Studies in Systems, Decision and Control

Volume 402

Series Editor

Janusz Kacprzyk, Systems Research Institute, Polish Academy of Sciences, Warsaw, Poland

The series "Studies in Systems, Decision and Control" (SSDC) covers both new developments and advances, as well as the state of the art, in the various areas of broadly perceived systems, decision making and control–quickly, up to date and with a high quality. The intent is to cover the theory, applications, and perspectives on the state of the art and future developments relevant to systems, decision making, control, complex processes and related areas, as embedded in the fields of engineering, computer science, physics, economics, social and life sciences, as well as the paradigms and methodologies behind them. The series contains monographs, textbooks, lecture notes and edited volumes in systems, decision making and control spanning the areas of Cyber-Physical Systems, Autonomous Systems, Sensor Networks, Control Systems, Energy Systems, Automotive Systems, Biological Systems, Vehicular Networking and Connected Vehicles, Aerospace Systems, Automation, Manufacturing, Smart Grids, Nonlinear Systems, Power Systems, Robotics, Social Systems, Economic Systems and other. Of particular value to both the contributors and the readership are the short publication timeframe and the world-wide distribution and exposure which enable both a wide and rapid dissemination of research output.

Indexed by SCOPUS, DBLP, WTI Frankfurt eG, zbMATH, SCImago.

All books published in the series are submitted for consideration in Web of Science.

More information about this series at https://link.springer.com/bookseries/13304

Piotr Kulczycki · Józef Korbicz · Janusz Kacprzyk
Editors

Fractional Dynamical Systems: Methods, Algorithms and Applications

Springer

Editors
Piotr Kulczycki
Systems Research Institute
Polish Academy of Sciences
Warsaw, Poland

Faculty of Physics and Applied Computer Science
AGH University of Science and Technology
Kraków, Poland

Janusz Kacprzyk
Systems Research Institute
Polish Academy of Sciences
Warsaw, Poland

Józef Korbicz
Institute of Control and Computation Engineering
University of Zielona Góra
Zielona Góra, Poland

ISSN 2198-4182 ISSN 2198-4190 (electronic)
Studies in Systems, Decision and Control
ISBN 978-3-030-89971-4 ISBN 978-3-030-89972-1 (eBook)
https://doi.org/10.1007/978-3-030-89972-1

This Springer imprint is published by the registered company Springer Nature Switzerland AG
The registered company address is: Gewerbestrasse 11, 6330 Cham, Switzerland

Introduction

Our world is composed of, and maybe even boils down, in the sense of structures and behavior, to various kinds of systems, and even systems of systems, which involve all aspects of life around us. The omnipresence and importance of systems have clearly implied a growing interest in science of the analysis, synthesis, modeling, and control of these systems. Firstly, just for gaining knowledge as well as understanding and finally to be able to use the tools and techniques developed to make them operate in a way that would be useful or beneficial to humans.

The huge research effort that has occurred over the centuries to attain the above intentions and aims has triggered first the development of new approaches, techniques, and tools as well as their implementations for practical solutions. This has happened across many disciplines of science and R&D, both "soft," represented by social sciences, as well as "hard," exemplified by control theory and engineering. Moreover, particular attention has been paid to more complex settings, illustrated by a necessity to account for multiple criteria, many stakeholders (agents) and as is relevant for the present volume, dynamics.

Among the large variety of approaches, tools and techniques, for our purposes one can certainly mention models based on the state space approach, differential and difference equations. All of these approaches have traditionally been characterized by a natural, yet traditional assumption of an integer order, for instance linear, that is of order 1, quadratic, that is of order 2, etc. This has made it possible to fully use the power of many strong analytic and algorithmic results obtained over the centuries.

Recently, however, it has become more and more obvious that the above integer order-type descriptions fail in some cases to adequately, effectively, and efficiently reflect all kind of system properties and behavior. Therefore, qualitatively new models have been proposed in which the order of respective tools has no longer been assumed to be integer, i.e., 1, 2, 3, etc., but has been allowed to be factional, e.g., 2/3 or π. This assumption has implied far-reaching consequences, both analytic and algorithmic, because in general, properties of traditional integer order systems cannot be directly extended by a straightforward generalization to fractional order. Fractional order dynamical systems, said to have fractional dynamics, are basically meant as those that can be modeled by fractional order differential equations which involve derivatives of a non-integer order. They are very useful for describing and analyzing, for instance, anomalies in the behavior of various systems or chaotic behavior, just to mention a couple.

This volume is meant to be a fully fledged, comprehensive presentation of many aspects related to the broadly perceived fractional order dynamical systems from both an analytic and theoretical points of view, algorithmic implementation of tools as well as techniques, and, finally, some examples of relevant and successful practical applications. It is worth noticing that Springer's book series *Studies in Systems, Decision and Control*, in which this volume appears, has been for many years the place chosen by numerous authors and editors for the publishing of monographs dedicated to fractional order dynamical systems, with the number of such books exceeding a dozen.

The volume is the next step in a novel initiative of the Committee on Automatic Control and Robotics of the Polish Academy of Sciences, which gathers the most prominent and active experts in the fields related to the scope of its activity. This goal of this publishing initiative is to prepare a high-quality and comprehensive volume containing works on an up-to-date and relevant topic. These works are requested from the best-known experts in the field and contain both fully fledged and critical state of the art, as well as presentations of new and highly original results. The chapters are prepared by the members of the Committee as the main authors and after with their foreign coauthors. This volume is dedicated to the presentation of approaches, methods, algorithms, implementations, and applications of the broadly perceived fractional order dynamical systems, an area that attracts much attention from the research and scholarly community.

In this edited book, a natural division into parts is adopted which has been implied by the above-mentioned intention to cover theory and applications. We start with mathematical foundations, then provide an account of theoretical analyses as well as computational approaches and finish with real-world applications.

Part "Foundations" has a very important purpose. Firstly, introducing novice readers to the topic and providing a comprehensive view of what has been done in the field. Secondly, it provides a point of departure for the next chapters by delimiting problem classes and the main terminology.

Rafał Stanisławski ("Fractional Systems: State-of-the-Art") starts with a brief history of fractional calculus, followed by an insightful, critical as well as constructive assessment of the main ideas, concepts, and approaches designed, mainly from the point of view of the fractional order systems, notably for modeling and control. More specifically, major developments and advances in four subareas related to the fractional order systems, including foundations, implementations, applications, and control, are surveyed. Particular attention is paid to the influence of Polish researchers in the development of this discipline. The presentation is illustrative and can be useful for both novice as well as advanced readers.

Piotr Ostalczyk and Ewa Pawłuszewicz ("Fractional Systems: Theoretical Foundations") present an illustrative introduction to fractional calculus. First, elements of linear differential equations are briefly reviewed and then their equivalent state space description is given. Next, an equivalence of the descriptions considered is shown via fractional order transfer functions. As an illustration, simple equations describing the fractional order integrator and the inertial element are dealt with. Some remarks concerning the stability of solutions are given.

Part "Modeling, Behavior and Properties" is the main and most voluminous part of this book, which is concerned with various theoretical aspects related to fractional order dynamical systems, notably with various models, their forms, and solution approaches.

Stefan Domek ("Mixed Logical Dynamical Modeling of Discrete-Time Hybrid Fractional Systems") presents a novel method for the modeling of discrete-time systems of the non-integer order, which involve real and integer variables and their interactions. The proposed fractional order mixed logical dynamical approach provides a generalization of many models applicable to hybrid systems, sequential logical systems, as well as some classes of discrete event systems, fractional order linear systems with constraints, and fractional order nonlinear systems, the nonlinearities of which can be approximated by piecewise linear functions. Such systems often occur in real life as well as various fields of science and technology. In the chapter, the method proposed for integer order hybrid systems is employed and its fractional order version is proposed, which can be used to design new fractional order control algorithms. Some numerical examples are shown for illustration.

Andrzej Dzieliński, Dominik Sierociuk, Wiktor Malesza, Michał Macias, Michał Wiraszka, and Piotr Sakrajda ("Fractional Variable-Order Derivative and Difference Operators and Their Applications to Dynamical Systems Modelling") present a comprehensive and critical overview of some specific derivative and difference operators of fractional variable order, their properties, equivalent forms, and applications. When fundamental properties of a system or its structure change over time, a variation of the system's order may be observed and in such a case time-dependent variable order operators are employed. Recently, the case of time-varying order has attracted interest and popularity. The description and analysis of such variable order systems are, however, much more complicated than for the case of constant order systems, mainly due to the existence of many definitions proposed in the literature. In order to give a deeper insight into fractional variable order calculus, the authors provide an alternative, intuitive description of some particular variable order operators in the form of equivalent switching schemes. According to such a schematic interpretation of the variable order operators, the analysis of variable order systems can be simpler and more effective than on the basis of purely analytic definitions. The switching strategies introduced, given unambiguously, classify and identify ways of changing the order of derivatives. Based on those switching schemes, it is possible to categorize fractional order derivatives according to their behavior and intrinsic properties. The duality between the analytical solutions of chosen variable order operators of linear differential equations can be effectively derived. In turn, through the matrix representation of the variable orders operators, also presented in the chapter, numerical solutions to linear variable order control systems can be obtained. Moreover, the possibility of including initial conditions in some types of fractional variable order operators makes it possible to obtain solutions to non-zero initial problems. Examples of applications of these operators in automatic control and modeling of a heat transfer process in specific grid holes and two-dimensional fractal like structure media with a changing geometry are presented here.

Adam Czornik, Pham The Anh, Artur Babiarz, and Stefan Siegmund ("Asymptotic Behavior of Discrete Fractional Systems") investigate discrete linear fractional systems with variable coefficients, considering the forward and backward equations with the Caputo and Riemann–Liouville operators. For the backward equations, two types of the Caputo operators and two types of the Riemann–Liouville operators, depending on whether the sum of the fractional order that appears in the definition of the particular operator includes the initial condition or not, are dealt with. The work presents the existing results on asymptotic properties of the considered equations, their extension, supplementation, and a comparison. The basic research method used here is the transformation of fractional order equations into the appropriate convolutional-type Volterra equations.

Piotr Ostalczyk ("Variable-, Fractional-Order Linear System State-State Description Transformation") presents an analysis of the linear time variant SISO (single-input single-output) systems described by difference equations with variable, fractional order Grünwald–Letnikov backward differences. A state space like form, similar to the observability matrix is proposed. The solution of the linear time-variant fractional order SISO system is derived. For a special case, a similarity like transformation to a diagonal form is defined.

Rafał Stanisławski, Marek Rydel, and Krzysztof J. Latawiec ("Balanced Truncation Model Reduction in Approximation of Nabla Difference-Based Discrete-Time Fractional-Order Systems") present a comprehensive and critical survey of new results in the application of balanced truncation model reduction methods to the approximation of discrete-time non-commensurate fractional order systems. As the fractional order difference is used, the nabla difference based fractional order difference employing the Grünwald–Letnikov definition is used. The method uses the Fourier-based decomposition of the fractional order system and the balanced truncation model order reduction. The main advantage of the algorithm presented here is a specific representation of the fractional order system which makes it possible to obtain a simple, analytical formula for the determination of the Gramians. This contributes to a significant improvement of the computational efficiency of the balanced truncation reduction method. The simulation experiments confirm the effectiveness and efficiency of the method introduced in terms of a high modeling accuracy and a low computational cost.

Ewa Pawłuszewicz, Andrzej Koszewnik, and Piotr Burzyński ("State Feedback Law for Discrete-Time Fractional Order Nonlinear Systems") are concerned with an important problem where one knows the inputs and measurements of an investigated process and a relationship between them is needed. That is, there is a question about possible systems that provide a good description of the input–output behavior observed, which is the crucial idea of the realization problem. The realization of an input–output map that describes the system's behavior means to find a dynamical state space system with input and output which can reproduce the given behavior. A theoretical approach to this problem for both the continuous time and discrete-time linear systems has been generalized and extended to any time domain and an extension to a more general case of the differential/difference order, i.e., to systems defined by fractional order operators has also been presented. There are many processes in nature more accurate models using fractional differentiation integration operators and a rapid development of computational techniques has made possible to efficiently model the real phenomena with the generalizations of the n-th order differences to their fractional forms and the state space equations of control systems in discrete time. Stability of the closed-loop system given by a fractional order transfer function has also been studied.

Tadeusz Kaczorek and Łukasz Sajewski ("Some Specific Properties of Positive Standard and Fractional Interval Systems") analyze some specific properties of positive standard and fractional linear systems with interval state matrices. The stability, positivity, and transfer matrices of positive different orders fractional continuous-time linear systems are discussed. New necessary and sufficient conditions for the asymptotic stability of positive different orders fractional linear systems are proposed. It is shown that the transfer matrices of positive asymptotically stable different orders fractional linear systems have only nonnegative coefficients. New conditions for the interval stability of positive standard and fractional linear systems are given. It is shown that the adjoin matrix of the singular Metzler matrix with zero sum of entries of each row (column) has all equal entries.

Part "Stability and Controllability" is concerned with the main properties of dynamical systems: stability and controllability. Roughly, stability is a property of many systems, including the ones considered in this volume, in that a system is stable if it resists small efforts to change its direction or position. On the other hand, controllability is a property of many systems which boils down to an ability to change the system's states by changing its input.

Tadeusz Kaczorek ("Global Stability of Nonlinear Fractional Dynamical Systems") proposes new sufficient conditions for the global stability of different classes of nonlinear fractional feedback systems. The linear parts of the systems are positive systems with interval state matrices. The nonlinear parts are described by static nonlinear characteristics located in the first and third quarters of the plane. The feedbacks are described in general case by matrices with positive entries. The sufficient conditions for global stability are given for the following classes of nonlinear systems: positive interval continuous-time feedback nonlinear systems, fractional positive interval continuous-time feedback nonlinear systems, positive interval discrete-time feedback nonlinear systems, descriptor nonlinear feedback discrete-time systems, and positive nonlinear electrical circuits. Procedures are presented for the calculation of gain matrices of the characteristics of nonlinear elements of the systems. The efficiency of the procedures are demonstrated on numerical examples of nonlinear systems.

Jerzy Klamka ("Controllability of Fractional Linear Systems with Delays in Control") considers linear, fractional, continuous-time, finite-dimensional, dynamical control systems with multiple variable point delays, and distributed delay in admissible control described by linear ordinary differential state equations. Using notations, theorems and methods taken directly from functional analysis and linear controllability theory, necessary and sufficient conditions for global relative controllability in a given finite-time interval are formulated and proved. The main result of the chapter is to show, that global relative controllability of fractional linear systems, with different types of delays in admissible control is equivalent to non-singularity of a suitably defined relative controllability matrix. In the proofs of the main results, methods and concepts taken from the theory of linear bounded operators in Hilbert spaces are used. Applying a relative controllability matrix for relative controllable systems steering admissible control is proposed, which steers the fractional system from the given initial complete state to the desired final relative state. Some remarks and comments on the existing controllability results for linear fractional dynamical system with delays are also presented.

Part "Applications" is a very important, useful, and illustrative part, in which some real life applications of some of the tools and techniques presented in this volume are shown so that the readers can see the very procedure of how elements of fractional order dynamical systems can be employed in real-life cases.

Wojciech Mitkowski, Marek Długosz, and Paweł Skruch ("Selected Engineering Applications of Fractional-Order Calculus") show several examples of applications for using fractional order calculus in some selected engineering tasks. It is shown then that some real systems can be better mathematically described by using fractional order differential equations. The focus is on ladder network structures with fractional order elements to model both electrical and nonelectrical systems with distributed parameters. Examples of modeling of supercapacitors, batteries, a chain of vehicles functioning in an adaptive cruise control mode and thermal processes inside buildings are provided. The effectiveness of the proposed modeling approach is verified by both simulation and experiments.

Krzysztof Oprzędkiewicz and Wojciech Mitkowski ("Fractional Order State Space Models of the One-Dimensional Heat Transfer Process") present some fractional order, state space models of the one dimensional heat transfer process. The proposed models are based on the known semigroup state space model of a parabolic system with distributed control and observation. The first model presented is a time-continuous model using the Caputo definition of the fractional derivative with respect to time and the Riesz definition to express the fractional derivative with respect to length. For this model, the spectrum decomposition is analyzed and time response formulas are given. The second model proposed is a discrete-time model employing the discrete Grünwald–Letnikov operator. For this model, its accuracy and convergence are analyzed. Additionally, the stability of this model needs to be analyzed due to the fact that too big a size of finite-dimensional approximation of the model causes a loss of stability. The sufficient and necessary stability condition for this model is proposed and proved. The last discrete-time fractional order model proposed uses a new, memory effective method of solution for the discrete fractional state equation proposed by the authors. The proposed method uses the continuous fraction expansion (CFE) approximant to express the fractional derivative. For this model, the accuracy and convergence are analyzed and suitable conditions are proposed. All the presented results are verified using real experimental heat plant.

We hope that this volume will be interesting and useful for a wide audience who will be able to find a comprehensive, insightful, critical, and constructive information about fractional order dynamical systems, their analytic, and computational aspects, as well as implementations and real application. This can greatly facilitate the use of tools and techniques in this new and powerful modeling approach.

We would like to thank first of all the authors of the chapters for their extremely valuable works, presenting in an insightful but accessible way both the state of the art of modern knowledge as well as the latest trends and tendencies, sometimes too innovative and too early to enter the standard canon of applications. Our gratitude is also due to the peer reviewers whose insightful and constructive remarks and suggestions have helped the authors improve their contributions.

And last but not least, we wish to thank Dr. Thomas Ditzinger, Dr. Leontina di Cecco, and Mr. Holger Schaepe from Springer Nature for their dedication and help to implement and finish this important publication project on time, while maintaining the highest publication standards.

Warsaw, Poland
July 2021

Piotr Kulczycki
Józef Korbicz
Janusz Kacprzyk

Contents

Foundations

Fractional Systems: State-of-the-Art

Rafał Stanisławski

Abstract This chapter presents the state-of-the-art in the fields of the theory and applications of fractional-order systems. Since this book is edited under the auspices of the Committee on Automatic Control and Robotics of Polish Academy of Sciences, the main focus is on Polish contributions in this area. At the beginning, a brief history of fractional calculus is outlined and quantitative analysis of contributions to the field are given. The state-of-the-art is surveyed in the advances within four subareas related to fractional-order systems including foundations, implementations, applications and control systems.

1 Introduction

The birth of noninteger- or fractional-order systems can be traced back to 1695 when Leibnitz and L'Hospital communicated with each other regarding the possibility of a derivative to assume a noninteger/fractional value, e.g. 1/2. In their correspondence Leibnitz used for the first time, the notation $d^{1/2}y$ in 1697 [96, 159]. However the first work formally considering a fractional-order derivative was written over a hundred years later by Lacroix in 1819 [90, 159]. The book is devoted to differential and integral calculus but the author has also touched upon the fractional-order derivative problem. In that proposition, the fractional-order derivative has been described by using the Legendre's Gamma function. At the same time, fractional calculus was considered by Abel in 1823, where it was delivered in a mechanical problem related to the isochrone curve [1]. In that article, Abel proposed a fractional-order integration formula known as the Riemann-Liouville fractional integral and fractional-order differentiation known as the Caputo fractional derivative [1]. Note that Abel was only 21 years old at the time of the publication of that paper. After he died in 1829, his contribution remained in the shadow of other contributions. As can be read in

R. Stanisławski (✉)
Department of Electrical, Control and Computer Engineering, Opole University of Technology, ul. Prószkowska 76, 45-758 Opole, Poland
e-mail: r.stanislawski@po.edu.pl

P. Kulczycki et al. (eds.), *Fractional Dynamical Systems: Methods, Algorithms and Applications*, Studies in Systems, Decision and Control 402,
https://doi.org/10.1007/978-3-030-89972-1_1

Ref. [149], the reason for this was the limited accessibility of the article, which was published in Danish in a Norwegian journal. Abel's first and second works have been translated and published for a broader audience in 1881 and 1839, respectively [1].

A little later, works regarding the fractional calculus problem were provided by Liouville. He proposed the first broadly known formal definitions of the fractional-order integrator [100, 101]. Specifically, he delivered two definitions for the fractional-order derivative and integrator. The second definition is similar to the Riemann-Liouville definition used nowadays. Already as a student, Bernhard Riemann had explored the continuation investigation under the fractional-order derivative. In particular, Riemann proposed his definition of the fractional-order integrator [156]. Nevertheless, Liouville's and Riemann's definitions were different, both from each other and from the Lacroix proposition. To solve this, the above mentioned discrepancies led to several works including Cauchy's integral formula. The first results were proposed by Sonin in 1869 and then extended by Letnikov in 1872 [98, 182]. Ultimately, the problem was solved by Laurent in Ref. [95], where the current Riemann-Liouville definition was proposed. Under mild conditions, the proposed generalized definition is equivalent to the Riemann, Liouville, and Lacroix propositions. Other work by Letnikov published in Ref. [97] and the results of another German mathematician Grünwald [52], have also led to an alternative definition of the fractional-order derivative, called the Grünwald-Letnikov definition.

The foundations of fractional calculus, including currently explored definitions of fractional-order operators in terms of the Riemann-Liouville, Caputo, and Grünwald-Letnikov definitions, were proposed at the end of 19th century. It is worth noting that fractional calculus has been investigated with intrigue by scientists right from its birth. Many well-known mathematicians and engineers have put their research stamps on the fractional calculus area, to mention Laplace [158], Riemann [156], Laurent [95], Heavyside [54] or later Riesz [157].

Note that some more comprehensive, interesting historical surveys in the fractional calculus area are presented in Refs. [96, 102, 149, 158, 159].

The fundamental and holistic work summarizing the achievements in the fractional calculus area is the book [128] published in 1974. From that time on, we can observe an increasing attention to fractional calculus in science and technology. In particular, we can observe the extensive exploration in this field during the last decade of the 20th century. This fact can be observed in Fig. 1, which presents the number of publications indexed in the Web of Sciences Core Collection, that have been published in particular years with the phrase 'fractional-order' or 'noninteger-order' in their titles, keywords or abstracts. Fractional-order systems successfully impact several areas of science. This can be observed from Fig. 2, where the areas of published papers are presented.

We can see from Fig. 2 that the topic of fractional-order systems falls across the majority of science and engineering, including various areas of mathematics and physics, electrical, control and computer engineering, mechanics, optics, telecommunication, medicine, measurement and many other areas. Note that in 2020 papers considering fractional-order systems represented some 1.5 to 1.6 % of all papers in the sections of 'Automation control systems' and 'Mathematics-applied' included in the Web of Science system.

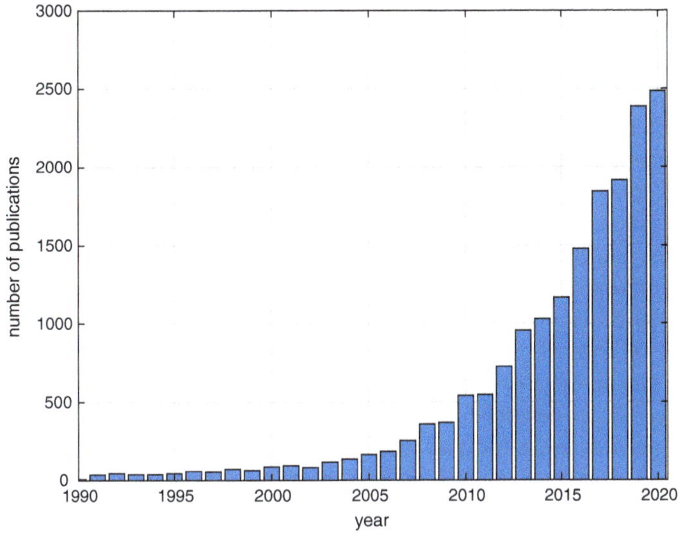

Fig. 1 Number of publications indexed in Web of Sciences in the fractional-order area in particular years

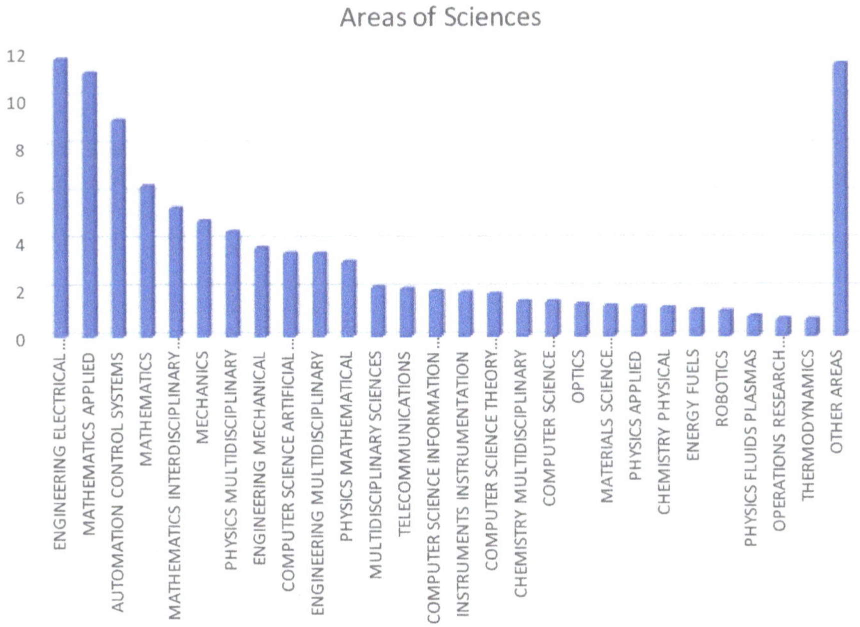

Fig. 2 Fractional calculus in areas of sciences and engineering

The popularity of fractional calculus results from the superposition of four main aspects:

(1) Fractional-order systems are a specific class of generalization, which need specific, often very complex methods for analysis and synthesis. Moreover, the introduction of this concept opens new perspectives on the system nonlinearity, parameters of time variability, positivity, etc.
(2) The specific nature of fractional-order derivatives/integrals generates numerical problems related to their implementation, which, in general, can be made in an approximate way only.
(3) Fractional-order models can more accurately describe various phenomena in (not only) physical processes than integer-order models. Also, fractional-order models can be more effective in system identification.
(4) Fractional-order generalizations of various control strategies can lead to higher control performances for both fractional- and integer-order systems.

These four main viewpoints will be addressed in separated sections to follow. Taking into account that this book is edited under the auspices of the Committee on Automatic Control and Robotics of the Polish Academy of Sciences, we focus on the (significant) Polish contributions in this area.

2 Foundations

One of the main streams in fractional-order systems is that fractional-order derivatives/integrators, when incorporated in dynamical systems, affect the system's properties. Therefore one of the main streams in this area is the analysis of fundamental properties of the system. Some selected contributions to this field are described below, starting with stability issues.

In the continuous-time case, the first simple stability test was proposed by Matignon [110]. The result is offered for linear time-invariant commensurate-order systems and it tests the stability on the basis of arguments of zeros of the characteristic pseudopolynomial (sometimes called pseudopoles or f-poles). In the test, the fractional-order incorporated in the system transforms the left-hand side of the complex plane into an order-dependent subplane. Several papers have been proposed on this basis, considering nonlinear and time-varying systems. Also, the Matignon stability result was generalized to the systems with rational, fractional-order derivatives [146]. Another approach in this context was introduced by Busłowicz in Ref. [22]. That paper is devoted to the stability analysis of a specific class of time-delayed systems. The paper shows that some stability results designed for classical (integer-order) systems in terms of Mikhailov stability criterion, can be effectively used for the class of fractional-order systems. That approach was continued recently in Ref. [113], which extends the Matignon theorem to commensurate-order systems. Moreover, the presented works show the high potential of the concept of stability analysis. The same origin-based Nyquist stability methods have successfully been implemented for commensurate- and noncommensurate fractional-order systems and also in nonlinear and stochastic cases [196, 210]. The author of this chapter claims that various current stability problems in both fractional- and integer-order systems can be solved using the Mikhailov method. However in the stability analysis of noncommensurate-order systems, it is the Lyapunov methods that are most often used [99], even though they are much more difficult to implement than e.g. the Mikhailov or Nyquist methods.

Similarly to the above methods, stability results for discrete-time fractional-order systems have been developed. The first stability results in this area were proposed by Dzieliński and Sierociuk in Ref. [46]. Sufficient stability conditions introduced in the paper, are proposed for a discrete-time commensurate fractional-order state-space system under the forward-shifted Grünwald-Letnikov difference. The results are based on pseudopoles of the characteristic pseudopolynomial defined in the $\mathscr{Z}$-domain. These sufficient stability conditions are conservative and need improvement, so the pseudopoles-based methodology has been explored. As a result, similar stability conditions have been proposed [25, 139, 187] based on the so-called stability areas. Also, in Ref. [188], analytically-driven conditions for this class of systems have been proposed. Note that, in discrete-time cases, the stability conditions are dependent on the definition of the fractional-order difference. In the backward Grünwald-Letnikov difference (often the so-called nabla difference), we have yet another stability condition. The pseudopoles stability areas for this class of systems are developed independently in Refs. [140, 160, 161, 186]. Comparing the stability results for delta (forward-shifted) and nabla (backward) difference presents, the definition of the difference operator affects the system stability. Also, in the case of discretization of the continuous-time system, the operator used in the discretization process affects the stability of the obtained discrete-time system. For the three most often used operators in Euler, Tustin and Al-Alaoui, this fact is considered in Ref. [192]. Alternative stability conditions are also obtained for the discrete-time systems using the Caputo definition [17, 203] and also in the variable-order case [119, 122]. As in the discrete-time case, the stability analysis of the noncommensurate systems is much more challenging than for the commensurate ones. In this area we can find some papers with hardly implementable analytical results (see [203]). However, it has been shown that the modified Mikhailov stability method leads to simple stability conditions of noncommensurate-order systems under the nabla difference [189]. As in the continuous-time case, in several publications there are considered various, specific subclasses of discrete-time fractional-order systems in the case of time-delay systems [162], interconnected systems [53], time-varying systems [138] and others. Also, the stability analysis of discrete-time nonlinear systems is more challenging and most results in this case are based on the Lyapunov stability [7, 204].

Let us shift now to other important system properties in terms of controllability and observability of fractional-order systems. This area's first result was proposed for delta difference-based discrete-time systems [63, 70]. The discrete-time results have been extended to time-delay systems in Ref. [78]. Generalizations have also been provided for discrete-time semilinear systems [79] and a more general class of nonlinear systems under various definitions of fractional-order difference [120]. Another discrete-time approach for discrete-time systems under the Caputo fractional operator with specified magnitude constraints is presented in Ref. [144]. Observability of discrete-time fractional-order systems has been analyzed in some papers [145], including nonlinear cases [117].

It has been mentioned in the introduction to this chapter, that the development of fractional-order systems opens a new possibility of accounting for time-varying parameters in terms of variable fractional orders. Note that in the case of the discrete-time Grünwald-Letnikov difference, the variable order can be implemented in various ways. In the literature we can find four definitions of the variable order operator in terms of type $\mathscr{A}$, $\mathscr{B}$, $\mathscr{D}$ and $\mathscr{E}$. The differences of type $\mathscr{A}$ and $\mathscr{B}$ are immediate results from the Grünwald-Letnikov definition but $\mathscr{D}$- and $\mathscr{E}$-type differences are based on a specific representation of the fractional-order difference [177]. The aforementioned definitions produce different results. It is interesting that all types of variable-order operator are practical, finite-length implementations. In Refs. [173, 175], there are presented equivalent, electric ladder circuits for all the above mentioned variable-order definitions. Time-varying fractional-order differences can be easily implemented in fractional-order systems [176] and generalized Kallman filter [174]. Properties of time-varying fractional-order operators can be found in [106]. Alternative results for the Caputo-based discrete-time variable-order difference are presented in Refs. [121, 195].

Another class of fractional-order systems, in terms of positive systems has attracted a significant research attention. Several leading results in this area have been due to Tadeusz Kaczorek. Note that positive fractional-order systems have unique properties. Rather surprisingly, the positive systems' stability proprieties are similar to those for the integer-order systems [24, 68, 71]. Also, stability conditions for various classes of positive fractional-order systems have quite strange properties [70]. In this area we can find many simple, analytically driven results for positive systems, including switching systems [71], nonlinear systems [74, 77], feedback nonlinear systems [76], time-delay systems [167], two-dimensional systems [66, 67] and many other subclasses. In Refs. [166, 167], stabilization results have been shown in the case of use of decentralized control of positive fractional systems, both in continuous- and discrete-time cases. Equivalent results for positive discrete-time systems with the forward-shifted Grünwald-Letnikov fractional-order difference are given in Refs. [23, 75, 166]. Reachability and controllability to zero of positive fractional systems have also been obtained by using a generalization of the Cayley-Hamilton theorem [64]. Equivalent results for the continuous-time systems case under the Riemann-Liouville derivative have been proposed in Ref. [65]. Some other generalizations of the above results for a class of noncommensurate-order systems are shown in Ref. [165]. That paper also contains the proposition of a minimum energy control scheme for the considered class of systems.

A distinct subclass of fractional-order dynamical systems are the so-called switching systems. In fractional control science, this subclass of systems is particularly important in sliding mode control, which has been given considerable research attention. Analysis and synthesis of fractional-order generalizations of switching systems are very challenging and therefore the issue has not been deeply explored. We can find only a few papers in the context of formulations of fractional-order switching systems [40, 41] and realization of their various control strategies [39, 56]. In the author's opinion, this area still has potential for exploration in the future.

In addition to the above mainstream contributions to fractional-order systems, we can find some other investigations into various interesting subclasses of this area. Fractional nonlinear equations with sequential derivatives under the Caputo definition are considered in Ref. [118]. In the comprehensive study of Ref. [107], the generalization of the Lagrangian variational problems using the fractional calculus can provide several new results. On this basis, the authors propose new optimality Euler–Lagrange conditions for the fundamental and isoperimetric problems, transversality conditions and the Noether symmetry theorems. Nice fractional-order generalization results concern multiagent systems. In this context special attention is paid to (fractional) consensus problems, where we can find several propositions [4–6]. A specific subclass of successfully investigated fractional-order systems is the fractional generalization of the Sturm-Liouville equation. The Sturm-Liouville theory's generalization is introduced in Ref. [80] and explored in Refs. [81, 82]. Taking into account that many differential equations in terms of Hermite, Laguerre, Jacobi and Legendre can be considered as Sturm-Liouville problems, the above result impacts several areas of fractional-order systems.

3 Implementations

One of the most important problems related to fractional-order systems is that implementation of fractional-order derivatives/differences may lead to computational explosion. Therefore, both in continuous- and discrete-time systems, various concepts have been proposed to approximate/implement fractional-order systems. These concepts are mainly based on two types of applications, where (1) approximators are used to model the fractional-order derivative/difference involved in a fractional-order system and (2) a rational, integer-order approximator models the whole fractional-order system.

In the continuous-time case, the most popular approach in modeling the fractional-order derivative is the Oustaloup approximator. The Oustaloup model is a rational, integer-order IIR filter class, described by the transfer function of arbitrary order. The Oustaloup approximation convergence is quite high, requiring a relatively low model length to obtain satisfactory modeling performances. The Oustaloup model also has an elegant, simple form and can be easily obtained with respect to an arbitrary (finite) frequency range [142]. Therefore this approach is implemented in the majority of numerical environments devoted to fractional-order systems and controls [12, 198]. Similarly we can obtain approximations based on the finite-length implementation of the Atangana–Baleanu fractional derivative operator [130]. Another concept of implementing the fractional-order derivative is the application of the subinterval-based method [184]. In the discrete-time case, the conceptually simplest but most numerically involved is the finite-length implementation of the Grünwald-Letnikov difference [136, 137, 147]. More numerically effective in this approach are various discrete-time integer-order IIR-class filters, which can be obtained in many ways. The useful concepts in this field are the application of the iterative methods in terms of continuous fraction expansion and the Muir recursion method based on the Tustin and Al-Alaoui discretization operators. An alternative IIR-based approach is the application of an orthonormal basis function-based model to the approximation of the Grünwald-Letnikov difference [185]. Comparative analysis of these approaches is presented in Ref. [190]. Other finite-length Grünwald-Letnikov-based approximation is based on Horner's algorithm [20]. In the area of discrete-time systems, the discrete-time equivalent to the Oustaloup approach has also been used [10, 11].

It is important to note that the use of all the above approximations in fractional-order systems, both in the continuous- and discrete-time cases, affects the proprieties of the systems. Exemplary stability conditions for these systems differ from the criteria for systems based on actual derivatives/differences and depend on the approximator used. This stability analysis problem for Oustaloup approximations has been studied in Ref. [10]. For the discrete-time case, stability analysis of models of fractional-order systems are developed in several Refs. [131, 170, 192].

The approximation concept presented above is widely considered in modeling of fractional-order systems [93]. In the case of system identification based on fractional-order models as an approximate implementation of fractional-order derivative/difference, the integer-order implementations have also been used both in continuous- [60], and discrete-time systems [94].

As mentioned above, the second way of modeling fractional-order systems is by an approximation of the system as a whole using integer-order approximations. One of the effective methods is to apply model order reduction algorithms within this approach. A few papers consider this concept, both in continuous- and discrete-time cases. In the continuous-time case, this approach has been presented in Refs. [87, 88], where the Oustaloup-based, expanded state-space system has been reduced. Similar approaches for the discrete-time case are presented in Refs. [163, 191]. In Ref. [191] the balanced truncation method to the specific FIR class model of the fractional-order system is applied. The main advantage of the presented methodology is the computational simplicity of the reduction process. In Ref. [163] the cross-gramian-based balanced truncation method is applied to a specific form of the discrete-time fractional-order system. The FIR-based approachis also proposed in [9], which applies the Laguerre polynomials to modeling of the fractional-order system. Another approach is presented in Ref. [27], where the use of the Loewner framework has been applied in modeling of continuous-time systems. As a result we obtain a model in the descriptor integer-order state-space system form.

Note that the use of 'explicit' approximations of fractional-order derivatives/ differences or integrators is usually not mandatory in the case of finite-time applications. Therefore several papers consider a general form of fractional equations in the context of numerical implementation. The specific matrix approach for the numerical implementation of the discrete-time fractional-order system under the Grünwald-Letnikov difference, as well as in the case of variable order with the problem of including initial conditions in fractional-order derivatives, is presented in Ref. [178]. Various numerical implementations of the Inverse Laplace Transform for fractional-order systems are given in Ref. [21]. This complete study also presents a comparative analysis of various methods in the considered inverse problem.

The separate, important issue related to fractional-order systems is the implementation problem within the industrial environment. This practically-oriented problem is crucial in both signal prediction of industrial process variables and implementation of fractional-order controllers. The problem is that the models are usually implemented in PLC controllers. Considering that the controllers are of relatively low computational performance and memory, the approximations' simplicity is crucial in this area. The PLC implementation of models based on the Atangana–Baleanu fractional derivative operator, the CFE-based and PSE-based approximations are presented in Ref. [129–131].

Since implementation of fractional-order systems is quite complicated, there are various fractional-order tools available for their implementation, The majority of software for implementing fractional-order systems has been designed in the Matlab/Simulink environment. One of most popular is the FOMCON Toolbox for Matlab [197, 198], which is focused on continuous-time models. This toolbox offers a wide spectrum of functionalities to analyze and implement the systems based on the Oustaloup approximation. The fractional-order PID controller is also implemented with methods for tuning its parameters. A cross-platform software tool for implementation of fractional-order systems and PID controllers is SoftFrac [12, 15]. The tool is implemented in Matlab, Python, and C distributions. This approach has been especially useful for implementation of fractional-order models in embedded systems. A more specific Matlab/Simulink tool for implementing variable-order discrete-time systems is proposed in [171]. Also, a specific Matlab tool for implementation of fractional-order derivative using the subinterval method has been developed in [183, 184].

4 Applications

One of the main reasons for the significant research attention paid to fractional-order systems is that this class of systems leads to better modeling performances in various physical processes than its integer-order counterparts. The first known practical application of fractional calculus was delivered by Abel in a mechanical problem related to the isochrone curve [159]. Note that, as it has been presented at the beginning of the chapter, along the way, Abel introduced some of the foundations of fractional calculus. Since that time, a plethora of applications of fractional-order approaches have been reported.

One of the most popular is the application in various formulations of diffusion processes, including probabilistic models based on random walks [51, 114], stochastic Brownian motion models [34, 62, 108], stochastic models based on master and Fokker-Planck equations [34], Sturm-Liouville problem [80–82], and several other diffusion problems [58, 82, 104, 105, 164, 193]. Therefore, fractional-order systems can be effectively used to model various "diffusion-related" processes.

One of the diffusion-related processes for which fractional-order models can be effectively used is heat conduction. In this area we can find various approaches, including different propositions of fractional-order models and approximation methods in modeling the conduction of various materials [18, 19, 89, 133, 134, 150, 151]. Other applications are related to heat conduction in thermal systems [13, 30, 103, 127].

Since there are analogies between diffusion processes and conduction in specific RC electrical circuits, both can be described by similar fractional-order models [179]. Therefore diffusion processes can be modeled by analog models in terms of the RC ladder network and vice versa. Moreover, the RC ladder network can be considered as the analog model's fractional-order derivative [115, 116, 172], as well as for various definitions of variable order derivative [173, 178]. On the other hand, it is well known that fractional-order models can effectively approximate some electrical processes in terms of charging and discharging supercapacitors [45, 47] and we can find various modeling approaches to this element using fractional-order equations [49, 84, 115]. The main problem in this area however, is that the model parameters, in this case, depend on the current direction in charging/discharging circuits [85, 86]. In general, in the modeling of supercapacitor-based circuits, we may need to use fractional-order nonlinear models. The existence of electrical components effectively modeled by fractional-order equations opens the door to a new idea of generalized, fractional-order circuits, where the capacitors and inductors may be, in general, considered as fractional-order elements [69, 72, 73]. This interesting concept has however brought some controversies regarding possible misinterpretation of physical units of fractional capacity and inductance [181]. This problem has also been studied in Refs. [92, 141], where this modern empirical approach of employing fractional-order models in electric circuit theory and practice is strongly advocated. It is worth mentioning that in the electrical engineering area, the application of fractional-calculus has a long history. In this area fractional calculus was first applied by Heavyside in 1890 to extend the theory of electromagnetic waves [54].

Fractional-order models can also be effectively used in other "diffusion-related" processes in many other plants, i.e., electrochemical processes [61, 169, 194], transportation systems [114], transmission and acoustics [111] and many other areas of science and engineering. Fractional-order derivatives are also commonly used in continuum and statistical mechanics for viscoelasticity problems [2, 8, 29, 36]. In this case, the fractional viscoelastic equation can be more appropriate than the classical one when modeling some types of viscoelastic damped structures. The approach has also been used for flexible robots [200]. Another novel approach is the application of fractional calculus in the modeling of biochemical phenomena related to tumorous disruptive pathologies [125]. In general, this class of model has also been used in synthesizing various processes in medicine [57, 83].

Note that fractional calculus can be effectively applied in many other approaches unrelated to the mathematical modeling approach. One of these is the application in image processing. One of the fundamental problems in image processing is edge detection, which can be solved by the use of fractional-order equations [109, 207]. Some other applications related to image processing, are signature verification [199], processing of geographical data [31, 209] and Great Salt Lake elevation modeling [168].

A challenging application for fractional-order models is the control science and engineering areas, which is presented in the next section.

5 Control Systems

Since fractional calculus has been an extensively explored field in science and technology over the past three decades, fractional-order generalizations of various control strategies have attracted considerable research attention. The specific properties of fractional-order derivatives/integrators may lead to improved control performances including accuracy, robustness, etc. This fractional generalization affects both classical control strategies such as PID controllers and advanced, model-based ones such as predictive/adaptive controllers for example. It is important to note that fractional-order controllers may not necessarily be related to the plant's fractional nature and these control concepts can also be useful in the case of classical, integer-order plants.

Definitely the most popular 'fractional generalization' in the control area is the fractional-order Proportional- Integral-Differential (PID) controller. The huge popularity of this control concept is reflected in the Web of Science Core Collection system where we can find over 1600 papers as a result of the search phrase "fractional-order PID". The first paper related to a fractional-order PID controller is Ref. [148]. In fractional-order PID controllers, we have two more parameters to tune in terms of the two fractional orders, which can significantly improve the control performance. In order to obtain an effective controller however, we have to tune as many as five parameters. Moreover, the controller's orders are mutually influenced by the gains, so tuning the fractional-order PID controller's parameters is quite challenging. To solve this optimization problem, various strategies have been implemented in the tunning process, including particle swarm optimization [153], genetic algorithms[143, 155], Fibonacci-search methods [55], differential evolution methods [143], extensions of classical tuning methods for PID controllers [16] and other techniques [124, 201, 206]. Alternative methods are proposed for time-varying controller parameters [130]. The second problem encountered in fractional-order PID controllers is the computational complexity of derivatives and integrators. This problem is particularly crucial because fractional-order PID controllers have to be used in industrial environments on hardware of usually low computational performance. To solve this problem, in the continuous-time case, the Oustaloup approximation of fractional-order derivative/integrator is most often used. In the discrete-time case, finite-length implementations of the Grünwald-Letnikov fractional-order derivative/integrator based on power series expansion and continuous fraction expansion [32, 131, 135], as well as the discrete-time counterpart of the Oustaloup approach are most commonly used. Note that fractional-order PID controllers have been implemented in various open software tools, i.e. FOMCON [198] or SoftFrac [12]. The simplicity and effectiveness of fractional-order PID controllers results in several practical applications in control loops for various processes, i.e., a two-mass drive system [48], a 3-DOF parallel manipulator [42], a radar-guided missile [206], a pneumatic position servo mechanism [91, 154], a magnetic levitation system [14, 15, 123], an inverted pendulum [132], a DC-motor in various systems [35, 197], a pressurized water nuclear reactor [152], the rotational speed control of an unmanned aerial vehicle's propulsion unit [50], a nonlinear water tank system [202] and many other plants.

The second field of fractional calculus-based control involves more advanced control techniques using fractional-order models of the processes. We consider predictive/adaptive model based control, linear-quadratic or linear-quadratic-Gaussian regulation, internal-model-control and many other control algorithms in this context.

In the design of predictive control, there are two main directions in the control of fractional-order systems. The first, more popular strategy is based on implementing a prediction control algorithm for a finite-length approximation of a fractional-order difference incorporated into the fractional-order system. In this case, we design a control strategy based on approximate fractional-order predictors and obtain a sort of a 'fractional' predictive controller [37]. In the second case, a control algorithm is based on integer-order approximation of the whole fractional-order system. In this case, the prediction scheme is similar to those based on a regular, integer-order system. Note that both prediction strategies can be used to implement fractional-order systems in the prediction process and various models of predictive control strategies can be used [37–39].

The optimal control problem for fractional discrete-time systems in terms of the linear quadratic regulator is considered in Refs. [43, 44, 180]. The proposed techniques are generalizations of the well-known methods designed for discrete-time fractional-order systems based on the Grünwald-Letnikov difference. An alternative approach for continuous-time Caputo-based systems is proposed in Ref. [59]. Other applications of these control problems include the control of a class of star graph system [112] and artificial intelligence based systems [33].

A number of other control strategies have been implemented, including various fractional-order generalizations of a sliding mode control concept, control of switching systems, or multiagent systems, also in the case of nonlinear systems [208], applications in a four-wheeled steerable mobile robot [205], a control strategy for therapy of HIV infection [28], control of Air-Breathing Hypersonic Vehicles [26], for Robotic Manipulators With Backlash Hysteresis [3], for a vehicle suspension system [126] and many other plants.

6 Conclusion

This chapter has outlined the state-of-the-art in the fields of the theory and applications of fractional-order systems, as seen from Polish contributions within the fractional framework. Firstly, a brief history of fractional calculus has been presented and a short quantitative analysis of contributions to the field has been provided. The state-of-the-art has been structured in four subareas related to fractional-order systems in terms of foundations, implementations, applications and control systems. In this chapter it has been indicated that Polish contributions to the field are a noteworthy developement. However the author would like to express his deepest apology for possibly omitting some valuable positions in the reference list, which can by no means be considered complete or exhaustive, owing to the vastness of the subject area.

References

1. Abel, N.H.: Solution de quelques problémes á l'aide d'intégrales definies. In: Ouvres Complétes, pp. 16–18. Christiania (1881)
2. Adolfsson, K., Enelund, I., Olsson, P.: On the fractional order model of viscoelasticity. Mech. Time-Depend. Mater. **9**, 15–34 (2005)
3. Ahmed, S., Wang, H., Tian, Y.: Adaptive high-order terminal sliding mode control based on time delay estimation for the robotic manipulators with backlash hysteresis. IEEE Trans. Syst. Man Cybern. Syst. **51**(2), 1128–1137 (2021). https://doi.org/10.1109/TSMC.2019.2895588
4. Almeida, R., Kamocki, R., Malinowska, A.B., Odzijewicz, T.: On the existence of optimal consensus control for the fractional Cucker-Smale model. Arch. Control Sci. **30**(4), 625–651 (2020)
5. Almeida, R., Kamocki, R., Malinowska, A.B., Odzijewicz, T.: Optimal leader-following consensus of fractional opinion formation models. J. Comput. Appl. Math. **381**, 112,996 (2021). https://doi.org/10.1016/j.cam.2020.112996
6. Almeida, R., Malinowska, A.B., Odzijewicz, T.: Optimal leader-follower control for the fractional opinion formation model. J. Optim. Theory Appl. **182**, 1171–1185 (2019)
7. Antoulas, A.: Approximation of Large-Scale Dynamical System. Society for Industrial and Applied Mathematics, Philadelphia (2005)
8. Bagley, R.L., Calico, R.A.: Fractional-order state equations for the control of viscoelastic damped structures. J. Guidance Control Dyn. **14**(2), 304–311 (1991)
9. Bania, P., Baranowski, J.: Laguerre polynomial approximation of fractional order linear systems. In: Advances in the Theory and Applications of Non-integer Order Systems. Lecture Notes in Electrical Engineering, vol. 257, pp. 171–182. Springer, Dordrecht, Netherlands (2013)
10. Baranowski, J., Bauer, W., Zagorowska, M.: Stability properties of discrete time-domain oustaloup approximation. In: Theoretical Developments and Applications of Non-Integer Order Systems, Lecture Notes in Electrical Engineering. Springer (2016)
11. Baranowski, J., Bauer, W., Zagorowska, M., Dziwinski, T., Piatek, P.: Time-domain oustaloup approximation. In: 20th IEEE International Conference on Methods and Models in Automation and Robotics. Miedzyzdroje, Poland (2015). https://doi.org/10.1109/MMAR.2015.7283857
12. Baranowski, J., et al.: Softfrac Project (2021). http://non-integer.pl
13. Battaglia, J.L., Cois, O., Puigsegur, L., Oustaloup, A.: Solving an inverse heat conduction problem using a noninteger identified model. J. Heat Mass Transf. **44**(14), 2671–2680 (2001)
14. Bauer, W., Baranowski, J.: Fractional (pid)-d-lambda controller design for a magnetic levitation system. Electronics **9**(12) (2020). https://doi.org/10.3390/electronics9122135
15. Bauer, W., Baranowski, J., Tutaj, A., Piatek, P., Bertsias, P., Kapoulea, S., Psychalinos, C.: Implementing fractional PID control for maglev with softfrac. In: 43rd International Conference on Telecommunications and Signal Processing (TSP), pp. 435–438. Brno, Czech Rep. (2020)
16. Bazanella, A.S., Pereira, L.F.A., Parraga, A.: A new method for PID tuning including plants without ultimate frequency. IEEE Trans. Control Syst. Technol. **25**(2), 637–644 (2017). https://doi.org/10.1109/TCST.2016.2557723

17. Brandibur, O., Kaslik, E., Mozyrska, D., Wyrwas, M.: Stability results for two-dimensional systems of fractional-order difference equations. Mathematics **8**(1751) (2020). https://doi.org/10.3390/math8101751
18. Brociek, R., Słota, D., Król, M., Matula, G., Kwaśny, W.: Comparison of mathematical models with fractional derivative for the heat conduction inverse problem based on the measurements of temperature in porous aluminum. Int. J. Heat Mass Transf. **143**, 118,440 (2019). https://doi.org/10.1016/j.ijheatmasstransfer.2019.118440
19. Brociek, R., Slota, D.: Reconstruction of the robin boundary condition and order of derivative in time fractional heat conduction equation. Math. Model. Nat. Phenom. **13**(1), 5 (2018). https://doi.org/10.1051/mmnp/2018008
20. Brzeziński, D.W.: Fractional order derivative and integral computation with a small number of discrete input values using Grünwald–Letnikov formula. Int. J. Comput. Methods **17**(05), 1940,006 (2020). https://doi.org/10.1142/S0219876219400061
21. Brzeziński, D.W., Ostalczyk, P.: Numerical calculations accuracy comparison of the inverse Laplace transform algorithms for solutions of fractional order differential equations. Nonlinear Dyn. **84**(1), 65–77 (2016). https://doi.org/10.1007/s11071-015-2225-8
22. Busłowicz, M.: Stability of linear continuous-time fractional order systems with delays of the retarded type. Bull. Pol. Acad. Sci. Tech. Sci. **56**(4), 319–324 (2008)
23. Busłowicz, M.: Robust stability of positive discrete-time linear systems of fractional order. Bull. Pol. Acad. Sci. Tech. Sci. **58**(4), 567–572 (2010)
24. Busłowicz, M., Kaczorek, T.: Simple conditions for practical stability of positive fractional discrete-time linear systems. Int. J. Appl. Math. Comput. Sci. **19**(2), 263–269 (2009)
25. Busłowicz, M., Ruszewski, A.: Necessary and sufficient conditions for stability of fractional discrete-time linear state-space systems. Bull. Pol. Acad. Sci. Tech. Sci. **61**(4), 779–786 (2013)
26. Cao, L., Tang, S., Zhang, D.: Fractional-order sliding mode control of air-breathing hypersonic vehicles based on linear-quadratic regulator. J. Aerosp. Eng. **31**(3), 04018,022 (2018). https://doi.org/10.1061/(ASCE)AS.1943-5525.0000852
27. Casagrande, D., Krajewski, W., Viaro, U.: The integer-order approximation of fractional-order systems in the Loewner framework. IFAC-PapersOnLine **52**(3), 43–48 (2019). https://doi.org/10.1016/j.ifacol.2019.06.008 15th IFAC Symposium on Large Scale Complex Systems LSS 2019
28. Chen, S.B., Rajaee, F., Yousefpour, A., Alcaraz, R., Chu, Y.M., Gómez-Aguilar, J., Bekiros, S., Aly, A.A., Jahanshahi, H.: Antiretroviral therapy of HIV infection using a novel optimal type-2 fuzzy control strategy. Alex. Eng. J. **60**(1), 1545–1555 (2021). https://doi.org/10.1016/j.aej.2020.11.009
29. Coimbra, C.: Mechanics with variable-order differential operators. Annalen der Physik **12**(11–12), 692–703 (2003)
30. Cois, O., Oustaloup, A., Battaglia, E., Battaglia, J.: Non integer model from modal decomposition for time domain identification. In: Proceedings of the 41st IEEE Conference on Decision and Control. Las Vegas, USA (2002)
31. Cooper, G.R.J., Cowan, D.R.: Filtering using variable order vertical derivatives. Comput. Geosci. **30**, 455–459 (2004)
32. da Costa, J.S.: An Introduction to Fractional Control. Control, Robotics & Sensors. Institution of Engineering and Technology (2012)
33. Dabri, A., Nazari, M., Butcher, E.A.: Adaptive neural-fuzzy inference system to control dynamical systems with fractional order dampers. In: American Control Conference (ACC), pp. 1972–1977. Seattle, WA (2017)
34. Denisov, S.I., Hanggi, P., Kantz, H.: Parameters of the fractional Fokker-Planck equation. EPL **85**(4) (2009). Paper ID: 40007
35. Dimeas, I., Petras, I., Psychalinos, C.: New analog implementation technique for fractional-order controller: a DC motor control. AEU - Int. J. Electron. Commun. **78**, 192–200 (2017). https://doi.org/10.1016/j.aeue.2017.03.010
36. Doehring, T.C., Freed, A.D., Carew, E.O., Vesely, I.: Fractional order viscoelasticity of the aortic valve cusp: an alternative to quasilinear viscoelasticity. J. Biomech. Eng. **127**(4), 700–708 (2005)

37. Domek, S.: Fuzzy predictive control of fractional-order nonlinear discrete-time systems. Acta Mechanica et Automatica **5**(2), 23–26 (2011)
38. Domek, S.: Switched state model predictive control of fractional-order nonlinear discrete-time systems. Asian J. Control **15**(3), 658–668 (2013). https://doi.org/10.1002/asjc.703
39. Domek, S.: Switched fractional state-space predictive control methods for non-linear fractional systems. In: Conference on Non-integer Order Calculus and Its Applications, Lecture Notes in Electrical Engineering, vol. 559, pp. 113–127. Springer (2019)
40. Domek, S.: Discrete-time switched models of non-linear fractional-order systems. In: Advanced, Contemporary Control, pp. 1176–1188. Springer (2020)
41. Domek, S.: Switched models of non-integer order. In: Kulczycki, P., Korbicz, J., Kacprzyk, J. (eds.) Automatic Control, Robotics, and Information Processing. Studies in Systems, Decision and Control, vol. 296. Springer (2021)
42. Dumlu, A., Erenturk, K.: Trajectory tracking control for a 3-DOF parallel manipulator using fractional-order $PI^{\lambda}D^{\mu}$ control. IEEE Trans. Ind. Electron. **61**(7), 3417–3426 (2014). https://doi.org/10.1109/TIE.2013.2278964
43. Dzieliń ski, A., Czyronis, P.M.: Fixed final time and free final state optimal control problem for fractional dynamic systems – linear quadratic discrete-time case. Bull. Pol. Acad. Sci. Tech. Sci. **61**(3), 681–690 (2013)
44. Dzieliński, A., Czyronis, P.M.: Optimal control problem for fractional dynamic systems – linear quadratic discrete-time case. In: Advances in the Theory and Applications of Non-integer Order Systems, Lecture Notes in Electrical Engineering, vol. 257, pp. 87–97. Springer (2013)
45. Dzieliński, A., Sierociuk, D., Sarwas, G.: Some applications of fractional order calculus. Bull. Pol. Acad. Sci. Tech. Sci. **58**(4), 583–592 (2009)
46. Dzieliński, A., Sierociuk, D.: Stability of discrete fractional order state-space systems. J. Vib. Control **14**(9–10), 1543–1556 (2008)
47. Dzieliński, A., Sierociuk, D.: Ultracapacitor modelling and control using discrete fractional order state-space model. Acta Montanistica Slovaca **13**(1), 136–145 (2008)
48. Erenturk, K.: Fractional-order $PI^{\lambda}D^{\mu}$ and active disturbance rejection control of nonlinear two-mass drive system. IEEE Trans. Ind. Electron. **60**(9), 3806–3813 (2013). https://doi.org/10.1109/TIE.2012.2207660
49. Freeborn, T.J., Maundy, B., Elwakil, A.S.: Measurement of supercapacitor fractional-order model parameters from voltage excited step response. IEEE J. Emerg. Sel. Top. Circuits Syst. **3**(3), 367–376 (2013)
50. Giernacki, W., Sadalla, T.: Comparison of tracking performance and robustness of simplified models of multirotor UAV's propulsion unit with CDM and PID controllers (with anti-windup compensation). CEAI **19**(3), 31–40 (2017)
51. Gorenflo, R., Mainardi, F.: Random walks models for space-fractional diffusion processes. Fract. Calc. Appl. Anal. **1**(2), 167–191 (1998)
52. Grünwald, A.: Ueber begrenzte derivationen und deren anwendung. Zeitschrift für angewandte Mathematik und Physik **12**, 441–480 (1867)
53. Grzymkowski, L., Trofimowicz, D., Stefański, T.P.: Stability analysis of interconnected discrete-time fractional-order LTI state-space systems. Int. J. Appl. Math. Comput. Sci. **30**(4), 649–658 (2020). https://doi.org/10.34768/amcs-2020-0048
54. Heaviside, O.: Electrical Papers. The Macmillan Company (1892)
55. Horla, D., Sadalla, T.: Optimal tuning of fractional-order controllers based on Fibonacci-search method. ISA Trans. **104**, 287–298 (2020). https://doi.org/10.1016/j.isatra.2020.05.022
56. Hosseinnia, S.H., Tejado, I., Vinagre, B.M., Sierociuk, D.: Boolean-based fractional order SMC for switching systems: application to a DC-DC buck converter. Signal Image Video Process. **6**, 445–451 (2012)
57. Hu, S., Liao, Z., Chen, W.: Sinogram restoration for low-dosed X-Ray computed tomography using fractional-order Perona-Malik diffusion. Math. Probl. Eng. **2012** (2012). Paper ID: 391050

58. Huang, F., Liu, F.: The time-fractional diffusion equation and fractional advection-dispersion equation. Aust. N. Z. Ind. Appl. Math. J. **46**, 1–14 (2005)
59. Idczak, D., Walczak, S.: On a linear-quadratic problem with Caputo derivative. Opuscula Mathematica **31**(1), 49–68 (2016). https://doi.org/10.7494/OpMath.2016.36.1.49
60. Jakowluk, W.: Optimal input signal design for fractional-order system identification optimal input signal design for fractional-order system identification. Bull. Pol. Acad. Sci. Tech. Sci. **67**(1), 37–44 (2019). https://doi.org/10.24425/bpas.2019.127336
61. Jesus, I.S., Machado, J.A.T.: Application of integer and fractional models in electrochemical systems. Math. Probl. Eng. **2012** (2012). Paper ID: 248175
62. Jumarie, G.: A Fokker-Planck equation of fractional order with respect to time. J. Math. Phys. **33**(4), 3536–3542 (1992)
63. Kaczorek, T.: Reachability and controllability to zero of cone fractional linear systems. Arch. Control Sci. **17**(4), 357–367 (2007)
64. Kaczorek, T.: Reachability and controllability to zero of positive fractional discrete-time systems. In: 2007 European Control Conference (ECC), pp. 1708–1712 (2007). https://doi.org/10.23919/ECC.2007.7068247
65. Kaczorek, T.: Fractional positive continuous-time linear systems and their reachability. Int. J. Appl. Math. Comput. Sci. **18**(2), 223–228 (2008)
66. Kaczorek, T.: Practical stability and asymptotic stability of positive fractional 2d linear systems. Asian J. Control **12**(2), 200–207 (2010)
67. Kaczorek, T.: Practical stability of positive fractional 2d linear systems. Multidimension. Syst. Signal Process. **21**(3), 231–238 (2010)
68. Kaczorek, T.: New stability tests of positive standard and fractional linear systems. Circuits Syst. **2**(4), 261–268 (2011)
69. Kaczorek, T.: Positivity and reachability of fractional electrical circuits. Acta Mechanica et Automatica **5**(2) (2011)
70. Kaczorek, T.: Selected Problems of Fractional Systems Theory. Springer, Berlin, Germany (2011)
71. Kaczorek, T.: Stability of positive fractional switched continuous-time linear systems. Bull. Pol. Acad. Sci. Tech. Sci. **61**(2), 349–352 (2013)
72. Kaczorek, T.: Zeroing of state variables in fractional descriptor electrical circuits by state-feedbacks. Arch. Electr. Eng. **63**(3) (2014)
73. Kaczorek, T.: Standard and positive electrical circuits with zero transfer matrices. Poznan Univ. Technol. Acad. J. Electr. Eng. **85**, 11–28 (2016)
74. Kaczorek, T.: Absolute stability of a class of fractional positive nonlinear systems. Int. J. Appl. Math. Comput. Sci. **29**(1), 93–98 (2019)
75. Kaczorek, T.: Decentralized stabilization of fractional positive descriptor discrete-time linear systems. In: Non-Integer Order Calculus and its Applications. RRNR 2017, Lecture Notes in Electrical Engineering, vol. 496. Springer (2019)
76. Kaczorek, T.: Global stability of positive standard and fractional nonlinear feedback systems. Bull. Pol. Acad. Sci. Tech. Sci. **68**(2), 285–288 (2020)
77. Kaczorek, T., Borawski, K.: Stability of positive nonlinear systems. In: 22nd International Conference on Methods and Models in Automation and Robotics. Miedzyzdroje, Poland (2017)
78. Klamka, J.: Controllability of fractional discrete-time systems with delay. In: Zeszyty Naukowe Politechniki Ślaskiej. Seria: Automatyka, vol. 151 (2008)
79. Klamka, J.: Local controllability of fractional discrete-time semilinear systems. acta mechanica et automatica **5**(2), 55–58 (2011)
80. Klimek, M., Agrawal, O.: Fractional Sturm-Liouville problem. Comput. Math. Appl. **66**(5), 795–812 (2013). https://doi.org/10.1016/j.camwa.2012.12.011
81. Klimek, M., Ciesielski, M., Błaszczyk, T.: Exact and numerical solutions of the fractional Sturm-Liouville problem. Fract. Calc. Appl. Anal. **21**(1), 45–71 (2018). https://doi.org/10.1515/fca-2018-0004

82. Klimek, M., Malinowska, A.B., Odzijewicz, T.: Applications of the fractional Sturm-Liouville problem to the space-time fractional diffusion in a finite domain. Fract. Calc. Appl. Anal. **19**(2), 516–550 (2016). https://doi.org/10.1515/fca-2016-0027
83. Ko, L.T., Chen, J.E., Shieh, Y.S., Scalia, M., Sung, T.Y.: A novel fractional-discrete-cosine-transform-based reversible watermarking for healthcare information management systems. Comput. Math. Methods Med. **2012** (2012). Paper ID: 757018
84. Kopka, R.: Estimation of supercapacitor energy storage based on fractional differential equations. Nanoscale Res. Lett. **12**(1), 636 (2017). https://doi.org/10.1186/s11671-017-2396-y
85. Kopka, R.: Discrepancy between derivative orders in fractional supercapacitor models for charging and discharging cycles. In: 2018 23rd International Conference on Methods Models in Automation Robotics (MMAR), pp. 567–572 (2018). https://doi.org/10.1109/MMAR.2018.8486079
86. Kopka, R.: Changes in derivative orders for fractional models of supercapacitors as a function of operating temperature. IEEE Access **7**, 47674–47681 (2019). https://doi.org/10.1109/ACCESS.2019.2909708
87. Krajewski, W., Viaro, U.: A method for the integer-order approximation of fractional-order systems. J. Frankl. Inst. **351**(1), 555–564 (2014)
88. Krajewski, W., Viaro, U.: A new method for the integer order approximation of fractional order models. In: Theoretical Developments and Applications of Non-Integer Order Systems, Lecture Notes in Electrical Engineering, vol. 357, pp. 81–92. Springer (2016)
89. Kukla, S., Siedlecka, U.: An analytical solution to the problem of time-fractional heat conduction in a composite sphere. Bull. Pol. Acad. Sci. Tech. Sci. **65**(2), 179–186 (2017)
90. Lacroix, S.F.: Traité du calcul Différentiel et du Calcul Intégral. Tome troisieme (1819)
91. Laski, P.A.: Fractional-order feedback control of a pneumatic servo-drive. Bull. Pol. Acad. Sci. Tech. Sci. **67**(1), 53–59 (2019)
92. Latawiec, K.J., Stanisławski, R., Łukaniszyn, M., Czuczwara, W., Rydel, M.: Fractional-order modeling of electric circuits: modern empiricism vs. classical science. In: 2017 Progress in Applied Electrical Engineering (PAEE), pp. 1–4 (2017). https://doi.org/10.1109/PAEE.2017.8008998
93. Latawiec, K.J., Stanisławski, R., Łukaniszyn, M., Rydel, M., Szkuta, B.R.: FFLD-based modeling of fractional-order state space LTI MIMO systems. In: Applied Physics, System Science and Computers, Proc. 1st Int. Conf. on Appl. Phys., Syst. Sci. and Comp. (APSAC2016), Lecture Notes in Electrical Engineering, vol. 428, Springer (2017). https://doi.org/10.1007/978-3-319-53934-8
94. Latawiec, K.J., Stanisławski, R., Łukaniszyn, M., Rydel, M., Szkuta, B.R.: Grunwald-Letnikoy-Laguerre modeling of discrete-time noncommensurate fractional-order state space LTI MIMO systems. In: Non-integer Order Calculus and Its Applications, Lecture Notes in Electrical Engineering, vol. 496, pp. 74–83. Springer (2018)
95. Laurent, H.: Sur le calcul des dérivées á indicies quelconques. Nouv. Annales de Mathématiques **3**(3), 240–252 (1884)
96. Lazarević, M.P., Rapaić, M.R., Śekara, T.B.: Introduction to fractional calculus with brief historical background. In: Advanced Topics on Applications of Fractional Calculus on Control Problems, System Stability and Modeling. WSEAS Press (2014)
97. Letnikov, A.V.: Theory of Differentiation with an Arbitrary Index. Moskow Matem, Sbornik (1869)
98. Letnikov, A.V.: An explanation of the concepts of the theory of differentiation of arbitrary index. Moskow Matem. Sbornik **6**, 413–445 (1872)
99. Li, Y., Chen, Y., Podlubny, I.: Stability of fractional-order nonlinear dynamic systems: Lyapunov direct method and generalized Mittag-Leffler stability. Comput. Math. Appl. **59**(5), 1810–1821 (2010)
100. Lioubille, J.: Méemoire sur le théoréme des fonctions complémentaires. J. fur reine und angew. Math. **11**, 1–19 (1934)
101. Liouville, J.: Mémoire sur l'intégration de l'équation $(mx^2 + nx + p)\frac{d^2y}{dx^2} + (qx + r)\frac{dy}{dx} + sy = 0$ á l'aide des différentielles á indices quelconques. ournal d l'Ecole Polytechnique **21**, 163–186 (1832)

102. Lützen, J.: Differentiation of arbitrary order. In: Studies in the History of Mathematics and Physical Sciences. Springer, New York (1990)
103. Maachou, A., Malti, R., Melchior, P., Battaglia, J.L., Hay, B.: Thermal system identification using fractional models for high temperature levels around different operating points. Nonlinear Dyn. **70**(2), 941–950 (2012)
104. Machado, J.A.T.: Fractional control of heat diffusion systems. Nonlinear Dyn. **54**(3), 263–282 (2012)
105. Mainardi, F.: Fractional relaxation-oscilation and fractional diffusion-wave phenomena. Chaos Solitons Fractals **7**, 1461–1477 (1996)
106. Malesza, W., Sierociuk, D.: Duality properties of variable-type and -order differences. In: Non-Integer Order Calculus and its Applications. RRNR 2017, Lecture Notes in Electrical Engineering, vol. 496. Springer (2018)
107. Malinowska, A.B., Odziejewicz, T., Torres, D.F.: Advanced Methods in the Fractional Calculus of Variations. Springer (2015)
108. Mandelbrot, B.B., Van Ness, J.W.: Fractional Brownian motion, fractional noises and applications. SIAM Rev. **10**(4), 422–437 (1968)
109. Mathieu, B., Melchior, P., Oustaloup, A., Ceyral, C.: Fractional differentiation for edge detection. Signal Process. - Spec. Issue: Fract. Signal Process. Appl. Arch. **83**(11), 2421–2432 (2003)
110. Matignon, D.: Stability properties for generalized fractional differential systems. ESAIM Proc. **5**, 145–158 (1998)
111. Matignon, D., d'Andrea-Novel, B., Depalle, P., Oustaloup, A.: Viscothermal Losses in Wind Instruments: A Non-integer Mode. Academic Verlag, Berlin, Germany (1994)
112. Mehandiratta, V., Mehra, M., Leugering, G.: Fractional optimal control problems on a star graph: optimality system and numerical solution. Math. Control Relat. Fields **11**, 189 (2021). https://doi.org/10.3934/mcrf.2020033
113. Mendiola-Fuentes, J., Melchor-Aguilar, D.: Modification of Mikhailov stability criterion for fractional commensurate order systems. J. Franklin Inst. **355**, 2779–2790 (2018)
114. Metzler, R., Klafter, J.: The restaurant at the end of the random walk: recent developments in the description of anomalous transport by fractional dynamics. J. Phys. A **37**(31), 161–208 (2004)
115. Mitkowski, W., Bauer, W., Zagorowska, M.: RC-ladder networks with supercapacitors. Arch. Electr. Eng. **67**(2), 377–389 (2018)
116. Mitkowski, W., Skruch, P.: Fractional-order models of the supercapacitors in the form of RC ladder networks. Bull. Pol. Acad. Sci. Tech. Sci. **61**(3), 581–587 (2013)
117. Mozyrska, D., Bartosiewicz, Z.: On observability of nonlinear discrete-time fractional-order control systems. In: New Trends in Nanotechnology and Fractional Calculus Applications, pp. 305–312. Springer (2010)
118. Mozyrska, D., Girejko, E., Wyrwas, M.: Fractional nonlinear systems with sequential operators. Cent. Eur. J. Phys. **11**(10), 1295–1303 (2013)
119. Mozyrska, D., Oziablo, P., Wyrwas, M.: Stability of fractional variable order difference systems. Fract. Calc. Appl. Anal. **22**(3) (2019). https://doi.org/10.1515/fca-2019-0044
120. Mozyrska, D., Pawluszewicz, E.: Local controllability of nonlinear discrete-time fractional order systems. Bull. Pol. Acad. Sci. Tech. Sci. **61**(1), 251–256 (2013)
121. Mozyrska, D., Wyrwas, M.: Stability of linear discrete-time systems with the Caputo fractional-, variable-order h-difference operator of convolution type. In: Proceedings of International Conference on Fractional Differentiation and its Applications (ICFDA) 2018 (2018). https://doi.org/10.2139/ssrn.3270846
122. Mozyrska, D., Wyrwas, M.: Stability of linear systems with Caputo fractional-, variable-order difference operator of convolution type. In: 2018 41st International Conference on Telecommunications and Signal Processing (TSP), pp. 1–4 (2018)
123. Mughees, A., Mohsin, S.A.: Design and control of magnetic levitation system by optimizing fractional order PID controller using ant colony optimization algorithm. IEEE Access **8**, 116,704–116,723 (2020). https://doi.org/10.1109/ACCESS.2020.3004025

124. Muñoz, J., Monje, C.A., Nagua, L.F., Balaguer, C.: A graphical tuning method for fractional order controllers based on iso-slope phase curves. ISA Trans. **105**, 296–307 (2020). https://doi.org/10.1016/j.isatra.2020.05.045
125. Neto, J.P., Coelho, R.M., Valério, D., Vinga, S., Sierociuk, D., Malesza, W., Macias, M., Dzieliński, A.: Simplifying biochemical tumorous bone remodeling models through variable order derivatives. Comput. Math. Appl. **75**(9), 3147–3157 (2018). https://doi.org/10.1016/j.camwa.2018.01.037
126. Nguyen, S.D., Lam, B.D., Choi, S.B.: Smart dampers-based vibration control – part 2: Fractional-order sliding control for vehicle suspension system. Mech. Syst. Signal Process. **148**, 107,145 (2021). https://doi.org/10.1016/j.ymssp.2020.107145
127. Nowak, T.K., Duzinkiewicz, K., Piotrowski, R.: Numerical investigation of nuclear reactor kinetic and heat transfer fractional model with temperature feedback. In: 20th IIEEE international Conference on Methods and Models in Automation and Robotics, pp. 585–590. Miedzyzdroje, Poland (2015)
128. Oldham, K., Spanier, J.: The Fractional Calculus. Academic Press, Orlando, FL (1974)
129. Oprzedkiewcz, K., Gawin, E., Mitkowski, W.: A plc implementation of PSE approximant for fractional order operator. In: Non-integer Order Calculus and Its Applications, Lecture Notes in Electrical Engineering, vol. 496, pp. 102–112. Springer (2019)
130. Oprzedkiewcz, K., Mitkowski, W.: Accuracy estimation of the approximated Atangana-Baleanu operator. J. Appl. Math. Comput. Mech. **18**(4), 53–62 (2019). https://doi.org/10.17512/jamcm.2019.4.05
131. Oprzedkiewcz, K., Mitkowski, W., Gawin, E.: The PLC Implementation of Fractional-Order Operator Using CFE Approximation. Springer (2017)
132. Oprzedkiewcz, K., Rosol, M., Zeglen, J.: Fractional order (pid beta)-d-alpha controller for the inverted pendulum. In: Automation 2020: Towards Industry of the Future, Advances in Intelligent Systems and Computing, vol. 1140, pp. 170–181. Springer (2020)
133. Oprzedkiewicz, K., Dziedzic, K., Wieckowski, L.: Non integer order, discrete, state space model of heat transfer process using Grunwald-Letnikov operator. Bull. Pol. Acad. Sci. Tech. Sci. **67**(5), 905–914 (2019). https://doi.org/10.24425/bpasts.2019.130873
134. Oprzedkiewicz, K., Mitkowski, W., Gawin, E., Dziedzic, K.: The Caputo vs. Caputo-Fabrizio operators in modeling of heat transfer process. Bull. Pol. Acad. Sci. Tech. Sci. **66**(4), 501–507 (2019)
135. Oprzedkiewicz, K., Stanisławski, R., Gawin, E., Mitkowski, W.: A new algorithm for a CFE-approximated solution of a discrete-time non integer-order state equation. Bull. Pol. Acad. Sci. Tech. Sci. **65**(4), 429–437 (2017)
136. Ostalczyk, P.: The non-integer difference of the discrete-time function and its application to the control system synthesis. Int. J. Syst. Sci. **31**(12), 1551–1561 (2000)
137. Ostalczyk, P.: A note on the Grünwald-Letnikov fractional-order backward-difference. Physica Scripta **136** (2009). Paper ID: 014036
138. Ostalczyk, P.: Stability analysis of a discrete-time system with a variable-fractional-order controller. Bull. Pol. Acad. Sci. Tech. Sci. **58**(4), 613–619 (2010)
139. Ostalczyk, P.: Equivalent descriptions of a discrete-time fractional-order linear system and its stability domains. Int. J. Appl. Math. Comput. Sci. **22**(3), 533–538 (2012)
140. Ostalczyk, P.: Discrete Fractional Calculus. Applications in Control and Image Processing. World Scientific (2016)
141. Ostalczyk, P.: However: "differential-integral fractional-order calculus". riposte to ryszard sikora's article: "fractional derivatives in electrical circuit theory – critical remarks" (in polish). Przegląd Elektrotechniczny **93**(3), 175–180 (2017)
142. Oustaloup, A., Levron, F., Nanot, F.: Frequency band complex non integer differentiator: characterization and synthesis. IEEE Trans. Circuits Syst. I: Fundam. Theory Appl. **47**(1), 25–40 (2000)
143. Pan, I., Das, S., Gupta, A.: Handling packet dropouts and random delays for unstable delayed processes in NCS by optimal tuning of $PI^{\lambda}D^{\mu}$ controllers with evolutionary algorithms. ISA Trans. **50**(4), 557–572 (2011). https://doi.org/10.1016/j.isatra.2011.04.002

144. Pawluszewicz, E.: Constrained controllability of the h-difference fractional control systems with caputo type operator. Discrete Dyn. Nat. Soc. **2015**(638420) (2015). https://doi.org/10.1155/2015/638420
145. Pawluszewicz, E.: Perfect observers for fractional discrete-time linear systems. Kybernetika **52**(6), 914–928 (2016)
146. Petráš, I.: Stability of fractional-order systems with rational orders: a survey. Fract. Calc. Appl. Anal. **12**(3), 269–298 (2009)
147. Podlubny, I.: Fractional Differential Equations. Academic Press, Orlando, FL (1999)
148. Podlubny, I.: Fractional-order systems and $PI^{\lambda}D^{\mu}$ controllers. IEEE Trans. Autom. Control **44**(1), 208–214 (1999). https://doi.org/10.1109/9.739144
149. Podlubny, I., Magin, R.L., Trymorush, I.: Niels Henrik Abel and the birth of fractional calculus. Fract. Calc. Appl. Anal. **20**(5), 1068–1075 (2017). https://doi.org/10.1515/fca-2017-0057
150. Povstenko, Y.: Fractional heat conduction in an infinite medium with a spherical inclusion. Entropy **15**(10), 4122–4133 (2013)
151. Povstenko, Y.Z.: Fractional heat conduction in infinite one-dimensional composite medium. J. Therm. Stresses **36**(4), 351–363 (2013). https://doi.org/10.1080/01495739.2013.770693
152. Puchalski, B., Rutkowski, T.A., Duzinkiewicz, K.: Fuzzy multi-regional fractional PID controller for pressurized water nuclear reactor. ISA Trans. **103**, 86–102 (2020). https://doi.org/10.1016/j.isatra.2020.04.003
153. Ramezanian, H., Balochian, S.: Optimal design a fractional- order PID controller using particle swarm optimization algorithm. Int. J. Control Autom. **6**(4), 55–67 (2013)
154. Ren, H., Fan, J., Kaynak, O.: Optimal design of a fractional-order proportional-integer-differential controller for a pneumatic position servo system. IEEE Trans. Ind. Electron. **66**(8), 6220–6229 (2019). https://doi.org/10.1109/TIE.2018.2870412
155. Ren, H.P., Zheng, T.: Optimization design of power factor correction converter based on genetic algorithm. In: International Conference on Genetic and Evolutionary Computation, pp. 293–296 (2011). https://doi.org/10.1109/ICGEC.2010.79
156. Riemann, B.G.: Versuch einer auffassung der integration und differentiation. In: Gesannnelte Werke, pp. 331–344 (1876)
157. Riesz, M.: L'intégrales de riemann-liouville et le probléme de cauchy. Acta Math. **81** (1949)
158. Ross, B.: The developement of fractional calculus 1695–1900. Histroia Mathematica **4**, 75–89 (1977)
159. Ross, B.E.: Fractional Calculus and Its Applications; Proceedings of the International Conference Held at the University of New Haven. Springer, London, UK (1975)
160. Ruszewski, A.: Practical and asymptotic stability of fractional discrete-time scalar systems described by a new model. Arch. Control Sci. **26**(4), 441–452 (2016)
161. Ruszewski, A.: Stability analysis for the new model of fractional discrete-time linear state-space systems. In: Theory and Applications of Non-integer Order Systems, Lectures Notes on Electrical Engineering, vol. 407, pp. 381–389. Springer (2017)
162. Ruszewski, A.: Stability of discrete-time fractional linear systems with delays. Arch. Control Sci. **29**(3), 549–567 (2019). https://doi.org/10.24425/acs.2019.130205
163. Rydel, M.: New integer-order approximations of discrete-time non-commensurate fractional-order systems using the cross Gramian. Adv. Comput. Math. **45**, 631–653 (2019). https://doi.org/10.1007/s10444-018-9633-5
164. Sabatier, J., Aoun, M., Oustaloup, A., Gregoire, G., Ragot, F., Roy, P.: Fractional system identification for lead acid battery state of charge estimation. Signal Process. **86**(10), 2645–2657 (2006)
165. Sajewski, L.: Reachability, observability and minimum energy control of fractional positive continuous-time linear systems with two different fractional orders. Multidimension. Syst. Signal Process. **27**(1), 27–41 (2016)
166. Sajewski, L.: Decentralized stabilization of descriptor fractional positive continuous-time linear systems with delays. In: 22nd International Conference on Methods and Models in Automation and Robotics. Miedzyzdroje, Poland (2017)

167. Sajewski, L.: Stabilization of positive descriptor fractional discrete-time linear systems with two different fractional orders by decentralized controller. Bull. Pol. Acad. Sci. Tech. Sci. **65**(5), 709–714 (2017)
168. Sheng, H., Chen, Y.Q.: Farima with stable innovations model of great salt lake elevation time series. Signal Process. **91**(3), 553–561 (2011)
169. Sheng, H., Chen, Y.Q., Qiu, T.S., et al.: Analysis of biocorrosion electrochemical noise using fractional order signal processing techniques. In: Sabatier, J. (ed.) Fractional Processes and Fractional-Order Signal Processing Signals and Communication Technology, pp. 189–202. Springer, Dordrecht, Netherlands (2012)
170. Siami, M., Tavazoei, M.S., Haeri, M.: Stability preservation analysis in direct discretization of fractional order transfer functions. Signal Process. **91**(3), 508–512 (2011)
171. Sierociuk, D.: Fractional Variable Order Derivative Simulink Toolkit (2019). https://www.mathworks.com/matlabcentral/fileexchange/38801-fractional-variable-order-derivative-simulink-toolkit
172. Sierociuk, D., Dzieliński, A.: New method of fractional order integrator analog modeling for orders 0.5 and 0.25. In: Proceedings of the 16th International Conference on Methods and Models in Automation and Robotics, Miedzyzdroje, Poland, pp. 137–141 (2011)
173. Sierociuk, D., Macias, M., Malesza, W.: Analog realization of fractional variable-type and -order iterative operator. Appl. Math. Comput. **336**, 138–147 (2018). https://doi.org/10.1016/j.amc.2018.04.047
174. Sierociuk, D., Macias, M., Malesza, W., Sarwas, G.: Dual estimation of fractional variable order based on the unscented fractional order Kalman filter for direct and networked measurements. Circuits Syst. Signal Process. **35**(6), 2055–2082 (2019). https://doi.org/10.1007/s00034-016-0255-1
175. Sierociuk, D., Macias, M., Malesza, W., Wiraszka, M.S.: Analog realization of a fractional recursive variable-type and order operator for a particular switching strategy. Electronics **9**(5) (2020)
176. Sierociuk, D., Malesza, W.: Fractional variable order discrete-time systems, their solutions and properties. Int. J. Syst. Sci. **48**(14), 3098–3105 (2017). https://doi.org/10.1080/00207721.2017.1365969
177. Sierociuk, D., Malesza, W., Macias, M.: Derivation, interpretation, and analog modelling of fractional variable order derivative definition. Appl. Math. Model. **39**(13), 3876–3888 (2015). https://doi.org/10.1016/j.apm.2014.12.009
178. Sierociuk, D., Malesza, W., Macias, M.: Numerical schemes for initialized constant and variable fractional-order derivatives: matrix approach and its analog verification. J. Vib. Control **22**(8), 2032–2044 (2016). https://doi.org/10.1177/1077546314565438
179. Sierociuk, D., Skovranek, T., Macias, M., Podlubny, I., Petras, I., Dzielinski, A., Ziubinski, P.: Diffusion process modeling by using fractional-order models. Appl. Math. Comput. **257**, 2–11 (2015). https://doi.org/10.1016/j.amc.2014.11.028
180. Sierociuk, D., Vinagre, B.M.: Infinite horizon state-feedback LQR controller for fractional systems. In: Proceedings of the 49th IEEE Conference on Decision and Control (CDC), Atlanta, GA, pp. 10,824–10,829 (2010)
181. Sikora, R.: Fractional derivatives in electrical circuit theory - critical remarks. Arch. Electr. Eng. **66**(1), 155–163 (2017)
182. Sonin, N.Y.: On differentiation with arbitrary index. Mo pp. 1–38 (1869)
183. Sowa, M.: Subival. https://msowascience.com/
184. Sowa, M.: Application of subival in solving initial value problems with fractional derivatives. Appl. Math. Comput. **319**, 86–103 (2018). https://doi.org/10.1016/j.amc.2017.01.047
185. Stanisławski, R.: New Laguerre filter approximators to the Grünwald-Letnikov fractional difference. Math. Probl. Eng. **2012**, 1–21 (2012). Article ID: 732917
186. Stanisławski, R.: New results in stability analysis for LTI SISO systems modeled by GL-discretized fractional-order transfer functions. Fract. Calc. Appl. Anal. **20**(1), 243–259 (2017). https://doi.org/10.1515/fca-2017-0013

187. Stanisławski, R., Latawiec, K.J.: Stability analysis for discrete-time fractional-order LTI state-space systems. Part I: new necessary and sufficient conditions for asymptotic stability. Bull. Pol. Acad. Sci. Tech. Sci. **61**(2), 353–361 (2013)
188. Stanisławski, R., Latawiec, K.J.: Stability analysis for discrete-time fractional-order LTI state-space systems. Part II: new stability criterion for FD-based systems. Bull. Pol. Acad. Sci. Tech. Sci. **61**(2), 362–370 (2013)
189. Stanisławski, R., Latawiec, K.J.: A modified Mikhailov stability criterion for a class of discrete-time noncommensurate fractional-order systems. Commun. Nonlinear Sci. Numer. Simul. **96**, 105,697 (2021). https://doi.org/10.1016/j.cnsns.2021.105697
190. Stanisławski, R., Latawiec, K.J., Łukaniszyn, M.: A comparative analysis of Laguerre-based approximators to the Grünwald-Letnikov fractional-order difference. Math. Probl. Eng. **2015**, 1–10 (2015). Article ID: 512104
191. Stanisławski, R., Rydel, M., Latawiec, K.J.: Modeling of discrete-time fractional-order state space systems using the balanced truncation method. J. Frankl. Inst. **354**(7), 3008–3020 (2017)
192. Stanisławski, R., Rydel, M., Latawiec, K.J.: New stability tests for discretized fractional-order systems using the Al-Alaoui and Tustin operators. Complexity **2018**(2036809), 1–9 (2018)
193. Sun, H., Chen, W., Chen, Y.: Variable-order fractional differential operators in anomalous diffusion modeling. Physica A: Stat. Mech. Appl. **388**(21), 4586–4592 (2009). https://doi.org/10.1016/j.physa.2009.07.024
194. Sun, H.H., Onaral, B., Tsao, Y.: Application of the positive reality principle to metal electrode linear polarization phenomena. IEEE Trans. Biomed. Eng. **31**(10), 664–674 (1984)
195. Tavares, D., Almeida, R., Torres, D.F.: Caputo derivatives of fractional variable order: numerical approximations. Commun. Nonlinear Sci. Numer. Simul. **35**, 69–87 (2016). https://doi.org/10.1016/j.cnsns.2015.10.027
196. Tavazoei, M., Asemani, M.H.: On robust stability of incommensurate fractional-order systems. Commun. Nonlinear Sci. Numer. Simul. **90**, 105,344 (2020)
197. Tepljakov, A., Gonzalez, E.A., Petlenkov, E., Belikov, J., Monje, C.A., Petráš, I.: Incorporation of fractional-order dynamics into an existing PI/PID DC motor control loop. ISA Trans. **60**, 262–273 (2016). https://doi.org/10.1016/j.isatra.2015.11.012
198. Tepljakov, A., Petlenkov, E., Belikov, J.: FOMCON Toolbox (2011). http://www.fomcon.net/
199. Tseng, C.C.: Design of variable and adaptive fractional order FIR differentiators. Signal Process. **86**(10), 2554–2566 (2006)
200. Valério, D., Sa da Costa, J.: Non-integer order control of a flexible robot. In: Proceedings of the IFAC Workshop on Fractional Differentiation and its Applications, FDA'04. Bordeaux, France (2004)
201. Verma, S.K., Yadav, S., Nagar, S.K.: Optimization of fractional order PID controller using grey wolf optimizer. J. Control Autom. Electr. Syst. **28**(3), 314–322 (2017). https://doi.org/10.1007/s40313-017-0305-3
202. Wiraszka, M.S., Wierzchowski, M., Wojciuk, M.: State-dependent fractional-order PI control strategy for a nonlinear water tank system. In: 2019 20th International Carpathian Control Conference (ICCC), pp. 1–6 (2019)
203. Wyrwas, M., Mozyrska, D., Girejko, E.: Stability of discrete fractional-order nonlinear systems with the Nabla Caputo difference. IFAC Proc. Vol. **46**(1), 167–171 (2013). 6th IFAC Workshop on Fractional Differentiation and Its Applications
204. Wyrwas, M., Pawluszewicz, E., Girejko, E.: Stability of nonlinear h -difference systems with n fractional orders. Kybernetika **51**(1), 112–136 (2015)
205. Xie, Y., Zhang, X., Meng, W., Zheng, S., Jiang, L., Meng, J., Wang, S.: Coupled fractional-order sliding mode control and obstacle avoidance of a four-wheeled steerable mobile robot. ISA Trans. **108**, 282–294 (2021). https://doi.org/10.1016/j.isatra.2020.08.025
206. Yaghi, M., Önder Efe, M.: H_2/H_∞-neural-based FOPID controller applied for radar-guided missile. IEEE Trans. Ind. Electron. **67**(6), 4806–4814 (2020). https://doi.org/10.1109/TIE.2019.2927196
207. Yang, H.: A novel fractional-order signal processing based edge detection method. In: Proceedings of the International Conference on Control Automation Robotics & Vision, pp. 1122–1127. Nanjing, China (2010)

208. Yang, Y., Tan, J., Yue, D., Xie, X., Yue, W.: Observer-based containment control for a class of nonlinear multiagent systems with uncertainties. IEEE Trans. Syst. Man Cybern. Syst. **51**(1), 588–600 (2021). https://doi.org/10.1109/TSMC.2018.2875515
209. Zha, D.: Underwater 2-D source localization based on fractional order correlation using vector hydrophone. In: Proceedings of the 2008 Congress on Image and Signal Processing, pp. 31–34 (2008)
210. Zhang, S., Liu, L., Xue, D.: Nyquist-based stability analysis of non-commensurate fractional-order delay systems. Appl. Math. Comput. **377**, 125,111 (2020)

Fractional Systems: Theoretical Foundations

Piotr Ostalczyk and Ewa Pawluszewicz

Abstract An overview of the fractional order integro-differential, methods and tools of fractional calculus that are commonly used in control and automation are presented. The presented results mainly concern the fractional differential operators Caputo, Riemann–Liouville and Grünwald–Letnikov, as they are some of the most used in the fields mentioned. Moreover, for zero initial value these operators coincides.

1 Introduction

In recent years fractional calculus has been viewed as a power tool for describing the behavior of real systems, see for example [8, 39, 45, 59, 68] and references therein. The term *fractional* basically means all non-integer numbers. In fact, in nature, there are many processes that can be more accurately modelled using fractional differintegrals, see for example [5, 8, 37, 42, 59, 86], references therein and others.

Although many believe that this is a new field of science, its foundations date back to the 17th century. Fractional order derivatives and integrals have a long standing theoretical foundation. The beginning of fractional calculus is dated to 1695. The ground for a fractional calculus was given by Liouville, Holmgren and Riemann, Grünwald, Letnikov, Riemann and others, see for example [41, 57]. At the same time, when theoretical beginnings were given, applying fractional calculus to various problems developed. Neils Abel was the first to notice and show that the application of fractional calculus could be more economical and useful than classical integer order calculus [57].

P. Ostalczyk
Faculty of Electrical Engineering, Institute of Control and Industrial Electronics, Warsaw University of Technology, Warsaw, Poland
e-mail: piotr.ostalczyk@ee.pw.edu.pl

E. Pawluszewicz (✉)
Institute of Robotics and Mechatronics, Bialystok University of Technology, Bialystok, Poland
e-mail: e.pawluszewicz@pb.edu.pl

P. Kulczycki et al. (eds.), *Fractional Dynamical Systems: Methods, Algorithms and Applications*, Studies in Systems, Decision and Control 402,
https://doi.org/10.1007/978-3-030-89972-1_2

A significant increase in interest in non-integer calculus, both from the theoretical and application point of view, took place in the second half of the 20th century. Undoubtedly, its practical applications were greatly influenced by the development of computer and IT techniques and tools. This progression contributed to the fact that today fractional calculus has become a power tool in descriptions of behaviours of real systems, see for example [28, 39, 42, 51, 52, 62, 77] and references therein. In modeling real phenomena, authors emphatically use generalizations of nth order differences to their fractional forms. The first steps in this topic were made in [47, 49].

On the other hand, one can find several definitions/notations of fractional derivatives among which the most popular are the Caputo, Riemman–Louville and Grünwald–Letnikov operators. Properties of the Caputo and Riemman–Louville integro-differential operators were developed by [3, 9, 15, 21, 41, 43, 45]. One can also consider a Grünwald–Letnikov fractional operator as a natural extension of the classical derivative in fractional calculus, see for example in [10, 39, 40, 68].

Applications of fractional calculus can be found today in many fields: for example in physics and mechanics [15, 37, 72], in the description and analysis of diffusion processes [29, 72], viscoelasticity [6, 7], economics [81], robotics [69], heat transferal [58, 77] and image processing [62]. Fractional order calculus has also proved to be an important tool for modeling control and dynamic systems [38, 51, 79], controllability and observability of systems [43, 55, 56, 74], identification processes, stability and stabilizability of systems [12, 71, 85], optimization [4, 27] and regulation [75, 86]. In automatic regulation and its industrial applications to process controlling, the most popular and commonly used are PID controllers. It is known however, that controllers of fractional orders (FOPID) are in many cases more robust and provide better optimal preferences than the classical ones, see [42, 79, 86]. These properties follow from the fact that such controllers have more tuning freedom. However the usage of FOPID controllers usually requires some approximations which makes their applications more complex [83, 86].

The purpose of this chapter is to review the basic non-integer order methods and tools that are used in automatic control. For this reason, in Sect. 2 the basic notation needed in the next parts of the chapter is given. Also the three most commonly used fractional order differential operators in control and automation, i.e. Caputo, Riemman–Louville and Grünwald–Letnikov fractional integro–differentials are recalled. It can be mentioned that the fractional order difference operators are omitted but an interested reader may refer for example to [46, 47, 56, 85]. Also fractional differential operators with variable order are missing here, but one may refer for example to [17, 44, 65, 76, 84]. Unilateral (one-side) Laplace transformation of the fractional order integral and derivative is discussed in Sect. 3. In the next step, in Sect. 4 fractional-order single-input-single-output systems are consider. The numerical approach to finding the solution is presented. Since zero initial conditions are assumed, the presented approach is valid for equations with Caputo, Riemman–Louville and Grünwald–Letnikov fractional differentials. Properties and applications of the fractional-order transfer function are discussed in Sect. 5. The Fourier transform is also described in this section. As a consequence, frequency characteristics of the considered systems and relation between fractional-order and inertial element are discussed. In Sect. 6 non-linear fractional-order systems are presented and studied based on the linear approximation their properties. In Sect. 7 stability conditions of SISO systems are given.

2 Fractional Derivatives and Integrals Definitions

2.1 Notation

The following notation will be used:

- $\mathbb{N}$—the set of natural numbers (including zero). Its elements will be in general denoted by letters i, j, k, m, n, p, q;
- $\mathbb{Q}$—the set of rational numbers. Its elements will be in general denoted by Greek letters $\nu = \frac{m}{n}$, $\mu = \frac{p}{q}$ for integers m, n, p, g, etc;
- $\mathbb{R}$—the set of real numbers. Its elements in general will be denoted by Greek letters ν, μ;
- $\Gamma(\cdot)$—gamma Euler function that is defined as

$$\Gamma(z) = \int_0^\infty (\tau)^{z-1} e^{\tau} d\tau \tag{1}$$

for complex numbers with $Re(z) > 0$ or by

$$\Gamma(z) = \lim_{n\to\infty} \frac{1}{z} \prod_{n=1}^{\infty} \frac{(1+\frac{1}{n})^z}{1+\frac{z}{n}}; \tag{2}$$

- $\mathbf{1}(\cdot)$—the Heaviside'a (unit step) function;
- the two parameters Mittag–Leffler function

$$E_{\alpha,\beta}(z) = \sum_{i=0}^{+\infty} \frac{z^i}{\Gamma(\alpha i + \beta)}. \tag{3}$$

Properties of this function in details are describe for example in [16, 68].
- For a discrete variable $k \in \mathbb{N}$ and a bounded order function $\nu \in \mathbb{R}$ a kernel function is defined as:

$$a^{\nu}(k) = \begin{cases} 1 & \text{for } k = 0 \\ (-1)^k \frac{\nu(\nu-1)\cdots(\nu-k+1)}{k!} & \text{for } k \in \mathbb{N} \end{cases}, \tag{4}$$

or equivalently

$$a^{\nu}(k) = (-1)^k \frac{\Gamma(\nu)}{\Gamma(k+1)\Gamma(\nu-k+1)}. \tag{5}$$

The function defined above will be called an *oblivion* or *decay function* for $\nu \in \mathbb{R}_+$ and a "collection" one for $\nu \in \mathbb{R}_-$. In order to standardize the notation, the order ν will be limited to positive values only. For a fractional-order the notation $-\nu < 0$ will be used. Properties of this function are given in [1, 53, 54]. Solutions of fractional differential equations with variable order and theirs application are discussed in [70, 78].

Fractional order calculus is based on generalization of differentiation and integration to an arbitrary order ν. It has led to the introduction of the basic continuous differintegral operator, see [50, 57]:

$$_aD_t^{\nu} = \begin{cases} \frac{d^{\nu}}{dt^{\nu}}, & \text{for } Re(\nu) > 0; \\ 1, & \text{for } Re(\nu) = 0; \\ \int_0^t (d\tau)^{\nu}, & \text{for } Re(\nu) < 0. \end{cases} \tag{6}$$

where a is a constant connected with initial conditions, the order ν can be rational, irrational or even complex.

There is an array of definitions of differintegral in the literature, see for example [41, 45, 73]. The three most frequently used in automatic and control definitions will be presented below. All presented results will be given without proofs.

2.2 Riemann–Liouville Fractional-Order Left-Sided Derivative/Integral

Let $[t_0, t]$, where $-\infty < t_0 < t < +\infty$, be a continuous real variable function $f(t)$ and let $\nu > 0$. Then the *Riemann–Liouville left-sided fractional-order integral* is defined as follows

$$^{RL}_{t_0}I_t^{\nu} f(t) = \frac{1}{\Gamma(\nu)} \int_{t_0}^{t} \frac{f(\tau)}{(t-\tau)^{1-\nu}} d\tau. \tag{7}$$

On a base of the Riemann–Liouville fractional-order integral its counterpart—derivative, i.e. the *Riemann–Liouville left-sided fractional-order derivative* can be defined as

$$^{RL}_{t_0}D_t^{\nu} f(t) = \left(\frac{d}{dt}\right)^n {}^{RL}_{t_0}I_t^{n-\nu} f(t) = \frac{1}{\Gamma(n-\nu)} \left(\frac{d}{dt}\right)^n \int_{t_0}^{t} \frac{f(\tau)}{(t-\tau)^{1+\nu-n}} d\tau \tag{8}$$

for any $n-1 < \nu < n$, $n \in \mathbb{N}$. For $n = 1$

Table 1 Riemann–Liouville fractional-order derivatives with $t_0 = 0$

$f(t)$	${}_0^{RL}D_t^0 f(t)\ t > 0,\ \nu \in \mathbb{R}$
$\delta(t)$	$\frac{t^{-\nu-1}}{\Gamma(-\nu)}$
$\delta(t-\tau)$	$\frac{(t-\tau)^{-\nu-1}}{\Gamma(-\nu)}$
$\mathbf{1}(t)$	$\frac{t^{-\nu}}{\Gamma(1-\nu)}$
$\mathbf{1}(t-\tau)$	$\begin{cases} \frac{(t-\tau)^{-\nu}}{\Gamma(1-\nu)} & \text{for} \quad t > \tau \\ 0 & \text{for } 0 \leqslant t \leqslant \tau \end{cases}$
$t^{\alpha}\mathbf{1}(t)$	$\frac{\Gamma(\alpha+1)}{\Gamma(\alpha+1-\nu)} t^{\alpha-\nu}$ for $\alpha > -1$
$e^{\alpha t}\mathbf{1}(t)$	$t^{-\nu} E_{1,1-\nu}(\alpha t)$
$cosh\left(\sqrt{\alpha t}\right)$	$t^{-\nu} E_{2,1-\nu}\left(\alpha t^2\right)$
$\frac{sinh(\sqrt{\alpha t})}{\sqrt{\alpha t}}$	$t^{1-\nu} E_{2,2-\nu}\left(\alpha t^2\right)$

$$ {}_{t_0}^{RL}D_t^{\nu} f(t) = \frac{d^{\nu}}{dt^{\nu}} = \frac{1}{\Gamma(1-\nu)} \frac{d}{dt} \int_{t_0}^{t} \frac{f(\tau)}{(t-\tau)^{\nu}} d\tau. \tag{9} $$

In Table 1 the Riemann–Liouville fractional-order derivatives with $t_0 = 0$ of the chosen, the most common functions in automatic and control applications are presented.

For $\nu, \mu \in \mathbb{R}_+$ the following results are true

$$ {}_{t_0}^{RL}I_t^{\nu}\left[{}_{t_0}^{RL}I_t^{\mu} f(t)\right] = {}_{t_0}^{RL}I_t^{\mu}\left[{}_{t_0}^{RL}I_t^{\nu} f(t)\right] = {}_{t_0}^{RL}I_t^{\nu+\mu} f(t), \tag{10} $$

$$ {}_{t_0}^{RL}D_t^{\nu}\left[{}_{t_0}^{RL}D_t^{\mu} f(t)\right] = {}_{t_0}^{RL}D_t^{\mu}\left[{}_{t_0}^{RL}D_t^{\nu} f(t)\right] = {}_{t_0}^{RL}D_t^{\nu+\mu} f(t), \tag{11} $$

$$ {}_{t_0}^{RL}D_t^{\nu}\left[{}_{t_0}^{RL}I_t^{\nu} f(t)\right] = f(t). \tag{12} $$

For $n-1 \leqslant \nu < n$ it holds

$$ {}_{t_0}^{RL}I_t^{\nu}\left[{}_{t_0}^{RL}D_t^{\nu} f(t)\right] = f(t) - \sum_{i=1}^{n} \left[{}_{t_0}^{RL}D_t^{\nu-i}\right]_{t=t_0} \frac{(t-t_0)^{\nu-i}}{\Gamma(\nu-i+1)}. \tag{13} $$

For $0 \leqslant n-1 \leqslant \nu < n$ and $\nu, \mu \geqslant 0$ it follows that

$$ {}_{t_0}^{RL}I_t^{\mu}\left[{}_{t_0}^{RL}D_t^{\nu} f(t)\right] = {}_{t_0}^{RL}D_t^{\nu-\mu} f(t) - \sum_{i=1}^{n} \left[{}_{t_0}^{RL}D_t^{\nu-i}\right]_{t=t_0} \frac{(t-t_0)^{\mu-i}}{\Gamma(\mu-i+1)}. \tag{14} $$

The results presented in this section are valid for integer orders $\nu = n$ and $\mu = m$. The considered fractional-order derivatives have similar properties to the classical ones [60].

2.3 Grünwald–Letnikov Fractional-Order Left-Sided Derivative/Integral

For a continuous real variable t function $f(t)$ defined over interval $[t_0, t]$, where $-\infty < t_0 < t < +\infty$, one defines

$$h = \frac{t - t_0}{k} > 0. \tag{15}$$

The *Grünwald–Letnikov fractional-order left-hand derivative* is defined as a following limit, see [41, 61, 68]

$${}_{t_0}^{GL}D_t^{\nu} f(t) = \lim_{\substack{h \to 0 \\ t - t_0 = kh}} \left[\frac{1}{h^{\nu}} \sum_{i=0}^{k} a^{\nu}(i) f(k - i) \right]. \tag{16}$$

The *Grünwald–Letnikov fractional-order left-hand integral* is defined as a following limit

$${}_{t_0}^{GL}D_t^{\nu} f(t) = \lim_{\substack{h \to 0 \\ t - t_0 = kh}} \left[h^{\nu} \sum_{i=0}^{k} a^{\nu}(i) f(k - i) \right]. \tag{17}$$

Remark 1 The Riemann–Liouville fractional-order operator can be used successfully in practical issues related to a non-zero initial conditions, see for example [26, 34] and references therein. The most common reason is that in many cases the past values of real phenomena should be memorized, see for example in [10, 26, 68]. In practice, the memory of the considered phenomena has influence on the present values of the process and on its future. The initialized fractional order Riemann–Liouville derivative is defined in the following way, see [10, 26]

$$\begin{aligned} {}_{t_0}^{RL}D^{\nu} f(t) &= \frac{1}{\Gamma(n - \alpha)} \frac{d^n}{dt^n} \int_a^t \frac{f(\tau)}{(t - \tau)^{1+\nu-n}} d\tau, \quad t > t_0, \\ f(t) &:= \begin{cases} \phi(t), & \text{for } t_0 < t \leq t_0, \\ 0, & \text{for } t \leq t_0. \end{cases} \end{aligned} \tag{18}$$

with $n - 1 < \nu \leq n, n \in \mathbb{N}$, and f is function of the class C^{n-1}. In (18) function $\phi(t)$ represents the initial history of the process described by f. The practical usefulness of this operator is due to the fact that its value depends on all past values of the fractionally derived function, so the history or memory of the process is naturally included in the analysis. Also, it provides a recursive solution in time and hence reduces computing time [2].

2.4 *Caputo Fractional-Order Left-Sided Derivative/Integral*

For $[t_0, t]$, where $-\infty < t_0 < t < +\infty$, and classical derivatives $f^{(n)}(t)$ of the function $f(t)$ one defines the *Caputo fractional-order derivative* in the following way

$$ {}_{t_0}^{C}D_t^{\nu} f(t) = \frac{1}{\Gamma(n-\nu)} \int_{t_0}^{t} \frac{f^{(n)}(\tau)}{(t-\tau)^{1+\nu-n}} d\tau \tag{19} $$

for $n - 1 < \nu < n$. If $n = 1$ and $0 < \nu < 1$, then

$$ {}_{t_0}^{C}D_t^{\nu} f(t) = \frac{1}{\Gamma(1-\nu)} \int_{t_0}^{t} \frac{f'(\tau)}{(t-\tau)^{1+\nu-n}} d\tau \tag{20} $$

The relation between the Riemann–Liouville fractional-order derivative and the Caputo fractional-order derivative can be stated as follows, see for example [10, 41]:

$$ {}_{t_0}^{C}D_t^{\nu} f(t) = {}_{t_0}^{RL}D_t^{\nu} \left[f(t) - \sum_{i=0}^{n-1} \frac{f^{(i)}(t_0)}{i!} (t-t_0)^i \right] \tag{21} $$

or

$$ {}_{t_0}^{C}D_t^{\nu} f(t) = {}_{t_0}^{RL}D_t^{\nu} f(t) - \sum_{i=0}^{n-1} \frac{f^{(i)}(t_0)}{i!} {}_{t_0}^{RL}D_t^{\nu} \left[(t-t_0)^i \right]. \tag{22} $$

Taking into account equality

$$ {}_{t_0}^{C}D_t^{\nu} \left[(t-a)^{\beta} \right] = \frac{\Gamma(1+\beta)}{\Gamma(1+\beta-\nu)} (t-a)^{\beta-\nu} \tag{23} $$

from (22) one gets

$$ {}_{t_0}^{C}D_t^{\nu} f(t) = {}_{t_0}^{RL}D_t^{\nu} f(t) - \sum_{i=0}^{n-1} \frac{f^{(i)}(t_0)}{i!} \frac{\Gamma(1+i)}{\Gamma(1+i-\nu)} (t-a)^{i-\nu}. \tag{24} $$

3 Unilateral (One-Sided) Laplace Transform of the Fractional-Order Integral and Derivative

It is well known that the Laplace transform is an integral transform that converts a function of a real variable t (often time) to a function of a complex variable s (complex frequency). Suppose that $t_0 = 0$ and that f must be locally integrable for any real $t \geq 0$. For locally integrable functions that decay at infinity or are of exponential

Table 2 The one-sided Laplace transform of selected functions

$f(t)$	$\mathcal{L}\left\{{}_0^{RL}I_t^{\nu} f(t)\right\}$ $t > 0,\ \nu \in \mathbb{R}_+$
$\delta(t)$	$\frac{1}{s^{\nu}}$
$\mathbf{1}(t)$	$\frac{1}{s^{\nu+1}}$
$t^{\alpha}\mathbf{1}(t)$	$\frac{\Gamma(\alpha+1)}{s^{\alpha+\nu+1}}$ $\alpha > -1$
$e^{\alpha t}\mathbf{1}(t)$	$\frac{1}{s^{\nu}(s-\alpha)}$
$t^{\beta-1}e^{\alpha t}\mathbf{1}(t)$	$\frac{\Gamma(\beta)}{s^{\nu}(s-\alpha)^{\beta}}$ $\alpha > 0$
$\sin(\alpha t)\mathbf{1}(t)$	$\frac{1}{s^{\nu}(s^2+\alpha^2)}$
$\cos(\alpha t)\mathbf{1}(t)$	$\frac{\alpha}{s^{\nu-1}(s^2+\alpha^2)}$

order $\alpha \in \mathbb{R}_+$ i.e. there exist such constants $\beta >$ and $\alpha \geq 0$ that

$$|f(t)| \leq \beta e^{\alpha t}, \tag{25}$$

the *Laplace transform* of a function $f(t)$ is a unilateral (one-sided) transform defined by

$$F(s) = \mathcal{L}\{f(t)\} = \int_0^{+\infty} e^{-st} f(t)dt. \tag{26}$$

For ${}_0^{RL}I_t^{\nu} f(t)$ a fractional-order integral of a function $f(t)$ the unilateral Laplace transform is

$$\mathcal{L}\left\{{}_0^{RL}I_t^{\nu} f(t)\right\} = \frac{1}{s^{\nu}} F(s). \tag{27}$$

The unilateral Laplace transform of the fractional-order integral $\mathcal{L}\left\{{}_0^{RL}I_t^{\nu} f(t)\right\}, t > 0$, for the most commonly used in control and automatic functions are collected in Table 2.

The unilateral transform of the fractional-order derivative contains the initial values of consecutive fractional orders of $f(t)$ for $n-1 \leqslant \nu < n$

$$\begin{aligned}\mathcal{L}\left\{{}_0^{RL}D_t^{\nu} f(t)\right\} &= s^{\nu}F(s) - \sum_{i=0}^{n-1} s^i \left[{}_0^{RL}D_t^{\nu-i-1} f(t)\right]_{t=0} \\ &= s^{\nu}F(s) - \left[1\ s\ s^2\ \cdots\ s^{n-1}\right] \begin{bmatrix} \left[{}_0^{RL}D_t^{\nu-1} f(t)\right]_{t=0} \\ \left[{}_0^{RL}D_t^{\nu-2} f(t)\right]_{t=0} \\ \left[{}_0^{RL}D_t^{\nu-3} f(t)\right]_{t=0} \\ \vdots \\ \left[{}_0^{RL}D_t^{\nu-n} f(t)\right]_{t=0} \end{bmatrix}. \end{aligned} \tag{28}$$

One should mention that formulae (27) and (28) are also valid for integer order integration and differentiation.

3.1 *The Unilateral Laplace Transform of the Caputo Fractional-Order Left-Side Derivative*

One can show that for $n-1 \leqslant \nu < n$ the following formula is valid

$$\mathcal{L}\left\{{}_0^C D_t^\nu f(t)\right\} = s^\nu F(s) - \left[s^{\nu-1}\ s^{\nu-2}\ s^{\nu-3}\ \cdots\ s^{\nu-n}\right] \begin{bmatrix} [f(t)]_{t=0} \\ \left[f^{(1)}(t)\right]_{t=0} \\ \left[f^{(2)}(t)\right]_{t=0} \\ \vdots \\ \left[f^{(n-1)}(t)\right]_{t=0} \end{bmatrix}. \tag{29}$$

Remark 2 The main advantage of the formula (29) over (28) is the occurrence of consecutive integer-order derivative $\left[f^{(i)}(t)\right]_{t=0}$ for $i = 0, 1, \ldots, n-1$, evaluated for $t = 0$. They can be more easily interpreted physically than the terms $\left[{}_0^{RL} D_t^{\nu-i} f(t)\right]_{t=0}$ for $i = 1, 2, \ldots, n$, in the second mentioned formula.

3.2 *The Unilateral Laplace Transform of the Grünwald–Letnikov Fractional-Order Left-Side Derivative*

For the case when $0 \leqslant \nu < 1$ the result (29) is valid for $\left[f^{(i)}(t)\right]_{t=0}$ where $i = 0, 1, \ldots, n-1$. For $\nu \geqslant 1$ one can split the fractional-order derivative into the form ${}_{t_0}^{GL} D_t^{\nu} f(t) = {}_{t_0}^{GL} D_t^{\nu-1} g(t)$ where $g(t) = {}_{t_0}^{GL} D_t^{1} f(t)$.

4 Linear Fractional-Order SISO Systems

The linear fractional-order SISO (singe-input single-output) system can be described by the equation of the following form

$$\sum_{i=0}^{n} A_i(t)_{t_0} D_t^{\nu_i} y(t) = \sum_{j=0}^{m} B_j(t)_{t_0} D_t^{\mu_j} u(t), \tag{30}$$

where $\nu_n > \nu_{n-1} > \ldots \nu_1 > \nu_0 \geq 0$ and $\mu_m > \mu_{m-1} > \ldots > \mu_1 > \mu_0 \geq 0$ are real numbers, A_i, B_j, $i = 0, 1, \ldots, n$, $j = 0, 1, \ldots, m$, are real coefficients with $A_n \neq 0$. Assuming that $t_0 = 0$, one has ${}_{t_0=0}D_t^{\delta} = D_t^{\delta}$ where D_t^{δ} can denote the Riemann–Liouville, Caputo or Grünwald–Letnikov fractional-order left-side derivative. It is known that fractional-order differential equations in terms of the Riemann–Liouville derivatives require initial conditions expressed in terms of initial values of fractional derivatives of the unknown function [36, 67, 73]. In the case of the Caputo fractional-order derivative, initial conditions are expressed in terms of initial values of integer order derivatives. It is known that in fact for $t_0 = 0$ these derivatives and also Grünwald–Letnikov coincides [36, 67, 73]. This allows us to work with the Caputo fractional-order derivative or use the Riemann–Liouville derivative, but avoid the problem of initial values by treating only the case of zero initial conditions, see [5]. Generally, expressing initial conditions in terms of fractional derivatives of a function is not a big problem. Examples confirming this are discussed in [36].

The orders of the system (30) can be arbitrary complex quantities. This type of orders is named a *non-commensurate order*. In practice there can occur two interesting cases when the orders are *commensurate*, i.e. when

$$\nu_k, \mu_k = k\delta, \quad \delta \in \mathbb{R}_+ \tag{31}$$

or

$$\nu_k, \mu_k = k\delta, \text{and} \quad \delta = \frac{1}{q}; \quad q \in \mathbb{Z}_+. \tag{32}$$

In the case $q = 1$ the derivatives are of integer order and the system becomes an ordinary differential equation. More about systems with commensurate and non-commensurate fractional orders and comparison between them can be found in [8, 35, 59] Assuming that

$$\begin{aligned} \nu_i &\in \mathbb{Q}_+, \ \text{for } i = 0, 1, \ldots, n, \\ \mu_j &\in \mathbb{Q}_+, \ \text{for } j = 0, 1, \ldots, m, \end{aligned} \tag{33}$$

every order can be expressed as a quotient of two integers

$$\begin{aligned} \nu_i &= \frac{n_i}{q}, \ \text{for } i = 0, 1, \ldots, n, \\ \mu_j &= \frac{m_i}{q}, \ \text{for } j = 0, 1, \ldots, m, \end{aligned} \tag{34}$$

where q is the least common denominator of all orders. Denoting now

$$\nu = \frac{1}{q} \leqslant 1 \tag{35}$$

one can write Eq. (30) in the form

$$\sum_{i=0}^{n} A_i(t) {}_{t_0}^{GL}D_t^{n_i \nu} y(t) = \sum_{j=0}^{m} B_j(t) {}_{t_0}^{GL}D_t^{m_j \nu} u(t). \tag{36}$$

In the equation given above, all orders are a multiplicity of a fractional order ν (when $q \neq 1$) such equation will be called a commensurate one.

Remark 3 Every fractional-order non-commensurate differential equation with rational orders can be transformed into the commensurate one.

4.1 *Numerical Solution of the Linear, Time-Invariant Differential Equation with Grünwald–Letnikov Fractional-Order Left-Hand Derivatives*

Without the loss of generality one assumes $t_0 = 0$. In a time invariant differential equation all coefficients A_i, B_j, $i = 1, 1, \ldots, n$, $j = 0, 1, \ldots, m$, are constant. Equation (36) can be expressed in a vector form

$$\begin{bmatrix} A_n & A_{n-1} & \cdots & A_0 \end{bmatrix} \begin{bmatrix} {}_{t_0}^{GL}D_t^{n_n \nu} y(t) \\ {}_{t_0}^{GL}D_t^{n_{n-1} \nu} y(t) \\ \vdots \\ {}_{t_0}^{GL}D_t^{n_0 \nu} y(t) \end{bmatrix} = \begin{bmatrix} B_m & B_{m-1} & \cdots & B_0 \end{bmatrix} \begin{bmatrix} {}_{t_0}^{GL}D_t^{m_m \mu} u(t) \\ {}_{t_0}^{GL}D_t^{m_{m-1} \mu} u(t) \\ \vdots \\ {}_{t_0}^{GL}D_t^{m_0 \mu} u(t) \end{bmatrix}. \tag{37}$$

Different schemes for fractional-order differential equations can be used [24]. For $t = kh$ the vectors of derivatives can be approximated [22, 30, 82] as follows

$$\begin{bmatrix} {}_{t_0}^{GL}D_t^{n_n \nu} y(kh) \\ {}_{t_0}^{GL}D_t^{n_{n-1} \nu} y(kh) \\ \vdots \\ {}_{t_0}^{GL}D_t^{n_1 \nu} y(kh) \\ {}_{t_0}^{GL}D_t^{n_0 \nu} y(kh) \end{bmatrix} = \begin{bmatrix} \frac{1}{h^{n_n \nu}} & 0 & \cdots & 0 \\ 0 & \frac{1}{h^{n_{n-1} \nu}} & \cdots & 0 \\ \vdots & \vdots & & \vdots \\ 0 & 0 & \cdots & 0 \\ 0 & 0 & \cdots & \frac{1}{h^{n_0 \nu}} \end{bmatrix} \cdot$$

$$\cdot \begin{bmatrix} a^{n_n \nu}(0) & a^{n_n \nu}(1) & \cdots & a^{n_n \nu}(k) \\ a^{n_{n-1} \nu}(0) & a^{n_{n-1} \nu}(1) & \cdots & a^{n_{n-1} \nu}(k) \\ \vdots & \vdots & & \vdots \\ a^{n_1 \nu}(0) & a^{n_1 \nu}(1) & \cdots & a^{n_1 \nu}(k) \\ a^{n_0 \nu}(0) & a^{n_0 \nu}(1) & \cdots & a^{n_0 \nu}(k) \end{bmatrix} \begin{bmatrix} y(kh) \\ y(kh-h) \\ \vdots \\ y(h) \\ y(0) \end{bmatrix},$$

$$\begin{bmatrix} {}_{t_0}^{GL}D_t^{m_m\nu}y(kh) \\ {}_{t_0}^{GL}D_t^{m_{m-1}\nu}y(kh) \\ \vdots \\ {}_{t_0}^{GL}D_t^{m_1\nu}y(kh) \\ {}_{t_0}^{GL}D_t^{m_0\nu}y(kh) \end{bmatrix} = \begin{bmatrix} \frac{1}{h^{m_m\nu}} & 0 & \cdots & 0 \\ 0 & \frac{1}{h^{m_{m-1}\nu}} & \cdots & 0 \\ \vdots & \vdots & & \vdots \\ 0 & 0 & \cdots & 0 \\ 0 & 0 & \cdots & \frac{1}{h^{m_0\nu}} \end{bmatrix} \cdot \begin{bmatrix} a^{m_m\nu}(0) & a^{m_m\nu}(1) & \cdots & a^{m_m\nu}(k) \\ a^{m_{m-1}\nu}(0) & a^{m_{m-1}\nu}(1) & \cdots & a^{m_{m-1}\nu}(k) \\ \vdots & \vdots & & \vdots \\ a^{m_1\nu}(0) & a^{m_1\nu}(1) & \cdots & a^{m_1\nu}(k) \\ a^{m_0\nu}(0) & a^{m_0\nu}(1) & \cdots & a^{m_0\nu}(k) \end{bmatrix} \begin{bmatrix} u(kh) \\ u(kh-h) \\ \vdots \\ u(h) \\ u(0) \end{bmatrix} \tag{38}$$

Substitution (38) to (37) gives

$$\begin{bmatrix} \frac{A_n}{h^{n_n\nu}} & \frac{A_{n-1}}{h^{n_{n-1}\nu}} & \cdots & \frac{A_0}{h^{n_0\nu}} \end{bmatrix} \begin{bmatrix} a^{n_n\nu}(0) & a^{n_n\nu}(1) & \cdots & a^{n_n\nu}(k) \\ a^{n_{n-1}\nu}(0) & a^{n_{n-1}\nu}(1) & \cdots & a^{n_{n-1}\nu}(k) \\ \vdots & \vdots & & \vdots \\ a^{n_1\nu}(0) & a^{n_1\nu}(1) & \cdots & a^{n_1\nu}(k) \\ a^{n_0\nu}(0) & a^{n_0\nu}(1) & \cdots & a^{n_0\nu}(k) \end{bmatrix} \begin{bmatrix} y(kh) \\ y(kh-h) \\ \vdots \\ y(h) \\ y(0) \end{bmatrix} =$$

$$\begin{bmatrix} \frac{B_m}{h^{m_m\nu}} & \frac{B_{m-1}}{h^{m_{m-1}\nu}} & \cdots & \frac{B_0}{h^{m_0\nu}} \end{bmatrix} \begin{bmatrix} a^{m_m\nu}(0) & a^{m_m\nu}(1) & \cdots & a^{m_m\nu}(k) \\ a^{m_{m-1}\nu}(0) & a^{m_{m-1}\nu}(1) & \cdots & a^{m_{m-1}\nu}(k) \\ \vdots & \vdots & & \vdots \\ a^{m_1\nu}(0) & a^{m_1\nu}(1) & \cdots & a^{m_1\nu}(k) \\ a^{m_0\nu}(0) & a^{m_0\nu}(1) & \cdots & a^{m_0\nu}(k) \end{bmatrix} \cdot \begin{bmatrix} u(kh) \\ u(kh-h) \\ \vdots \\ u(h) \\ u(0) \end{bmatrix}. \tag{39}$$

Now, one defines the following coefficients

$$\begin{bmatrix} a_0 \; a_1 \; \cdots \; a_k \end{bmatrix} =$$

$$= \left[\tfrac{A_n}{h^{n_n \nu}} \; \tfrac{A_{n-1}}{h^{n_{n-1} \nu}} \; \cdots \; \tfrac{A_0}{h^{n_0 \nu}} \right] \begin{bmatrix} a^{n_n \nu}(0) & a^{n_n \nu}(1) & \cdots & a^{n_n \nu}(k) \\ a^{n_{n-1} \nu}(0) & a^{n_{n-1} \nu}(1) & \cdots & a^{n_{n-1} \nu}(k) \\ \vdots & \vdots & & \vdots \\ a^{n_1 \nu}(0) & a^{n_1 \nu}(1) & \cdots & a^{n_1 \nu}(k) \\ a^{n_0 \nu}(0) & a^{n_0 \nu}(1) & \cdots & a^{n_0 \nu}(k) \end{bmatrix} \tag{40}$$

$$\begin{bmatrix} b_0 \; b_1 \; \cdots \; b_k \end{bmatrix} =$$

$$= \left[\tfrac{B_m}{h^{m_m \nu}} \; \tfrac{B_{m-1}}{h^{m_{m-1} \nu}} \; \cdots \; \tfrac{B_0}{h^{m_0 \nu}} \right] \begin{bmatrix} a^{m_m \nu}(0) & a^{m_m \nu}(1) & \cdots & a^{m_m \nu}(k) \\ a^{m_{m-1} \nu}(0) & a^{m_{m-1} \nu}(1) & \cdots & a^{m_{m-1} \nu}(k) \\ \vdots & \vdots & & \vdots \\ a^{m_1 \nu}(0) & a^{m_1 \nu}(1) & \cdots & a^{m_1 \nu}(k) \\ a^{m_0 \nu}(0) & a^{m_0 \nu}(1) & \cdots & a^{m_0 \nu}(k) \end{bmatrix} \tag{41}$$

By definition

$$a^{n_i \nu}(0) = 1 \quad \text{for } i = 0, 1, \ldots, n. \tag{42}$$

Hence,

$$a_0 = \frac{A_n}{h^{n\nu}} + \frac{A_{n-1}}{h^{(n-1)\nu}} + \cdots + \frac{A_0}{h^{0\nu}}. \tag{43}$$

One can always find $h > 0$ such that $a_0 \neq 0$. Then, Eq. (37) is transformed to the form

$$\begin{bmatrix} a_0 \; a_1 \; \cdots \; a_k \end{bmatrix} \begin{bmatrix} y(kh) \\ y(kh-h) \\ \vdots \\ y(h) \\ y(0) \end{bmatrix} = \begin{bmatrix} b_0 \; b_1 \; \cdots \; b_k \end{bmatrix} \begin{bmatrix} u(kh) \\ u(kh-h) \\ \vdots \\ u(h) \\ u(0) \end{bmatrix} \tag{44}$$

The equation given above is also valid for $k-1, k-2, \ldots, 1, 0$. Hence, for $k-1$ one can write

$$\begin{bmatrix}0 \; a_0 \cdots a_{k-1}\end{bmatrix}\begin{bmatrix} y(kh) \\ y(kh-h) \\ \vdots \\ y(h) \\ y(0) \end{bmatrix} = \begin{bmatrix}0 \; b_0 \cdots b_{k-1}\end{bmatrix}\begin{bmatrix} u(kh) \\ u(kh-h) \\ \vdots \\ u(h) \\ u(0) \end{bmatrix} \tag{45}$$

and further for $k-2$

$$\begin{bmatrix}0 \; 0 \; a_0 \cdots a_{k-2}\end{bmatrix}\begin{bmatrix} y(kh) \\ y(kh-h) \\ \vdots \\ y(h) \\ y(0) \end{bmatrix} = \begin{bmatrix}0 \; 0 \; b_0 \cdots b_{k-2}\end{bmatrix}\begin{bmatrix} u(kh) \\ u(kh-h) \\ \vdots \\ u(h) \\ u(0) \end{bmatrix} \tag{46}$$

Collecting all such equations, one has, see [31]

$$\begin{bmatrix} a_0 & a_1 & a_2 & \cdots & a_k \\ 0 & a_0 & a_1 & \cdots & a_{k-1} \\ 0 & 0 & a_0 & \cdots & a_{k-2} \\ \vdots & \vdots & \vdots & & \vdots \\ 0 & 0 & 0 & \cdots & a_0 \end{bmatrix}\begin{bmatrix} y(kh) \\ y(kh-h) \\ \vdots \\ y(h) \\ y(0) \end{bmatrix} = \begin{bmatrix} b_0 & b_1 & b_2 & \cdots & b_k \\ 0 & b_0 & b_1 & \cdots & b_{k-1} \\ 0 & 0 & b_0 & \cdots & b_{k-2} \\ \vdots & \vdots & \vdots & & \vdots \\ 0 & 0 & 0 & \cdots & b_0 \end{bmatrix}\begin{bmatrix} u(kh) \\ u(kh-h) \\ \vdots \\ u(h) \\ u(0) \end{bmatrix} \tag{47}$$

By (42) the left-hand matrix in (47) is always non-singular. Hence, the numerical solution [32] of Eq. (36) has the form

$$\begin{bmatrix} y(kh) \\ y(kh-h) \\ \vdots \\ y(h) \\ y(0) \end{bmatrix} = \begin{bmatrix} a_0 & a_1 & a_2 & \cdots & a_k \\ 0 & a_0 & a_1 & \cdots & a_{k-1} \\ 0 & 0 & a_0 & \cdots & a_{k-2} \\ \vdots & \vdots & \vdots & & \vdots \\ 0 & 0 & 0 & \cdots & a_0 \end{bmatrix}^{-1}\begin{bmatrix} b_0 & b_1 & b_2 & \cdots & b_k \\ 0 & b_0 & b_1 & \cdots & b_{k-1} \\ 0 & 0 & b_0 & \cdots & b_{k-2} \\ \vdots & \vdots & \vdots & & \vdots \\ 0 & 0 & 0 & \cdots & b_0 \end{bmatrix}\begin{bmatrix} u(kh) \\ u(kh-h) \\ \vdots \\ u(h) \\ u(0) \end{bmatrix} \tag{48}$$

In general to solve the fractional-order differential equation one can use more sophisticated numerical procedures [19–21, 33, 51].

5 Fractional-Order Transfer Function

Fractional-order differential equation (36) may describe a linear time-invariant dynamical system. Suppose that $y(\cdot)$ and $u(\cdot)$ are original functions. Assuming zero initial conditions, one can apply the unilateral Laplace transform to both sides of the Eq. (36)

$$\mathcal{L}\left\{\sum_{i=0}^{n} A_{i\,t_0}^{\,GL} D_t^{n_i \nu} y(t)\right\} = \mathcal{L}\left\{\sum_{j=0}^{n} B_{j\,t_0}^{\,GL} D_t^{m_j \nu} u(t)\right\}. \tag{49}$$

By linearity of Laplace transform one has

$$\sum_{i=0}^{n} A_i \mathcal{L}\left\{{}_{t_0}^{GL} D_t^{n_i \nu} y(t)\right\} = \sum_{j=0}^{n} B_j \mathcal{L}\left\{{}_{t_0}^{GL} D_t^{m_j \nu} u(t)\right\}. \tag{50}$$

Denoting $Y(s) = \mathcal{L}\{y(t)\}$ and $U(s) = \mathcal{L}\{u(t)\}$ one defines the fractional order transfer function

$$G(s) = \frac{Y(s)}{U(s)} = \frac{B_m s^{\mu_m} + B_{m-1} s^{\mu_{m-1}} + \cdots + B_0 s^{m_0}}{s^{\nu_n} + A_{n-1} s^{\nu_{n-1}} + \cdots + A_0 s^{\nu_0}}. \tag{51}$$

If (36) is of the commensurate order equation, then one has

$$G(s) = \frac{Y(s)}{U(s)} = \frac{B_m s^{m\nu} + B_{m-1} s^{(m-1)\nu} + \cdots + B_0 s^{0\nu}}{s^{n\nu} + A_{n-1} s^{(n-1)\nu} + \cdots + A_0 s^{0\nu}}, \tag{52}$$

Now, introducing a new complex variable

$$p = s^{\nu} \tag{53}$$

one can expressed fractional-order transfer function (52) in the form

$$G(p) = \frac{Y(p)}{U(p)} = \frac{B_m p^m + B_{m-1} p^{m-1} + \cdots + B_0 p^0}{p^n + A_{n-1} s^{n-1} + \cdots + A_0 p^0}. \tag{54}$$

In fact, (54) is the transfer function of the integer order associated with the fractional-order transfer function (51). Applying classical methods the transfer function (54) can be expressed as

$$G(p) = \frac{Y(p)}{U(p)} = \frac{\prod_{j=1}^{m}(p - z_j)}{\prod_{i=1}^{n}(p - p_i)}, \tag{55}$$

where z_j is a pseudo-zero of the fractional-order transfer function, p_i is a pseudo-pole. They may be real or complex conjugate, distinct or multiple. From this moment forward the considerations will be restricted to the case of distinct poles. Applying common procedures the transfer function (55) may be represented as a sum of fractions

$$G(p) = \frac{Y(p)}{U(p)} = \sum_{i=1}^{n} \frac{C_i}{p - p_i}. \tag{56}$$

Now, one considers a simple term where without generality, one can put $C = 1$. Returning to the variable s one has

$$\frac{1}{p - p_i} = \frac{1}{s^{\nu} - p_i} = \sum_{j=1}^{q} \frac{p^{j-1}}{s^{j\nu - 1}(s - p^q)}. \tag{57}$$

Then the inverse Laplace transform of (57) is

$$\mathcal{L}^{-1}\left\{\frac{1}{s^{\nu} - p_i}\right\} = \sum_{j=1}^{q} p_i^{j-1} E_t(j\nu - 1, p_i^q), \tag{58}$$

where $E_t(x, y) = t^x \sum_{i=0}^{+\infty} \frac{(yt)^i}{\Gamma(x+i+1)}$ for $x \in \mathbb{R}_+$ and $y \in \mathbb{C}$. Formula (58) helps in evaluation of the response of the linear time-invariant fractional-order system.

5.1 Fourier Transform of the Fractional-Order Differential Equation

Substituting $s = j\omega$ where $j^2 = -1$, one has

$$s^{\nu} = (j\omega)^{\nu} = \left(\omega e^{j\frac{\pi}{2}}\right)^{\nu} = \omega^{\nu}\left[\cos\left(\nu\frac{\pi}{2}\right) + j\sin\left(\nu\frac{\pi}{2}\right)\right]. \tag{59}$$

Substitution (59) into (52) leads to a frequency characteristic

$$\begin{aligned} G(j\omega) = \frac{Y(j\omega)}{U(j\omega)} &= \frac{B_m(j\omega)^{\mu_m} + B_{m-1}(j\omega)^{\mu_{m-1}} + \cdots + B_0(j\omega)^{m_0}}{(j\omega)^{\nu_n} + A_{n-1}(j\omega)^{\nu_{n-1}} + \cdots + A_0(j\omega)^{\nu_0}} \\ &= P(\omega) + jQ(\omega), \end{aligned} \tag{60}$$

where $P(\omega)$ and $Q(\omega)$ are real and imaginary parts of $G(j\omega)$, respectively.

5.2 Fractional-Order Integrator

For $n = 1$ with $A_0 = 0$ and for $m = 0$ with $B_0 \neq 0$ from (36) one gets the particular case of the fractional order differential equation

$${}_{t_0}^{GL}D_t^{\nu} y(t) = B_0 u(t). \tag{61}$$

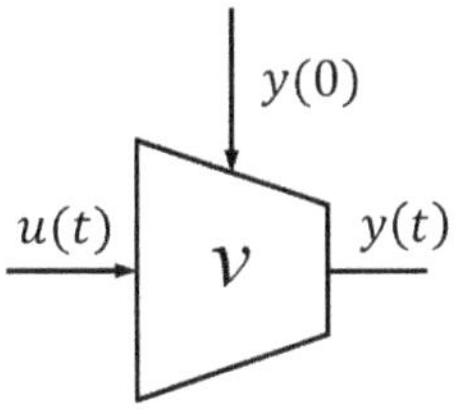

Fig. 1 Block diagram of a fractional order integrator

In the automatic control Eq. (61) describes the so called *fractional-order integrator*. Its fractional-order transfer function is as follows

$$G_I(s) = \frac{Y(s)}{U(s)} = \frac{B_0}{s^\nu} = \frac{1}{(T_u s)^\nu}, \tag{62}$$

where T_u is an element time constant. Its block diagram is presented in Fig. 1.

Remark 4 For the classical integer (first) order integrator a triangle is usually used as a block. To emphasise the fractionality of the integrator of order ν a trapezoid is applied. Inside the trapezoid a value of the order is put. The integrators may be performed due to different types of derivatives.

The impulse response of the ideal fractional-order integrator with $B_0 = 1$ is calculated according to the formula

$$g_I^\nu(t) = \mathcal{L}^{-1}\left\{\frac{1}{(T_u s)^\nu}\right\} = \frac{t^{\nu-1}}{T_u^\nu \Gamma(\nu)}. \tag{63}$$

The unit step response of the ideal fractional-order integrator is calculated according to the formula (Figs. 2 and 3)

$$h_I^\nu(t) = \mathcal{L}^{-1}\left\{\frac{1}{(T_u s)^{\nu+1}}\right\} = \frac{t^{\nu}}{T_u^\nu \Gamma(\nu+1)}. \tag{64}$$

To evaluate a response $y(t)$ of the fractional-order integrator (61) to any input signal $u(t)$ such that $\mathcal{L}\{u(t)\} = U(s)$ one can apply a convolution theorem. Then the response may be described by the following formula

$$\begin{aligned} y(t) = \mathcal{L}^{-1}\left\{G_I^\nu(s)U(s)\right\} &= \mathcal{L}^{-1}\left\{G_I^\nu(s)\right\} * \mathcal{L}^{-1}\left\{U(s)\right\} \\ &= g_I^\nu * u(t) = \int_0^t g_I^\nu u(t-\tau)d\tau. \end{aligned} \tag{65}$$

Substituting (63) into (65) leads to (Figs. 4 and 5)

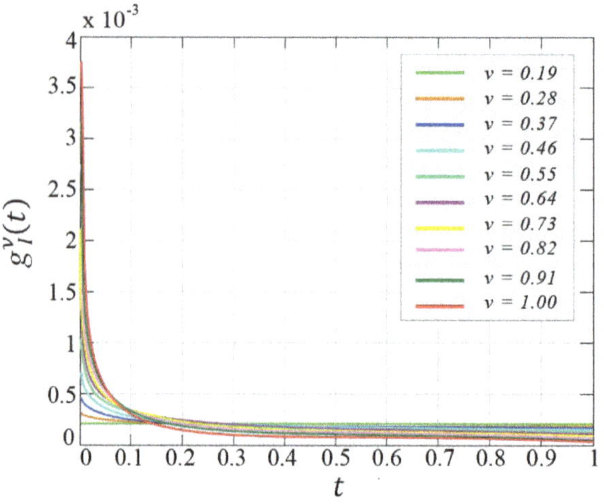

Fig. 2 Dirac impulse response of the fractional-order ideal integrator for $\nu \in [0.19, 0.28, \ldots, 0.91, 1.00]$

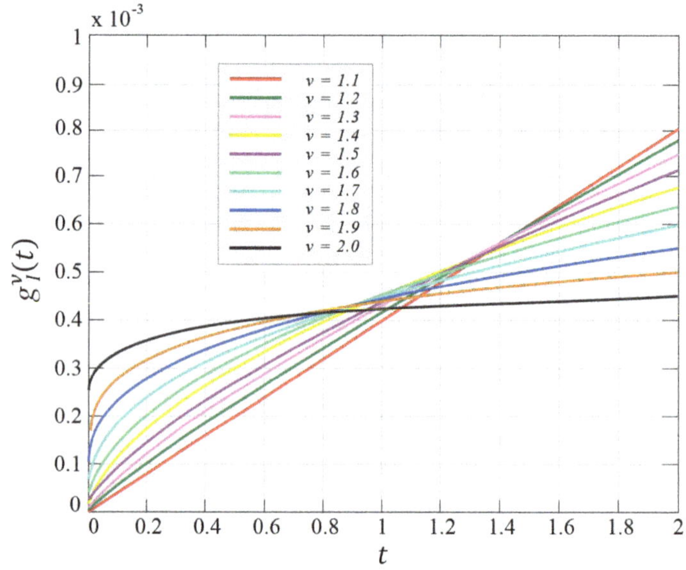

Fig. 3 Dirac impulse response of the fractional-order ideal integrator for $\nu \in [1.1, 1.2, \ldots, 1.9, 2.0]$

$$y(t) = \frac{1}{T_u^\nu \Gamma(\nu)} \int_0^t \tau^{\nu-1} u(t-\tau) d\tau = \frac{1}{T_u^\nu \Gamma(\nu)} \int_0^t u(\tau)(t-\tau)^{\nu-1} d\tau. \quad (66)$$

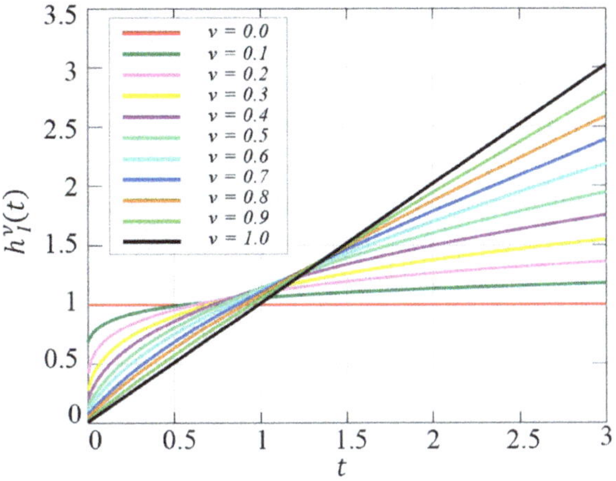

Fig. 4 Discrete unit step response of the fractional-order ideal integrator for $\nu \in [0.0, 0.1, \ldots, 0.9, 1.0]$

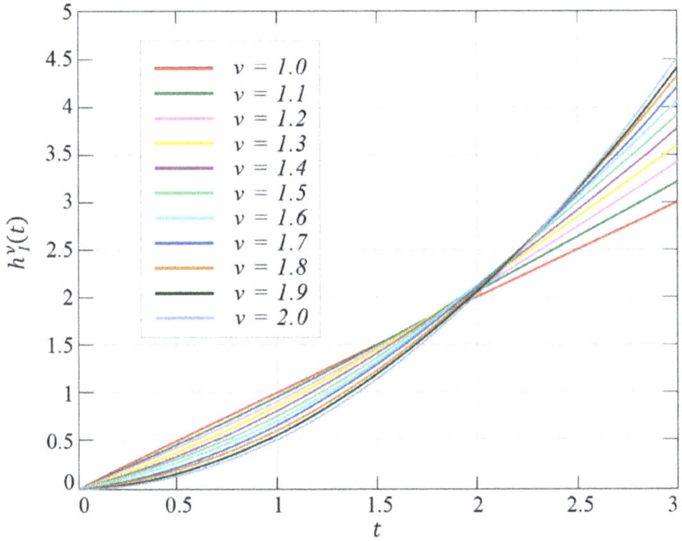

Fig. 5 Discrete unit step response of the fractional-order ideal integrator for $\nu \in [1.0, 1.1, \ldots, 1.9, 2.0]$

5.2.1 Frequency Characteristics of the Fractional-Order Integrator

Substitution of equality (59) into (62) gives

$$G_I^\nu(j\omega) = \frac{1}{(T_u j\omega)^\nu} = \frac{1}{T_u^\nu(\omega e^{j\omega\frac{\pi}{2}})} = \frac{1}{(T_u\omega)^\nu}\frac{1}{\cos\left(\nu\frac{\pi}{2}\right) + j\sin\left(\nu\frac{\pi}{2}\right)}. \tag{67}$$

After elementary transformations one extracts real and imaginary parts of the frequency characteristic

$$G_I^\nu(j\omega) = P_I^\nu(\omega) + jQ_I^\nu(\omega), \tag{68}$$

where

$$P_I^\nu(\omega) = \frac{1}{(T_u\omega)^\nu}\cos\left(\nu\frac{\pi}{2}\right), \tag{69}$$

$$Q_I^\nu(\omega) = -\frac{1}{(T_u\omega)^\nu}\sin\left(\nu\frac{\pi}{2}\right). \tag{70}$$

From (69) and (70) one immediately gets a Nyquist plot

$$Q_I^\nu = -\tan\left(\nu\frac{\pi}{2}\right)P_I^\nu. \tag{71}$$

Remark 5 Formula (71) reveals that the Nyquist plot of the fractional-order ideal integrator for $\omega \in [0, +\infty)$ is a half-line with a slope depending on the order $-\nu\frac{\pi}{2}$. The half-line parameters are independent on the time constant T_u.

In Fig. 6 the Nyquist plots of the fractional-order integrator with chosen orders are presented.

Remark 6 For $\nu = 1$ one has

$$P_I^1(\omega) = \frac{1}{(T_u\omega)^\nu}\cos\left(\frac{\pi}{2}\right) = 0, \tag{72}$$

$$Q_I^1(\omega) = -\frac{1}{(T_u\omega)^\nu}\sin\left(\frac{\pi}{2}\right) = \frac{1}{(T_u\omega)\nu} \tag{73}$$

and it is a commonly known frequency characteristic of a classical integrator.

From relations (72) and (73) one evaluates the magnitude characteristics

$$\left|G_I^\nu(j\omega)\right| = \sqrt{\left[P_I^\nu(\omega)\right]^2 + \left[Q_I^\nu(\omega)\right]^2} = \frac{1}{(T_u\omega)^\nu}. \tag{74}$$

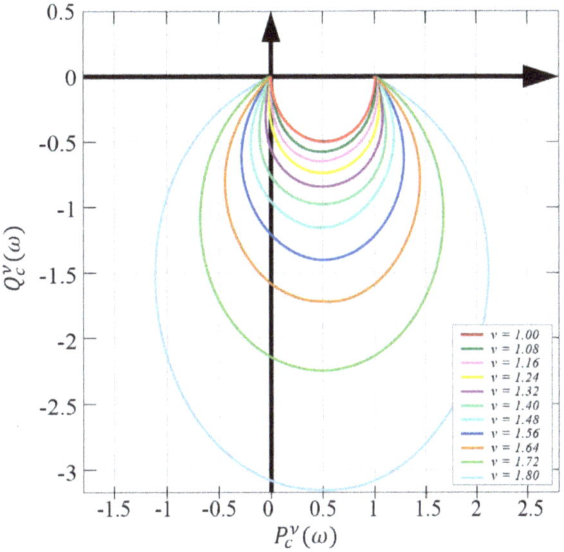

Fig. 6 Nyquist plots of the fractional-order ideal integrator for selected orders ν

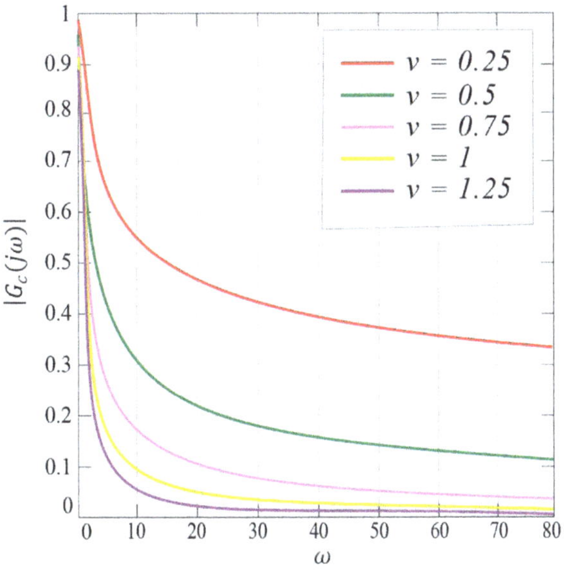

Fig. 7 Magnitude plots of the fractional-order ideal integrator for selected orders ν

It is presented in Figs. 7 and 8. The appropriate phase frequency characteristic given in Fig. 8 has the specially simple form

$$\phi_I^\nu(\omega) = \tan^{-1}\left(\frac{Q_I^\nu(\omega)}{P_I^\nu(\omega)}\right) = -\tan^{-1}\left[\tan\left(\nu\frac{\pi}{2}\right)\right] = -\nu\frac{\pi}{2}. \tag{75}$$

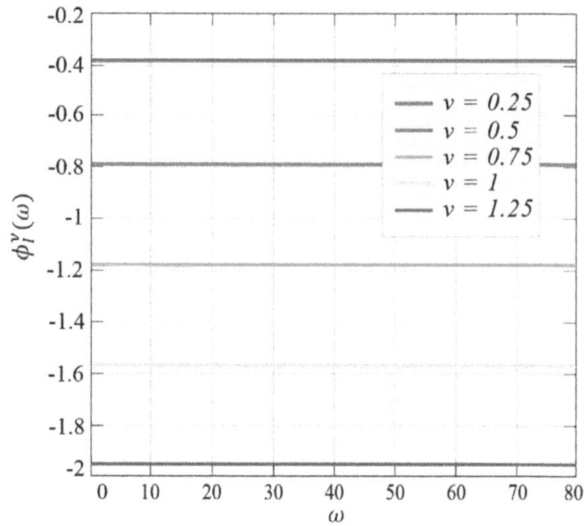

Fig. 8 Magnitude plots of the fractional-order ideal integrator for selected orders ν

Finally, the Bode plots will be considered. From Eq. (74) one derives

$$m_I(\omega) = 20\log\left|G_I^\nu(j\omega)\right| = -20\log(T_u\omega)^\nu = -20\nu\log(T_u\omega) \tag{76}$$

Remark 7 The phase shift is negative and independent from frequency ω.

Bode characteristics related to formula (76) are given in Fig. 9.

Remark 8 The transfer function of the open-loop system allows the use of Bode's ideal transfer function method in order to determine controller parameters. For details see [42].

Now, based on Eq. (76) one evaluates a considered element gain drop due to the frequency. One takes into consideration one decade of frequency $[\omega_0, 10\omega_0]$ for a fixed ω_0. Hence,

$$\begin{aligned} m_I^\nu(\omega_0, T_u) &= -20\nu\log(T_u\omega_0), \\ m_I^\nu(10\omega_0, T_u) &= -20\nu\log(T_u 10\omega_0) \end{aligned} \tag{77}$$

and

$$\Delta m_I^\nu(\omega_0) = -20\nu\log(T_u 10\omega_0) + 20\nu\log(T_u\omega_0) = -20\nu\log(10) = -20\nu. \tag{78}$$

A gain drop equals to $-20\nu\left[\frac{dB}{dec}\right]$. Finally, one analyses the influence of the time instant to the Bode plot. It is easy to check that for

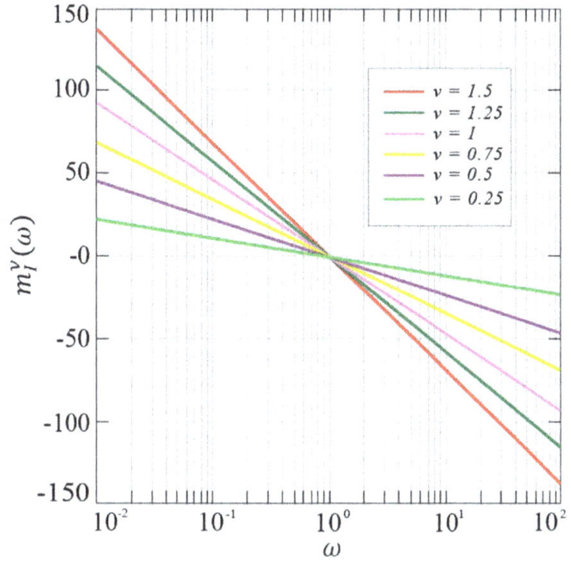

Fig. 9 Bode plots of the fractional-order ideal integrator for selected orders ν

$$\omega = \frac{1}{T_u} \tag{79}$$

one has

$$m_I\left(\frac{1}{T_u}\right) = -20\log\left(T_u\frac{1}{T_u}\right) = -20\log(1) = 0. \tag{80}$$

This means that the Bode plot crosses the vertical axis at $\omega = \frac{1}{T_u}$. This is plotted in Fig. 10.

5.3 *Fractional-Order Inertial Element*

The fractional-order inertial element is formed by a unity negative feedback of the fractional-order integrator with the proportional element representing the value of a time instant. Described configuration is presented in Fig. 11.

From the block diagram in Fig. 11 one derives the following relation between input signal $u(t)$ and output $y(t)$ signal:

$$ {}_{t_0}^{GL}D_t^\nu y(t) + \frac{1}{T_u^\nu}y(t) = \frac{1}{T_u^\nu}u(t). \tag{81}$$

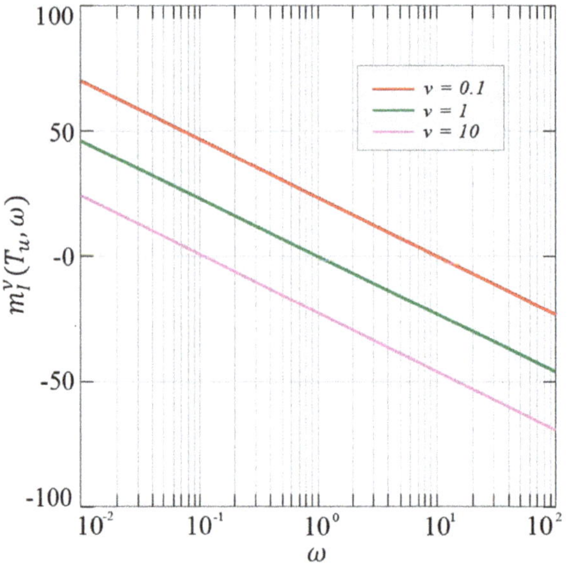

Fig. 10 Bode plots of the fractional-order ideal integrator for selected time constants $T_u \in \{0.1, 1.0, 10.0\}$

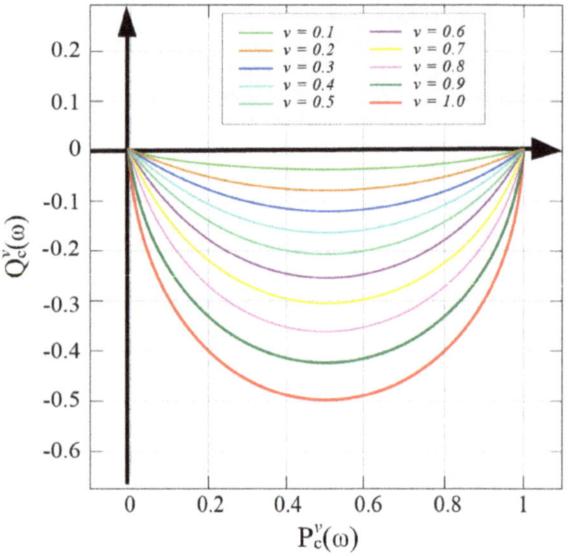

Fig. 11 Block diagram of the fractional-order inertial element

Assuming zero initial conditions and applying the unilateral Laplace transform to both sides of Eq. (81) one obtains

$$s^\nu Y(s) + \frac{1}{T_u^\nu} Y(s) = \frac{1}{T_u^\nu} U(s), \tag{82}$$

where $Y(s) = \mathcal{L}\{y(t)\}$ and $U(s) = \mathcal{L}\{u(t)\}$. Further

$$G_c(s) = \frac{Y(s)}{U(s)} = \frac{\frac{1}{T_u^\nu}}{s^\nu + \frac{1}{T_u^\nu}} = \frac{1}{s^\nu T_u^\nu + 1}. \tag{83}$$

Then the impulse response of the inertial fractional-order element is given by formula (58) with $p_i = -\frac{1}{T_u^\nu}$. The unit step response can be calculated as the classical integral of the impulse response.

Now, define a new parameter related to the time instant T_u as

$$\omega_u = \frac{1}{T_u} \tag{84}$$

and assume that $0 < \nu < 2$. Then the frequency transfer function is of the form

$$G_c^\nu(j\omega) = \frac{1}{\left(\frac{j\omega}{\omega_u}\right)^\nu + 1} = \frac{1}{\left(\frac{\omega}{\omega_u}\right)^\nu \left[\cos\left(\nu\frac{\pi}{2}\right) + j\sin\left(\nu\frac{\pi}{2}\right)\right] + 1}. \tag{85}$$

After elementary operations one obtains

$$G_c^\nu(j\omega) = P_c^\nu(\omega) + jQ_c^\nu(\omega), \tag{86}$$

where

$$P_c^\nu(j\omega) = \frac{1 + \left(\frac{\omega}{\omega_u}\right)^\nu \cos\left(\nu\frac{\pi}{2}\right)}{\left(\frac{\omega}{\omega_u}\right)^{2\nu} + 2\left(\frac{\omega}{\omega_u}\right)^\nu \cos\left(\nu\frac{\pi}{2}\right) + 1}, \tag{87}$$

$$Q_c^\nu(j\omega) = \frac{\left(\frac{\omega}{\omega_u}\right)^\nu \sin\left(\nu\frac{\pi}{2}\right)}{\left(\frac{\omega}{\omega_u}\right)^{2\nu} + 2\left(\frac{\omega}{\omega_u}\right)^\nu \cos\left(\nu\frac{\pi}{2}\right) + 1}. \tag{88}$$

Functions $P_c^\nu(\omega)$ and $Q_c^\nu(\omega)$ are real and imaginary parts of $G_c^\nu(j\omega)$, respectively. The elimination of frequency ω from (87) and (88) gives the relation in which the graph gives the Nyquist plot

$$\left[P_c^\nu - \frac{1}{2}\right]^2 + \left[Q_c^\nu - \frac{1}{2}\tan^{-1}\left(\nu\frac{\pi}{2}\right)\right]^2 = \left[\frac{1}{2\sin\left(\nu\frac{\pi}{2}\right)}\right]^2. \tag{89}$$

It is easy to note that (89) is in fact the equation of a circle. Moreover, the relation (89) is independent to the frequency ω. The radius and the circle center given by (89) depend only on order ν. Hence, the Nyquist plot evaluated for $[0, +\infty)$ is a circle arch. All arches cross points $(0, j0)$ and $(1, j0)$ on the complex plane.

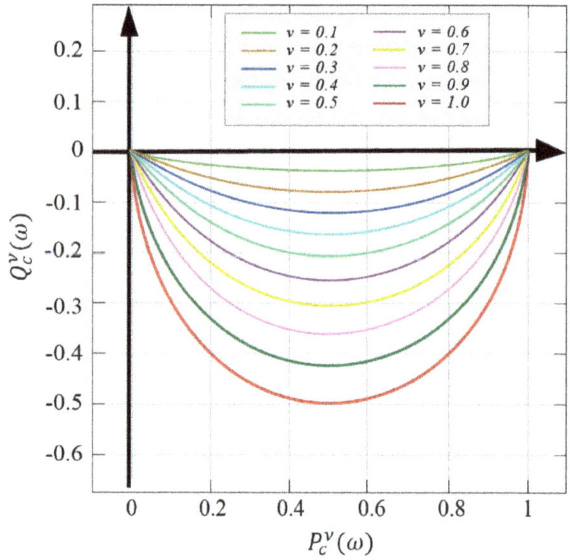

Fig. 12 Frequency characteristics of a fractional-order inertial element for $\nu \in \{0.1, 0.2, \ldots, 0.9, 1.0\}$

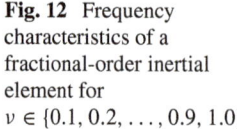

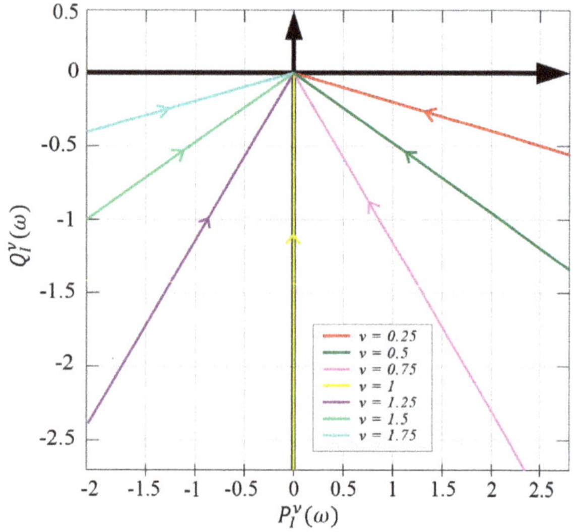

Fig. 13 Frequency characteristics of a fractional-order inertial element for $\nu \in \{1.00, 1.08, \ldots, 1.72, 1.80\}$

In Figs. 12 and 13 the Nyquist plots related to different orders ν are given. Moreover, the Nyquist plots in the forms of circle arches are characterized by the following parameters:

1. the circle centers (with arches) are placed in points

$$\left(\frac{1}{2}, \frac{1}{2}\tan^{-1}\left(\nu\frac{\pi}{2}\right)\right); \tag{90}$$

2. the circle radii equal to

$$r(\nu) = \frac{1}{2\sin\left(\nu\frac{\pi}{2}\right)}. \tag{91}$$

Consider now the magnitude characteristics of the fractional-order inertial element. From (86)–(88) one has

$$\left|G_c^\nu(j\omega)\right| = \frac{1}{\left(\frac{\omega}{\omega_u}\right)^{2\nu} + 2\left(\frac{\omega}{\omega_u}\right)^{\nu}\cos\left(\nu\frac{\pi}{2}\right) + 1}. \tag{92}$$

The magnitude plots family is presented in Figs. 14 for $0 < \nu \leq 1$ and 15 for $1 < \nu < 2$, respectively.

The analysis of Figs. 14 and 15 reveals that for $0 < \nu \leqslant 1$ the graphs show monotonically decreasing functions. For $1 < \nu < 2$ there is one maximum in all plots. The maxima values depend on ν, i.e.

$$\left|G_c^\nu(j\omega)\right|_{max} = 2r(\nu) = \frac{1}{\sin\left(\nu\frac{\pi}{2}\right)}. \tag{93}$$

Relation (93) is plotted in Fig. 16. It can be proven that the maximum $\left|G_c^\nu(j\omega)\right|$ is achieved at *resonant frequency*

$$\omega_r = \omega_u\sqrt{\cos\left(\nu\frac{\pi}{2}\right)}. \tag{94}$$

Then,

$$P_c^\nu(\omega_r) = 1, \quad Q_c^\nu(\omega_r) = -\tan^{-1}\left(\nu\frac{\pi}{2}\right), \tag{95}$$

which means that the points related to the resonant frequencies lie on the half line defined by

$$P_c^\nu(\omega_r) = 1, \quad Q_c^\nu(\omega_r) \leqslant 0. \tag{96}$$

The next characteristic points are

$$P_c^\nu(\omega_u) = \frac{1}{2}, \quad Q_c^\nu(\omega_u) = -r(\nu) + \frac{1}{2}\tan^{-1}\left(\nu\frac{\pi}{2}\right). \tag{97}$$

Points related to the frequency ω_u lie on half line defined by

$$P_c^\nu(\omega_u) = \frac{1}{2}, \; Q_c^\nu(\omega_u) \leqslant 0, \tag{98}$$

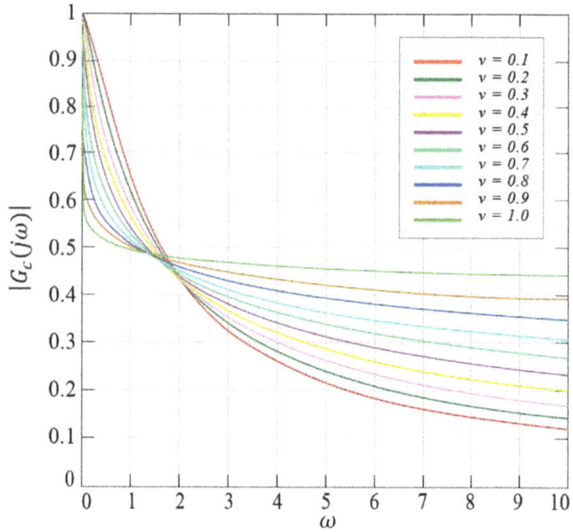

Fig. 14 Magnitude characteristics of a fractional-order inertial element for $\nu \in \{0.1, 0.2, \ldots, 0.9, 1.0\}$

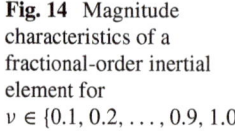

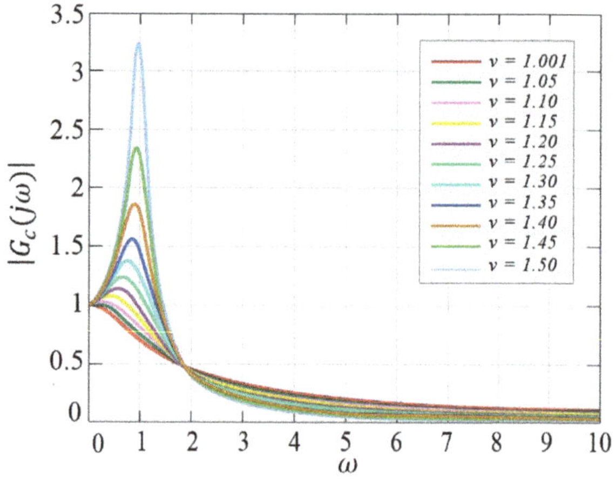

Fig. 15 Magnitude characteristics of a fractional-order inertial element for $\nu \in \{1.50, 1.55, \ldots, 1.65, 1.70\}$

Finally, the third characteristic frequency, denoted by ω_0, related to crossing the imaginary axis by the Nyquist plot of the fractional-order inertial element is defined by

$$P_c^{\nu}(\omega_0) = 0, \quad Q_c^{\nu}(\omega_0) = Q_c^{\nu}(\omega_r). \tag{99}$$

Solving Eq. (99) for ω_o one gets

$$\omega_o = \omega_u \frac{1}{\left[-\cos\left(\nu\frac{\pi}{2}\right)\right]} \tag{100}$$

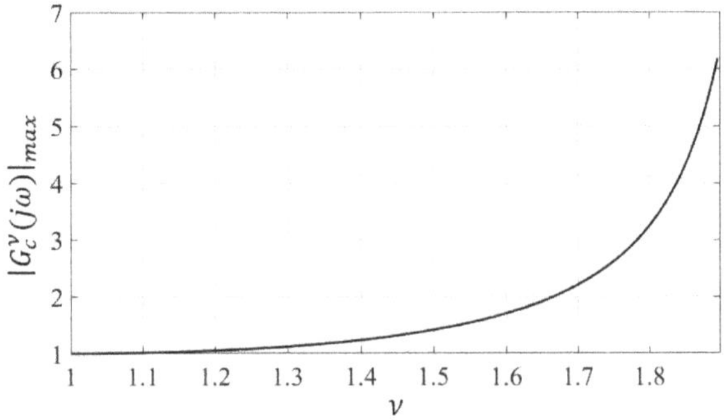

Fig. 16 Maximal values of magnitude characteristics of a fractional-order inertial element versus order $1 < \nu < 2$

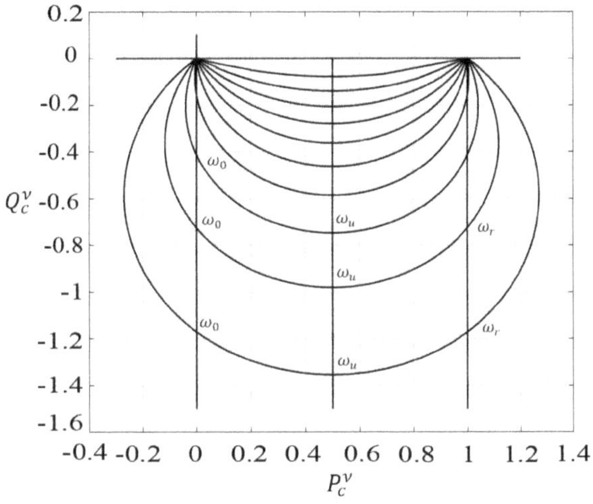

Fig. 17 Nyquist plots of the fractional-order inertial elements with characteristic frequencies

an appropriate half line is characterised by conditions

$$P_c^\nu(\omega_0) = 0, \quad Q_c^\nu(\omega_0) \leqslant 0. \tag{101}$$

Half lines characterized by (96), (98) and (101) are plotted in Fig. 17. The Nyquist plot contains information concerning the phase shift of the element. From real and imaginary parts of $G_c\nu(\cdot)$ given by (87), (88) one obtains the phase frequency

$$\varphi_c(\omega) = -\tan^{-1}\left[\frac{\left(\frac{\omega}{\omega_u}\right)^\nu \sin\left(\nu\frac{\pi}{2}\right)}{1+\left(\frac{\omega}{\omega_u}\right)^\nu \cos\left(\nu\frac{\pi}{2}\right)}\right]. \tag{102}$$

The phase frequency characteristics of the fractional inertial element are presented in Fig. 18 for $0 < \nu \leq 1$ and in Fig. 19 for $1 < \nu < 2$, respectively.

Further information provides logarithmic frequency characteristics. From (92) one gets

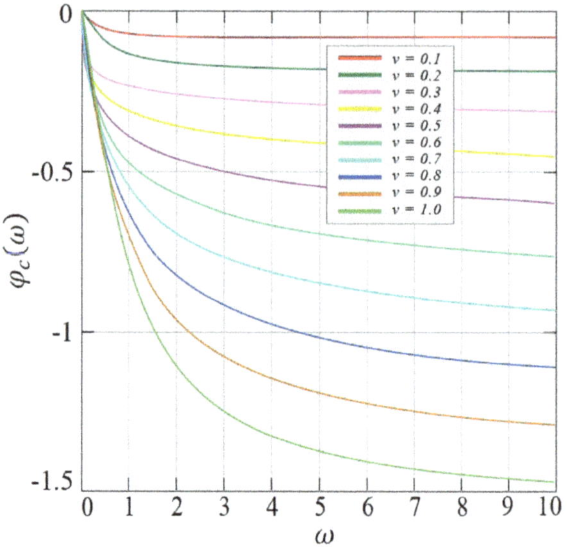

Fig. 18 Phase frequency characteristics of the fractional-order inertial element for orders $\nu \in \{0.1, 0.2, \ldots, 0.9, 1.0\}$

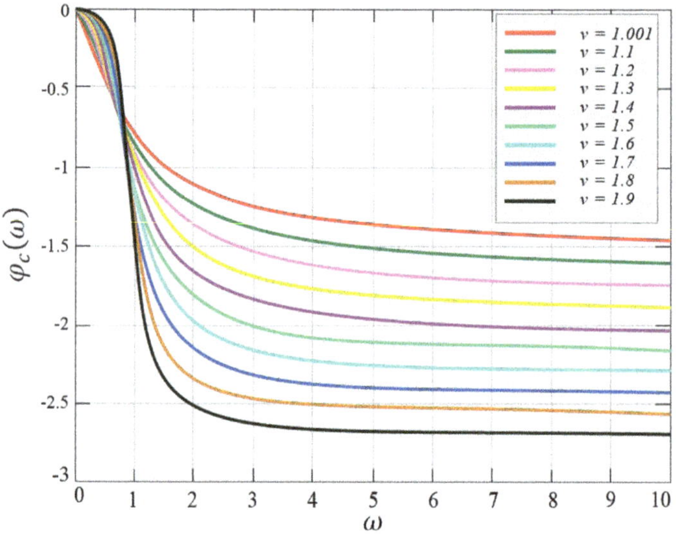

Fig. 19 Phase frequency characteristics of the fractional-order inertial element for orders $\nu \in \{1.0, 1.1, 1.2, \ldots, 1.9\}$

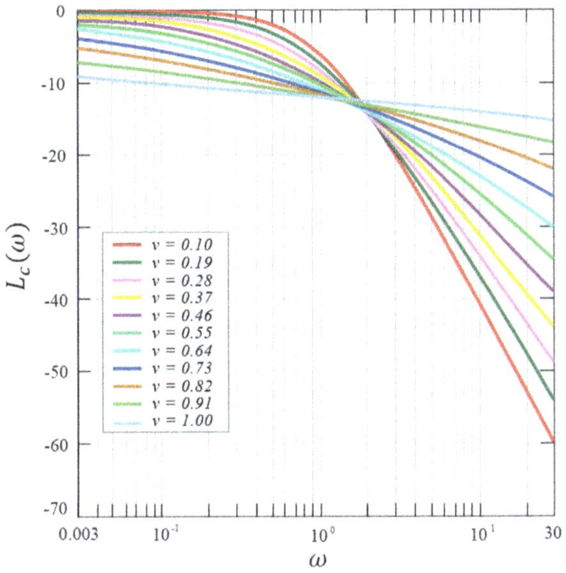

Fig. 20 Logarithmic magnitude frequency characteristics of the fractional-order inertial element for orders $\nu \in \{0.10, 0.19, \ldots, 0.91, 1.00\}$

$$L_c(\omega) = 20\log_{10}\left|G_c^{\nu}(j\omega)\right|$$

$$= -20\log_{10}\left[\left(\frac{\omega}{\omega_u}\right)^{2\nu} + 2\left(\frac{\omega}{\omega_u}\right)^{\nu}\cos\left(\nu\frac{\pi}{2}\right) + 1\right]. \tag{103}$$

Appropriate logarithmic magnitude frequency characteristics of the fractional-order inertial element are presented in Figs. 20 for $0 < \nu \leq 1$ and 21 for $1 < \nu < 2$, respectively.

5.3.1 Relation Between Fractional-Order Integrator and Inertial Element Nyquist Plots

For $\omega \in [0, +\infty)$ one can prove that the real and imaginary parts of the fractional-order integrator and inertial element are related by the following formula

$$\tan\left[\phi(\omega)\right] = \left|\frac{1 - P_I^{\nu}(\omega)}{Q_I^{\nu}(\omega)}\right| = \left|\frac{1 - P_c^{\nu}(\omega)}{Q_c^{n}u\omega)}\right|. \tag{104}$$

The formula (104) has simple geometrical interpretation. One can derive the fact that the Nyquist plots of the fractional-order integrator and inertial elements are tangent for $\omega = +\infty$. This is shown in Fig. 22.

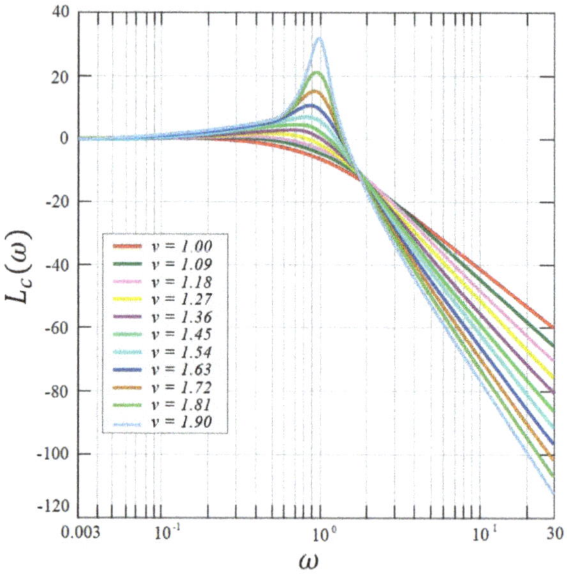

Fig. 21 Logarithmic magnitude frequency characteristics of the fractional-order inertial element for orders $\nu \in \{1.00, 1.09, \ldots, 1.81, 1.90\}$

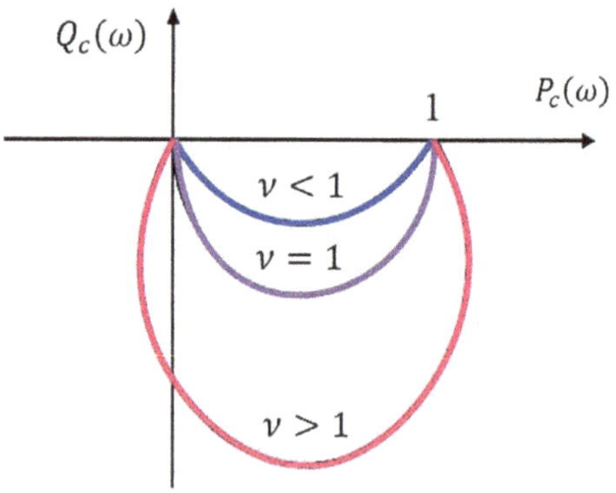

Fig. 22 Tangential relation between the Nyquist plots of the fractional-order ideal integrator and inertial element

Remark 9 The relations shown in Fig. 22, Fig. 23 also include the relation between classical (i.e. for $\nu = n = 1$) ideal integrator and first order inertial element. Here, the Nyquist plot of the integrator coincides with the negative part of the imaginary axis.

6 Elements of the State-Space Equivalence of the Fractional-Order Differential Equation

Considers a set of non-linear fractional-order differential equations that can be transformed to the system of a commensurate order equations:

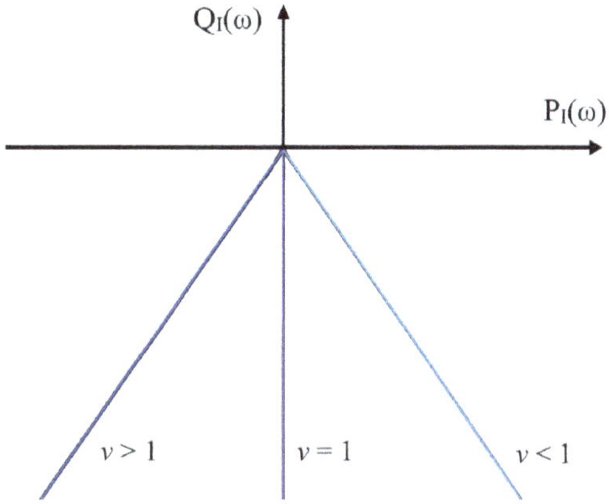

Fig. 23 Tangential relation between the Nyquist plots of the fractional-order ideal integrator and inertial element

$$\begin{aligned} {}_{t_0}^{RL}D_t^{\nu}\mathbf{x}(t) &= \mathbf{f}\left[\mathbf{x}(t), u(t)\right], \\ y(t) &= \mathbf{g}\left[\mathbf{x}(t), u(t)\right], \end{aligned} \tag{105}$$

where a state vector and its fractional order derivative are denoted as

$$\mathbf{x}(t) = \begin{bmatrix} x_1(t) \\ x_2(t) \\ \vdots \\ x_n(t) \end{bmatrix}, \quad {}_{t_0}^{RL}D_t^{\nu}\mathbf{x}(t) = \begin{bmatrix} {}_{t_0}^{RL}D_t^{\nu}x_1(t) \\ {}_{t_0}^{RL}D_t^{\nu}x_2(t) \\ \vdots \\ {}_{t_0}^{RL}D_t^{\nu}x_n(t) \end{bmatrix}. \tag{106}$$

Note that in (106) the Riemann–Liouville fractional order derivative is used but the presented results are also valid for Grünwald–Letnikov and Caputo derivatives. The function $\mathbf{f}\left[\mathbf{x}(t), u(t)\right]$ defining the dynamics of system (105) is of the form

$$\mathbf{f}\left[\mathbf{x}(t), u(t)\right] = \begin{bmatrix} f_1\left[\mathbf{x}(t), u(t)\right] \\ f_2\left[\mathbf{x}(t), u(t)\right] \\ \vdots \\ {}_{t_0}^{RL}D_t^{m_j\nu}\mathbf{x}(t) = \mathbf{f}\left[\mathbf{x}(t), u(t), t\right] \\ f_n\left[\mathbf{x}(t), u(t)\right] \end{bmatrix}$$

$$= \begin{bmatrix} f_1 \left[x_1(t), x_2(t), \ldots, x_n(t), u(t) \right] \\ f_2 \left[x_1(t), x_2(t), \ldots, x_n(t), u(t) \right] \\ \vdots \\ f_n \left[x_1(t), x_2(t), \ldots, x_n(t), u(t) \right] \end{bmatrix}, \tag{107}$$

$$\mathbf{g} \left[\mathbf{x}(t), u(t) \right] = \left[g_1 \left[\mathbf{x}(t), u(t) \right] \right]. \tag{108}$$

Assume that for function $u(t) = u_0 = \text{const}$ one gets a non-linear algebraic equation

$$\mathbf{0} = \mathbf{f} \left[\mathbf{x}_0, u_0 \right]. \tag{109}$$

Applying Implicit Function Theorem (see for example in [3, 56]) and solving for $\mathbf{x}_0$ the Eq. (109) one gets the so called *steady state*. Expanding function $\mathbf{f}$ into the Taylor series around the steady state $\mathbf{x}_0$ and u_0 and limiting to the first terms, one gets a linear approximation of the system (105).

Remark 10 One should note that in the linearization procedure

$$\mathbf{x}(t) = \mathbf{x}_0 + \delta \mathbf{x}(t) \tag{110}$$

is valid for relatively small perturbations of function $x(t)$ around its steady-state value x_0.

Then the linear approximation of the system (105) is of the following form:

$$\begin{bmatrix} f_1 \left[\delta x_1(t), \delta x_2(t), \ldots, \delta x_n(t), \delta u(t) \right] \\ f_2 \left[\delta x_1(t), \delta x_2(t), \ldots, \delta x_n(t), \delta u(t) \right] \\ \vdots \\ f_n \left[\delta x_1(t), \delta x_2(t), \ldots, \delta x_n(t), \delta u(t) \right] \end{bmatrix}$$

$$= \begin{bmatrix} a_{11} & a_{12} & \cdots & a_{1n} \\ a_{21} & a_{22} & \cdots & a_{2n} \\ \vdots & \vdots & & \vdots \\ a_{n1} & a_{n2} & \cdots & a_{nn} \end{bmatrix} \begin{bmatrix} \delta x_1(t) \\ \delta x_2(t) \\ \vdots \\ \delta x_n(t) \end{bmatrix} + \begin{bmatrix} b_{11} \\ b_{21} \\ \vdots \\ b_{n1} \end{bmatrix} \delta u(t). \tag{111}$$

Hence, one gets a linear state-space description [25]

$$ {}_{t_0}^{RL}D_t^{\nu}\delta\mathbf{x}(t) = \mathbf{A}\delta\mathbf{x}(t) + \mathbf{b}\delta u(t). \tag{112} $$

Analogous procedure performed on the nonlinear function (109) gives

$$ \mathbf{g}\left[\mathbf{x}(t), u(t)\right] = \begin{bmatrix} c_{11} & c_{12} & \cdots & c_{1n} \end{bmatrix} \begin{bmatrix} \delta x_1(t) \\ \delta x_2(t) \\ \vdots \\ \delta x_n(t) \end{bmatrix} + [d_{11}]\,\delta u(t), \tag{113} $$

The output equation is as follows

$$ y(t) = \mathbf{c}\delta\mathbf{x}(t) + \mathbf{d}\delta u(t) \tag{114} $$

Further on, the prefix δ in terms δx and δu will be omitted.

6.1 Solution of the Fractional-Order State-Space Equation

Consider state-space equations with the Caputo derivatives

$$ \begin{aligned} {}_{t_0}^{C}D_t^{\nu}\mathbf{x}(t) &= \mathbf{A}\mathbf{x}(t) + \mathbf{b}u(t), \\ y(t) &= \mathbf{c}\mathbf{x}(t) + \mathbf{d}u(t). \end{aligned} \tag{115} $$

Suppose that $0 < \nu < 1$, $t_0 = 0$ and a vector $\mathbf{x}(t_0) = \mathbf{x}(0)$ is a vector of initial conditions, $u(t)$ is a known function. Note that the dynamics of the system (115), i.e. the equation ${}_{t_0}^{C}D_t^{\nu}\mathbf{x}(t) = \mathbf{A}\mathbf{x}(t) + \mathbf{b}u(t)$ is a linear non-homogenous equation. Since the solution of homogenous linear equation

$$ {}_{t_0}^{C}D_t^{\nu}\Phi_0(t) = \mathbf{A}\Phi_0(t) \tag{116} $$

is, see [39, 40]

$$ \Phi_0(t) = \sum_{i=0}^{+\infty} \frac{t^{i\nu}}{\Gamma(i\nu+1)}\mathbf{A}^i. \tag{117} $$

Then the solution of dynamics is as follows

$$ \mathbf{x}(t) = \Phi_0(t)\mathbf{x}(0) + \int_0^t \Phi(t-\tau)\mathbf{b}u(\tau)d\tau, \tag{118} $$

where

$$\Phi(t) = \sum_{i=0}^{+\infty} \frac{t^{(i+1)\nu-1}}{\Gamma[(i+1)\nu]} \mathbf{A}^i. \quad (119)$$

The matrix (117) is a solution of the following differential equation

$$ {}_{t_0}^{C}D_t^{\nu} \Phi_0(t) = \mathbf{A}\Phi_0(t). \quad (120)$$

6.2 Relation Between the State-Space Description of Fractional-Order System and Its Transfer Function

Consider a fractional-order dynamical system described by the state-space equations

$$\begin{aligned} {}_{t_0}^{RL}D_t^{\nu}\mathbf{x}(t) &= \mathbf{A}\mathbf{x}(t) + \mathbf{b}u(t), \\ y(t) &= \mathbf{c}\mathbf{x}(t) + \mathbf{du(t)}. \end{aligned} \quad (121)$$

The block diagram of this system is presented in Fig. 24. Assuming zero initial conditions and applying the unilateral Laplace transform to the Eqs. (121) with zero initial conditions one gets

$$\begin{aligned} s^{\nu}\mathbf{X}(s) &= \mathbf{A}\mathbf{X}(s) + \mathbf{b}U(s), \\ Y(s) &= \mathbf{c}\mathbf{X}(s) + \mathbf{d}U(s). \end{aligned} \quad (122)$$

Then elementary operations on Eqs. (122) lead to the fractional-order transfer function

$$G(s) = \frac{Y(s)}{U(s)} = \mathbf{c}\left(s^{\nu}\mathbf{1} - \mathbf{A}\right)^{-1}\mathbf{b} + \mathbf{d}. \quad (123)$$

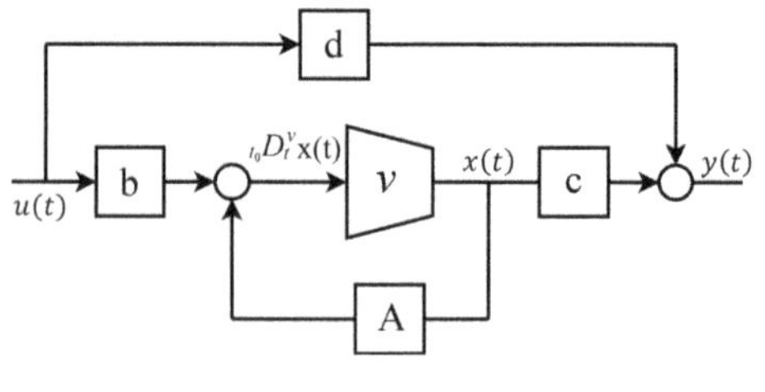

Fig. 24 Block diagram of a state-space realization

6.3 Realizations of the Fractional-Order Transfer Functions

Assuming that $\mathbf{d} = \mathbf{0}$ and taking into account Remark 3 the given fractional-order commensurate dynamical system can be described by the transfer function being a proper rational function with nominator of degree m and denominator of degree n (i.e. $m < n$). Dividing the numerator and denominator by $s^{n\nu}$ gives

$$G(s) = \frac{Y(s)}{U(s)} = \frac{B_m s^{(-n+m)\nu} + B_{m-1} s^{(-n+m-1)\nu} + \cdots + B_0 s^{-n\nu}}{1 + A_{n-1} s^{(-1)\nu} + \cdots + A_0 s^{-n\nu}}. \tag{124}$$

Then, one can write

$$Y(s) = (B_m s^{(-n+m)\nu} + B_{m-1} s^{(-n+m-1)\nu} + \cdots + B_0 s^{-n\nu}) E(s) \tag{125}$$

where

$$E(s) = \frac{U(s)}{1 + A_{n-1} s^{(-1)\nu} + \cdots + A_0 s^{-n\nu}}, \tag{126}$$

or equivalently

$$E(s) = U(s) - (A_{n-1} s^{(-1)\nu} + \cdots + A_0 s^{-n\nu}) E(s). \tag{127}$$

If $X_i(s) = \mathcal{L}[x(t)]$ for $i = 1, 2, \ldots, n$, then the state-variables are related as follows

$$\begin{aligned} s^{\nu} X_1(s) &= X_2(s) \\ s^{\nu} X_2(s) &= X_3(s) \\ &\vdots \\ s^{\nu} X_{n-1}(s) &= X_n(s) \\ s^{\nu} X_n(s) &= -A_0 X_1(s) - A_1 X_2(s) - \cdots - A_{n-1} X_n(s) + U(s) \\ Y(s) &= B_0 X_1(s) + B_1 X_2(s) + \cdots + B_{m-1} X_m(s) + B_m X_{m+1}(s). \end{aligned} \tag{128}$$

Applying an inverse Laplace transform to the equations presented above one gets

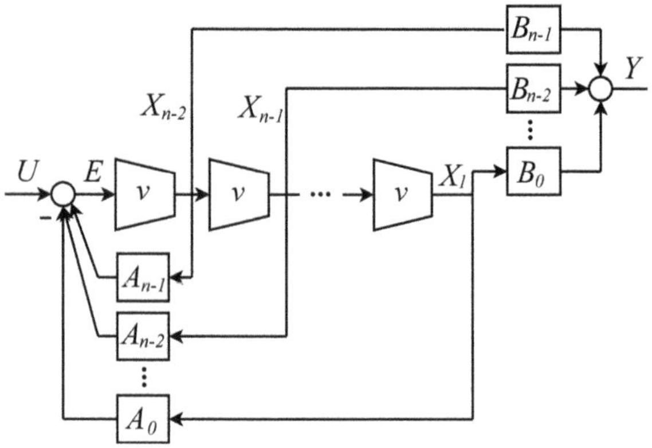

Fig. 25 Block diagram of a realization of the fractional-order transfer function

$$
\begin{aligned}
{}_{t_0}^{RL}D_t^{\nu}x_1(t) &= x_2(t)\\
{}_{t_0}^{RL}D_t^{\nu}x_2(t) &= x_3(t)\\
&\vdots\\
{}_{t_0}^{RL}D_t^{\nu}x_{n-1}(t) &= x_n(t)\\
{}_{t_0}^{RL}D_t^{\nu}x_n(t) &= -A_0x_1(t) - A_1x_2(t) - \cdots - A_{n-1}x_n(t) + u(t)\\
y(t) &= B_0x_1(t) + B_1x_2(t) + \cdots + B_{m-1}x_m(t) + B_mx_{m+1}(t).
\end{aligned}
\tag{129}
$$

System of Eqs. (129) can be written in the vector-matrix form (121) with

$$
\mathbf{A} = \begin{bmatrix} 0 & 1 & 0 & \cdots & 0 & 0 \\ 0 & 0 & 1 & \cdots & 0 & 0 \\ \vdots & \vdots & \vdots & & \vdots & \vdots \\ 0 & 0 & 0 & \cdots & 0 & 1 \\ -A_0 & -A_1 & -A_2 & \cdots & -A_{n-2} & -A_{n-1} \end{bmatrix}, \quad \mathbf{b} = \begin{bmatrix} 0 \\ 0 \\ \vdots \\ 0 \\ 1 \end{bmatrix}, \tag{130}
$$

$$
\mathbf{c} = \begin{bmatrix} B_0 & B_1 & \cdots & B_{m-1} & B_m & 0 & \cdots & 0 \end{bmatrix}, \quad \mathbf{d} = [0]. \tag{131}
$$

In Fig. 25 the block diagram realizing the state-space Eq. (122) is presented.

6.4 Similarity Transformations of Matrices Describing the Fractional-Order Systems

Now, one considers a transformation of the state $\mathbf{x}(t)$

$$
\mathbf{x}(t) = \mathbf{T}\mathbf{w}(t), \tag{132}
$$

where $\mathbf{w}(t) = [w_1(t), w_2(t), \ldots, w_n(t)] \in \mathbb{R}^n$ is a new n components state vector and $\mathbf{T} \in \mathbb{R}^{n\times n}$ is a non-singular transformation matrix. Substitution (132) into (121) leads to

$$\begin{aligned} {}_{t_0}^{RL}D_t^{\nu}\left[\mathbf{Tw}(t)\right] &= \mathbf{ATw}(t) + \mathbf{b}u(t) \\ y(t) &= \mathbf{cTw}(t) + \mathbf{du(t)}. \end{aligned} \tag{133}$$

Since the operator ${}_{t_0}^{RL}D_t^{\nu}$ is linear, then the left-hand side multiplication of the first equations of (133) gives

$$\begin{aligned} {}_{t_0}^{RL}D_t^{\nu}\mathbf{w}(t) &= \mathbf{T}^{-1}\mathbf{ATw}(t) + \mathbf{T}^{-1}\mathbf{b}u(t) \\ y(t) &= \mathbf{cTw}(t) + \mathbf{du(t)}, \end{aligned} \tag{134}$$

or

$$\begin{aligned} {}_{t_0}^{RL}D_t^{\nu}\mathbf{w}(t) &= \mathbf{A}_w\mathbf{w}(t) + \mathbf{b}_w u(t) \\ y(t) &= \mathbf{c}_w\mathbf{w}(t) + \mathbf{du(t)} \end{aligned} \tag{135}$$

where

$$\mathbf{A}_w = \mathbf{T}^{-1}\mathbf{AT}, \quad \mathbf{b}_w = \mathbf{T}^{-1}\mathbf{b}, \quad \mathbf{c}_w = \mathbf{cT}. \tag{136}$$

By the way of example as a transformation matrix, one considers a transformation involutory matrix

$$\mathbf{T} = \begin{bmatrix} 0\,0\,0 \cdots 0\,1 \\ 0\,0\,1 \cdots 1\,0 \\ \vdots\,\vdots\,\vdots \quad\; \vdots\,\vdots \\ 0\,1\,0 \cdots 0\,0 \\ 1\,0\,0 \cdots 0\,0 \end{bmatrix}. \tag{137}$$

Then, appropriate matrices take the forms

$$\mathbf{A}_w = \begin{bmatrix} -A_{n-1} & -A_{n-2} & -A_{n-2} & \cdots & -A_2 & -A_1 & -A_0 \\ 1 & 0 & 0 & \cdots & 0 & 0 & 0 \\ \vdots & \vdots & \vdots & & \vdots & \vdots & \vdots \\ 0 & 0 & 0 & \cdots & 1 & 0 & 0 \\ 0 & 0 & 0 & \cdots & 0 & 1 & 0 \end{bmatrix}, \ \mathbf{b}_w = \begin{bmatrix} 1 \\ 0 \\ \vdots \\ 0 \\ 0 \end{bmatrix}, \tag{138}$$

$$\mathbf{c}_w = \begin{bmatrix} 0 \cdots 0 \ B_m \ B_{m-1} \cdots B_1 \ B_0 \end{bmatrix}, \ \mathbf{d} = [0]. \tag{139}$$

7 Stability of the Fractional-Order Linear Systems

Consider the transfer function (54) of the integer order associated with the fractional-order transfer function (51). In general case the characteristic polynomial of the system with transfer function (54), i.e. polynomial

$$w(s) = s^{\nu_n} + A_{n-1}s^{\nu_{n-1}} + \ldots + A_0 s^{\nu_0} \tag{140}$$

is a multifunction of the complex variable with Riemann surface as the domain. This surface has an infinite number of leafs, so the fractional polynomial (141) has an infinite number of zeros. A finite number of zeros will be in the main leaf of the Riemann surface [61, 66], see Fig. 26. From a stability point of view only the main leaf defined for $\pi < arg(s) < \pi$ is important. The Riemann surface has a finite number of leafs only for commensurate order polynomial

$$\bar{w}(s) = s^{n\nu} + A_{n-1}s^{(n-1)\nu} + \cdots + A_0. \tag{141}$$

Substituting $p = s^{\nu}$ in polynomial $w(s)$ one obtains

$$\tilde{w}(s) = p^n + A_{n-1}p^{(n-1)} + \cdots + A_0 \tag{142}$$

which is the characteristic polynomial of the integer order associated with the polynomial (141). The fractional order system of commensurate order is stable (in the sense bounded input—bounded output stability) if and only if [13, 61, 66]

$$w(s) \neq 0 \quad \text{for} \quad Re(s) \geq 0. \tag{143}$$

One can show that [11]

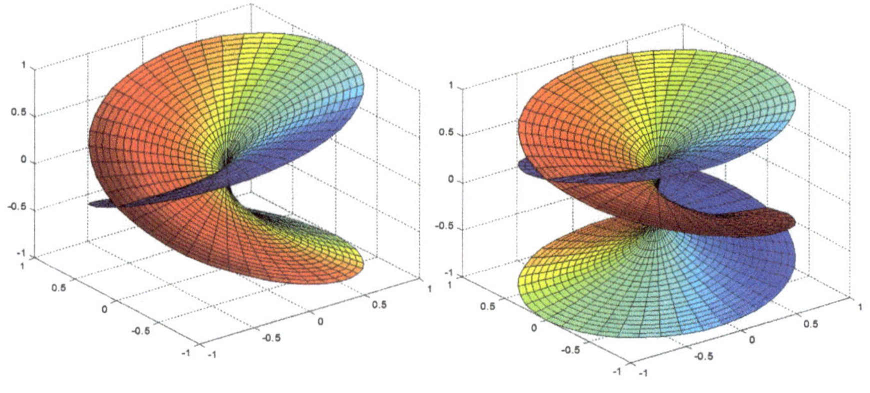

Fig. 26 Riemann surface for $w(s) = s^{\frac{1}{2}}$, and $w(s) = s^{\frac{1}{3}}$

Fig. 27 Stability region (in grey)

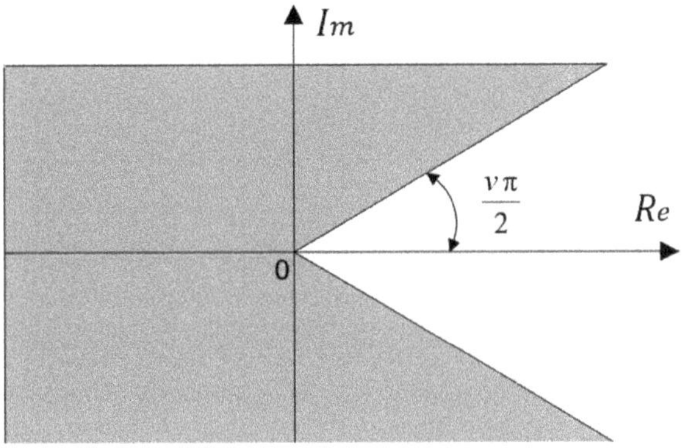

$$\det\left[s^{\nu}\mathbf{I}-\mathbf{A}\right]=\det\left[p\mathbf{I}-\mathbf{A}\right]=\det\left[p\mathbf{I}-\mathbf{A}_w\right]=\prod_{i=1}^{n}(p-\sigma_i)\,, \tag{144}$$

where $\sigma_i = p_i$ and p_i are the same as in the denominator in formula (52). Elements σ_i will be further called *pseudo-poles* of the fractional-order systems.

Assume that the order of the commensurate differential equation satisfies the condition

$$0 < \nu \leqslant 1. \tag{145}$$

The linear time-invariant dynamical systems described by the fractional-order differential equations (30) are asymptotically stable (in the sense bounded input—bounded output) [12, 48] if and only if

$$|\arg(\sigma_i)| > \nu\frac{\pi}{2} \quad \text{for } i = 1, 2, \ldots, n. \tag{146}$$

Conditions (143) and (146) are equivalent, see for example [13].

A configuration of pseudo-poles of any asymptotically stable system is given in Fig. 27. For $\nu = 1$ a stability region is a classical open left half complex plane. This is presented in Fig. 28. Finally, for $1 < \nu < 2$ the system is asymptotically stable if and only if the pseudo-poles lie in the shadowed area presented in Fig. 29. One should mention that if there is σ_i $i, \in \{1, 2, \ldots, n\}$, then

$$|\arg(\sigma_i)| = \nu\frac{\pi}{2} \tag{147}$$

for any σ_i, $i \in \{1, 2, \ldots, n\}$, then the given system is on a stability border.

Fig. 28 Stability

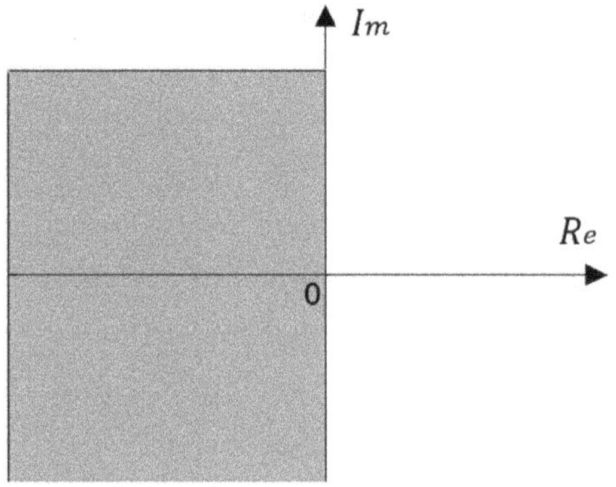

Fig. 29 Stability

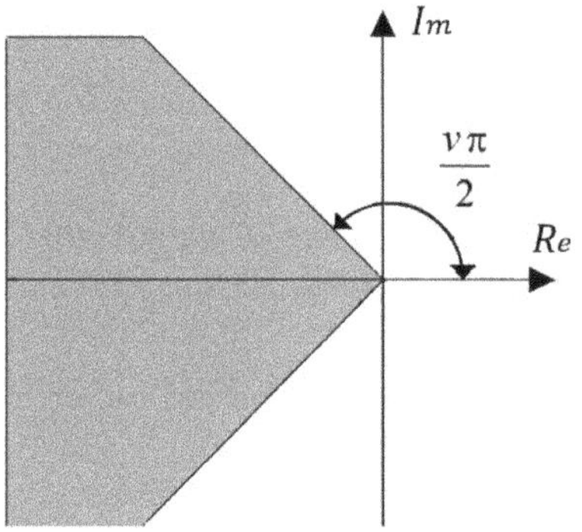

Remark 11 The given fractional order system could be stable even when the polynomial $\tilde{w}(s)$ given by (142) is not stable. But it can be shown that fractional order system (30) with $0 < \nu \leq 1$ is stable if the polynomial (142) is asymptotically stable.

One can shown that the dynamics of fractional order system (121), i.e. system

$$ {}_{t_0}^{RL}D_t^{\nu}\mathbf{x}(t) = \mathbf{A}\mathbf{x}(t) + \mathbf{b}u(t) \tag{148}$$

with commensurate orders of derivatives with characteristic polynomial $w(s) = det[s^{\nu}\mathbf{I} - \mathbf{A}]$ is stable if and only if

$$\varDelta \arg w(j\omega) = n\frac{\pi}{2} \tag{149}$$

where $w(j\omega) = \det(\mathbf{I}(j\omega) - \mathbf{A})$ and is defined by $\mathbf{I}(j\omega) = (j\omega)^{\nu}\mathbf{I}_{n\times n}$ and $\mathbf{I}_{n\times n}$ is the n-dimensional identity matrix. Condition (148) means that the plot of (149) starts for $\omega = 0$ at the point $\omega(0) = \det(-\mathbf{A})$ and with ω increasing from zero to infinity, turns strictly counter-clockwise and goes through n quadrants of the complex plane. The plot of function (149) will be called the generalized (to the class of fractional order systems) Mikhailov plot. The method of checking the condition (148) is described in [14].

8 Final Remarks

In this chapter elements of fractional calculus important from the point of application in control and automatics were shortly reviewed. Considering how wide the area of research, applications, and even industrial applications of fractional calculus is, we have only touched upon this topic in this chapter. The presented results have been restricted to only continuous-time systems.

From an application point of view important for analysis and synthesis of fractional order systems one can use packets of Matlab/Simulink. More on this can be found in [18, 80]. Additionally, for designing of Crone regulators (see for example [63, 64]) one can use *Crone toolbox for Matlab*. However, taking into account the limitation of computational resources, computational complexity should be as low as possible. There are works addressed to the optimization of the number steps in approximation used in computing of the Grünwald–Letnikov fractional order operator [2, 80] and references therein.

Acknowledgements The work of P. Ostalczyk was supported by the National Science Centre Poland Grant Number 2016/23/B/ST7/03686. The work of E. Pawluszewicz is supported with University Work No WZ/WM-IIM/1/2019 Faculty of Mechanical Engineering, Bialystok University of Technology.

References

1. Akgül, A., Inc, M., Baleanu, D.: On solutions of variable-order fractional differential equations. Int. J. Optim. Control: Theor. Appl. **7**(1), 112–116 (2017)
2. Alagoz, B.B., Tepljakov, A., Ates, A.: Time-domain identification of one noninteger order plus time delay models from step response measurements. Int. J. Model. Simul. Sci. Comput. **10**(1), 1941011–1–1941011–22 (2019)
3. Allendoerfer, C.B.: Theorems About Differentiable Functions. Calculus of Several Variables and Differentiable Manifolds. Macmillan, New York (1974)
4. Almeida, R., Kamocki, R., Malinowska, A.B., Odzijewicz, T.: On the existence of optimal consensus control for the fractional Cucker-Smale model. Arch. Control Sci. **30**(4), 625–651 (2020)
5. Ambroziak, L., Lewon, D., Pawluszewicz, E.: The use of fractional order operators in modeling of RC-electrical systems. Control Cybern. **45**(3), 275–288 (2016)
6. Bagley, R.L., Torvik, P.J.: Fractional calculus in the transient analysis of viscoelasticity damped structures. AIAA J. **23**(6), 918–925 (1985)
7. Bagley, R.L., Torvik, P.J.: On the fractional calculus model of viscoelastic behaviour. J. Rheol. **30**(1), 133–155 (1986)
8. Bąkała, M., Duch, P., Machado, J.A.T., Ostalczyk, P., Sankowski, D.: Commensurate and non-commensurate fractional-order discrete models of an electric individual-wheel drive on an autonomous platform. Entropy **22**(3) (2020). https://doi.org/10.3390/e22030300

9. Baleanu, D., Diethelm, K., Trujillo, J.J.: Fractional Calculus: Models and Numerical Methods. Nonlinearity and Chaos. World Scientific, Singapore, Series on Complexity (2012)
10. Bandyopadhyay, B., Kamal, S.: Stabilization and Control of Fractional Order Systems: a Sliding Mode Approach. Lecture Notes in Electrical Engineering, vol. 317, pp. 55–90. Springer International Publishing, New York (2015)
11. Busłowicz, M.: Frequency domain method for stability analysis of linear continuous-time fractional systems. In: Malinowski, K., Rutkowski, L. (eds.) Recent Advances in Control and Automation. Academic Publishing House EXIT, Warsaw (2008)
12. Busłowicz, M.: Stability of linear continuous-time fractional order systems of commensurate order. J. Autom. Mob. Robot. Intell. Syst. **2**, 15–21 (2009)
13. Busłowicz, M.: Wybrane zagadnienia z zakresu liniowych ciȩgłych układów niecałowitego rzdu. Pomiary Automatyka Kontrola **2**, 93–114 (2010). (in Polish)
14. Busłowicz, M.: Stability of continuous-time linear systems described by state equation with fractional commensurate orders of derivatives. Acta Mech. Autom. **6**(4), 17–20 (2012)
15. Carpinteri, A., Mainardi, F.: Fractals and Fractional Calculus in Continuous Mechanics. Springer, New York (1997)
16. Chak, A.M.: A generalization of the Mittag-Leffler function. Mat. Vesn. **19**(4), 257–262 (1967)
17. Coimbra, C.: Mechanics with variable-order differential operators. Annu. Phys. **12**, 692–703 (2003)
18. Copot, C., Ionescu, C.M., Muresan, C.I.: Image-Based and Fractional Order Control for Mechatronic System. Theory and Applications with Matlab. Springer Publisher, New York (2020)
19. Deng, W.: Short memory principle and a predictor-corrector approach for fractional differential equations. J. Comput. Appl. Math. **206**, 174–188 (2007)
20. Diethelm, K.: An algorithm for the numerical solution of differential equations of fractional order. Electron. Trans. Numer. Anal. **5**, 1–6 (1997)
21. Diethelm, K.: A predictor - corrector approach for the numerical solution of fractional differential equations. Nonlinear Dyn. **29**, 3–22 (2002)
22. Ditzian, Z.: Fractional derivatives and best approximation. Acta Math. Hug. **81**, 323–348 (1998)
23. Djennoune, S., Bettayeb, M., Al-Saggaf, U.M.: Synchronization of fractional-order discrete-time chaotic systems by an delayed reconstructor: application to secure communication. Int. J. Appl. Math. Comput. Sci. **29**(1), 179–194 (2019)
24. Dorčák, L.: Numerical models for simulation the fractional-order control systems. In: UEF-04-94. The Academy of Science, Institute of Experimental Physics, Kosice, pp. 1–12 (1994)
25. Dorčák, L., Petráš, G., Koštial, I., Trepák, J.: Fractional-order state space models. In: International Carpathian Control Conference ICCC', Czech Republic, pp. 193–198 (2002)
26. Du, M., Wang, Z.: Correcting the initialization of models with fractional derivatives via history-depend conditions. Acta Mech. Sin. 320–325 (2016)
27. Dzieliński, A.: Optimal control for discrete fractional systems. In: Babiarz, A., Czornik, A., Klamka, J., Niezabitowski, M. (eds.) Theory and Applications of Non-integer Order Systems. Lecture Notes in Electrical Engineering, vol. 407. Springer International Publishing AG, Cham, pp. 175–185 (2017)
28. Dzieliński, A., Sarwas, G., Sierociuk, D.: Time domain validation of ultracapacitor fractional order model. In: 49th IEEE Conference on Decision and Control DCD, pp. 3730–3735 (2010)
29. Evangelista, L.R., Lenzi, E.K.: (Autor): Fractional Diffusion Equations and Anomalous Diffusion. Cambridge University Press, Cambridge (2018)
30. Farid, H.: Discrete-time fractional differentiation form integer derivatives. TR2004-528, Dartmouth College, Computer Science, pp. 1–9 (2004)
31. Fiedler, M.: Special Matrices and Their Applications in Numerical Mathematics. Martinus Nijhoff Publishers, Dordrecht (1986)
32. Ford, N.J., Simpson, A.C.: The numerical solution of fractional differential equations: speed versus accuracy. Numer. Algorithms **26**, 333–346 (2001)
33. Gorenflo, R.: Fractional Calculus: Some Numerical Methods. CISM Courses Lect. **378**, 277–290 (2001)

34. Hartley, T.T., Lorenzo, C.F.: Control of initialized fractional-order systems. NASA/TM-2002-211377/Rev1 Raport, Glenn Research Center, pp. 1–40 (2002)
35. Hcheichi, K., Bouani, F.: Comparison between commensurate and non-commensurate fractional systems. Int. J. Adv. Comput. Sci. Appl. **9**(11), 685–691 (2018)
36. Heymens, N., Podlubny, I.: Physical interpretation of initial conditions for fractional differential equations with Riemann - Liouville fractional derivatives. Rheol. Acta **45**, 765–771 (2006)
37. Hilfer, R.: Applications of Fractional Calculus in Physics. World Scientific, Singapore (2000)
38. Jifeng, W., Yuankai, L.: Frequency domain analysis and applications for fractional-order control systems. J. Phys.: Conf. Ser. **b13**, 268–273 (2005)
39. Kaczorek, T.: Selected Problems of Fractional System Theory. Springer, Heidelberg (2011)
40. Kaczorek, T., Rogowski, K.: Fractional Linear Systems and Electrical Circuits. Studies in Systems, Decision and Control, vol. 13. Springer, Heidelberg (2015)
41. Kilbas, A.A., Srivastawa, H.M., Trujilo, J.J.: Theory and Applications of Fractional Differential Equations. North-Holland Mathematics Studies, Amsterdam (2006)
42. Koszewnik, A., Ostaszewski, M., Pawluszeicz, E.: Experimental studies of the fractional PID and TID controllers for industrial process. Int. J. Control Autom. Syst. **19**, 1847–1862 (2021)
43. Lakshmikantham, V., Leela, S., Devi, J.V.: Theory of Fractional Dynamical Systems. Cambridge/Academic, Cambridge (2009)
44. Lorenzo, C.F., Hartley, T.T.: Variable order and distributed order fractional operators. Nonlinear Dyn. **29**, 57–98 (2002)
45. Loverro, A.: Fractional calculus: history, definitions and applications for the engineer. Department of Aerospace and Mechanical Engineering. University of Notre Dame, USA (2004)
46. Lubich, C.H.: Discretized fractional calculus. SIAM J. Math. Anal. **17**, 704–719 (1986)
47. Machado, J.A.T.: Theory analysis and design of fractional - order digital control systems. J. Syst. Anal. - Model. - Simul. **27**, 107–122 (1997)
48. Matignon, D.: Stability properties for generalized fractional differential system. In: ESIM: Proceedings Fractional Differential Systems: Models, Methods and Applications, pp. 145–158 (1998)
49. Miller, K.S., Ross, B.: Proceedings of the International Symposium on Univalent Functions, Fractional Calculus and Their Applications, Nihon University, Kōriyama, Japan Fractional Difference Calculus (1988), pp. 139–152
50. Miller, K.S., Ross, B.: An Introduction to the Fractional Calculus and Fractional Differential Equations. Wiley, New York (1993)
51. Momani, S., Odibat, Z.: Numerical approach to differential equations of fractional order. J. Comput. Appl. Math. **207**(11), 96–110 (2007)
52. Monie, C.A., Chen, Y.Q., Vinagre, B.M., Xue, B.M., Feliu, V.: Fractional-Order Systems and Controls. Fundamentals and Applications. Advances in Industrial Control. Springer, London (2010)
53. Mozyrska, D., Ostalczyk, P.: Generalized Fractional-Order Discrete-Time Integrator. Complexity **2017**, Article ID 3452409, 11 pp
54. Mozyrska, D., Ostalczyk, P.: Variable-, fractional-order oscillation element. In: Babiarz, A., Czornik, A., Klamka, J., Niezabitowski, M. (eds.) Theory and Applications of Non-integer Order Systems. Lecture Notes in Electrical Engineering, vol. 407, pp. 65–75. Springer International Publishing AG, Cham (2017)
55. Mozyrska, D., Pawluszewicz, E.: Local controllability of nonlinear discrete-time fractional order systems. Bull. Pol. Acad. Sci. Tech. Sci. **61**(1), 251–256 (2013)
56. Mozyrska, D., Pawluszewiczb, E., Wyrwas, M.: Local observability and controllability of nonlinear discrete-time fractional order systems based on their linearization. Int. J. Syst. Sci. **48**(4), 788–794 (2017)
57. Oldham, K.B., Spanier, J.: The Fractional Calculus. Academic, New York (1974)
58. Oprzedkiwicz, K.: Fractional order, discrete model of heat transfer process using time and spatial Grünwald-Letnikov operator. Bull. Pol. Acad. Sci.: Tech. Sci. **69**(1), 1–10 (2021)
59. Ortigueira, M.D., Bengochea, G.: Non-commensurate fractional linear systems: new results. J. Adv. Res. **25**, 11–17 (2020)

60. Osler, T.J.: Fractional derivatives and Leibniz rule. Am. Math. Mon. Taylor & Francis, Ltd **78**(6), 645–649 (1971)
61. Ostalczyk, P.: Epitome of the fractional calculus. Theory and its applications in automatics. Publishing Department of Technical University of Lodz, Lodz (2008)
62. Ostalczyk, P.: Discrete Fractional Calculus. Some Applications in Control and Image Processing. Series in Computer Vision, vol. 4. World Scientific Publishing Co Pte Ltd., Singapore (2016)
63. Oustaloup, A.: La commande CRONE. Éditions Hermès, Paris (1991)
64. Oustaloup, A.: Diversity and Non-integer Differentiation for System Dynamics. Wiley, Hoboken (2014)
65. Pawluszewicz, E., Koszewnik, A., Burzynski, P.: On Grünwald-Letnikov fractional operator with measurable order on continuous-discrete-time scale. Acta Mech. Autom. **14**(3), 161–165 (2020)
66. Petras, I.: Fractional-Order Nonlinear Systems Modeling. Analysis and Simulation. Springer, Berlin (2011)
67. Podlubny, I.: Fractional-Order Systems and - Controllers. IEEE Trans. Autom. Control **44**(1), 208–214 (1999)
68. Podlubny, I.: Fractional Differential Equations. Academic, San Diego (1999)
69. Poty, A., Melchior, P., Oustaloup, A.: Dynamic path planning for mobile robot using fractional potential field. In: IEEE First International Symposium on Control, Communications and Signal Processing, pp. 557–561 (2004)
70. Razminia, A., Dizaji, A.F., Majda, V.J.: Solution existence for non-autonomous variable-order fractional differential equations. Math. Comput. Model. **55**, 1106–1117 (2012)
71. Ruszewski, A.: Stability of discrete-time fractional linear systems with delays. Arch. Control Sci. **29**(3), 549–567 (2019)
72. Sabatier, J., Agrawal, O., Machado, J.A.T. (eds.): Advances in Fractional Calculus. Theoretical Developments and Applications in Physics and Engineering. Springer, London (2007)
73. Samko, S., Kilbas, A., Marichew, O.: Fractional Integrals and Derivatives. Theory and Applications. Gordon & Breach Sci. Publishers, New York (1987)
74. Shamardan, A.B., Moubarak, M.R.A.: Controllability and Observability for Fractional Control Systems. J. Fract. Calc. **15**, 25–34 (1999)
75. Si, X., Yang, H., Ivanov, I.G.: Conditions and a computation method of the constrained regulation problem for a class of fractional-order nonlinear continuous-time systems. Int. J. Appl. Math. Comput. Sci. **31**(1), 17–28 (2021)
76. Sierociuk, D., Malesza, W.: Fractional variable order discrete-time systems, their solutions and properties. Int. J. Syst. Sci. **48**(14), 3098–3105 (2017)
77. Sierociuk, D., Dzieliński, A., Sarwas, G., Petráš, I., Podlubny, I., Skovranek, T.: Modeling heat transfer in heterogenous media using fractional calculus. Philos. Trans. Math. Phys. Eng. Sci. **371** (2013)
78. Sun, H.H., Chang, A., Zhang, Y., Chen, W.: Review on variable-order fractional differential equations: mathematical foundations, physical models, numerical methods and applications. Fract. Calc. Anal. **22**, 27–57 (2019)
79. Teplajkov, A.: Fractional Order Modelling and Control of Dynamic Systems. Springer, Berlin (2017)
80. Tepljakov, A., Petlekov, E., Belikov, J.: A flexible MATLAB tool for optimal fractional-order PID controller design subject to specifications. In: Proceedings of the 31st Chinese Control Conference, pp. 4698–4703 (2012)
81. Traore, A., Sene, N.: Model of economic growth in the context of fractional derivative. Alex. Eng. J. **59**, 4843–4850 (2020)
82. Tuan, V.K., Gorenflo, R.: Extrapolation to the limit for numerical fractional differentiation. Z. Angew. Math. Mech. **75**(8), 646–648 (1995)
83. Valério, D., Costa, S.: Tuning-rules for fractional PID controllers. In: Proceedings of the 2nd IFAC Workshop on Fractional Differentiation and Its Applications, Porto, pp. 89–94 (2004)

84. Valerio, D., da Costa, J.S.: Variable-order fractional derivatives and their numerical approximations. Signal Process. **91**(3), 470–483 (2011)
85. Wyrwas, M., Pawluszewicz, E., Girejko, E.: Stability of nonlinear H-difference systems with n fractional orders. Kybernetika **51**(1), 112–136 (2015)
86. Zhao, C., Xue, D., Chen, Y.Q.: A fractional order PID tuning algorithm for a class of fractional order plants. In: Proceedings of the IEEE, International Conference on Mechatronics & Automation, pp. 216–221 (2005)

Modeling, Behavior and Properties

Mixed Logical Dynamical Modeling of Discrete-Time Hybrid Fractional Systems

Stefan Domek

Abstract In this paper a method for modeling discrete-time systems of non-integer order, involving real and integer variables and their interactions is introduced. The proposed fractional-order mixed logical dynamical (FO MLD) approach generalizes a wide set of models applicable to fractional-order linear hybrid systems, sequential logical systems, some classes of discrete event systems, FO linear systems with constraints and FO nonlinear systems, the nonlinearities of which can be approximated by piecewise linear functions (FO PWL). Such systems occur often in various fields of life, science and technology. The paper adopts the method proposed for integer-order hybrid systems (MLD). The proposed FO MLD can be used to design new fractional-order control algorithms. Some numerical examples are also given.

1 Introduction

Models play an important role in automation. They are used, among others things, for simulation experiments and analysis of control systems, for the synthesis and optimization of control algorithms, for tuning of controllers, in signal estimation, as well as directly to determine controls in advanced discrete control systems, e.g. in model predictive control. Hence, their efficiency i.e. correct model behavior under diverse states of the environment, in conjunction with their numerical and implementation simplicity, is of great importance.

This is particularly crucial in the design of safety–critical systems, such as new trans-portation systems, advanced industrial technologies or health care monitoring, which deal with heterogeneous components exhibiting a variety of different behaviors, and consequently, where different mathematical representations have to be mixed to analyze the overall behavior of the controlled system [7, 8, 11, 34].

S. Domek (✉)
Department of Automatic Control and Robotics, West Pomeranian University of Technology Szczecin, Al. Piastów 17, 70-310 Szczecin, Poland
e-mail: stefan.domek@zut.edu.pl

P. Kulczycki et al. (eds.), *Fractional Dynamical Systems: Methods, Algorithms and Applications*, Studies in Systems, Decision and Control 402,
https://doi.org/10.1007/978-3-030-89972-1_3

The mathematical models of the vast majority of the processes surrounding us are traditionally associated with differential equations, typically derived from physical laws governing the dynamics of the system under consideration [9, 35, 40]. Additionally, in many applications, they display a non-linear nature, which is often so strong that it is not enough to use linearized models to adequately describe their properties. In turn, determining and the subsequent use of non-linear models is usually very difficult [24, 55].

The systems to be controlled also contain the real-valued discrete signals satisfying Boolean relations, if–then-else conditions, on/off conditions, etc. [5, 10, 16, 21, 46, 54]. Hence the most effective description method for such complex systems are hybrid models that describe in a common framework, the dynamics of real-valued variables, the dynamics of discrete variables and their interaction. These systems and models have been the subject of intensive studies for decades because of the interesting theoretical problems as well as the relevance to practical applications [12, 27, 29, 45, 47, 51, 59]. One of the most popular method for modeling such hybrid systems is the so-called MLD approach [1, 3, 5, 32, 33, 52].

Unfortunately, the research work conducted worldwide shows that even such advanced hybrid models as mentioned above, are often insufficient for an effective description of many phenomena from various fields of life, science and technology, for example biological, social, socio-cognitive, economic, transport uses, worldwide information and many technological processes and systems, not to mention control and monitoring systems including solutions based on artificial intelligence [17–20, 41, 43, 44, 49]. This applies to, for example, the dynamics of aircraft, population restricted growth, the distribution of parameters in charge transfer, the diffusion mechanism in batteries, the general behavior of an epidemics (SEIR), etc. [25, 37, 39, 57], where the most effective modeling methods are those using non-integer order differential calculus [13, 15, 21, 42, 48]. If such systems also include discrete-valued signals, a special modeling approach may be necessary.

In this paper we propose a new hybrid fractional-order mixed dynamical logical model (FO MLD) based on an effective integer-order MLD model [2, 4–6]. As in the original work, we assume the system to be specified in discrete time because of the digital nature of computer controllers and powerful mathematical programming software available in practice [5, 56]. Thus discrete-time hybrid systems, whose continuous dynamics of which is described by linear fractional-order difference equations and discrete dynamics are described by finite state machines, both synchronized by the same clock, will be considered.

The work is structured as follows: in Sect. 2 preliminary assumptions of fractional-order difference calculus and, based on them, discrete-time dynamic state space models of fractional-order are recalled. Section 3 outlines the fundamentals of the concept of discrete-time hybrid MLD models. In Sect. 4 the new discrete-time fractional-order hybrid MLD state space model is defined. Next, in Sect. 5, some examples of discrete-time FO MLD models are given. Finally, the whole is summarized in Conclusions.

The following notation will be used: $\mathbb{R}(\mathbb{R}_+)$ stand for a set of all (non-negative) real numbers; $\mathbb{Z}(\mathbb{Z}_+)$ stand for a set of all (non-negative) integers; $\mathbb{R}^n$ and $\mathbb{R}^{n\times m}$

refer to the n-dimensional Euclidean space and the set of all $n \times m$ real matrices, respectively; $\{0, 1\}^n$ refers to the n-dimensional set of binary (logical) elements 0 or 1 only.

2 Preliminaries

2.1 *Fractional-Order Difference*

The generalized differential operator of $\alpha \in \mathbb{R}$ order of the function $\varphi(t), t \in \mathbb{R}_+$ on the interval $[t_0, t], 0 \leq t_0 < t$, can be written as ${}_{t_0}D_t^{\alpha}\varphi(t)$ with the remark that the generalized integral operator on the interval $[t_0, t]$ is written as ${}_{t_0}I_t^{\alpha}\varphi(t)$, where ${}_{t_0}I_t^{\alpha}\varphi(t) = {}_{t_0}D_t^{-\alpha}\varphi(t)$. There are several known definitions of the operator ${}_{t_0}D_t^{\alpha}\varphi(t)$ proposed by various researchers, which differ in properties and/or the range of applicability [22, 41, 43, 44, 50, 53]. However the Grünwald-Letnikov definition of the fractional-order derivative is particularly popular for reasons of application, especially for digital control systems, where it is natural to use discretized function values $\varphi(t)$ taken with a sampling period for the purpose of computations T_s [36]:

Definition 1 A derivative of fractional order $\alpha \in \mathbb{R}$ of function $\varphi(t), t \in \mathbb{R}_+$ is defined according to Grünwald and Letnikov as follows:

$$ {}_{t_0}^{GL}D_t^{\alpha}\varphi(t) = \lim_{h \to 0} T_s^{-\alpha} \sum_{j=0}^{\lceil \frac{t-t_0}{T_s} \rceil} c_j^{\alpha}\varphi(t - jT_s), \tag{1} $$

where the symbol $\lceil k \rceil$ denotes the largest integer less than or equal to k, and

$$ c_j^{\alpha} = \begin{cases} 1 & \text{for } \ j = 0 \\ (-1)^j \frac{\alpha(\alpha-1)....(\alpha-j+1)}{j!} & \text{for } \ j > 0 \end{cases} \tag{2} $$

or in recursive manner

$$ c_j^{\alpha} = \left(1 - \frac{1+\alpha}{j}\right)c_{j-1}^{\alpha}, \quad c_0^{\alpha} = 1. \tag{3} $$

All of the above considerations cover the case of continuous time, but in practice they can be mapped into the discrete time case as well. As already mentioned above, if computer control systems are used, i.e. discrete control algorithms and discrete models of controlled plants are applied, discrete functions defined at discrete time instants $t \in \mathbb{Z}_+$ must be considered. Without losing the generality of further considerations, it can be assumed the sampling period $T_s = 1$.

In such a case, the fractional-order difference calculus represents an equivalent to the fractional-order differential calculus. Hence, based upon (1), the following definition may be introduced [36, 43]:

Definition 2 A discrete fractional-order difference of a discrete function is defined by:

$$ {}_{t_0}\Delta_t^\alpha \varphi(t) = \sum_{j=0}^{t-t_0} c_j^\alpha \varphi(t-j), \quad \alpha \in \mathbb{R}, \quad t \in \mathbb{Z}, \tag{4}$$

with the most commonly adopted simplified notation $t_0 = 0$ as:

$$\Delta^\alpha \varphi(t) = \sum_{j=0}^{t} c_j^\alpha \varphi(t-j). \tag{5}$$

An accurate implementation of the fractional-order difference (5), especially at the bounded resource platforms, requires many summations. However many integer-order finite-length discrete-time approximations also exist. The best known of them are based on PSE (Power Series Expansion) and CFE (Continuous Fraction Expansion) and allow one to approximate a fractional-order difference with the use of digital FIR or IIR filters. [14, 20, 50, 53].

2.2 *Discrete-Time Fractional-Order State Space Models*

Dynamic systems of non-integer order can be modeled in many ways, with transfer function and state space descriptions being the most popular.

The nonlinear discrete-time fractional-order state space model can be introduced on the basis of the integer-order model [14]:

$$x(t+1) = f(x(t), u(t)), \quad t \in \mathbb{Z}, \quad x(0) = x_0. \tag{6}$$

Denoting

$$f_d(x(t), u(t)) = f(x(t), u(t)) - x(t) \tag{7}$$

we get

$$\Delta^1 x(t+1) = f_d(x(t), u(t)), \quad t \in \mathbb{Z}, \quad x(0) = x_0, \tag{8}$$

where the individual vectors $x(t) \in \mathrm{X} \subseteq \mathbb{R}^n$, $u(t) \in \mathrm{U} \subseteq \mathbb{R}^m$, $y(t) \in \mathrm{Y} \subseteq \mathbb{R}^p$ denote the model state, input and output, respectively, x_0 is the initial state and $t \in \mathbb{Z}$ denotes a discrete-time independent variable (consecutive sample instants).

Hence, by analogy, we can write:

Definition 3 A nonlinear discrete-time fractional-order α state space model is given by nonlinear state and output equations.

$$\Delta^\alpha x(t+1) = f_d(x(t), u(t)), \quad t \in \mathbb{Z}, \quad x(0) = x_0, \tag{9}$$

$$y(t) = g(x(t), u(t)). \tag{10}$$

In the linear case, by analogy with integer-order models, it may be introduced the definition of linear discrete-time model of fractional-order α in state space [14, 36]:

Definition 4 A linear discrete-time fractional-order α state space model is given by the state and output equations.

$$\Delta^\alpha x(t+1) = A_d x(t) + Bu(t), \quad t \in \mathbb{Z}, \quad x(0) = x_0, \tag{11}$$

$$y(t) = Cx(t) + Du(t), \tag{12}$$

where $A \in \mathbb{R}^{n \times n}$ is the state matrix, $B \in \mathbb{R}^{n \times m}$, $C \in \mathbb{R}^{p \times n}$ are the input and output matrices, matrix $D \in \mathbb{R}^{p \times m}$ is equal to zero in the most common case,

$$A_d = A - I_n \tag{13}$$

is the so-called complemented state matrix and $I_n \in \mathbb{R}^{n \times n}$ denotes the identity matrix.

According to (5), we can write the linear discrete-time model of fractional-order α in state space (11) as:

$$x(t+1) = A_d x(t) + Bu(t) - \sum_{j=1}^{t+1} c_j^\alpha x(t+1-j), \quad x(0) = x_0. \tag{14}$$

It should be noted that in hybrid systems the parameters of the nonlinear model (9), (10), i.e. the vector functions f_d and/or g, or the linear model (11)–(13), (14), i.e. the matrices A_d, B, C, D and/or the fractional-order α, respectively, can vary in time.

3 Fundamentals of the Concept of Discrete-Time, Integer-Order Hybrid MLD State Space Models

In hybrid systems, two types of dynamic elements coexist and affect each other: continuous components, most often at the lower level of the hierarchical system, described by linear difference equations; and logical/discrete components, most often at the higher level of the hierarchical system, the dynamics of which are described by finite machines. Therefore comprehensive modeling of such systems is a difficult, but also an increasingly visible subject of scientific research [5, 10, 28, 33, 45, 47, 52]. The need for an adequate description of hybrid systems in automation was recognized several decades ago, e.g. in mechanical and power systems with on/off switches or valves, in transport systems with gears or speed selectors, as well as in control systems where discontinuous two-position or three-position controllers and continuous PID controllers with variable structure were used.

Thereafter, several modeling formalisms have been developed to describe hybrid systems, e.g. the so-called Piecewise Affine (PWA) models [26, 27, 55, 60], Linear Complementarity (LC) models [31] or discrete-time hybrid automatons (DHA) [38]. In 1999, a systematic approach to discrete-time mathematical modeling of dynamic systems which involve the interaction of physical laws, logic rules and constraints was proposed [5].

The key idea of combining, in one MLD model, a description of the linear dynamic part described by linear difference equations subject to possible constraints on physical states, continuous inputs, and continuous auxiliary variables, with the logical part including discrete states, binary or logical inputs and binary auxiliary variables, taking into account their possible constraints, is to embed the logical part in the extended state equations by transforming Boolean variables into integers {0, 1} and expressing the constraints and logical relations as mixed number-linear inequalities [2, 4, 5].

3.1 Boolean Algebra and Equivalent Linear Integer Inequalities

Let us assume a standard logical notation, i.e. X_i represents a logic statement that may be of value TRUE or FALSE. In Boolean algebra, a logic statement can be combined in compound statements by means of connectives, e.g. NOT, AND, OR, EXCLUSIVE OR, IMPLIES, IF AND ONLY IF. Connectives are defined by means of the so-called truth table and satisfy several well-known properties, which can be used to transform compound statements into equivalent statements involving different connectives, and simplify complex statements. Note that all connectives can be defined in terms of a subset of them, which is said to be a complete set of connectives [5, 58]. Suppose also that literal statement X_i can be associated with a

Table. 1 Some examples of equivalence of logical expressions and linear inequalities

Logical operations	Symbols	Equivalent linear inequalities
NOT	$\neg X_1$	$\delta_1 = 0$
AND	$X_1 \wedge X_2$	$\delta_1 = 1, \ \delta_2 = 1$
OR	$X_1 \vee X_2$	$\delta_1 + \delta_2 \geq 1$
EXCLUSIVE OR	$X_1 \oplus X_2$	$\delta_1 + \delta_2 = 1$
IMPLIES	$X_1 \rightarrow X_2$	$\delta_1 - \delta_2 \leq 0$
IF AND ONLY IF	$X_1 \leftrightarrow X_2$	$\delta_1 - \delta_2 = 0$

logical $\delta_i \in \{0, 1\}$, which has a value of either 1 if value X_i is TRUE, or 0 if value X_i is FALSE.

Then, as in [2, 59], it can be shown that any logic problem in which a given Boolean relation $S(X_1, X_2, \ldots, X_i, \ldots, X_n) = \text{TRUE}$ must be proved, can be translated into linear integer inequalities involving logical variables δ_i, and solved by means of a linear integer program. For example, it can be easily shown that the logical propositions and linear constraints presented in Table 1 are equivalent.

In general, the procedure for transformation from given Boolean logic relation $S(X_1, X_2, \ldots, X_i, \ldots, X_n) = \text{TRUE}$ to linear inequalities is as follow:

convert compound statements into equivalent statements Conjunctive Normal Form

$$\underset{\substack{j=1 \\ P_j, N_j \subseteq \{1,2,\ldots,n\}}}{\overset{m}{\wedge}} \left[\underset{i \in P_j}{\vee} X_i \underset{i \in N_j}{\vee} \neg X_i \right] = \text{TRUE}; \tag{15}$$

transform the obtained Conjunctive Normal Form into set of linear inequalities

$$\begin{cases} \sum\limits_{i \in P_1} \delta_i + \sum\limits_{i \in N_1} (1 - \delta_i) \geq 1 \\ \vdots \\ \sum\limits_{i \in P_m} \delta_i + \sum\limits_{i \in N_m} (1 - \delta_i) \geq 1 \end{cases} . \tag{16}$$

Note that the obtained set of linear inequalities (16) corresponds to the affiliation of the vector of numeric variables $\delta \in \{0, 1\}^n$ to the polyhedron $\Omega\delta \leq \omega$ [2].

Example Let us consider the following logic relation [5]:

$$S(X_1, X_2, X_3) = [X_3 \leftrightarrow X_1 \wedge X_2]. \tag{17}$$

It can be converted to the following Conjunctive Normal Form:

$$[X_3 \vee \neg X_1 \vee \neg X_2] \wedge [X_1 \vee \neg X_3] \wedge [X_2 \vee \neg X_3]. \tag{18}$$

This Conjunctive Normal Form can be transformed into the following set of linear inequalities:

$$\begin{cases} \delta_3 + (1-\delta_1) + (1-\delta_2) \geq 1 \\ \delta_1 + (1-\delta_3) \geq 1 \\ \delta_2 + (1-\delta_3) \geq 1 \end{cases} . \tag{19}$$

Products of binary variables.

When considering discrete/logical systems, it can be noticed that in many cases there are products of binary variables in them. To describe them with linear inequalities, it is necessary to introduce the so-called logical auxiliary variables [5, 59]:

$$\delta_{aux} \triangleq \delta_1 \delta_2. \tag{20}$$

It is easy to see that (20) is equivalent to

$$[\delta_{aux} = 1] \leftrightarrow [\delta_1 = 1] \wedge [\delta_2 = 1] \tag{21}$$

and therefore, according to Table 1, logical statement (21) can be transformed into the following set of linear inequalities:

$$\begin{cases} -\delta_1 + \delta_{aux} \leq 0 \\ -\delta_2 + \delta_{aux} \leq 0 \\ \delta_1 + \delta_2 - \delta_{aux} \leq 1 \end{cases} . \tag{22}$$

Note, that the obtained set of linear inequalities (19) corresponds to the affiliation of the vector of numeric variables $\delta_j \in \{0, 1\}$ to the polyhedron $\Omega\delta \leq \omega$ of the form

$$\begin{bmatrix} -1 & 0 & 1 \\ 0 & -1 & 1 \\ 1 & 1 & -1 \end{bmatrix} \begin{bmatrix} \delta_1 \\ \delta_2 \\ \delta_{aux} \end{bmatrix} \leq \begin{bmatrix} 0 \\ 0 \\ 1 \end{bmatrix}. \tag{23}$$

Products of binary and continuous variables.

When considering hybrid systems, it can be noticed that in many cases there are products of a binary variable $\delta \in \{0, 1\}$ and a continuous variable $\varphi(x, u, t) \in \mathbb{R}$ in them. To describe them with linear inequalities, it is necessary to introduce the auxiliary variable [5, 59]:

$$\psi(x, u, t) \triangleq \delta\varphi(x, u, t), \quad u \in \mathrm{U}, x \in \mathrm{X}. \tag{24}$$

It may be seen that (24) is equivalent to

$$[\delta = 0] \rightarrow [\psi(x, u, t) = 0], \quad [\delta = 1] \rightarrow [\psi(x, u, t) = \varphi(x, u, t)]. \tag{25}$$

If we assume also that for the linear function $\varphi(x, u, t)$ the following bounded values exist

$$\varphi_m \triangleq \min_{u \in \mathrm{U},\, x \in \mathrm{X}} \varphi(x, u, t), \tag{26}$$

$$\varphi_M \triangleq \max_{u \in \mathrm{U},\, x \in \mathrm{X}} \varphi(x, u, t), \tag{27}$$

therefore, logical statements (22) can be transformed into the following set of linear inequalities:

$$\begin{array}{c} \psi(x, u, t) \leq \varphi_M \delta \\ \psi(x, u, t) \geq \varphi_m \delta \\ \psi(x, u, t) \leq \varphi(x, u, t) - \varphi_m(1 - \delta) \\ \psi(x, u, t) \geq \varphi(x, u, t) - \varphi_M(1 - \delta) \end{array}. \tag{28}$$

3.2 *Functional Components of Discrete-Time Hybrid MLD State Space Models*

Generally, in a functional sense, the MLD model can be interpreted as a discrete hybrid automaton including an Event Generator, a Finite State Machine, a Mode Selector and a Switched Piecewise Affine System [3, 24, 54], as shown in Fig. 1.

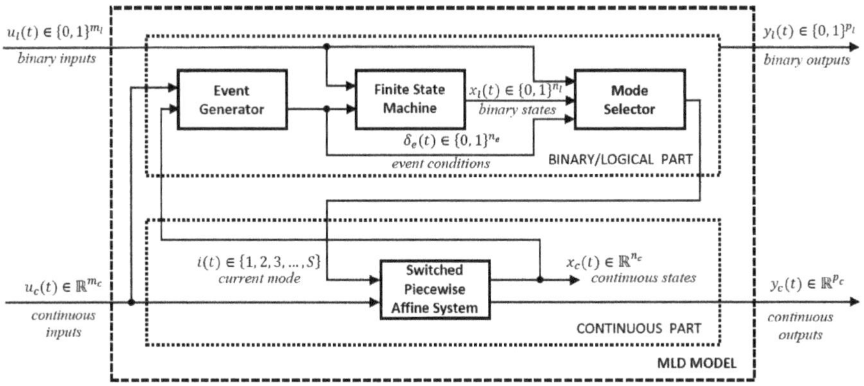

Fig. 1 Functional scheme of the MLD model

Taking into account the previous considerations, all significant components of the model can be described by conventional state equations, binary or continuously respectively, with constraints in the form of linear matrix inequalities. Below are some collected comments on the functional components of the model.

Event Generator.

In the Event Generator the vector $\delta_e(t) \in \{0, 1\}^{n_e}$ of event variables $\delta_{e,j}(t) \in \{0, 1\}$ is generated by linear threshold conditions over continuous states $x_c(t) \in \mathbb{R}^{n_c}$, continuous inputs $u_c(t) \in \mathbb{R}^{m_c}$, and discrete time, separately for each j-th event variable:

$$\left[\delta_{e,j}(t) = 1\right] \leftrightarrow \left[H_j x_c(t) + K_j u_c(t) \leq W_j\right], \tag{29}$$

where $H_j \in \mathbb{R}^{n_c \times n_c}$, $K_j \in \mathbb{R}^{n_c \times m_c}$, $W_j \in \mathbb{R}^{n_e}$ are appropriately chosen real matrices and real vector.

Moreover, since the model is to be specified in discrete time, it is assumed that the relation $\neg[\varphi(x, u, t) \leq 0]$ to be true if and only if $\varphi(x, u, t) \geq \varepsilon$, where $\varepsilon > 0$ is a small tolerance (usually the accuracy of discrete arithmetic). This is equivalent to the assumption that $-\varepsilon < \varphi(x, u, t) < 0$ cannot occur due to the finite number of bits used to represent real numbers in the arithmetic unit [1, 5].

In Table 2 a few of the most common event conditional expressions are represented as an equivalent set of linear inequalities. Other examples of equivalences can be found in [2, 5].

Finite State Machine.

The inputs for the binary state of the Finite State Machine are the binary vector $u_l(t) \in \{0, 1\}^{m_l}$ and the binary event conditions vector $\delta_e(t) \in \{0, 1\}^{n_e}$. The state $x_l(t) \in \{0, 1\}^{n_l}$ evolves according to a Boolean state updates function:

Table. 2 Some equivalences between logic statements and sets of linear inequalities

Logic statements	Linear inequalities
$[\varphi(x, u, t) \leq 0] \wedge [\delta(t) = 1]$	$\varphi(x, u, t) - \delta(t) \leq$ $-1 + \varphi_m(1 - \delta(t))$
$[\varphi(x, u, t) \leq 0] \vee [\delta(t) = 1]$	$\varphi(x, u, t) \leq \varphi_M \delta(t)$
$[\varphi(x, u, t) \geq 0] \leftrightarrow [\delta(t) = 1]$	$-\varphi_m \delta(t) \leq \varphi(x, u, t) - \varphi_m$ $-(\varphi_M + \varepsilon)\delta(t) \leq$ $\varphi(x, u, t) - \varepsilon$
$[\varphi(x, u, t) \leq 0] \leftrightarrow [\delta(t) = 1]$	$\varphi(x, u, t) \leq \varphi_M(1 - \delta(t))$ $\varphi(x, u, t) \geq \varepsilon + (\varphi_m - \varepsilon)\delta(t)$
IF $[\delta(t) = 1]$ THEN $z(t) =$ $\varphi_1(x, u, t)$ ELSE $z(t) = \varphi_2(x, u, t)$	$(\varphi_{2m} - \varphi_{1M})\delta(t) + z(t) \leq$ $\varphi_2(x, u, t)$ $(\varphi_{1m} - \varphi_{2M})\delta(t) - z(t) \leq$ $-\varphi_2(x, u, t)$ $(\varphi_{1m} - \varphi_{2M})(1 - \delta(t)) +$ $z(t) \leq \varphi_1(x, u, t)$ $(\varphi_{2m} - \varphi_{1M})(1 - \delta(t)) -$ $z(t) \leq -\varphi_1(x, u, t)$

$$x_l(t+1) = f_l(x_l(t), u_l(t), \delta_e(t)), \tag{30}$$

for example $x_l(t+1) = \neg\delta_e(t) \vee (x_l(t) \oplus u_l(t))$.

Mode Selector.

The inputs for the binary Mode Selector are the current binary states $x_l(t) \in \{0,1\}^{n_l}$, the binary vector $u_l(t) \in \{0,1\}^{m_l}$ and the binary event conditions vector $\delta_e(t) \in \{0,1\}^{n_e}$. The active mode is selected by an affine expression depending on the value of the Boolean function of the current binary inputs:

$$i(t) = f_M(x_l(t), u_l(t), \delta_e(t)). \tag{31}$$

Example Let us assume that in a MLD model the Mode Selector depends only on the current values of $x_l(t)$ and $\delta_e(t)$, according to the following Boolean functions:

$$i(t) = \begin{bmatrix} x_l(t) \vee \neg\delta_e(t) \\ x_l(t) \wedge \delta_e(t) \end{bmatrix}. \tag{32}$$

It is easy to see that such a system has $S = 3$ modes, as shown in Table 3:

Switched Piecewise Affine System.

In switched piecewise affine systems a switching signal defines the instantaneous degree of activity of each dynamic local model [2, 3, 24, 27]. The state equation can be rewritten as a combination of affine terms and IF–THEN-ELSE conditions:

Table. 3 Logic statements and selection of modes

$x_l(t)$	$\delta_e(t)$	$x_l(t) \vee \neg\delta_e(t)$	$x_l(t) \wedge \delta_e(t)$	Mode $i(t)$
0	0	1	0	$i(t) = \begin{bmatrix} 1 \\ 0 \end{bmatrix}$
0	1	0	0	$i(t) = \begin{bmatrix} 0 \\ 0 \end{bmatrix}$
1	0	1	0	$i(t) = \begin{bmatrix} 1 \\ 0 \end{bmatrix}$
1	1	1	1	$i(t) = \begin{bmatrix} 1 \\ 1 \end{bmatrix}$

$$x_c^1(t) = \begin{cases} A_1 x_c(t) + B_1 u_c(t) + f_1 & \text{if} \quad i(t) = 1 \\ 0 & \text{otherwise} \end{cases}$$
$$\vdots \qquad , \tag{33}$$
$$x_c^S(t) = \begin{cases} A_S x_c(t) + B_S u_c(t) + f_S & \text{if} \quad i(t) = S \\ 0 & \text{otherwise} \end{cases}$$

$$x_c(t+1) = \sum_{i=1}^{S} x_c^i(t), \quad x_c^i(t) \in \mathbb{R}^{n_c}. \tag{34}$$

The output equation assumes a similar form

$$y^1(t) = \begin{cases} C_1 x_c(t) + D_1 u_c(t) + g_1 & \text{if} \quad i(t) = 1 \\ 0 & \text{otherwise} \end{cases}$$
$$\vdots \qquad , \tag{35}$$
$$y^S(t) = \begin{cases} C_S x_c(t) + D_S u_c(t) + g_S & \text{if} \quad i(t) = S \\ 0 & \text{otherwise} \end{cases}$$

$$y_c(t) = \sum_{i=1}^{S} y^i(t), \quad y^i(t) \in \mathbb{R}^{p_c}. \tag{36}$$

3.3 *Discrete-Time Integer-Order Hybrid MLD State Space Models*

Based on the previous considerations, in conclusion, the following definition can be introduced [1, 5, 33]:

Definition 5 A linear discrete-time integer-order hybrid state space MLD model is given by.

$$x(t+1) = A_1 x(t) + B_1 u(t) + B_2 \delta(t) + B_3 z(t) + B_4, \quad x(0) = x_0, \tag{37}$$

$$y(t) = C_1 x(t) + D_1 u(t) + D_2 \delta(t) + D_3 z(t) + D_4, \tag{38}$$

$$E_2 \delta(t) + E_3 z(t) \le E_1 u(t) + E_4 x(t) + E_5, \tag{39}$$

where

$$x(t) = \begin{bmatrix} x_c(t) \\ x_l(t) \end{bmatrix} \in \mathbb{R}^{n_c} \times \{0, 1\}^{n_l}, n = n_c + n_l, \tag{40}$$

$$u(t) = \begin{bmatrix} u_c(t) \\ u_l(t) \end{bmatrix} \in \mathbb{R}^{m_c} \times \{0, 1\}^{m_l}, m = m_c + m_l, \tag{41}$$

$$y(t) = \begin{bmatrix} y_c(t) \\ y_l(t) \end{bmatrix} \in \mathbb{R}^{p_c} \times \{0, 1\}^{p_l}, p = p_c + p_l, \tag{42}$$

are the state, input and output vectors respectively, composed of appropriate vectors of the real and discrete components of the model. A_1, $B_1 - B_3$, C_1, $D_1 - D_3$, $E_1 - E_4$ are real valued matrices with appropriate dimensions, and B_4, D_4, E_5 denote real vectors. Furthermore, $z(t) \in \mathbb{R}^{r_c}$ and $\delta \in \{0, 1\}^{r_l}$ represent two auxiliary continuous and binary vectors respectively, additionally introduced for translating propositional logic or PWA functions into linear inequalities.

This kind of modeling was named as mixed logical dynamical because it allows one to describe the dynamic properties of such processes as [5]:

- sequential logical systems (Finite State Machines, Automata);
- some classes of discrete event systems;
- linear systems without or with constraints;
- nonlinear dynamic systems, where the nonlinearity can be expressed through combinational logic;
- nonlinear systems, the nonlinearities of which can be suitably approximated by piecewise linear functions;
- linear hybrid systems.

The special simulation language HYSDEL (HYbrid Systems DEscription Language) was developed in [11, 56] to obtain MLD models from a high level textual description of the hybrid dynamics. Examples of real-world applications that can be naturally modeled within the MLD framework are reported in [6, 8, 30, 32, 33, 37, 47, 52].

4 MLD Model of Discrete-Time Fractional-Order Hybrid Systems

According to techniques used in the integer-order case, here we propose discrete-time fractional-order hybrid mixed logical dynamical models. The introduced models include binary functional components, the same as described by Definition 5 and a continuous part consisting of fractional-order models (11)–(14). This is summarized by the following definition:

Definition 6 A linear discrete-time fractional-order hybrid state space mixed logical dynamical model (FO MLD) is given by.

$$
\begin{aligned}
x(t+1) = A_1x(t) + B_1u(t) - \sum_{i=1}^{t+1} A_{1,i}x(t+1-i) \\
+B_2\delta(t) + B_3z(t) - \sum_{i=1}^{t+1} B_{3,i}z(t+1-i) + B_4, \quad x(0) = x_0
\end{aligned}
, \tag{43}
$$

$$
y(t) = C_1x(t) + D_1u(t) + D_2\delta(t) + D_3z(t) + D_4, \tag{44}
$$

$$
E_2\delta(t) + E_3z(t) \leq E_1u(t) + E_4x(t) + E_5, \tag{45}
$$

where

$$
x(t) = \begin{bmatrix} x_c(t) \\ x_l(t) \end{bmatrix} \in \mathbb{R}^{n_c} \times \{0, 1\}^{n_l}, n = n_c + n_l, \tag{46}
$$

$$
u(t) = \begin{bmatrix} u_c(t) \\ u_l(t) \end{bmatrix} \in \mathbb{R}^{m_c} \times \{0, 1\}^{m_l}, m = m_c + m_l, \tag{47}
$$

$$
y(t) = \begin{bmatrix} y_c(t) \\ y_l(t) \end{bmatrix} \in \mathbb{R}^{p_c} \times \{0, 1\}^{p_l}, p = p_c + p_l, \tag{48}
$$

are the state, input and output vectors respectively, composed of appropriate vectors of the.

real and discrete components of the model. A_{d1}, $B_1 - B_3$, $B_{3,i}$, C_1, $D_1 - D_3$, $E_1 - E_4$ are real valued matrices with appropriate dimensions, and B_4, D_4, E_5 denote real vectors. The matrices $A_{1,i}$ and $B_{3,i}$ have, according to (2), (3) and (14), the block forms:

$$
A_{1,i} = \begin{bmatrix} (-1)^i \binom{\alpha}{i} I_{n_c} & 0_{n_c \times n_l} \\ 0_{n_l \times n_c} & 0_{n_l} \end{bmatrix} \in \mathbb{R}^{(n_c+n_l)\times(n_c+n_l)}, \tag{49}
$$

$$
B_{3,i} = \begin{bmatrix} B_{z,i} \\ 0_{n_l \times r_c} \end{bmatrix} \in \mathbb{R}^{(n_c+n_l)\times r_c}, \quad B_{z,i} \in \mathbb{R}^{n_c \times r_c}, \tag{50}
$$

where $0_n \in \mathbb{R}^{n \times n}$ and $0_{n_1 \times n_2} \in \mathbb{R}^{n_1 \times n_2}$ denote the zero matrices. Vectors $z(t) \in \mathbb{R}^{r_c}$ and $\delta \in \{0, 1\}^{r_l}$ represent two auxiliary continuous and binary variables, respectively.

The proposed FO MLD model, like the integer-order model (37)–(42), includes the following important classes of systems (with $A_{1,i}$, $B_{3,i} = 0$):

1. sequential logical systems (Finite State Machines, Automata) (if $n_c, m_c, p_c = 0$);
2. some classes of discrete event systems (if n_c, $p_c = 0$);
3. linear systems of integer-order (if $n_l, m_l, p_l, r_l, r_c = 0$);
4. linear systems of integer-order with constraints (if $n_l, m_l, p_l = 0$);

5. nonlinear dynamic systems of integer-order, where the nonlinearity can be expressed through combinational logic (if $n_l = 0$) or suitably approximated by piecewise linear functions (if $m_l, p_l = 0$);
6. linear hybrid systems of integer-order.

However, the proposed FO MLD model (43)–(48) for $A_{1,i} \neq 0$ and/or $B_{3,i} \neq 0$ can additionally describe or suitably approximate a large quantity of complex situations:

7. fractional-order linear systems (if $n_l, m_l, p_l, r_l, r_c = 0$);
8. fractional-order constrained linear systems (if $n_l, m_l, p_l = 0$);
9. fractional-order nonlinear dynamic systems, where the nonlinearity can be expressed through combinational logic (if $n_l = 0$) or suitably approximated by piecewise linear functions (if $m_l, p_l = 0$);
10. fractional-order linear hybrid systems.

The high versatility of the model (43)–(48) makes it possible to use it widely wherever a linear description of the complex phenomenon is desired, e.g. in the synthesis of control algorithms. It can be shown for example that the FO MLD models can easily be used to describe dynamic non-integer order nonlinear processes by using a piecewise linear approximation (PWL) [24]. In this way we obtain a simplified version of this model designated as fractional-order switched model FO PWL [16, 20–22] and we can directly determine the fractional-order optimal control or, for example, the predictive control algorithm for nonlinear processes FO NMPC [14, 15, 23].

On the other hand, a certain disadvantage of MLD models is the requirement to enumerate all possible combinations of binary states, binary inputs and binary auxiliary variables δ(t). Therefore, although most of such combinations lead usually to empty sets, it should be noted that the generality of the models (39)–(44) leads in some specific cases, to high application complexity where it may be necessary to use more simplified models [34, 41]. Fortunately, as it can be shown following [4], the dimension of some of the matrices and decision vectors in the FO MLD model may be reduced by appropriate selection of auxiliary variables $z(t) \in \mathbb{R}^{r_c}$ and $\delta \in \{0, 1\}^{r_l}$.

5 Some Examples of Discrete-Time FO MLD Models

The proposed generalized model (43)–(48) can describe various discrete processes and integer-order continuous processes previously mentioned in items 1–6. Many examples of them can be found in [1, 2, 5, 11, 52].

Some simple examples of fractional-order models listed above in items 7–10 will be shown below.

Example 1 *Fractional-order linear systems.*

Let us consider the following simple linear fractional-order system (FOS):

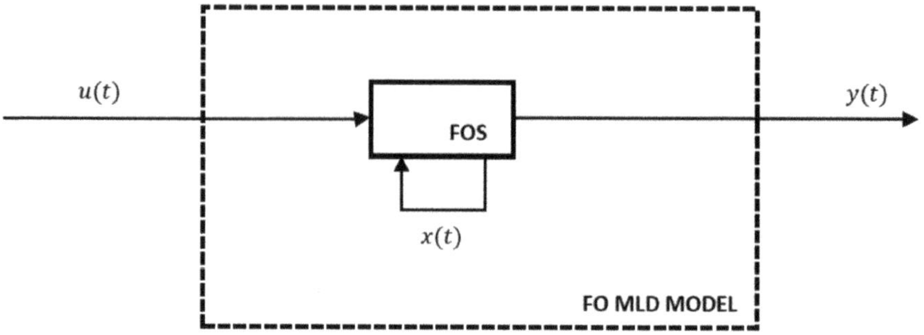

Fig. 2 The structure of the process in Example 1

$$\begin{aligned} x(t+1) &= 0.6x(t) + 2u(t) - \sum_{i=1}^{t+1}(-1)^i \binom{0.5}{i} x(t+1-i) \\ y(t) &= 0.75x(t) \end{aligned}, \tag{51}$$

where $A_d = 0.6$, $B = 2$, $C = 0.75$, $D = 0$, $\alpha = 0.5$.

For such system we have $n_l, m_l, p_l, r_l, r_c = 0, n_c, m_c, p_c = 1$ and

$$A_1 = 0.6, \mathrm{A}_{1,i} = (-1)^i \binom{0.5}{i}, \quad B_1 = 2, \quad C_1 = 0.75,$$

$$B_2 = B_3 = B_4 = 0, \quad B_{3,i} = 0, \quad D_i = 0, \quad E_i = 0.$$

Thus, the FO MLD model (43)–(48) takes the same form as the ordinary dynamic model of the fractional-order (14). Figure 2 shows the structure of the process.

Example 2 *Fractional-order linear system with constraints.*

Let us consider the same linear discrete-time fractional-order system as in Example 1. Additionally we assume the following constraints imposed on the state and input variables:

$$\begin{aligned} x(t) &\in \left[x_m, x_M\right] \\ u(t) &\in \left[u_m, u_M\right] \end{aligned}, \tag{52}$$

where these limiting conditions can be equivalently written as LMI:

$$\begin{aligned} \varepsilon &\le x(t) - x_m \\ \varepsilon &\le x_M - x(t) \\ u_m &\le u(t) \\ u(t) &\le u_M \end{aligned}. \tag{53}$$

and $\varepsilon > 0$ is the small tolerance (typically precision of arithmetic) above which the constraint "> 0" is considered satisfied. Figure 3 shows the structure of the process.

For such a system we have, as before $n_l, m_l, p_l, r_l, r_c = 0, \quad n_c, m_c, p_c = 1$ and

$$A_1 = 0.6, \mathrm{A}_{1,i} = (-1)^i \binom{0.5}{i}, \quad B_1 = 2, \quad C_1 = 0.75,$$

$$B_2 = B_3 = B_4 = 0, \quad B_{3,i} = 0, \quad D_i = 0, \quad E_2 = E_3 = 0.$$

Thus, the FO MLD model (39)–(44) takes the same form:

$$\begin{aligned} x(t+1) = 0.6x(t) + 2u(t) - \sum_{i=1}^{t+1}(-1)^i \binom{0.5}{i} x(t+1-i) \\ y(t) = 0.75x(t) \end{aligned}, \tag{54}$$

with the constraint matrix inequality

$$0 \leq E_1 u(t) + E_4 x(t) + E_5, \tag{55}$$

where

$$E_1 = \begin{bmatrix} 0 \\ 0 \\ 1 \\ -1 \end{bmatrix}, \quad E_4 = \begin{bmatrix} 1 \\ -1 \\ 0 \\ 0 \end{bmatrix}, \quad E_5 = \begin{bmatrix} -x_m - \varepsilon \\ x_M - \varepsilon \\ -u_m \\ u_M \end{bmatrix}. \tag{56}$$

Example 3 *Fractional-order nonlinear dynamic system.*

Let us consider a discrete-time nonlinear fractional-order dynamic systems, where the nonlinearity can be suitably approximated by two switched linear subsystems:

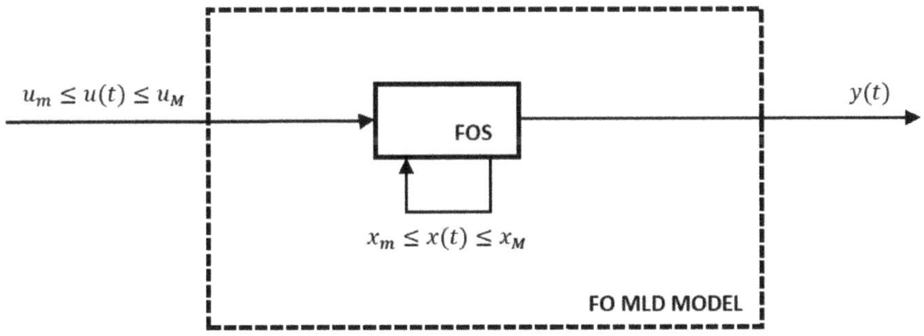

Fig. 3 The structure of the process in Example 2

$$\begin{aligned} x(t+1) &= -0.7x(t) + u(t) - \sum_{i=1}^{t+1} (-1)^i \binom{0.8}{i} x(t+1-i) \\ y(t) &= 0.9x(t) \end{aligned} \quad \text{for } x(t) < 0, \tag{57}$$

$$\begin{aligned} x(t+1) &= 0.4x(t) + u(t) - \sum_{i=1}^{t+1} (-1)^i \binom{0.6}{i} x(t+1-i) \\ y(t) &= 1.3x(t) \end{aligned} \quad \text{for } x(t) \geq 0, \tag{58}$$

where

$$A_{d\_1} = -0.7,\ B_{\_1} = 1,\ C_{\_1} = 0.9,\ D_{\_1} = 0, \alpha_{\_1} = 0.8,$$

$$A_{d\_2} = 0.4,\ B_{\_2} = 1,\ C_{\_2} = 1.3,\ D_{\_2} = 0,\ \alpha_{\_2} = 0.6.$$

Additionally we adopt the same constraints of state and input variables (52) as in Example 2. Figure 4 shows the structure of the process.

The model's switching condition may be associated with an auxiliary binary variable $\delta(t)$ in such a way that

$$[\delta(t) = 1] \leftrightarrow [x(t) \geq 0], \tag{59}$$

whereby the logical conditions of variable constraints can be written as LMI:

$$\begin{array}{c} -x_m\delta(t) \leq x(t) - x_m \\ -(x_M + \varepsilon)\delta(t) \leq -x(t) - \varepsilon, \quad \varepsilon > 0 \end{array}. \tag{60}$$

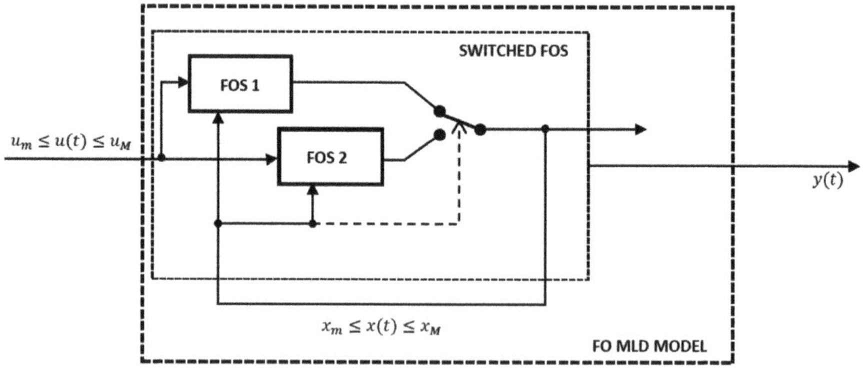

Fig. 4 The structure of the process in Example 3

By introducing an auxiliary vector

$$z(t) = \delta(t)x(t) \in \mathbb{R}^{n_c} \tag{61}$$

the state and output equations of a fractional-order hybrid model composed of the two previously mentioned switched models can be written as:

$$\begin{aligned} x(t+1) &= -0.7x(t) + u(t) - \sum_{i=1}^{t+1}(-1)^i\binom{0.8}{i}x(t+1-i) \\ &\quad + 1.1z(t) - \sum_{i=1}^{t+1}(-1)^i\left[\binom{0.6}{i} - \binom{0.8}{i}\right]z(t+1-i). \\ y(t) &= 0.9x(t) + 0.4z(t) \end{aligned} \tag{62}$$

For such a system we have $n_l, m_l, p_l = 0, \quad n_c, m_c, p_c, r_l, r_c = 1$ and

$$A_1 = -0.7, \quad \mathrm{A}_{1,i} = (-1)^i\binom{0.8}{i}, \quad B_{3,i} = (-1)^i\left[\binom{0.6}{i} - \binom{0.8}{i}\right],$$

$$B_1 = B_2 = B_3 = 1, \quad B_4 = 0, \quad C_1 = 0.9, \quad D_3 = 0.4, \quad D_1 = D_2 = D_4 = 0.$$

Additionally, taking into account the constraints imposed on the state variable, one can write for the auxiliary variable:

$$\begin{aligned} z(t) &\le x_M\delta(t) \\ z(t) &\ge x_m\delta(t) \\ z(t) &\le x(t) - x_m(1-\delta(t)) \\ z(t) &\ge x(t) - x_M(1-\delta(t)) \end{aligned} \tag{63}$$

and the LMI constraints will take the form:

$$\underbrace{\begin{bmatrix} -x_m \\ -(x_M+\varepsilon) \\ 0 \\ 0 \\ -x_M \\ x_m \\ -x_m \\ x_M \end{bmatrix}}_{E_2}\delta(t) + \underbrace{\begin{bmatrix} 0 \\ 0 \\ 0 \\ 0 \\ 1 \\ -1 \\ 1 \\ -1 \end{bmatrix}}_{E_3} z(t) \le \underbrace{\begin{bmatrix} 0 \\ 0 \\ 1 \\ -1 \\ 0 \\ 0 \\ 0 \\ 0 \end{bmatrix}}_{E_1} u(t) + \underbrace{\begin{bmatrix} 1 \\ -1 \\ 0 \\ 0 \\ 0 \\ 0 \\ 1 \\ -1 \end{bmatrix}}_{E_4} x(t) + \underbrace{\begin{bmatrix} -x_m \\ -\varepsilon \\ -u_m \\ u_M \\ 0 \\ 0 \\ -x_m \\ x_M \end{bmatrix}}_{E_5}. \tag{64}$$

Example 4 *Fractional-order nonlinear dynamic system with output saturation.*

Let us consider the same nonlinear discrete-time fractional-order switched dynamic systems as in Example 3, with output matrices $C_{\_1} = C_{\_2} = C = 1$. Additionally, we assume that this system is cascaded by the nonlinear saturation function:

$$sat(y(t)) = \begin{cases} y_m & \text{if } y(t) \le -1 \\ y(t) & \text{if } -1 < y(t) \le 1 \\ y_M & \text{if } 1 \le y(t) \end{cases}, \tag{65}$$

where

$$\begin{aligned} y_m &\triangleq \min_{x \in X} Cx(t) \\ y_M &\triangleq \max_{x \in X} Cx(t) \end{aligned}. \tag{66}$$

Figure 5 shows the structure of the process.

Taking into account the auxiliary binary variable $\delta(t)$ from Example 3, here denoted as $\delta_1(t)$, we introduce two additional auxiliary binary variables $\delta_2(t), \delta_3(t)$ and combine all variables into one auxiliary binary vector:

$$\delta(t) = \begin{bmatrix} \delta_1(t) \\ \delta_2(t) \\ \delta_3(t) \end{bmatrix} \in \{0, 1\}^{r_l}. \tag{67}$$

Therefore, the saturation function may be associated with δ_2, δ_3 in such a way, that:

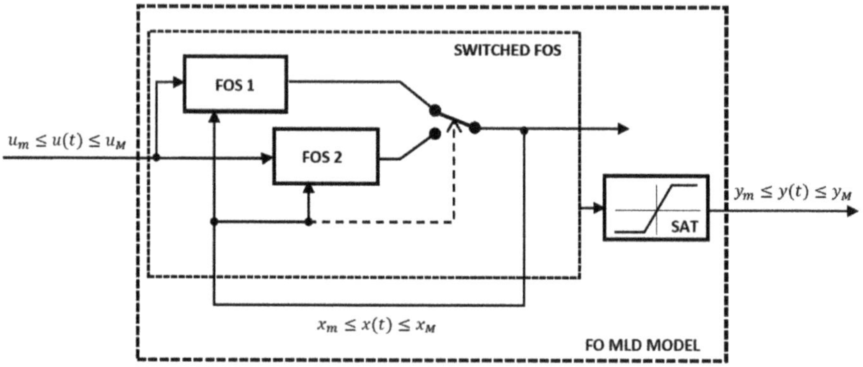

Fig. 5 The structure of the process in Example 4

$$\begin{array}{c}[Cx(t) < -1] \rightarrow [\delta_2(t) = 1] \\ [Cx(t) > -1] \rightarrow [\delta_2(t) = 0] \\ [Cx(t) < 1] \rightarrow [\delta_3(t) = 0] \\ [Cx(t) > 1] \rightarrow [\delta_3(t) = 1]\end{array}. \tag{68}$$

The logical conditions can be rewritten, respectively, as:

$$\begin{array}{c}Cx(t) - (y_m + 1)\delta_2(t) \geq -1 \\ Cx(t) - (1 + y_M)(1 - \delta_2(t)) \leq -1 \\ Cx(t) + (1 - y_m)(1 - \delta_3(t)) \geq 1 \\ -Cx(t) + (y_M - 1)\delta_3(t) \geq -1\end{array}. \tag{69}$$

For the saturation function it is also true:

$$\begin{array}{c}[\delta_2(t) = 1] \rightarrow [\delta_3(t) = 0] \\ [\delta_3(t) = 1] \rightarrow [\delta_2(t) = 0]\end{array}, \tag{70}$$

or, equivalently,

$$\begin{array}{c}\delta_2(t) - (1 - \delta_3(t)) \leq 0 \\ \delta_3(t) - (1 - \delta_2(t)) \leq 0\end{array}. \tag{71}$$

Taking into account the auxiliary vector $z(t)$ from Example 3, here denoted as $z_1(t)$, we introduce an additional auxiliary vector

$$z_2(t) = sat(Cx(t)) \tag{72}$$

and combine the two vectors into one auxiliary vector

$$z(t) = \begin{bmatrix} z_1(t) \\ z_2(t) \end{bmatrix} \in \mathbb{R}^{r_c}, \quad r_c = n_c + p_c. \tag{73}$$

So, we can write:

$$\begin{array}{c}[\delta_2(t) = 0] \rightarrow [z_2(t) \leq Cx(t)] \\ [\delta_3(t) = 0] \rightarrow [z_2(t) \geq Cx(t)]\end{array}, \tag{74}$$

and, equivalently,

$$\begin{array}{c}z(t) - (y_M - y_m)\delta_2(t) \leq Cx(t) \\ z(t) + (y_M - y_m)\delta_3(t) \geq Cx(t)\end{array}, \tag{75}$$

and that

$$\begin{array}{c} z(t) \geq -1 \\ z(t) - (y_M + 1)(1 - \delta_2(t)) \leq -1 \\ z(t) \leq 1 \\ z(t) + (1 - y_m)(1 - \delta_3(t)) \geq 1 \end{array}. \tag{76}$$

As a result of above considerations, the state and output equations of a fractional-order hybrid model composed of the two previously mentioned switched models can be written as:

$$\begin{aligned} x(t+1) &= -0.7x(t) + u(t) - \sum_{i=1}^{t+1} (-1)^i \binom{0.8}{i} x(t+1-i) \\ &\quad + 1.1z(t) - \sum_{i=1}^{t+1} B_{3,i} z(t+1-i). \\ y(t) &= D_3 z(t) \end{aligned} \tag{77}$$

For such a system we have $n_l, m_l, p_l = 0, \quad n_c, m_c, p_c = 1, \quad r_c = 2, r_l = 3$ and

$$A_1 = -0.7, \quad A_{1,i} = (-1)^i \binom{0.8}{i}, \quad B_{3,i} = \left[(-1)^i \left(\binom{0.6}{i} - \binom{0.8}{i} \right) \; 0 \right], \tag{78}$$

$$B_1 = 1, \quad B_2 = 0, \quad B_3 = 1.1, \quad B_4 = 0, \quad C_1 = 0, \quad D_1 = D_2 = D_4 = 0,$$

$$D_3 = \left[0 \; 1 \right].$$

Taking into account the constraints imposed on the state variable, as in Example 3, and on the saturated output variable, the LMI constraints will take the form:

$$
\underbrace{\begin{bmatrix} -x_m & 0 & 0 \\ -(x_M+\varepsilon) & 0 & 0 \\ 0 & 0 & 0 \\ 0 & 0 & 0 \\ -x_M & 0 & 0 \\ x_m & 0 & 0 \\ -x_m & 0 & 0 \\ x_M & 0 & 0 \\ 0 & 0 & 0 \\ 0 & y_M+1 & 0 \\ 0 & 0 & 0 \\ 0 & 0 & 1-y_m \\ 0 & -(y_M-y_m) & 0 \\ 0 & 0 & -(y_M-y_m) \end{bmatrix}}_{E_2}\delta(t)+\underbrace{\begin{bmatrix} 0 & 0 \\ 0 & 0 \\ 0 & 0 \\ 0 & 0 \\ 1 & 0 \\ -1 & 0 \\ 1 & 0 \\ -1 & 0 \\ 0 & -1 \\ 0 & 1 \\ 0 & 1 \\ 0 & -1 \\ 0 & 1 \\ 0 & -1 \end{bmatrix}}_{E_3} z(t)
$$

$$
\leq \underbrace{\begin{bmatrix} 0 \\ 0 \\ 1 \\ -1 \\ 0 \\ 0 \\ 0 \\ 0 \\ 0 \\ 0 \\ 0 \\ 0 \\ 0 \\ 0 \end{bmatrix}}_{E_1} u(t)+\underbrace{\begin{bmatrix} 1 \\ -1 \\ 0 \\ 0 \\ 0 \\ 0 \\ 1 \\ -1 \\ 0 \\ 0 \\ 0 \\ 0 \\ 1 \\ -1 \end{bmatrix}}_{E_4} x(t)+\underbrace{\begin{bmatrix} -x_m \\ -\varepsilon \\ -u_m \\ u_M \\ 0 \\ 0 \\ -x_m \\ x_M \\ 1 \\ y_M \\ 1 \\ y_m \\ 0 \\ 0 \end{bmatrix}}_{E_5}. \tag{79}
$$

Example 5 *Fractional-order linear hybrid systems.*

Let us consider a simple hybrid process containing the fractional-order system as in Example 2 and a Finite State Machine (FSM) [5]:

$$
\begin{gathered}
[x_l(t)=0] \wedge [x_c(t) \leq 0] \rightarrow [x_l(t+1)=0] \\
[x_l(t)=0] \wedge [x_c(t) > 0] \rightarrow [x_l(t+1)=1]. \\
[x_l(t)=1] \rightarrow [x_l(t+1)=0]
\end{gathered} \tag{80}
$$

The logical state of FSM $x_l(t)$ remains in 0 as long as the continuous state $x_c(t)$ is non-positive. If $x_c(t) > 0$ at some time-instant t, then FSM generates a digital

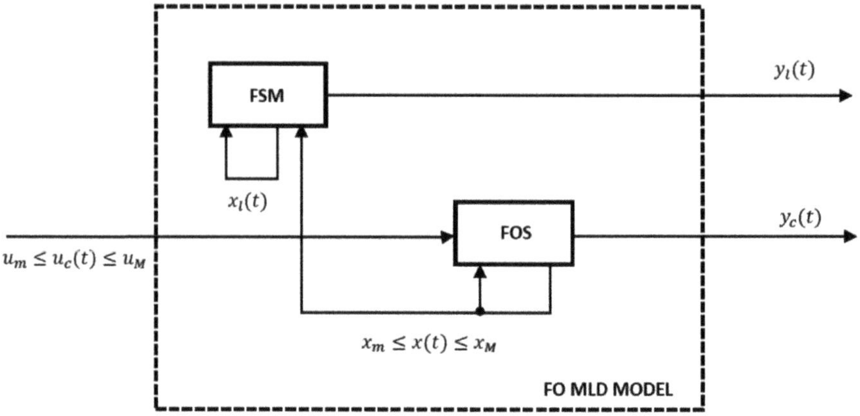

Fig. 6 The structure of the process in Example 5

impulse, i.e. $x_l(t+1) = 1$ and $x_l(t+2) = 0$. The FSM dynamics are hence driven by events generated by the underlying linear system. Figure 6 shows the structure of the process.

Let us introduce the auxiliary logical variables $\delta_1(t)$, $\delta_2(t)$ defined as

$$\begin{gathered}[\delta_1(t) = 1] \leftrightarrow [x_c(t) \le 0] \\ [\delta_2(t) = 1] \leftrightarrow [x_l(t) = 0] \wedge [\delta_1(t) = 0]\end{gathered} \tag{81}$$

and, equivalently,

$$\begin{gathered}x_c \le x_M(1 - \delta_1(t)) \\ x_c \ge \varepsilon + (x_m - \varepsilon)\delta_1(t) \\ \delta_2(t) \le (1 - \delta_1(t)) \\ \delta_2(t) \le (1 - x_l(t)) \\ \delta_2(t) \ge (1 - \delta_1(t)) + (1 - x_l(t)) - 1\end{gathered} \quad . \tag{82}$$

Therefore, the FSM as a logical part in the proposed FO MLD model is defined by mixed-integer linear inequality (45) along with the equalities:

$$\begin{gathered}x_l(t+1) = \delta_2(t) \\ y_l(t) = x_l(t)\end{gathered} \quad . \tag{83}$$

Additionally, according to results of Example 2., for such a system we have $n_c, m_c, p_c = 1, r_c = 0, n_l = p_l = 1, m_l = 0, r_l = 2$ and

$$A_1 = \begin{bmatrix} 0.6 & 0 \\ 0 & 0 \end{bmatrix}, \quad A_{1,i} = \begin{bmatrix} (-1)^i \binom{0.5}{i} & 0 \\ 0 & 0 \end{bmatrix}, \quad B_1 = \begin{bmatrix} 2 \\ 0 \end{bmatrix}, \quad C_1 = \begin{bmatrix} 0.75 & 0 \\ 0 & 1 \end{bmatrix},$$

$$B_2 = \begin{bmatrix} 0 & 0 \\ 0 & 1 \end{bmatrix}, \quad B_3 = B_{3,i} = B_4 = 0, \quad D_1 = D_2 = D_3 = D_4 = 0, \quad E_3 = 0,$$

as well as the same constraints imposed on the continuous state variable $x_c(t)$. Therefore, the FO MLD model (43)–(48) takes the form:

$$\begin{aligned} \begin{bmatrix} x_c(t+1) \\ x_l(t+1) \end{bmatrix} &= \begin{bmatrix} 0.6 & 0 \\ 0 & 0 \end{bmatrix} \begin{bmatrix} x_c(t) \\ x_l(t) \end{bmatrix} + \begin{bmatrix} 2 \\ 0 \end{bmatrix} u_c(t) + \begin{bmatrix} 0 & 0 \\ 0 & 1 \end{bmatrix} \begin{bmatrix} \delta_1(t) \\ \delta_2(t) \end{bmatrix} \\ &- \sum_{i=1}^{t+1} \begin{bmatrix} (-1)^i \binom{0.5}{i} & 0 \\ 0 & 0 \end{bmatrix} \begin{bmatrix} x_c(t+1-i) \\ x_l(t+1-i) \end{bmatrix}, \\ \begin{bmatrix} y_c(t) \\ y_l(t) \end{bmatrix} &= \begin{bmatrix} 0.75 & 0 \\ 0 & 1 \end{bmatrix} \begin{bmatrix} x_c(t) \\ x_l(t) \end{bmatrix} \end{aligned} \tag{84}$$

with the LMI constraints that define the FO MLD model in the form:

$$\underbrace{\begin{bmatrix} 0 & 0 \\ 0 & 0 \\ 0 & 0 \\ 0 & 0 \\ x_M & 0 \\ x_m - \varepsilon & 0 \\ 1 & 1 \\ 0 & 1 \\ -1 & -1 \end{bmatrix}}_{E_2} \begin{bmatrix} \delta_1(t) \\ \delta_2(t) \end{bmatrix} \leq \underbrace{\begin{bmatrix} 0 \\ 0 \\ 1 \\ -1 \\ 0 \\ 0 \\ 0 \\ 0 \\ 0 \end{bmatrix}}_{E_1} u_c(t) + \underbrace{\begin{bmatrix} 1 & 0 \\ -1 & 0 \\ 0 & 0 \\ 0 & 0 \\ -1 & 0 \\ 1 & 0 \\ 0 & 0 \\ 0 & -1 \\ 0 & 1 \end{bmatrix}}_{E_4} \begin{bmatrix} x_c(t) \\ x_l(t) \end{bmatrix} + \underbrace{\begin{bmatrix} -x_m - \varepsilon \\ x_M - \varepsilon \\ -u_m \\ u_M \\ x_M \\ -\varepsilon \\ 1 \\ 1 \\ -1 \end{bmatrix}}_{E_5}. \tag{85}$$

6 Conclusions

The paper describes a method for generalized modeling of non-integer order discrete systems involving real and integer variables and their interactions. A new discrete-time fractional-order mixed logical dynamical model (FO MLD) derived from integer-order models has been proposed. The model can describe a wide class of systems of both integer and non-integer order, those continuous, discrete and hybrid ones, as well as many non-linear systems, the nonlinearities of which can be approximated by Piecewise Affine (PWA) models. The model can be employed for process modeling but can also be applied directly to the synthesis of complex, effective control algorithms, optimal and predictive ones.

Essentially the idea underlying the proposed approach to process modeling consists of replacing process binary variables and logical conditions resulting, for example, from signal constraints by a system of linear matrix inequalities (LMI). Formally, it allows logical relationships to be included into the on-line control optimization problem, as is the case with, for example, the nonlinear, fractional-order model predictive control. However the generality of the FO MLD model leads, in some specific cases to its high implementation complexity. Thus, for a more complete opinion, a further analysis of the computational complexity of the proposed method is needed. Development of a Nonlinear Fractional-order MPC Toolbox using the proposed model will be of great help here.

References

1. Bemporad, A.: An efficient technique for translating mixed logical dynamical systems into piecewise affine systems. In: Proceedings of the 41st IEEE Conference on Decision and Control, Las Vegas, pp. 1970–1975 (2002)
2. Bemporad, A.: Modeling, control and reachability analysis of discrete time hybrid systems. Disc School on Hybrid systems. Technical Report. University of Siena, Italy (2003)
3. Bemporad, A.: Efficient algorithm for converting mixed logical dynamical systems into an equivalent piecewise affine form. IEEE Trans. Control (2004)
4. Bemporad, A., Giorgetti, N.: Logic-based solution methods for optimal control of hybrid systems. IEEE Trans. Autom. Control **51**(6), 963–976 (2006)
5. Bemporad, A., Morari, M.: Control of systems integrating logic, dynamics, and constraints. Automatica **35**, 407–427 (1999)
6. Bemporad, A., Torrisi, F.D., Morari, M.: Discrete-time hybrid modeling and verification of the batch evaporator process benchmark. Eur. J. Control **7**(4), 382–399 (2001)
7. Blesa, J., Puig, V., Bolea, Y.: Fault detection using interval LPV models in an open-flow canal. Control Eng. Pract. **18**, 460–470 (2010)
8. Borrelli, F., Bemporad, A., Morari, M.: Predictive Control for Linear and Hybrid Systems. Cambridge University Press (2017)
9. Bortolussi, L. Policriti, A.: Hybrid systems and biology: continuous and discrete modeling for systems biology. In: Bernardo, M., Degano, P. (eds.): SFM'08: Proceedings of the Formal Methods for the Design of Computer, Communication, and Software Systems, 8th International Conference on Formal Methods for Computational Systems Biology. Springer, 424–448 (2008)
10. Branicky, M.S., Borkar, V.S., Mitter, S.K.: A unified framework for hybrid control: model and optimal control theory. IEEE Trans. Autom. Control **43**(1), 31–45 (1998)
11. Carloni, L.P., Passerone, R., Pinto, A., Sangiovanni-Vincentelli, A.L.: Languages and Tools for Hybrid Systems Design. Foundations and Trends in Electronic Design Automation, vol. 1(1–2). Publishers inc., Boston – Delft (2006)
12. Cavalier, T.M., Pardalos, P.M., Soyster, A.L.: Modeling and integer programming techniques applied to propositional calculus. Comput. Oper. Res. **17**(6), 561–570 (1990)
13. Chen, Y.Q., Petráš, I., Xue, D.: Fractional order control. In: American Control Conference, St. Louis, pp. 1397–1410 (2009)
14. Domek, S.: Fractional-Order Calculus in Model Predictive Control (In Polish). West Pomeranian University of Technology Academic Press, Szczecin (2013)

15. Domek, S.: Switched state model predictive control of fractional-order nonlinear discrete-time systems. In: Pisano, A., Caponetto, R. (eds.): Advances in Fractional Order Control and Estimation. Asian J. Control, Special Issue **15**(3), 658–668 (2013)
16. Domek, S.: Piecewise affine representation of discrete in time, non-integer order systems. In: Mitkowski, W., Kacprzyk, J., Baranowski, J. (eds.): Advances in the Theory and Applications of Non-integer Order Systems. LN in Electrical Engineering, vol. 257, 149–160. Springer (2013)
17. Domek, S.: Multiple use of the fractional-order differential calculus in the model predictive control. In: Proceedings of 19th International Conference on Methods and Models in Automation and Robotics, Mi\c{e}dzyzdroje, pp. 359–362 (2014)
18. Domek, S.: Fractional-order model predictive control with small set of coincidence points. In: Latawiec, K., Łukaniszyn, M., Stanisławski, R. (eds.) Advances in Modelling and Control of Non-integer Order Systems. Lecture Notes in Electrical Engineering, vol. 320, pp. 135–144. Springer (2014)
19. Domek, S.: Model-plant mismatch in fractional order model predictive control. In: Domek, S., Dworak, P. (eds.) Theoretical Developments and Applications of Non-integer Order Systems. Lecture Notes in Electrical Engineering, vol. 357, pp. 281–291. Springer (2016)
20. Domek, S.: Approximation and stability analysis of some kinds of switched fractional linear systems. In: Mitkowski, W., Kacprzyk, J., Oprz\c{e}dkiewicz, K., Skruch, P. (eds.) Trends in Advanced Intelligent Control, Optimization and Automation. AISC, vol. 577, pp. 442–454. Springer (2017)
21. Domek, S.: Fractional linear systems with memory deficiency and their state-space integer-order approximation. In: Ostalczyk, P., Sankowski, D., Nowakowski, J. (eds.) Non-integer Order Calculus and Its Applications. Lecture Notes in Electrical Engineering, vol. 496, pp. 164–179. Springer (2018)
22. Domek, S.: Discrete-time switched models of non-linear fractional-order systems. In: Bartoszewicz, A., Kabziński, J., Kacprzyk, J. (eds): Advanced, Contemporary Control. Advances in Intelligent Systems and Computing, vol. 1196, pp. 1176–1188. Springer (2020)
23. Domek, S.: Switched fractional state-space predictive control methods for non-linear fractional systems. In: Malinowska A., Mozyrska, D., Sajewski, Ł. (eds): Advances in Non-Integer Order Calculus and Its Applications. RRNR 2018. Lecture Notes in Electrical Engineering, vol. 559, pp. 113–127. Springer (2020)
24. Domek, S.: Switched models of non-integer order. In: Kulczycki, P., Korbicz, J., Kacprzyk, J. (eds): Automatic Control, Robotics, and Information Processing. Studies in Systems, Decision and Control, vol. 296. Springer (2021)
25. Dzieliński, A., Sierociuk, D., Sarwas, G.: Some applications of fractional order calculus. Bull. Pol. Acad. Sci. - Tech. Sci. **58**(4), 583–592 (2010)
26. Fang, L., Lin, H., Antsaklis, P.J.: Stabilization and performance analysis for a class of switched systems. In: Proceedings of 43rd IEEE Conference on Decision Control, Atlantis, pp. 1179–1180 (2004)
27. Ferrari-Trecate, G., Cuzzola, F.A., Mignone, D., Morari, M.: Analysis of discrete-time piecewise affine and hybrid systems. Automatica **38**(12), 2139–2146 (2002)
28. Geyer, T., Torrisi, F.D., Morari, M.: Efficient mode enumeration of compositional hybrid systems. Int. J. Control **83**(2), 313–329 (2010)
29. Ghods, A.H., Fu, L., Rahimi-Kian, A.: An efficient optimization approach to real-time coordinated and integrated freeway traffic control. IEEE Trans. Intell. Transp. Syst. **11**(4), 873–888 (2010)
30. Hariprasad, K., Bhartiya, S.: An efficient and stabilizing model predictive control of switched systems. IEEE Trans. Autom. Control **62**(7), 3401–3407 (2017)
31. Heemels, W.P.M.H., Schutter, B.D., Bemporad, A.: Equivalence of hybrid dynamical models. Automatica **37**(7), 1085–1091 (2001)
32. Hejri, M., Giua, A.: Hybrid modeling and control of switching DC-DC converters via MLD systems. In: Proceedings of 7th IEEE International Conference on Automation Science and Engineering, Trieste, Italy (2011)

33. Hejri, M., Giua, A., Mokhtari, H.: On the complexity and dynamical properties of mixed logical dynamical systems via an automaton-based realization of discrete-time hybrid automaton. Int. J. Robust Nonlinear Control **28**, 4713–4746 (2018)
34. Hejri, M., Mokhtari, H.: Hybrid modeling and control of a DC-DC boost converter via extended mixed logical dynamical systems (EMLDs). In: Proceedings of 5th Annual International Power Electronics, Drive Sys and Technol Conf (PEDSTC), Tehran (2014)
35. Ionescu, C., Lopes, A., Copot, D., Machado, J.A.T., Bates, J.H.T.: The role of fractional calculus in modeling biological phenomena: a review. Commun. Nonlinear Sci. Numer. Simul. **51**, 141–159 (2017)
36. Kaczorek T.: Selected Problems of Fractional Systems Theory. Springer (2011)
37. Kim, Y., Kato, T., Okuma, S., Narikiyo, T.: Traffic network control based on hybrid dynamical system modeling and mixed integer nonlinear programming with convexity analysis. IEEE. Trans. Syst. Man Cybern. Syst. Hum. **38**(2), 346–357 (2008)
38. Lygeros, J., Johansson, K.H., Simic, S.N., Zhang, J., Sastry, S.S.: Dynamical properties of hybrid automata. IEEE Trans. Autom. Control **48**(1), 2–17 (2003)
39. Matsui, G., Tachibana, T., Kogiso, K., Sugimoto, K.: Dynamic resource management in optical grid. IEEE Trans. Control Syst. Technol. **22**(4), 1607–1614 (2014)
40. Mäkilä, P.M., Partington, J.R.: On linear models for nonlinear systems. Automatica **39**, 1–13 (2003)
41. Monje, C.A., Chen, Y.Q., Vinagre, B.M., Xue, D., Feliu, V.: Fractional Order Systems and Controls. Springer (2010)
42. Muddu Madakyaru, M., Narang, A., Patwardhan, S.C.: Development of ARX models for predictive control using fractional order and orthonormal basis filter parameterization. Ind. Eng. Chem. Res. **48**(19), 8966–8979 (2009)
43. Ostalczyk, P.: The non-integer difference of the discrete-time function and its application to the control system synthesis. Int. J. Syst. Sci. **31**(12), 1551–1561 (2000)
44. Podlubny, I.: Fractional Differential Equations. Academic Press, San Diego (1999)
45. Raman, R., Grossmann, I.E.: Relation between MILP modeling and logical inference for chemical process synthesis. Comput. Chem. Eng. **15**(2), 73–84 (1991)
46. Rihan, F.A., Al-Mdallal, Q.M., Al Sakaji, H.J., Hashish, A.: A fractional-order epidemic model with time-delay and nonlinear incidence rate. Chaos, Solitons Fractals **126**, 97–105 (2019)
47. Ripaccioli, G., Bemporad, A., Assadian, F., Dextreit, C., Di Cairano, S., Kolmanovsky, I.V.: Hybrid modeling, identification, and predictive control: an application to hybrid electric vehicle energy management. In: Majumdar, R., Tabuada, P. (eds.): Hybrid Systems: Computation and Control. 12th International Conference on HSCC 2009, San Francisco, CA, LNCS, vol. 5469, pp. 321–336 (2009)
48. Romero, M., De Madrid, Á.P., Mañoso, C., Vinagre, B.M.: Fractional-order generalized predictive control: formulation and some properties. In: Proceedings of 11th International Conference on Control, Automation, Robotics and Vision, Singapore, pp. 1495–1500 (2010)
49. Romero, M., Vinagre, B.M., De Madrid, Á.P.: GPC control of a fractional–order plant: improving stability and robustness. In: Proceedings of the 17th IFAC World Congress, Seoul, pp. 14266–14271 (2008)
50. Rydel, M., Stanisławski, R., Bialic, G., Latawiec, K.: Modeling of discrete-time fractional-order state space systems using the balanced truncation method. In: Domek, S., Dworak, P. (eds.) Theoretical Developments and Applications of Non-integer Order Systems. Lecture Notes in Electrical Engineering, vol. 357, pp. 119–127. Springer (2016)
51. Savkin, A.V., Evans, R.J.: Hybrid Dynamical Systems. Controller and Sensor Switching Problems. Birkhäuser, Boston (2002)
52. Sirmatel, I.I., Geroliminis, N.: Mixed logical dynamical modeling and hybrid model predictive control of public transport operations. Transp. Res. Part B: Methodol. **114**, 325–345 (2018)
53. Stanisławski, R., Latawiec, K.: Normalized finite fractional differences: the computational and accuracy breakthroughs. Int. J. Appl. Math. Comput. Sci. **22**(4), 907–919 (2012)
54. Sun, Z., Ge, S.S.: Switched Linear Systems. Control and Design. Springer (2005)

55. Szűcs, A., Kvasnica, M., Fikar, M.: Optimal piecewise affine approximations of nonlinear functions obtained from measurements. In: Proceedings of the 4th IFAC Conference on Analysis and Design of Hybrid Systems, Eindhoven, pp. 160–165 (2012)
56. Torrisi, F.D., Bemporad, A.: HYSDEL - a tool for generating computational hybrid models for analysis and synthesis problems. IEEE Trans. Control Syst. Technol. **12**(2), 235–249 (2004)
57. Vargas-De-Leó, C.: On the global stability of SIS, SIR and SIRS epidemic models with standard incidence. Chaos, Solitons Fractals **44**, 1106–1110 (2011)
58. Williams, H.P.: Logical problems and integer programming. Bull. Inst. Math. Appl. **13**, 18–20 (1977)
59. Williams, H.P.: Model Building in Mathematical Programming, 3rd edn. Wiley (1993)
60. Zhang, W., Abate, A., Hu, J.: Stabilization of discrete-time switched linear systems: a control-Lyapunov function approach. In: Majumdar, R., Tabuada, P. (eds.): Hybrid Systems: Computation and Control. In: Proceedings of 12th International Conference on HSCC 2009, San Francisco, CA, LNCS, vol. 5469, pp. 411–426 (2009)

Fractional Variable-Order Derivative and Difference Operators and Their Applications to Dynamical Systems Modelling

Andrzej Dzieliński, Dominik Sierociuk, Wiktor Malesza, Michał Macias, Michał Wiraszka, and Piotr Sakrajda

Abstract The chapter presents an overview of some particular derivative and difference operators of fractional variable order, their properties, equivalent forms, and applications. When fundamental properties of a system or its structure are changing in time, a variation of the system's order may be observed. In such a case, time-dependent variable order operators are taken into consideration. Recently, cases where order is time-varying, have began to be studied extensively. In order to give a deeper insight into fractional variable order calculus, alternative, intuitive descriptions of some particular variable order operators, in the form of equivalent switching schemes, is provided. According to such a schematic interpretation of variable order operators analysis of variable order systems can be simpler and more effective than on the basis of purely analytical definitions. Based on those switching schemes it is possible to categorize fractional order derivatives according to their behaviour and intrinsic properties. Thanks to this schematic description and duality property between chosen variable order operators, analytical solutions of variable order linear differential equations can be effectively derived. Examples of applications of these operators to automatic control and modelling of the heat transfer process in specific

---

A. Dzieliński (✉) · D. Sierociuk · W. Malesza · M. Macias · M. Wiraszka · P. Sakrajda
Faculty of Electrical Engineering, Institute of Control and Industrial Electronics, Warsaw University of Technology, Koszykowa 75, 00-662 Warszawa, Poland
e-mail: andrzej.dzielinski@ee.pw.edu.pl

D. Sierociuk
e-mail: dominik.sierociuk@ee.pw.edu.pl

W. Malesza
e-mail: wiktor.malesza@ee.pw.edu.pl

M. Macias
e-mail: michal.macias@ee.pw.edu.pl

M. Wiraszka
e-mail: michal.wiraszka@ee.pw.edu.pl

P. Sakrajda
e-mail: piotr.sakrajda@ee.pw.edu.pl

P. Kulczycki et al. (eds.), *Fractional Dynamical Systems: Methods, Algorithms and Applications*, Studies in Systems, Decision and Control 402,
https://doi.org/10.1007/978-3-030-89972-1_4

grid-holes and two-dimensional fractal-like structure media, of which the geometry is changing in time, are presented.

1 Introduction

Integral and differential calculus of non integer (fractional) order is a natural generalization of the well-known differential and integral calculus (integer order), where derivatives and integrals can be obtained not only for the integer orders but also non-integer or even complex ones.

The history of fractional order calculus is nearly as old as traditional differential calculus with the first possible definition of fractional order derivatives mentioned in 1695 by Leibniz and L'Hôspital. At the end of the 19th century Liouville and Riemann introduced the first definition of fractional derivative; however, only in late 1960's, did the idea draw the attention of engineers.

The first monograph about fractional order calculus was written in 1974 by K. B. Oldham and J. Spanier [22]. Other very important books on this topic are those of S. G. Samko at al. [32], I. Podlubny [29], and Miller-Ross [20].

Initially the fractional order differential calculus was used only by mathematicians and theoretical physicists; however, recently we can see intensive expansion of the use of fractional calculus in a wide range of scientific areas such as economics, bio-engineering and control engineering. Since a fractional derivative is not defined in a point, but on an interval (the same as integral for integer order case), the models based on fractional derivatives better describe complex dynamics. Thus, fractional calculus is applied in the description and modeling of various phenomena and systems of a complex nature or internal structures. An example of a complex dynamical system (described by partial differential equations) is the process of diffusion. In the solution of the equation describing this process, taking into account the simplest approach of an infinite heating rod and the description in the Laplace operator domain, there is a square root element of the complex variable s, i.e., $\sqrt{s}$, which can be interpreted as a derivative of order of 0.5. In [3], results of successful modeling of the heat transfer process in a solid material is presented. Moreover, in [34], similar results for heat transfer in heterogeneous materials, described by anomalous diffusion using a fractional order partial differential equation, are shown. Memory effect within a fractional calculus approach has been investigated in various areas, for example in nonlinear heat conduction [7] and viscoelastic [6]. Another example of successful exploitation of fractional calculus are ultracapacitors. Modeling of these devices based on anomalous diffusion (fractional order model) is presented in [2, 4, 5]. In [35], analysis of resonance phenomenon in a circuit with ultracapacitor is investigated. This analysis clearly shows that modelling of ultracapacitors based only on integer order calculus causes unacceptable errors in obtaining values of resonance frequency. It is also presented that the fractional order model can describe such a phenomenon where resonance frequency is different from maximum current frequency. Not only in modeling, fractional calculus was also found interesting in signal processing. For example in control theory, fractional calculus is used to obtain more efficient algorithms like fractional order PID controllers [24] and variable order PID controllers [1]. More theoretical background and applications of fractional calculus can be found in [20–23, 29, 32].

An interesting case is when the fractional order is time-varying. Then, the fundamental properties of a system or its structure are changing in time and a variation of the system's order is to be observed. This behavior can be described as state vector-dependent or time-dependent (or both time- and state vector-dependent). Below a time-dependent variable order case is taken into consideration. Recently, the case when order is time-varying, began to be studied extensively. Such a behavior can be met, e.g. in chemistry (when the properties of the system are changing due to chemical reactions), medicine, electrochemistry, material science and other areas. In [11], variable order differential equations are used to model the memory behavior of shape-memory polymer (SMP). The presented results clearly show that the variable order model is more suitable than constant order. In [44] usage of variable order fractional calculus to describe the evolution of protein lateral diffusion ability in cell membranes is presented. In [33], the process of an electrochemical device with parameters changing in time is presented. The authors have shown that the order of the model used depends on time, which implies that for describing long-time behavior, a variable order model has to be used. Results of very accurate modelling of the heat transfer process in a time-varying grid-hole structure are presented in [31].

A description and analysis of such variable order systems is much more complicated than is the case for constant order systems, mainly due to the variety of definitions given in the literature. In [13, 43], three general types of variable order derivative definitions have been given. These definitions are given mainly as further modifications of mathematical formulas—without derivation, intuitive interpretation and deep analysis of them.

In order to give a deeper insight into fractional variable order calculus, four switching schemes equivalent to four definitions of variable order derivatives (two reported in the literature and our two new introduced definitions) have been proposed, among others, in the applicant's works. Introduced switching strategies, given unambiguously, classify and identify ways of changing the order of derivatives (integrals). Based on those switching schemes, it is possible to categorize fractional order derivatives according to their behavior and intrinsic properties. Moreover, based on proposed switching schemes, it is possible to build analog models of variable order systems for different types of definitions. The obtained experimental results confirmed theoretical results of equivalence between obtained switching schemes and corresponding mathematical formulas (which are modifications of the well-known Grünwald-Letnikov formula).

2 Variable Order Operators

In this section, definitions of variable order operators (both derivatives and differences), together with corresponding equivalent switching schemes, are presented.

The following definition constitutes a starting point for generalization of constant fractional order difference operators onto a variable order case. A constant fractional order derivative and its discrete approximation (difference operator) are defined, respectively, in the following way

$$ {}_0\mathrm{D}_t^{\alpha} x(t) = \lim_{h \to 0} \frac{1}{h^{\alpha}} \sum_{j=0}^{\eta} (-1)^j \binom{\alpha}{j} x(t - jh), $$

where $\eta = \lfloor t/h \rfloor$, h is a step time, and

$$ {}_0\Delta_k^{\alpha} x_k \equiv \sum_{j=0}^{k} w(\alpha, j) x_{k-j}\,, \tag{1} $$

where

$$ w(\alpha, j) = \frac{1}{h^{\alpha}} (-1)^j \binom{\alpha}{j}, \tag{2} $$

and

$$ \binom{\alpha}{j} \equiv \begin{cases} 1 & \text{for} \quad j = 0\,, \\ \frac{\alpha(\alpha-1)\ldots(\alpha-j+1)}{j!} & \text{for} \quad j > 0\,. \end{cases} $$

Operators for variable order case presented below exhibit different behavior. Below are studied definitions of variable order derivatives and differences, as well as examples for Heaviside step function with order function (3) are presented (Fig. 1)

$$ \alpha(t) = \begin{cases} \alpha_1 = -1, & t \in [0, 1) \\ \alpha_2 = -2, & t \in [1, 2). \end{cases} \tag{3} $$

The first one is obtained by replacing a constant order α by variable order $\alpha(t)$.

$$ {}_0^{\mathscr{A}}\mathrm{D}_t^{\alpha(t)} x(t) = \lim_{h \to 0} \frac{1}{h^{\alpha(t)}} \sum_{j=0}^{\eta} (-1)^j \binom{\alpha(t)}{j} x(t - jh), $$

and its discrete approximation is given by

$$ {}_0^{\mathscr{A}}\Delta_k^{\alpha_k} x_k \equiv \sum_{j=0}^{k} {}^{\mathscr{A}} w(\alpha(\cdot), k, j) x_{k-j}\,, \tag{4} $$

where

$$ {}^{\mathscr{A}} w(\alpha(\cdot), k, j) = \frac{(-1)^j}{h^{\alpha_k}} \binom{\alpha_k}{j}. \tag{5} $$

The equivalent switching scheme to $\mathscr{A}$-type operator (studied and proved in [38]) is presented in Fig. 2, where

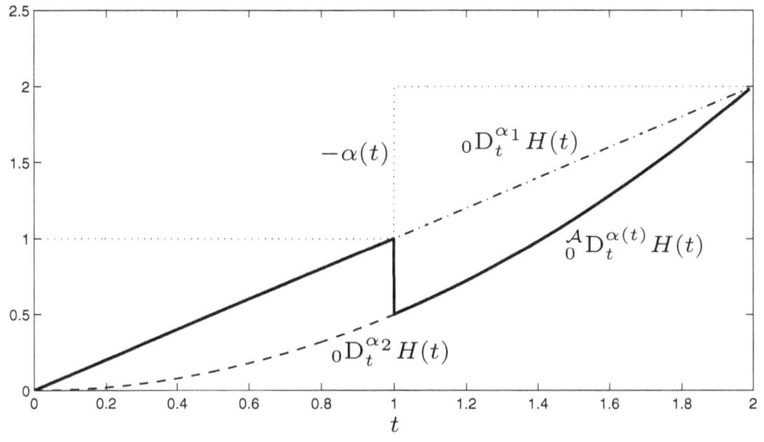

Fig. 1 Heaviside step function response of $\mathscr{A}$-type integrals for $\alpha(t)$ given by (3)

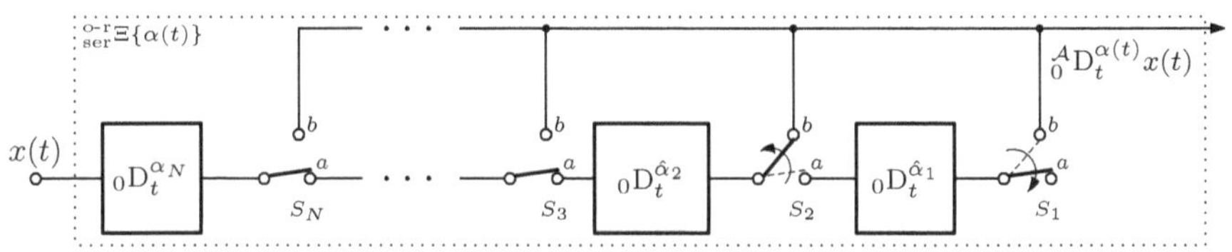

Fig. 2 Output-reductive switching scheme (switching from α_1 to α_2)

$$\alpha_j = \alpha_{j+1} + \hat{\alpha}_j, \qquad j = 1, \ldots, N-1,$$

and

$$S_i = \begin{cases} b & \text{for } t_{i-1} \le t < t_i, \\ a & \text{otherwise}, \end{cases} \qquad i = 1, \ldots, N.$$

The second definition assumes that coefficients for past samples are obtained for order that was present for these samples (Fig. 3)

$${}_0^{\mathscr{B}}\mathrm{D}_t^{\alpha(t)} x(t) = \lim_{h \to 0} \sum_{j=0}^{\eta} \frac{(-1)^j}{h^{\alpha(t-jh)}} \binom{\alpha(t-jh)}{j} x(t-jh).$$

Its discrete approximation is given by

$${}_0^{\mathscr{B}}\Delta_k^{\alpha_k} x_k \equiv \sum_{j=0}^{k} {}^{\mathscr{B}}w(\alpha(\cdot), k, j) x_{k-j}\,, \tag{6}$$

where

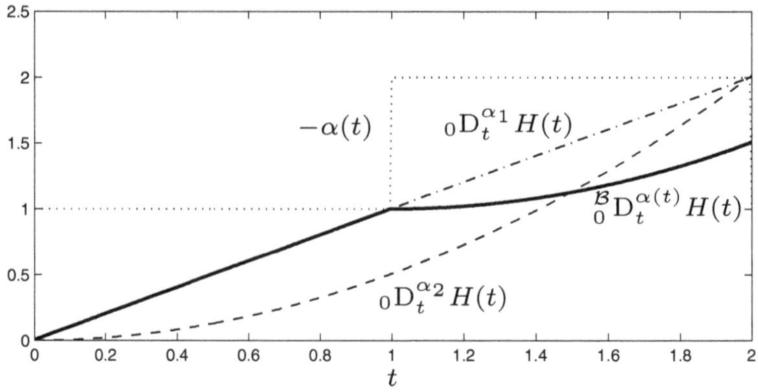

Fig. 3 Heaviside step function response of $\mathscr{B}$-type integrals for $\alpha(t)$ given by (3)

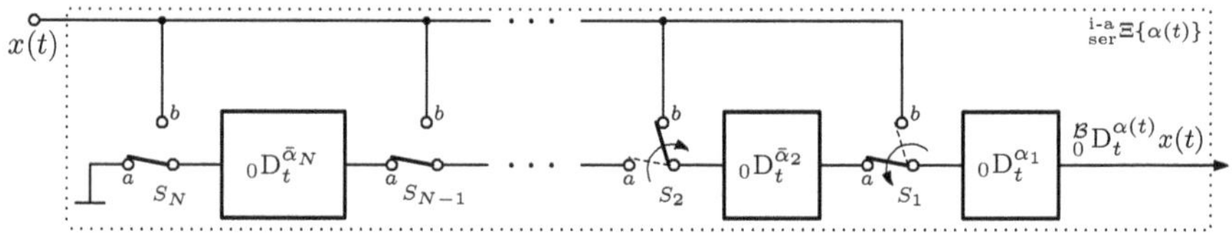

Fig. 4 Input-additive switching scheme (switching from α_1 to α_2)

$$ {}^{\mathscr{B}}w(\alpha(\cdot), k, j) = \frac{(-1)^j}{h^{\alpha_{k-j}}} \binom{\alpha_{k-j}}{j}. \tag{7} $$

The equivalent switching scheme to $\mathscr{B}$-type operator (studied and proved in [37]) is presented in Fig. 4.
For this switching system

$$ \alpha_j = \alpha_{j-1} + \bar{\alpha}_j, \qquad j = 2, \ldots, N, $$

and

$$ S_i = \begin{cases} b & \text{for } t_{i-1} \le t < t_i, \\ a & \text{otherwise,} \end{cases} \qquad i = 1, \ldots, N. $$

The third definition is less intuitive and assumes that coefficients for the newest samples are obtained, respectively, for the oldest orders

$$ {}_0^{\mathscr{C}}\mathrm{D}_t^{\alpha(t)} x(t) = \lim_{h \to 0} \frac{1}{h^{\alpha(jh)}} \sum_{j=0}^{\eta} (-1)^j \binom{\alpha(t)}{j} x(t - jh), $$

and its discrete approximation

$$\begin{pmatrix} {}_0^{\mathbb{T}}\mathrm{D}_0^{\alpha(0)}x(0) \\ {}_0^{\mathbb{T}}\mathrm{D}_h^{\alpha(h)}x(h) \\ {}_0^{\mathbb{T}}\mathrm{D}_{2h}^{\alpha(2h)}x(2h) \\ \vdots \\ {}_0^{\mathbb{T}}\mathrm{D}_{kh}^{\alpha(kh)}x(kh) \end{pmatrix} = \lim_{h\to 0} {}^{\mathbb{T}}W(\alpha,k) \begin{pmatrix} x(0) \\ x(h) \\ x(2h) \\ \vdots \\ x(kh) \end{pmatrix},$$

where matrix ${}^{\mathbb{T}}W(\alpha,k) \in \mathbb{R}^{(k+1)\times(k+1)}$ is different for each parameter $\mathbb{T}$ and is explicitly given (excluding $\mathscr{C}$-type) in [18, 37, 38]. For example, the matrix form of $\mathscr{D}$-type derivative is the following

$${}^{\mathscr{D}}W(\alpha,k) = \left(\begin{array}{c|c|c|c|c|c} h^{-\alpha_0} & 0 & 0 & \cdots & 0 & 0 \\ \hline \mathfrak{q}_{2,1} & h^{-\alpha_1} & 0 & \cdots & 0 & 0 \\ \hline \mathfrak{q}_{3,1} & \mathfrak{q}_{3,2} & h^{-\alpha_2} & \cdots & 0 & 0 \\ \hline \vdots & \vdots & \vdots & \ddots & \ddots & \vdots \\ \hline \mathfrak{q}_{k,1} & \mathfrak{q}_{k,2} & \mathfrak{q}_{k+1,3} & \cdots & h^{-\alpha_{k-1}} & 0 \\ \hline \mathfrak{q}_{k+1,1} & \mathfrak{q}_{k+1,2} & \mathfrak{q}_{k+1,3} & \cdots & \mathfrak{q}_{k+1,k} & h^{-\alpha_k} \end{array}\right),$$

where, for $i, j = 1, \ldots, k+1$,

$$\mathfrak{q}_{i,j} = \begin{cases} q_{i-1}(\mathfrak{q}_{1,j}, \ldots, \mathfrak{q}_{i-1,j})^T & \text{for } i > j, \\ h^{-\alpha_i} & \text{for } i = j, \\ 0 & \text{for } i < j, \end{cases}$$

and q_r, for $r = 1, \ldots, k$, is given by (12).

It is worth noticing that for 4-tuple $\mathscr{T} = (\mathscr{A}, \mathscr{B}, \mathscr{D}, \mathscr{E})$ the semi-group property does not hold, i.e.,

$${}_0^{\mathscr{T}_i}\mathrm{D}_t^{\alpha(t)}\left({}_0^{\mathscr{T}_i}\mathrm{D}_t^{-\alpha(t)}x(t)\right) \neq x(t).$$

However, define the 4-tuple $\tilde{\mathscr{T}} = (\mathscr{D}, \mathscr{E}, \mathscr{A}, \mathscr{B})$, and then the following duality property holds [36]

$${}_0^{\mathscr{T}_i}\mathrm{D}_t^{\alpha(t)}\left({}_0^{\tilde{\mathscr{T}}_i}\mathrm{D}_t^{-\alpha(t)}x(t)\right) = x(t). \tag{18}$$

3 Variable Order Control Systems

Using the variable orders operators presented in Sect. 2 dynamical control systems can be defined. Below two possible approaches for finding solutions to such kind of systems are presented, namely: numerical (described in [16, 18, 39]) and analytical (described in [17]).

3.1 Approximate Solutions of Variable Order Control Systems

Recall the 4-tuple $\mathscr{T} = (\mathscr{A}, \mathscr{B}, \mathscr{D}, \mathscr{E})$ and define other quadruple $\tilde{\mathscr{T}} = (\mathscr{D}, \mathscr{E}, \mathscr{A}, \mathscr{B})$, where $\mathscr{T}_\ell$ and $\tilde{\mathscr{T}}_\ell$ denote the ℓ-th elements of $\mathscr{T}$ and $\tilde{\mathscr{T}}$, respectively. We also define two n-tuples $\mathbb{T} = (\mathbb{T}^1, \dots, \mathbb{T}^n)$, where $\mathbb{T}^i \in \mathscr{T}$, and $\tilde{\mathbb{T}} = (\tilde{\mathbb{T}}^1, \dots, \tilde{\mathbb{T}}^n)$, where $\tilde{\mathbb{T}}^i \in \tilde{\mathscr{T}}$, in both cases $i = 1, \dots, n$, and such that if $\mathbb{T}^i = \mathscr{T}_\ell$ then $\tilde{\mathbb{T}}^i = \tilde{\mathscr{T}}_\ell$ for some $\ell \in \{1, \dots, 4\}$.

Now, consider a time-variant non-commensurate fractional variable order system

$$ {}_0^{\mathbb{T}}\mathrm{D}_t^{\alpha(t)} x = A(t)x + B(t)u, \quad x(0) = 0 \tag{19a}$$

$$ y = C(t)x + D(t)u, \tag{19b}$$

where ${}_0^{\mathbb{T}}\mathrm{D}_t^{\alpha(t)} x = \left({}_0^{\mathbb{T}^1}\mathrm{D}_t^{\alpha_1(t)} x_1(t), \dots, {}_0^{\mathbb{T}^n}\mathrm{D}_t^{\alpha_n(t)} x_n(t)\right)^T \in \mathbb{R}^n$, $x = x(t) \in \mathbb{R}^n$, $u = u(t) \in \mathbb{R}^m$, $y = y(t) \in \mathbb{R}^p$, $A(t) = [a_{ij}(t)] \in \mathbb{R}^{n \times n}$, $B(t) = [b_{ir}(t)] \in \mathbb{R}^{n \times m}$, $C(t) = [c_{si}(t)] \in \mathbb{R}^{p \times n}$, $D(t) = [d_{sr}(t)] \in \mathbb{R}^{p \times m}$, for $t \in \mathbb{R}$, $i, j = 1, \dots, n$, $r = 1, \dots, m$, $s = 1, \dots, p$; and $\mathbb{T}^i \in \mathscr{T}$ is a type of variable order derivative definition. We assume variable orders to be piece-wise constant functions, i.e., for $i = 1, \dots, n$

$$ \alpha_i(t) = \alpha_i^{\nu+1} \in \mathbb{R} \quad \text{for}\, t_\nu \le t < t_{\nu+1}, \nu = 0, \dots, N-1, $$

where $N \in \mathbb{N}$ denotes the number of time-intervals.

In turns, system (19) can be rewritten in the equivalent numerical matrix form

$$ {}^{\mathbb{T}}\mathbb{W}(\alpha)\hat{x} = \mathbb{A}\hat{x} + \mathbb{B}\hat{u} \tag{20a}$$

$$ \hat{y} = \mathbb{C}\hat{x} + \mathbb{D}\hat{u}, \tag{20b}$$

where ${}^{\mathbb{T}}\mathbb{W}(\alpha) \in \mathbb{R}^{n(k+1) \times n(k+1)}$ is

$$ {}^{\mathbb{T}}\mathbb{W}(\alpha) = \text{block diag}\left({}^{\mathbb{T}^1}W(\bar{\alpha}_1, k), \dots, {}^{\mathbb{T}^n}W(\bar{\alpha}_n, k)\right), $$

and

$$ \bar{\alpha}_i = (\alpha_i(0), \dots, \alpha_i(kh)), \quad i = 1, \dots, n, \; k \ge N, \tag{21}$$

and $\hat{x} = (\bar{x}_1, \dots, \bar{x}_n)^T \in \mathbb{R}^{n(k+1) \times 1}$, $\bar{x}_i = (x_i(0), \dots, x_i(kh)) \in \mathbb{R}^{1 \times (k+1)}$, $\hat{u} = (\bar{u}_1, \dots, \bar{u}_m)^T \in \mathbb{R}^{m(k+1) \times 1}$, $\bar{u}_r = (u_r(0), \dots, u_r(kh)) \in \mathbb{R}^{1 \times (k+1)}$, and

$$\mathbb{A} = \begin{pmatrix} \hat{a}_{11} & \cdots & \hat{a}_{1n} \\ \vdots & \ddots & \vdots \\ \hat{a}_{n1} & \cdots & \hat{a}_{nn} \end{pmatrix} \in \mathbb{R}^{n(k+1)\times n(k+1)}, \quad \hat{a}_{ij} = \operatorname{diag}\left(a_{ij}^0, \ldots, a_{ij}^k\right) \in \mathbb{R}^{(k+1)\times(k+1)};$$

$$\mathbb{B} = \begin{pmatrix} \hat{b}_{11} & \cdots & \hat{b}_{1m} \\ \vdots & \ddots & \vdots \\ \hat{b}_{n1} & \cdots & \hat{b}_{nm} \end{pmatrix} \in \mathbb{R}^{n(k+1)\times m(k+1)}, \quad \hat{b}_{ir} = \operatorname{diag}\left(b_{ir}^0, \ldots, b_{ir}^k\right) \in \mathbb{R}^{(k+1)\times(k+1)};$$

$$\mathbb{C} = \begin{pmatrix} \hat{c}_{11} & \cdots & \hat{c}_{1n} \\ \vdots & \ddots & \vdots \\ \hat{c}_{p1} & \cdots & \hat{c}_{pn} \end{pmatrix} \in \mathbb{R}^{p(k+1)\times n(k+1)}, \quad \hat{c}_{si} = \operatorname{diag}\left(c_{si}^0, \ldots, c_{si}^k\right) \in \mathbb{R}^{(k+1)\times(k+1)};$$

$$\mathbb{D} = \begin{pmatrix} \hat{d}_{11} & \cdots & \hat{d}_{1m} \\ \vdots & \ddots & \vdots \\ \hat{d}_{p1} & \cdots & \hat{d}_{pm} \end{pmatrix} \in \mathbb{R}^{p(k+1)\times m(k+1)}, \quad \hat{d}_{sr} = \operatorname{diag}\left(d_{sr}^0, \ldots, d_{sr}^k\right) \in \mathbb{R}^{(k+1)\times(k+1)}.$$

Using the duality properties (18), we have the following result.

Theorem 1 *Solution of* (20a), *and thereby approximated solution of* (19a), *is*

$$\hat{x} = \left(I_{n(k+1)} - {}^{\tilde{\mathbb{T}}}\mathbb{W}(-\alpha)\mathbb{A}\right)^{-1} {}^{\tilde{\mathbb{T}}}\mathbb{W}(-\alpha)\mathbb{B}\hat{u}, \tag{22}$$

where

$${}^{\tilde{\mathbb{T}}}\mathbb{W}(-\alpha) = \text{block diag}\left({}^{\tilde{\mathbb{T}}^1} W(-\bar{\alpha}_1, k), \ldots, {}^{\tilde{\mathbb{T}}^n} W(-\bar{\alpha}_n, k)\right) \in \mathbb{R}^{n(k+1)\times n(k+1)}, \tag{23}$$

and $I_{n(k+1)} \in \mathbb{R}^{n(k+1)\times n(k+1)}$ *stands for identity matrix.*

For time-invariant control system (19), i.e., where $A(t) = A, B(t) = B, C(t) = C$ and $D(t) = D$, we have the following result (special case of Theorem 1).

Corollary 1 *Approximated numerical solution of time-invariant control system* (19) *is*

$$\hat{x} = \left(I_{n(k+1)} - {}^{\tilde{\mathbb{T}}}\mathbb{W}(-\alpha)(A \otimes I_{k+1})\right)^{-1} {}^{\tilde{\mathbb{T}}}\mathbb{W}(-\alpha)(B \otimes I_{k+1})\hat{u}, \tag{24a}$$

$$\hat{y} = (C \otimes I_{k+1})\hat{x} + (D \otimes I_{k+1})\hat{u}, \tag{24b}$$

where $\otimes$ *denotes Kronecker product of matrices, and* ${}^{\tilde{\mathbb{T}}}\mathbb{W}(-\alpha)$ *is given by* (23).

Example 1 Let us consider system (19) with two state variables, i.e, $n = 2$, single input $u(t) = H(t)$, where $H(t)$ denotes a Heaviside step function, i.e., $m = 1$, two outputs, i.e., $p = 2$, and the matrices

$$A(t) = \begin{pmatrix} 0 & \lambda_1(t) \\ -\lambda_2(t) & -2\lambda_2(t) \end{pmatrix}, \quad B(t) = \begin{pmatrix} \lambda_1(t) \\ 10\lambda_2(t) \end{pmatrix},$$

$$C(t) = \begin{pmatrix} 1 & 0 \\ 0 & 1 \end{pmatrix}, \quad D(t) = \begin{pmatrix} 1 \\ 0 \end{pmatrix},$$

where

$$\lambda_1(t) = \begin{cases} \lambda_1^1 & \text{for } t \in [0, 1) \\ \lambda_1^2 & \text{for } t \in [1, 3]. \end{cases}, \quad \lambda_2(t) = \begin{cases} \lambda_2^1 & \text{for } t \in [0, 1) \\ \lambda_2^2 & \text{for } t \in [1, 3]. \end{cases}$$

The types of variable order derivatives are given by the 2-tuple $\mathbb{T} = (\mathscr{D}, \mathscr{B})$, and then ${}_0^{\mathbb{T}}\mathrm{D}_t^{\alpha(t)} x = \left({}_0^{\mathscr{D}}\mathrm{D}_t^{\alpha_1(t)} x_1(t), {}_0^{\mathscr{B}}\mathrm{D}_t^{\alpha_2(t)} x_2(t)\right)^T$. The piece-wise constant variable orders $\alpha_i(t)$, $i = 1, 2$, are defined on two time-intervals ($N = 2$), that is

$$\alpha_1(t) = \begin{cases} \alpha_1^1 & \text{for } t \in [0, 1) \\ \alpha_1^2 & \text{for } t \in [1, 3]. \end{cases}, \quad \alpha_2(t) = \begin{cases} \alpha_2^1 & \text{for } t \in [0, 1) \\ \alpha_2^2 & \text{for } t \in [1, 3]. \end{cases}$$

The solution of system (19) will be calculated from (22) and (20b) and is depicted in Fig. 10.

3.2 Including Initial Conditions

The initial conditions problem of fractional order systems is an important issue especially in real plant applications due to "long memory effect" of fractional order operators. In [12, 14] the difference between initial conditions for Riemann-Liouville and Caputo definitions are underlined and can not be taken into account in the same way. In [30] an interpretation of initial conditions based on the internal state of the frequency distributed fractional integrator model is given. Article [8] presents physical interpretation of the initial conditions for different practical examples of fractional differential equations with Riemann-Liouville fractional derivatives. The initialization issue of linear fractional differential equations has been studied in [42].

In order to consider the non-zero initial conditions problem in variable order dynamics, the variable order operators have to be reconsidered. In [39], the recursive $\mathscr{D}$-type operator is considered.

Theorem 2 *Fractional variable order $\mathscr{D}$-type derivative for initial conditions in the form of constant function* ${}_{-\infty}^{\mathscr{D}}\mathrm{D}_t^{\alpha(t)} x(t) = c = \text{const}$ *for* $t = (-\infty, 0)$ *is given by*

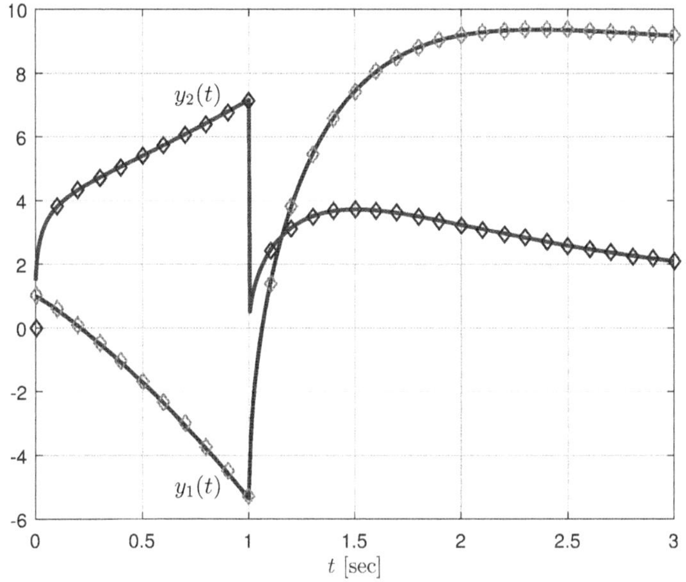

Fig. 10 Numerical solution (*solid lines*) of system from Example 1, compared with simulation results (*diamond lines*), for $\mathbb{T} = (\mathscr{D}, \mathscr{B})$, $\alpha_1(t) \neq \alpha_2(t)$, i.e., $\alpha_1^1 = 1$, $\alpha_1^2 = 0.25$, and $\alpha_2^1 = 0.5$, $\alpha_2^2 = 1$; and $\lambda_1^1 = -1$, $\lambda_1^2 = 2$, and $\lambda_2^1 = 3$, $\lambda_2^2 = 0.5$; for $h = 0.005$

the following relation

$$ {}_{-\infty}{}^{\mathscr{D}}\mathrm{D}_t^{\alpha(t)} x(t) = \lim_{h\to 0} \left[\frac{x(t)}{h^{\alpha(t)}} - \sum_{j=1}^{n} (-1)^j \binom{-\alpha(t)}{j} \left({}_{-\infty}{}^{\mathscr{D}}\mathrm{D}_{t-jh}^{\alpha(t)} x(t) - c \right) + c \right]. \tag{25} $$

The initial conditions are assumed to be in the form of a time-constant function.

In a matrix form, incorporating initial conditions is as follows.

Theorem 3 *Fractional order derivative given by* (25) *for initial conditions in the form of constant function is equivalent to the following matrix form*

$$ \begin{pmatrix} {}_{-\infty}{}^{\mathscr{D}}\mathrm{D}_h^{\alpha(h)} x(h) \\ {}_{-\infty}{}^{\mathscr{D}}\mathrm{D}_{2h}^{\alpha(2h)} x(2h) \\ \vdots \\ {}_{-\infty}{}^{\mathscr{D}}\mathrm{D}_{kh}^{\alpha(kh)} x(kh) \end{pmatrix} = \lim_{h\to 0} \mathfrak{Q}_1^k \begin{pmatrix} x(h) \\ x(2h) \\ \vdots \\ x(kh) \end{pmatrix} + C, \tag{26} $$

where $C = (c, \ldots, c)^T \in \mathbb{R}^{k\times 1}$,

$$ \mathfrak{Q}_1^k = \mathfrak{Q}(\alpha_k, k) \cdots \mathfrak{Q}(\alpha_1, 1), \tag{27} $$

and for $r = 1, \ldots, k$

$$\mathfrak{Q}(\alpha_r, r) = \left(\begin{array}{c|c|c} I_{r-1,r-1} & 0_{r-1,1} & 0_{r-1,k-r} \\ \hline q_r & h^{-\alpha_r} & 0_{1,k-r} \\ \hline 0_{k-r,r-1} & 0_{k-r,1} & I_{k-r,k-r} \end{array}\right) \in \mathbb{R}^{k \times k}, \tag{28}$$

where

$$q_r = (-\upsilon_{-\alpha_r,r-1}, -\upsilon_{-\alpha_r,r-2}, \ldots, -\upsilon_{-\alpha_r,1}) \in \mathbb{R}^{1 \times r-1},$$

and $\upsilon_{-\alpha_r,i} = (-1)^i \binom{-\alpha_r}{i}$, *for* $i = 1, \ldots, r-1$, *i.e., the* j*th element of* q_r *is*

$$(q_r)_j = -\upsilon_{-\alpha_r,r-j} = (-1)^{r-j} \binom{-\alpha_r}{r-j}, \quad j = 1, \ldots, r-1.$$

The numerical scheme for solving fractional-order differential equations, given in the matrix form, is depicted in the following example.

Example 2 Let us consider the variable order inertial system

$$y(t) = {}_{-\infty}{}^{\mathscr{D}}\mathrm{D}_t^{\alpha(t)} \left[a(t)(-y(t))\right], \quad y_0 = 0.2, \tag{29}$$

where (for the switching time $T = 0.7$ s.)

$$a(t) = \begin{cases} 20 & \text{for } t \leq 0.7, \\ 1.55 & \text{for } t > 0.7, \end{cases} \quad \alpha(t) = \begin{cases} -0.5 & \text{for } t \leq 0.7, \\ -0.25 & \text{for } t > 0.7. \end{cases} \tag{30}$$

Discrete approximation of (29) in matrix form, according to Theorem 3, is

$$y = (I_{k,k} + \mathfrak{Q}_1^k \Lambda)^{-1} C,$$

where $y = (y_1, \ldots, y_k)^T$, $y_0 = 0.2$, $C = (0.2, \ldots, 0.2)^T$, matrix $\mathfrak{Q}_1^k$ is calculated for variable order $\alpha(t)$ given by (30), and the matrix $\Lambda \in \mathbb{R}^{k \times k}$ of gain coefficients is

$$\Lambda = \mathrm{diag}\{\underbrace{20, \ldots, 20}_{\rho_1\text{-times}}, \underbrace{1.55, \ldots, 1.55}_{\rho_2\text{-times}}\}, \tag{31}$$

where $\rho_1 = 0.7/h$, $\rho_2 = 1.3/h$, such that $\rho_1 + \rho_2 = k$ (Fig. 11).

3.3 Analytical Solutions of Variable Order Control Systems

Methods for analytical solving of fractional variable order linear differential equations for different types of variable order derivative definitions (e.g. $\mathscr{A}$-type, $\mathscr{B}$-type, $\mathscr{D}$-type, $\mathscr{E}$-type) are established in [17].

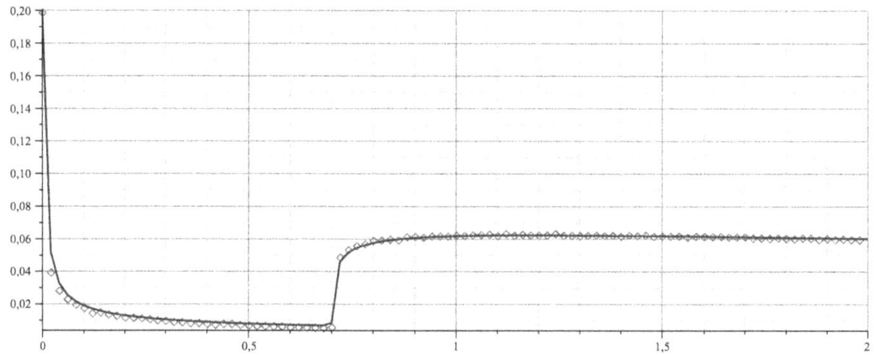

Fig. 11 Results of analog (*diamonds*) and numerical implementation (*continuous line*) of fractional variable order inertial system with initial condition 0.2 V; order switching from −0.5 to −0.25 at $t = 0.7$ s

Let us define two 4-tuples $\mathscr{T} = (\mathscr{D}, \mathscr{E}, \mathscr{A}, \mathscr{B})$ and $\tilde{\mathscr{T}} = (\mathscr{A}, \mathscr{B}, \mathscr{D}, \mathscr{E})$. Consider the following fractional $\mathscr{T}_\ell$-type, $\ell \in \{1, \ldots, 4\}$, variable order differential equation (shortly $\mathscr{T}_\ell$-differential equation)

$$ {}^{\mathscr{T}_\ell}_{0}\mathrm{D}^{\alpha(t)}_t x = \lambda(t)\left(a(t)\gamma(t)x + u\right), \quad x(0) = 0, \quad \ell \in \{1, \ldots, 4\}, \tag{32a}$$

$$ y(t) = \gamma(t)x(t), \tag{32b}$$

where $x = x(t)$ is the real valued unknown function, $u = u(t)$ is a real valued known function, and the following piece-wise constant functions, all defined on $t_i \le t < t_{i+1}$, for $i = 0, \ldots, N$, where $t_0 = 0$:

$$\alpha(t) = \alpha_i \in \mathbb{R}_+, \quad \lambda(t) = \lambda_i \in \mathbb{R}, \quad a(t) = a_i \in \mathbb{R}, \quad \gamma(t) = \gamma_i \in \mathbb{R}. \tag{33}$$

Using the duality property given in (18), i.e., ${}^{\tilde{\mathscr{T}}_\ell}_{0}\mathrm{D}^{-\alpha(t)}_t \left({}^{\mathscr{T}_\ell}_{0}\mathrm{D}^{\alpha(t)}_t x\right) = x$, applied to (32), we get

$$ y(t) = \gamma(t)\,{}^{\tilde{\mathscr{T}}_\ell}_{0}\mathrm{D}^{-\alpha(t)}_t \left(\lambda(t)(a(t)\gamma(t)x + u)\right), \quad \ell \in \{1, \ldots, 4\}. \tag{34}$$

Realization of (32), and thereby of (34), is depicted in Fig. 12, where the integration block ${}^{\tilde{\mathscr{T}}_\ell}_{0}\mathrm{D}^{-\alpha(t)}_t$ is replaced by switching scheme ${}^{\mathscr{X}_\ell}_{\mathscr{F}_j}/\{-\alpha(t)\}$, where the 4-tuple $\mathscr{X} =$ (o-r, i-a, i-r, o-a) stands for a type of switching scheme.
Below, only for $\mathscr{A}$-type differential equations the solution algorithm is presented.

We will use the following objects (for $i = 1, \ldots, N$, $j = 1, \ldots, N-1$):

$$ G_i(s) = \frac{1}{s^{\alpha_i} - a_i\lambda_i\gamma_i}, \quad \hat{G}_j(s) = \frac{1}{s^{\hat{\alpha}_j}} \tag{35}$$

and

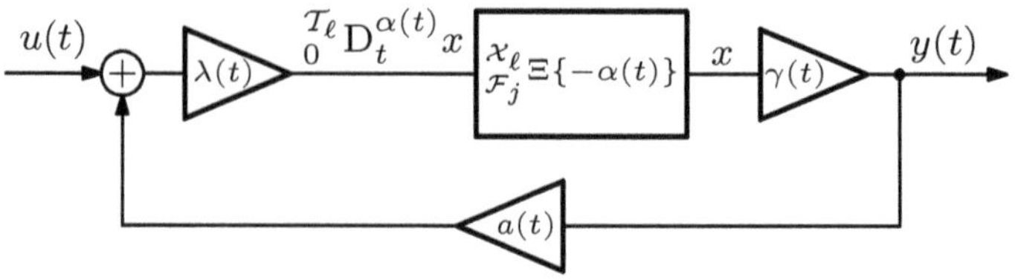

Fig. 12 Realization of $\mathscr{T}_\ell$-type differential equation (32)

$$w_i = H(t - t_{i-1}) - H(t - t_i), \quad \bar{w}_i = H(t) - H(t - t_i),$$

where s is a Laplace operator, $H(t)$ denotes Heaviside step function, and for $i = 1$ we have $w_1 = \bar{w}_1$.

Therefore, we have the following result.

Theorem 4 *Solution of $\mathscr{A}$-type differential equation (32) (for $\mathscr{T}_\ell = \mathscr{A}$) is the following*

$$y(t) = \sum_{i=1}^{N} \gamma_i y_i w_i, \tag{36}$$

where

$$\begin{aligned}
y_i(t) &= \mathscr{L}^{-1}\{G_i \mathscr{L}\{u_i(t)\}\}, \quad i = 1, \ldots, N, \\
u_i(t) &= u^{i-1}(t)\bar{w}_{i-1}(t) + \tilde{u}(t)w_i(t), \quad u_1 = \tilde{u}(t)w_i(t), \\
u^{i-1}(t) &= \hat{y}_{i-1}(t) - a_i\lambda_i\gamma_i y_{i-1}(t) \\
\hat{y}_{i-1}(t) &= \mathscr{L}^{-1}\{\hat{G}_{i-1}\mathscr{L}\{e_{i-1}(t)\bar{w}_{i-1}(t)\}\} \\
e_{i-1}(t) &= u_{i-1}(t) + a_{i-1}\lambda_{i-1}\gamma_{i-1}y_{i-1}(t),
\end{aligned}$$

where $\tilde{u}(t) = \lambda(t)u(t)$, and $\mathscr{L}^{-1}$ stands for inverse Laplace transform.

Example 3 Consider system (32), where $\mathscr{T}_\ell = \mathscr{A}$, $a(t) = 0$, $\lambda(t) = 1$, $\gamma(t) = 1$, and

$$\alpha(t) = \begin{cases} \frac{1}{2} & \text{for } t \in [0, 1), \\ 1 & \text{for } t \in [1, 2), \end{cases} \quad u(t) = \begin{cases} 1 & \text{for } t \in [0, 1), \\ 1 & \text{for } t \in [1, 2). \end{cases}$$

The solution, given by (36), is

$$y(t) = \begin{cases} \frac{2\sqrt{t}}{\sqrt{\pi}}, & \text{for } t \in [0, 1), \\ \frac{2}{\sqrt{\pi}} + t - 1, & \text{for } t \in [1, 2). \end{cases}$$

The presented methods providing analytical solutions, may lead, in general, to non-trivial symbolical computations. These methods can be helpful, e.g., in anal-

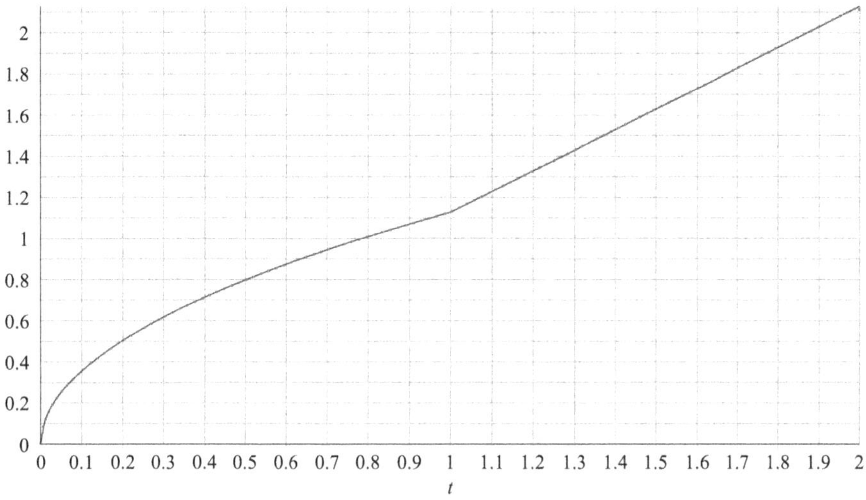

Fig. 13 Solution plots (*solid line*—analytical, *dotted line*—numerical (for step size 0.002)) of $\mathscr{A}$-type equation from Example 3

ysis of properties of variable order systems and, especially, to validate numerical algorithms for solving these types of equations (Fig. 13).

The solution methods of other types of differential equations are given in [17].

4 Application of Variable Order Calculus

4.1 Anti-Windup Algorithm

Nowadays, in steering processes, PID controllers are commonly used. Due to the many limitations in control signal values, modified and additional structures of PID controllers have to be applied. The most popular methods for improving the control process is using so-called anti-windup algorithms. The idea is to oppose increasing integration signal in the presence of saturated control signal. There exists many types of anti-windup algorithms, with back-calculation and reset integration the most popular among them. Fractional calculus has been recently intensively used for creating anti-windup methods [9, 25–28].

In the proposed method, presented in Fig. 14, the integrator's action is governed by changing (switching) the order of integration. We can distinguish precisely between two situations: if a control signal is not saturated, the integrator possesses some nominal order value (integer or fractional); otherwise, if the control signal achieves saturation limit, the order of integration switches to the zero value, serving in consequence as a gain action. It has been verified that only in the case of $\mathscr{D}$-type integrator are

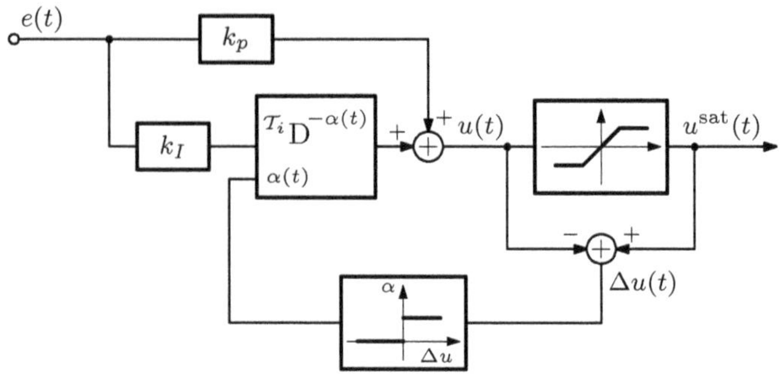

Fig. 14 Variable order anti-windup scheme

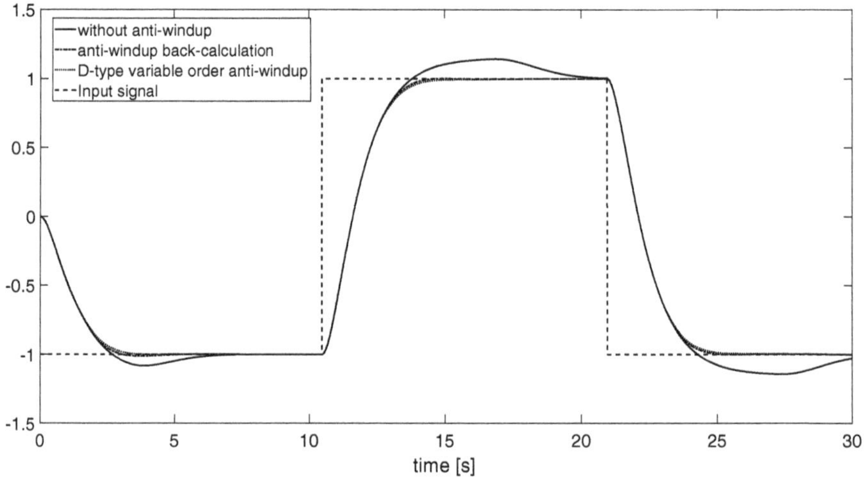

Fig. 15 Results for anti-windup fractional variable order $\mathscr{D}$-type and integer order controller

the results satisfactory, comparable with the classical back-calculation anti-windup algorithm.

In Fig. 15, results for $\mathscr{D}$-type derivative are presented, and as it can be noticed, this version of algorithm produces very similar results to the traditional anti-windup method. Zero steady error is obtained as well as similar dynamic behavior.

What is more, it seems to yield even better results than is seen in the control signal depicted in Fig. 16. The proposed method has the advantage over the classic one in that it does not require any parameters to be adjusted as it is in the back-calculation method.

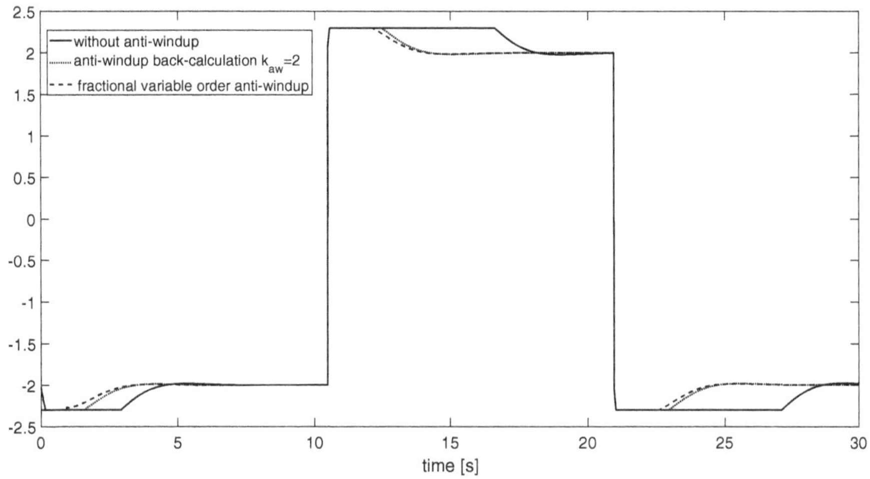

Fig. 16 Comparison of control signal $u^{\text{sat}}(t)$ without anti-windup, with anti-windup for $k_{\text{aw}} = 2$, and with proposed method

4.2 *Modelling of Heat Transfer Process*

Diffusion processes are common examples of real-world fractional order model applications [10, 19, 34, 40, 41]. In particular, studies have shown that such models can provide excellent approximation in the case of heat diffusion in heterogeneous media with complex structure. Fractional order models become especially useful in cases when traditional models fail, i.e. when the medium is porous or fractal-like [31].

The physical process of heat diffusion is described by the heat equation:

$$\frac{\partial T(x, y, t)}{\partial t} - \lambda \nabla^2 T(x, y, t) = Q(x, y, t), \tag{37}$$

where $T(x, y, t)$ is temperature at time t and point (x, y), λ is thermal diffusivity of considered material, and $Q(x, y, t)$ is a function describing heat loss.

Because the analytical description of the heat transfer process in heterogeneous medium can be extremely complicated, let us approximate the heterogeneous medium heat transfer process with a homogeneous one replacing the normal heat equation with the fractional order partial differential anomalous diffusion equation

$$\frac{\partial^\alpha T(x, y, t)}{\partial t^\alpha} - \lambda_\alpha \nabla^2 T(x, y, t) = Q(x, y, t), \tag{38}$$

where α is an order of anomalous diffusion, and λ_α is thermal diffusivity for anomalous diffusion model. Coefficient λ_α can differ from λ used in Eq. 37 describing the same problem, but in the classical way.

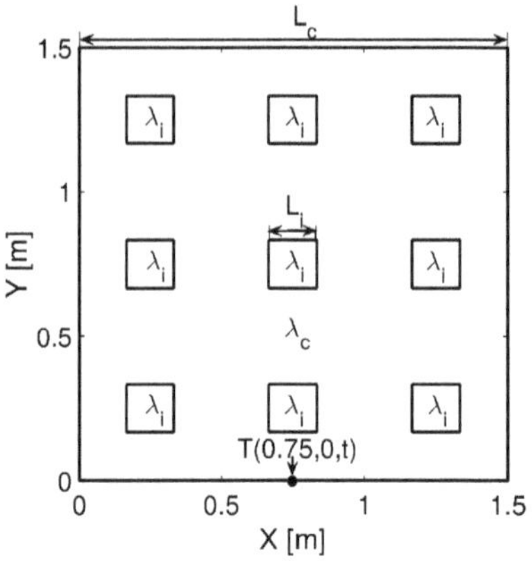

Fig. 17 Structure of investigated dynamically changing medium

For the purpose of this paper a basic case will be considered, i.e. a heated object is perfectly insulated ($Q(x, y, t) = 0$) and the experiment can be locally reduced to a one-dimensional problem (x can be omitted), so Eq. 38 is reduced to:

$$\frac{\partial^\alpha T(y,t)}{\partial t^\alpha} - \lambda_\alpha \frac{\partial^2 T(y,t)}{\partial y^2} = 0. \tag{39}$$

Knowing that heat flux is given by the equation

$$\Phi_H(y,t) = -\lambda \frac{\partial T(y,t)}{\partial y}$$

one can solve Eq. 39, eventually deriving relation between temperature and heat flux [34]:

$$T(y,t) = \frac{1}{\sqrt{\lambda_\alpha}} \frac{\partial^{-\alpha/2} \Phi_H(y,t)}{\partial t^{-\alpha/2}}. \tag{40}$$

Let us now consider the structures presented in Fig. 17. They are clearly heterogeneous, consisting of both a highly conductive external part and multiple insulated areas characterized with low thermal conductivity. In the experiment, the whole bottom edge ($y = 0$) of the plate is heated uniformly with a constant heat flux and so the problem can be approximated by a one-dimensional case. The output of the system is a temperature at the heated edge. Considered structures are characterized by μ parameter.

$$\mu = \frac{9L_i}{L_c} \tag{41}$$

This parameter linearly describes the dimension of insulated regions where $\mu = 3$ means that the whole plate is made of insulator and $\mu = 0$ means that there is no insulator at all. Values of μ corresponds with α order of a system. Results of applying

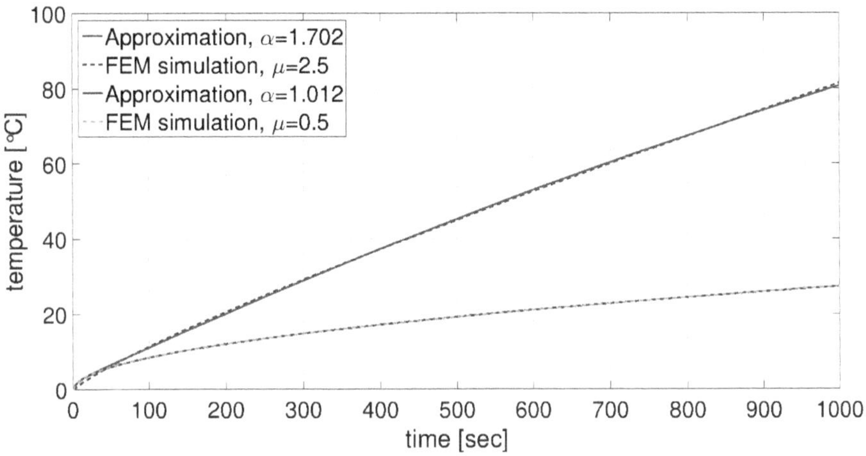

Fig. 18 Results of invariant order fractional order heat transfer model

Eq. 40 to the aforementioned structures in comparison with the finite element method simulations are shown in Fig. 18.

It is clear that for invariant structures the fractional order approximation is very accurate. However one can consider the problem where the structure changes over time during the heating process, e.g. in the plate with $\mu = 2.5$, the insulated regions diminish after time t_v transforming a structure into a plate with $\mu = 0.5$. In such a case α and λ_α are given by the functions of time:

$$\lambda_\alpha(t) = \begin{cases} 9.63 \cdot 10^{-4} & t < t_v \\ 2.41 \cdot 10^{-4} & t \geqslant t_v \end{cases},$$

$$\alpha(t) = \begin{cases} 1.702 & t < t_v \\ 1.012 & t \geqslant t_v \end{cases}.$$

The crucial decision is to provide a proper variable order definition. Three definition are considered: $\mathscr{B}$-, $\mathscr{D}$-, and $\mathscr{E}$-type:

$$T(0,t) = {}_0^{\mathscr{B}}\mathrm{D}_t^{-\alpha(t)/2} \frac{1}{\sqrt{\lambda_{\alpha(t)}}} \Phi_H(t) ,$$

$$T(0,t) = {}_0^{\mathscr{D}}\mathrm{D}_t^{-\alpha(t)/2} \frac{1}{\sqrt{\lambda_{\alpha(t)}}} \Phi_H(t) ,$$

$$T(0,t) = {}_0^{\mathscr{E}}\mathrm{D}_t^{-\alpha(t)/2} \frac{1}{\sqrt{\lambda_{\alpha(t)}}} \Phi_H(t) .$$

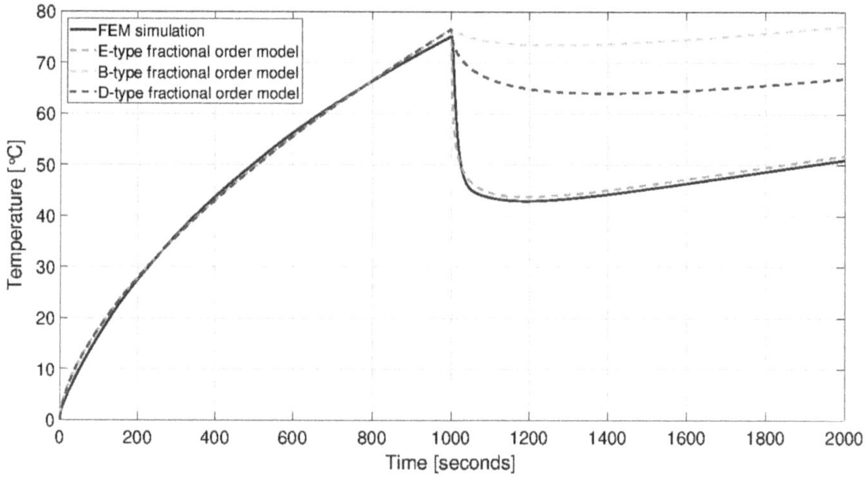

Fig. 19 Comparison between finite element method model and various types of variable order models

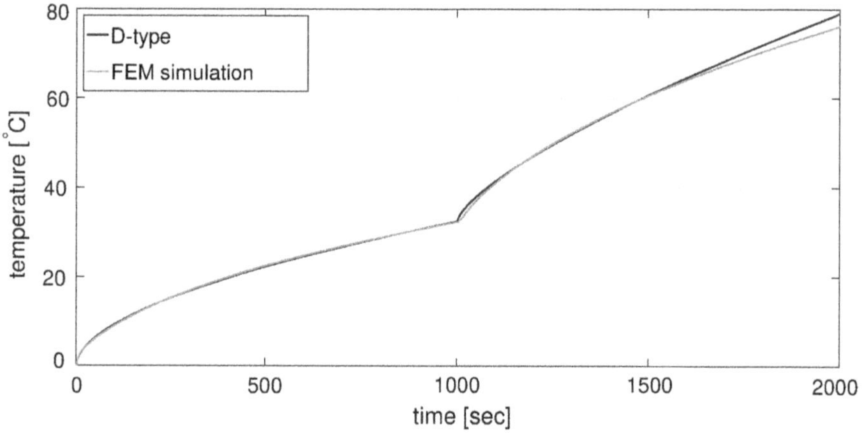

Fig. 20 Comparison of the results of the numerical FEM simulation and $\mathscr{D}$-type variable order model

Results of variable order models compared to FEM simulation are shown in Fig. 19. One can notice the effect of the fast drop in temperature at the moment of switch caused by a sudden rise of conductivity. The most interesting thing however, is that such a complex behaviour can be modeled so accurately by one of the variable order definitions.

On the other hand, in the situation where after the moment of switch insulated regions are increased, other definitions are more accurate. Obtained results are presented in Fig. 20 and as we can see the $\mathscr{D}$-type of definition for this manner of changing structure is the most appropriate.

5 Conclusions

In the chapter an alternative, intuitive description of variable order operators in the form of equivalent serial switching schemes has been presented. According to this schematic interpretation of variable order operators, analysis of variable order systems can be simpler and more effective than those based on pure analytical definitions. Thanks to this schematic description and the duality property between chosen variable order operators, analytical solutions of variable order control systems can be effectively derived. In turn, through matrix representation of variable orders operators, which is the other important achievement, numerical solutions of variable order control systems can be calculated. Moreover, the possibility of including initial conditions in fractional variable order operators allows obtaining of solutions of non-zero initial problems.

A variety of considered types of variable order operators allows the possibility to construct alternative control algorithms—such as anti-windup—considerably effective and comparable (and even better in some aspects) to integer order control methods.

A very interesting application of variable fractional order operators is the modelling of heat transfer in a variable time structure. As was presented, different types of variable order definitions can very accurately describe dynamics for different methods of structure changes. In the case where the holes in a structure increase their size, definition type $\mathscr{D}$ was the most accurate. More importantly, other definitions, in the same case give much worse results. On the other hand, in the case where the holes decrease in size, the definition $\mathscr{E}$ was the most accurate. This clearly suggests that considering many types of variable fractional order operators is very necessary. Moreover, thanks to the richness of definitions we are able to describe many patterns for order changing.

References

1. Dabiri, A., Moghaddam, B.P., Tenreiro Machado, J.A.: Optimal variable-order fractional PID controllers for dynamical systems. J. Comput. Appl. Math. (2018)
2. Dzieliński, A., Sarwas, G., Sierociuk, D.: Time domain validation of ultracapacitor fractional order model. In: 2010 49th IEEE Conference on Decision and Control (CDC), pp. 3730–3735 (2010)
3. Dzieliński, A., Sierociuk, D.: Fractional order model of beam heating process and its experimental verification. In: Baleanu, D., Guvenc, Z.B., Machado, J.A.T. (eds.) New Trends in Nanotechnology and Fractional Calculus Applications, pp. 287–294. Springer, Netherlands (2010)
4. Dzieliński, A., Sarwas, G., Sierociuk, D.: Comparison and validation of integer and fractional order ultracapacitor models. Adv. Differ. Equ. **2011**, 11 (2011)
5. Garrappa, R., Mainardi, F., Maione, G.: Models of dielectric relaxation based on completely monotone functions. Fract. Calc. Appl. Anal. **19**(5), 1105–1160 (2016)
6. Giusti, A.: On infinite order differential operators in fractional viscoelasticity. Fract. Calc. Appl. Anal. **20**(4), 854–867 (2017)

7. Harris, P.A., Garra, R.: Nonlinear heat conduction equations with memory: physical meaning and analytical results. J. Math. Phys. **58**(6), 063501 (2017)
8. Heymans, Nicole, Podlubny, Igor: Physical interpretation of initial conditions for fractional differential equations with Riemann-Liouville fractional derivatives. Rheol. Acta **45**(5), 765–771 (2006). Jun
9. Ikeda, F., Toyama, R., Toyama, S.: Anti-windup controller design by fractional calculus for linear control systems with input saturation. In: 2009 ICCAS-SICE, pp. 3317–3320 (2009)
10. Kosztolowicz, T.: Subdiffusion in a system with a thick membrane. J. Membr. Sci. **320**(1–2), 492–499 (2008)
11. Li, Z., Wang, H., Xiao, R., Yang, S.: A variable-order fractional differential equation model of shape memory polymers. Chaos, Solitons Fractals **102**, 473–485 (2017). Future Directions in Fractional Calculus Research and Applications
12. Lorenzo, C.F., Hartley, T.T.: Initialization in fractional order systems. In: The European Control Conference, Porto, Portugal, pp. 1471–1476 (2001)
13. Lorenzo, C.F., Hartley, T.T.: Variable order and distributed order fractional operators. Nonlinear Dyn. **29**(1–4), 57–98 (2002)
14. Lorenzo, C.F., Hartley, T.T.: Initialization of fractional differential equations: theory and application. In: The ASME 2007 International Design Engineering Technical Conferences, DETC2007-34814 Las Vegas, USA (2007)
15. Macias, M., Sierociuk, D.: An alternative recursive fractional variable-order derivative definition and its analog validation. In: Proceedings of International Conference on Fractional Differentiation and Its Applications, Catania, Italy (2014)
16. Malesza, W., Macias, M.: Numerical solution of fractional variable order linear control system in state-space form. Bull. Pol. Acad. Sci.: Tech. Sci. **65**(5) (Special Section on Multilevel Converters), 715–724 (2017)
17. Malesza, W., Macias, M., Sierociuk, D.: Analytical solution of fractional variable order differential equations. J. Comput. Appl. Math. **348**, 214–236 (2019)
18. Malesza, W., Macias, M., Sierociuk, D.: Matrix approach and analog modeling for solving fractional variable order differential equations. In: Latawiec, K.J., Łukaniszyn, M., Stanisławski, R. (eds.) Advances in Modelling and Control of Non-integer-Order Systems, pp. 71–80. Springer International Publishing, Cham (2015)
19. Metzler, Ralf, Klafter, Joseph: The random walk's guide to anomalous diffusion: a fractional dynamics approach. Phys. Rep. **339**(1), 1–77 (2000)
20. Miller, K.S., Ross, B.: An Introduction to the Fractional Calculus and Fractional Differential Equations. Wiley, New York (1993)
21. Monje, C.A., Chen, Y., Vinagre, B.M., Xue, D., Feliu, V.: Fractional-Order Systems and Controls. Springer (2010)
22. Oldham, K.B., Spanier, J.: The Fractional Calculus. Academic Press (1974)
23. Ortigueira, M.D., Tenreiro Machado, J.A.: What is a fractional derivative? J. Comput. Phys. **293**, 4–13 (2015). Fractional PDEs
24. Ostalczyk, P., Rybicki, T.: Variable-fractional-order dead-beat control of an electromagnetic servo. J. Vib. Control (2008)
25. Padula, F., Visioli, A., Pagnoni, M.: On the anti-windup schemes for fractional-order PID controllers. In: Proceedings of 2012 IEEE 17th International Conference on Emerging Technologies Factory Automation (ETFA 2012), pp. 1–4 (2012)
26. Pandey, S., Soni, N.K., Pandey, R.K.: Fractional order integral and derivative (FOID) controller with anti-windup for temperature profile control. In: 2015 2nd International Conference on Computing for Sustainable Global Development (INDIACom), pp. 1567–1573 (2015)
27. Pandey, S., Dwivedi, P., Junghare, A.: Anti-windup fractional order PI^{λ}-PD^{μ} controller design for unstable process: a magnetic levitation study case under actuator saturation. Arab. J. Sci. Eng. (2017)
28. Pandey, S., Dwivedi, P., Junghare, A.S.: A novel 2-DOF fractional-order PI^{α}-D^{μ} controller with inherent anti-windup capability for a magnetic levitation system. AEU-Int. J. Electron. Commun. **79**, 158–171 (2017)

29. Podlubny, I.: Fractional Differential Equations. Academic Press (1999)
30. Sabatier, Jocelyn, Merveillaut, Mathieu, Malti, Rachid, Oustaloup, Alain: How to impose physically coherent initial conditions to a fractional system? Commun. Nonlinear Sci. Numer. Simul. **15**(5), 1318–1326 (2010)
31. Sakrajda, P., Sierociuk, D.: Modeling heat transfer process in grid-holes structure changed in time using fractional variable order calculus. In: Babiarz, A., Czornik, A., Klamka, J., Niezabitowski, M. (eds.) Theory and Applications of Non-integer Order Systems, pp. 297–306. Springer International Publishing, Cham (2017)
32. Samko, S.G., Kilbas, A.A., Maritchev, O.I.: Fractional Integrals and Derivative. Theory and Applications. Gordon & Breach Science Publishers (1987)
33. Sheng, H., Sun, H., Coopmans, C., Chen, Y., Bohannan, G.W.: Physical experimental study of variable-order fractional integrator and differentiator. In: Proceedings of the 4th IFAC Workshop Fractional Differentiation and its Applications FDA'10 (2010)
34. Sierociuk, D., Dzieliński, A., Sarwas, G., Petras, I., Podlubny, I., Skovranek, T.: Modelling heat transfer in heterogeneous media using fractional calculus. Philos. Trans. R. Soc. A: Math. Phys. Eng. Sci. **371**, 2013 (1990)
35. Sierociuk, D., Sarwas, G., Twardy, M.: Resonance phenomena in circuits with ultracapacitors. In: Proceedings of 12th International Conference on Environment and Electrical Engineering (EEEIC), pp. 197–202 (2013)
36. Sierociuk, D., Twardy, M.: Duality of variable fractional order difference operators and its application in identification. Bull. Pol. Acad. Sci. Tech. Sci. **62**(4), 809–815 (2014)
37. Sierociuk, Dominik, Malesza, Wiktor, Macias, Michal: Derivation, interpretation, and analog modelling of fractional variable order derivative definition. Appl. Math. Model. **39**(13), 3876–3888 (2015). https://doi.org/10.1016/j.apm.2014.12.009
38. Sierociuk, Dominik, Malesza, Wiktor, Macias, Michal: On the recursive fractional variable-order derivative: equivalent switching strategy, duality, and analog modeling. Circuits Syst. Signal Process. **34**(4), 1077–1113 (2015). Apr
39. Sierociuk, Dominik, Malesza, Wiktor, Macias, Michal: Numerical schemes for initialized constant and variable fractional-order derivatives: matrix approach and its analog verification. J. Vib. Control **22**(8), 2032–2044 (2016)
40. Sierociuk, Dominik, Skovranek, Tomas, Macias, Michal, Podlubny, Igor, Petras, Ivo, Dzielinski, Andrzej, Ziubinski, Pawel: Diffusion process modeling by using fractional-order models. Appl. Math. Comput. **257**, 2–11 (2015)
41. Sun, HongGuang, Zhang, Yong, Baleanu, Dumitru, Chen, Wen, Chen, YangQuan: A new collection of real world applications of fractional calculus in science and engineering. Commun. Nonlinear Sci. Numer. Simul. **64**, 213–231 (2018)
42. Trigeassou, J.C., Maamri, N.: Initial conditions and initialization of linear fractional differential equations. Signal Process. **91**(3), 427–436 (2011). Advances in Fractional Signals and Systems
43. Valerio, D., Sa da Costa, J.: Variable-order fractional derivatives and their numerical approximations. Signal Process. **91**(3, SI), 470–483 (2011)
44. Yin, Deshun, Pengfei, Qu.: Variable-order fractional MSD function to describe the evolution of protein lateral diffusion ability in cell membranes. Phys. A: Stat. Mech. Appl. **492**, 707–714 (2018)

Asymptotic Behavior of Discrete Fractional Systems

Adam Czornik, Pham The Anh, Artur Babiarz, and Stefan Siegmund

Abstract This work comprehensively describes and extends the results on asymptotic properties of linear discrete time-varying fractional order systems with Caputo and Riemann-Liouville forward and backward difference operators. In our considerations we take into account various definitions from the literature of fractional difference operators and we compare the dynamic properties of the corresponding systems. These equations are studied by converting them to the corresponding Volterra convolution equations. The main results are: explicit formulas for solutions, results on asymptotic stability, rates of growth or decay of solutions and solution separation. The work also formulates a number of open questions that may be the subject of future research.

1 Introduction

Continuous fractional calculus has a long history and is nearly as old as the integer-order calculus. Nowadays fractional calculus is studied both for its theoretical interest as well as its use in applications. In spite of the existence of a substantial mathematical theory of continuous fractional calculus, there was no similar development of discrete fractional calculus until very recently [16]. Over the past decade, there has been an

A. Czornik (✉) · A. Babiarz
Department of Automatic Control and Robotics, Silesian University of Technology, Akademicka 16, 44-100 Gliwice, Poland
e-mail: adam.czornik@polsl.pl

A. Babiarz
e-mail: artur.babiarz@polsl.pl

P. T. Anh
Department of Mathematics, Le Quy Don Technical University, 236 Hoang Quoc Viet, Hanoi, Vietnam

S. Siegmund
Faculty of Mathematics, Technische Universität Dresden, 01062 Dresden, Germany
e-mail: stefan.siegmund@tu-dresden.de

P. Kulczycki et al. (eds.), *Fractional Dynamical Systems: Methods, Algorithms and Applications*, Studies in Systems, Decision and Control 402,
https://doi.org/10.1007/978-3-030-89972-1_5

increased interest in developing discrete fractional calculus and dynamical models described by discrete fractional calculus (see [17, 23] and the reference therein).

In this work we investigate discrete linear fractional systems with variable coefficients. We consider forward and backward equations with Caputo and Riemann-Liouville operators. For backward equations, we distinguish two types of Caputo operators and two types of Riemann-Liouville operators depending on whether the sum of the fractional order that appears in the definition of the operator includes the initial condition or not. The first works on backward operators were based on the definition of the sum containing the initial condition (see [18]), however later, the operator based on the definition of the sum without the initial condition was introduced, justifying it by the greater similarity of such operators to the case with continuous time (see [15]).

The main aim of this work is to collect the existing results on asymptotic properties of the considered equations, their extension, supplementation, and comparison. The basic research method used in this work is to transform the fractional order equations into the appropriate convolution-type Volterra equations. The work is organized as follows: The next section is devoted to fractional-order differences and the relationships between them. In Sect. 3, we define the equations considered in the work, as well as the initial value problem and the existence and uniqueness of its solution. In Sect. 4 we present for each of the fractional order equations considered, two Volterra equations that are equivalent to them. Section 5 is devoted to multidimensional systems with constant coefficients. The main results of this section are explicit formulas for solutions and conditions for stability. In Sect. 6 we present results on the stability and the rate of growth or decay for one-dimensional equations. The results of separation of solutions of the Volterra equations are discussed in Sect. 7. Finally, Sect. 8 contains conclusions, summaries and directions for further research. In the remainder of this section, we introduce notation and definitions of fractional sums.

Denote by $\mathbb{R}$ the set of real numbers, by $\mathbb{Z}$ the set of integers, by $\mathbb{N}$ the set $\{0, 1, 2, \dots\}$ of natural numbers including 0, and by $\mathbb{Z}_{\leq 0} := \{0, -1, -2, \dots\}$ the set of non-positive integers. For $a \in \mathbb{R}$ we denote by $\mathbb{N}_a := a + \mathbb{N}$ the set $\{a, a+1, \dots\}$ and for a function $x\colon \mathbb{N}_a \to \mathbb{R}^d$ we define $\Delta x\colon \mathbb{N}_a \to \mathbb{R}^d$ by $(\Delta x)(t) = x(t+1) - x(t)$ and $\nabla x\colon \mathbb{N}_{a+1} \to \mathbb{R}^d$ by $(\nabla x)(t) = x(t) - x(t-1)$.

The Euler Gamma function $\Gamma\colon \mathbb{R} \setminus \mathbb{Z}_{\leq 0} \to \mathbb{R}$ is defined by

$$\Gamma(\alpha) := \lim_{n\to\infty} \frac{n^\alpha n!}{\alpha(\alpha+1)\cdots(\alpha+n)}$$

For $x \in \mathbb{R}$ we write as usual $\lceil x \rceil := \min\{k \in \mathbb{Z}\colon k \geq x\}$ and $\lfloor x \rfloor := \max\{k \in \mathbb{Z}\colon k \leq x\}$. A reader who is familiar with fractional difference equations may skip the next definition, see e.g. [17, 27].

Definition 1 *(Basic notions of fractional calculus)* Let $s, \nu \in \mathbb{R}$.
(a) *Falling factorial power* $(s)^{(\nu)}$: If $s+1, s+1-\nu \notin \mathbb{Z}_{\leq 0}$

$$(s)^{(\nu)} := \frac{\Gamma(s+1)}{\Gamma(s+1-\nu)}.$$

(b) *Rising factorial power* $(s)^{\overline{(\nu)}}$: If $s,\, s+\nu \notin \mathbb{Z}_{\leq 0}$

$$(s)^{\overline{(\nu)}} := \frac{\Gamma(s+\nu)}{\Gamma(s)}.$$

(c) *Binomial coefficient* $\binom{s}{\nu}$: If $s+1,\, \nu+1,\, s+1-\nu \notin \mathbb{Z}_{\leq 0}$

$$\binom{s}{\nu} := \frac{(s)^{(\nu)}}{\Gamma(\nu+1)} = \frac{\Gamma(s+1)}{\Gamma(\nu+1)\Gamma(s+1-\nu)}.$$

We recall notation for fractional sums (see e.g. [17]) and relate it to various notions from the literature in a remark afterwards.

Definition 2 *(Fractional sum)* Let $a \in \mathbb{R}$, $\nu \in (0,\, 1)$ and $x\colon \mathbb{N}_a \to \mathbb{R}^d$. The function $\Delta_a^{-\nu} x\colon \mathbb{N}_{a+\nu} \to \mathbb{R}^d$ defined by

$$(\Delta_a^{-\nu} x)(t) := \frac{1}{\Gamma(\nu)} \sum_{k=a}^{t-\nu} (t-k-1)^{(\nu-1)} x(k)$$

is called ν*-th fractional sum of* x.

Remark 1 (a) The ν-th fractional sum of $x\colon \mathbb{N}_a \to \mathbb{R}^d$ satisfies

$$\begin{aligned}
(\Delta_a^{-\nu} x)(t) &= \frac{1}{\Gamma(\nu)} \sum_{k=a}^{t-\nu} (t-k-\nu+1)^{\overline{(\nu-1)}} \\
&= \frac{1}{\Gamma(\nu)} \sum_{k=a}^{t-\nu} \frac{\Gamma(t-k)}{\Gamma(t-k-\nu+1)} x(k) \\
&= \sum_{k=a}^{t-\nu} \binom{t-k-1}{t-k-\nu} x(k) \\
&= \sum_{k=a}^{t-\nu} (-1)^{t-k-\nu} \binom{-\nu}{t-k-\nu} x(k), \qquad t \in \mathbb{N}_{a+\nu},
\end{aligned}$$

or equivalently,

$$(\Delta_a^{-\nu} x)(n+a+\nu) = \sum_{k=a}^{n+a} (-1)^{n+a-k} \binom{-\nu}{n+a-k} x(k), \qquad n \in \mathbb{N}_a.$$

(b) Some authors (see e.g. [10, 18]) define the ν-th fractional sum $\overline{\Delta}_a^{-\nu} x\colon \mathbb{N}_a \to \mathbb{R}^d$ of $x\colon \mathbb{N}_a \to \mathbb{R}^d$ as follows

$$
\begin{aligned}
(\overline{\Delta}_a^{-\nu} x)(t) &:= \frac{1}{\Gamma(\nu)} \sum_{k=a}^{t} (t-k+1)^{\overline{(\nu-1)}} x(k) \\
&= \frac{1}{\Gamma(\nu)} \sum_{k=a}^{t} \frac{\Gamma(t-k+\nu)}{\Gamma(t-k+1)} x(k) \\
&= \sum_{k=a}^{t} \binom{t-k+\nu-1}{t-k} x(k) \\
&= \sum_{k=a}^{t} (-1)^{t-k} \binom{-\nu}{t-k} x(k), \qquad t \in \mathbb{N}_a.
\end{aligned}
$$

It is easy to check that

$$
(\Delta_a^{-\nu} x)(t+\nu) = (\overline{\Delta}_a^{-\nu} x)(t), \qquad t \in \mathbb{N}_a. \tag{1}
$$

(c) Some authors (see e.g. [14, 25]) exclude $x(a)$ from the definition of the fractional sum and define the ν-th fractional sum $\widetilde{\Delta}_a^{-\nu} x \colon \mathbb{N}_{a+1} \to \mathbb{R}^d$ of $x \colon \mathbb{N}_a \to \mathbb{R}^d$ as follows

$$
\begin{aligned}
(\widetilde{\Delta}_a^{-\nu} x)(t) &:= \frac{1}{\Gamma(\nu)} \sum_{k=a+1}^{t} (t-k+1)^{\overline{(\nu-1)}} x(k) \\
&= \frac{1}{\Gamma(\nu)} \sum_{k=a+1}^{t} \frac{\Gamma(t-k+\nu)}{\Gamma(t-k+1)} x(k) \\
&= \sum_{k=a+1}^{t} \binom{t-k+\nu-1}{t-k} x(k) \\
&= \sum_{k=a+1}^{t} (-1)^{t-k} \binom{-\nu}{t-k} x(k), \qquad t \in \mathbb{N}_{a+1}.
\end{aligned}
$$

(d) An extensive discussion about relationships between the fractional sums $\overline{\Delta}_a^{-\nu}$ and $\widetilde{\Delta}_a^{-\nu}$ is presented in [2], where it has been shown [2, Lemma 3.1] that

$$
(\overline{\Delta}_{a+1}^{-\nu} \, x|_{\mathbb{N}_{a+1}})(t) = (\widetilde{\Delta}_a^{-\nu} x)(t), \qquad t \in \mathbb{N}_{a+1},
$$

and

$$
(\overline{\Delta}_a^{-\nu} x)(t) = \binom{t-a+\alpha-1}{t-a} x(a) + (\widetilde{\Delta}_a^{-\nu} x)(t), \qquad t \in \mathbb{N}_a,
$$

where $x|_{\mathbb{N}_{a+1}}$ is the restriction of $x \colon \mathbb{N}_a \to \mathbb{R}^d$ to the set $\mathbb{N}_{a+1}$.

2 Fractional Differences

In this section we provide definitions of fractional differences and discuss the relationships between them.

Definition 3 *(Caputo and Riemann-Liouville forward and backward differences)* Let $\alpha \in (0, 1)$, $a \in \mathbb{R}$ and $x: \mathbb{N}_a \to \mathbb{R}^d$.
(a) *Caputo forward difference* ${}_{\mathrm{C}}\Delta_a^\alpha := \Delta_a^{-(1-\alpha)} \circ \Delta$:

$$ {}_{\mathrm{C}}\Delta_a^\alpha x: \mathbb{N}_{a+1-\alpha} \to \mathbb{R}^d, \qquad t \mapsto ({}_{\mathrm{C}}\Delta_a^\alpha x)(t) = (\Delta_a^{-(1-\alpha)}(\Delta x))(t). \tag{2}$$

(b) *Riemann-Liouville forward difference* ${}_{\text{R-L}}\Delta_a^\alpha := \Delta \circ \Delta_a^{-(1-\alpha)}$:

$$ {}_{\text{R-L}}\Delta_a^\alpha x: \mathbb{N}_{a+1-\alpha} \to \mathbb{R}^d, \qquad t \mapsto ({}_{\text{R-L}}\Delta_a^\alpha x)(t) = (\Delta(\Delta_a^{-(1-\alpha)} x))(t). \tag{3}$$

(c) *Caputo backward difference* ${}_{\mathrm{C}}\nabla_a^\alpha := \Delta_{a+1}^{-(1-\alpha)} \circ \nabla$:

$$ {}_{\mathrm{C}}\nabla_a^\alpha x: \mathbb{N}_{a+2-\alpha} \to \mathbb{R}^d, \qquad t \mapsto ({}_{\mathrm{C}}\nabla_a^\alpha x)(t) = (\Delta_{a+1}^{-(1-\alpha)}(\nabla x))(t). \tag{4}$$

(d) *Riemann-Liouville backward difference* ${}_{\text{R-L}}\nabla_a^\alpha := \nabla \circ \Delta_a^{-(1-\alpha)}$:

$$ {}_{\text{R-L}}\nabla_a^\alpha x: \mathbb{N}_{a+2-\alpha} \to \mathbb{R}^d, \qquad t \mapsto ({}_{\text{R-L}}\nabla_a^\alpha x)(t) = (\nabla(\Delta_a^{-(1-\alpha)} x))(t). \tag{5}$$

(e) The Caputo and Riemann-Liouville forward and backward differences with the fractional sum $\overline{\Delta}_a^{-\nu}$ instead of $\Delta_a^{-\nu}$ are defined as

$$ {}_{\mathrm{C}}\overline{\Delta}_a^\alpha x: \mathbb{N}_a \to \mathbb{R}^d, \quad t \mapsto ({}_{\mathrm{C}}\overline{\Delta}_a^\alpha x)(t) = (\overline{\Delta}_a^{-(1-\alpha)}(\Delta x))(t), \tag{6}$$
$$ {}_{\text{R-L}}\overline{\Delta}_a^\alpha x: \mathbb{N}_a \to \mathbb{R}^d, \quad t \mapsto ({}_{\text{R-L}}\overline{\Delta}_a^\alpha x)(t) = (\Delta(\overline{\Delta}_a^{-(1-\alpha)} x))(t), \tag{7}$$
$$ {}_{\mathrm{C}}\overline{\nabla}_a^\alpha x: \mathbb{N}_{a+1} \to \mathbb{R}^d, \quad t \mapsto ({}_{\mathrm{C}}\overline{\nabla}_a^\alpha x)(t) = (\overline{\Delta}_{a+1}^{-(1-\alpha)}(\nabla x))(t), \tag{8}$$
$$ {}_{\text{R-L}}\overline{\nabla}_a^\alpha x: \mathbb{N}_{a+1} \to \mathbb{R}^d, \quad t \mapsto ({}_{\text{R-L}}\overline{\nabla}_a^\alpha x)(t) = (\nabla(\overline{\Delta}_a^{-(1-\alpha)} x))(t). \tag{9}$$

(f) The Caputo and Riemann-Liouville backward differences with $\widetilde{\Delta}_a^{-\nu}$ instead of $\Delta_a^{-\nu}$ are

$$ {}_{\mathrm{C}}\widetilde{\nabla}_a^\alpha x: \mathbb{N}_{a+2} \to \mathbb{R}^d, \quad t \mapsto ({}_{\mathrm{C}}\widetilde{\nabla}_a^\alpha x)(t) = (\widetilde{\Delta}_{a+1}^{-(1-\alpha)}(\nabla x))(t),$$
$$ {}_{\text{R-L}}\widetilde{\nabla}_a^\alpha x: \mathbb{N}_{a+2} \to \mathbb{R}^d, \quad t \mapsto ({}_{\text{R-L}}\widetilde{\nabla}_a^\alpha x)(t) = (\nabla(\widetilde{\Delta}_a^{-(1-\alpha)} x))(t).$$

If $a = 0$ we write ${}_{\mathrm{C}}\Delta^\alpha$, ${}_{\text{R-L}}\Delta^\alpha$, ${}_{\mathrm{C}}\nabla^\alpha$, ${}_{\text{R-L}}\nabla^\alpha$, as well as ${}_{\mathrm{C}}\overline{\Delta}^\alpha$, ${}_{\text{R-L}}\overline{\Delta}^\alpha$, ${}_{\mathrm{C}}\overline{\nabla}^\alpha$, ${}_{\text{R-L}}\overline{\nabla}^\alpha$, and ${}_{\mathrm{C}}\widetilde{\nabla}^\alpha$, ${}_{\text{R-L}}\widetilde{\nabla}^\alpha$, respectively.

Remark 2 Using (1) in Remark 1(d), we get for $\alpha \in (0, 1)$, $a \in \mathbb{R}$ and $x: \mathbb{N}_a \to \mathbb{R}^d$ the following relationships between the fractional differences (2)–(5) and (6)–(9)

$$
\begin{aligned}
({}_{\mathrm{C}}\overline{\Delta}_a^{\alpha} x)(t) &= ({}_{\mathrm{C}}\Delta_a^{\alpha} x)(t+1-\alpha), \quad && t \in \mathbb{N}_a,\\
({}_{\text{R-L}}\overline{\Delta}_a^{\alpha} x)(t) &= ({}_{\text{R-L}}\Delta_a^{\alpha} x)(t+1-\alpha), \quad && t \in \mathbb{N}_a,\\
({}_{\mathrm{C}}\overline{\nabla}_a^{\alpha} x)(t) &= ({}_{\mathrm{C}}\nabla_{a+1}^{\alpha} x)(t+1-\alpha), \quad && t \in \mathbb{N}_{a+1},\\
({}_{\text{R-L}}\overline{\nabla}_a^{\alpha} x)(t) &= ({}_{\text{R-L}}\nabla_{a+1}^{\alpha} x)(t+1-\alpha), \quad && t \in \mathbb{N}_{a+1}.
\end{aligned}
$$

The following two lemmas enable us to rewrite fractional difference equations as Volterra convolution equations in the next section.

Lemma 1 (Sum representations of fractional operators)
Let $\alpha \in (0,1)$ and $x\colon \mathbb{N}_a \to \mathbb{R}^d$. Then

$$
\begin{aligned}
({}_{\mathrm{C}}\overline{\Delta}_a^{\alpha} x)(n) = &\sum_{k=a+1}^{n+1} (-1)^{n+1-k}\binom{\alpha}{n+1-k} x(k)\\
&- (-1)^{n-a}\binom{\alpha-1}{n-a} x(a), \quad n \in \mathbb{N}_a,
\end{aligned}
\tag{10}
$$

$$
({}_{\text{R-L}}\overline{\Delta}_a^{\alpha} x)(n) = \sum_{k=a}^{n+1} (-1)^{n+1-k}\binom{\alpha}{n+1-k} x(k), \quad n \in \mathbb{N}_a,
\tag{11}
$$

$$
\begin{aligned}
({}_{\mathrm{C}}\overline{\nabla}_a^{\alpha} x)(n) = &\sum_{k=a+1}^{n} (-1)^{n-k}\binom{\alpha}{n-k} x(k)\\
&- (-1)^{n-a-1}\binom{\alpha-1}{n-a-1} x(a), \quad n \in \mathbb{N}_{a+1},
\end{aligned}
\tag{12}
$$

$$
({}_{\text{R-L}}\overline{\nabla}_a^{\alpha} x)(n) = \sum_{k=a}^{n} (-1)^{n-k}\binom{\alpha}{n-k} x(k), \quad n \in \mathbb{N}_{a+1},
\tag{13}
$$

$$
\begin{aligned}
({}_{\mathrm{C}}\widetilde{\nabla}_a^{\alpha} x)(n) = &\sum_{k=a+2}^{n} (-1)^{n-k}\binom{\alpha}{n-k} x(k)\\
&- (-1)^{n-a-2}\binom{\alpha-1}{n-a-2} x(a+1), \quad n \in \mathbb{N}_{a+2},
\end{aligned}
\tag{14}
$$

$$
({}_{\text{R-L}}\widetilde{\nabla}_a^{\alpha} x)(n) = \sum_{k=a+1}^{n} (-1)^{n-k}\binom{\alpha}{n-k} x(k), \quad n \in \mathbb{N}_{a+2}.
\tag{15}
$$

Proof We prove only (10), the proofs of (11)–(15) are similar. Using the facts (see [27, pp. 158, 164]) that

$$\binom{s}{\ell} = (-1)^{\ell}\binom{\ell - s - 1}{\ell} \quad \text{and} \quad \binom{s}{\ell} - \binom{s-1}{\ell-1} = \binom{s-1}{\ell}, \qquad s \in \mathbb{R}, \ell \in \mathbb{N},$$

and Definition 3, we get

$$\begin{aligned}
({}_{\mathrm{c}}\overline{\Delta}_a^{\alpha} x)(n) &= (\overline{\Delta}_a^{-(1-\alpha)} \circ \Delta x)(n) = \sum_{k=a}^{n} \binom{n-k-\alpha}{n-k} \Delta x(k) \\
&= \sum_{k=a}^{n} \binom{n-k-\alpha}{n-k} (x(k+1) - x(k)) \\
&= \sum_{k=a}^{n} \binom{n-k-\alpha}{n-k} x(k+1) - \sum_{k=a}^{n} \binom{n-k-\alpha}{n-k} x(k) \\
&= \sum_{k=a+1}^{n+1} \binom{n-k+1-\alpha}{n-k+1} x(k) - \sum_{k=a}^{n} \binom{n-k-\alpha}{n-k} x(k) \\
&= \sum_{k=a+1}^{n} \left(\binom{n-k+1-\alpha}{n-k+1} - \binom{n-k-\alpha}{n-k} \right) x(k) + x(n+1) \\
&\quad - \binom{n-a-\alpha}{n-a} x(a) \\
&= \sum_{k=a+1}^{n} \binom{n-k-\alpha}{n-k+1} x(k) + x(n+1) - \binom{n-a-\alpha}{n-a} x(a) \\
&= \sum_{k=a+1}^{n+1} \binom{n-k-\alpha}{n-k+1} x(k) - \binom{n-a-\alpha}{n-a} x(a) \\
&= \sum_{k=a+1}^{n+1} (-1)^{n-k+1} \binom{\alpha}{n-k+1} x(k) - (-1)^{n-a} \binom{\alpha-1}{n-a} x(a),
\end{aligned}$$

which proves (10). □

The following lemma provides fractional forward and backward Taylor difference formulas.

Lemma 2 (Taylor formula)
Let $\alpha \in (0, 1)$, $a \in \mathbb{R}$ and $x\colon \mathbb{N}_a \to \mathbb{R}^d$. Then for each $n \in \mathbb{N}_a$

$$x(n+1) = x(a) + \sum_{k=a}^{n} (-1)^{n-k} \binom{-\alpha}{n-k} ({}_{\mathrm{C}}\overline{\Delta}_a^{\alpha} x)(n), \tag{16}$$

$$x(n) = x(a) + \sum_{k=a}^{n} (-1)^{n-k} \binom{-\alpha}{n-k} ({}_{\mathrm{C}}\overline{\nabla}_a^{\alpha} x)(k). \tag{17}$$

Proof See [4, Theorems 35.8 and 36.4]. □

3 Linear Fractional Forward and Backward Difference Equations

Let $A\colon \mathbb{N}_0 \to \mathbb{R}^{d\times d}, x\colon \mathbb{N}_0 \to \mathbb{R}^d$, and $\alpha \in (0,1)$. We investigate linear time-varying

$$\text{forward } (\widehat{\Delta}^{\alpha} x)(n) = A(n)x(n) \quad \text{and} \quad \text{backward } (\widehat{\nabla}^{\alpha} x)(n) = A(n)x(n)$$

fractional difference equations with $\widehat{\Delta}^{\alpha} \in \{{}_{\mathrm{C}}\overline{\Delta}^{\alpha}, {}_{\mathrm{R\text{-}L}}\overline{\Delta}^{\alpha}\}$, $\widehat{\nabla}^{\alpha} \in \{{}_{\mathrm{C}}\overline{\nabla}^{\alpha}, {}_{\mathrm{R\text{-}L}}\overline{\nabla}^{\alpha}, {}_{\mathrm{C}}\widetilde{\nabla}^{\alpha}, {}_{\mathrm{R\text{-}L}}\widetilde{\nabla}^{\alpha}\}$. According to Definition 3(e)–(f), the equations are defined on $\mathbb{N}_0$, $\mathbb{N}_1$ or $\mathbb{N}_2$. More precisely,

$$({}_{\mathrm{C}}\overline{\Delta}^{\alpha} x)(n) = A(n)x(n), \qquad n \in \mathbb{N}_0, \tag{18}$$

$$({}_{\mathrm{R\text{-}L}}\overline{\Delta}^{\alpha} x)(n) = A(n)x(n), \qquad n \in \mathbb{N}_0, \tag{19}$$

$$({}_{\mathrm{C}}\overline{\nabla}^{\alpha} x)(n) = A(n)x(n), \qquad n \in \mathbb{N}_1, \tag{20}$$

$$({}_{\mathrm{R\text{-}L}}\overline{\nabla}^{\alpha} x)(n) = A(n)x(n), \qquad n \in \mathbb{N}_1, \tag{21}$$

$$({}_{\mathrm{C}}\widetilde{\nabla}^{\alpha} x)(n) = A(n)x(n), \qquad n \in \mathbb{N}_2, \tag{22}$$

$$({}_{\mathrm{R\text{-}L}}\widetilde{\nabla}^{\alpha} x)(n) = A(n)x(n), \qquad n \in \mathbb{N}_2. \tag{23}$$

In order to define initial value problems, note that by Lemma 1,

$$({}_{\mathrm{C}}\overline{\Delta}^{\alpha} x)(0) = x(1) - x(0), \qquad ({}_{\mathrm{R\text{-}L}}\overline{\Delta}^{\alpha} x)(0) = x(1) - \alpha x(0),$$
$$({}_{\mathrm{C}}\overline{\nabla}^{\alpha} x)(1) = x(1) - x(0), \qquad ({}_{\mathrm{R\text{-}L}}\overline{\nabla}^{\alpha} x)(1) = x(1) - \alpha x(0),$$
$$({}_{\mathrm{C}}\widetilde{\nabla}^{\alpha} x)(2) = x(2) - x(1), \qquad ({}_{\mathrm{R\text{-}L}}\widetilde{\nabla}^{\alpha} x)(2) = x(2) - \alpha x(1),$$

and hence for the forward fractional difference equations (18)–(19) an initial value $x(0) \in \mathbb{R}^d$ determines $x(1)$ (and also $x(n)$ for $n \in \mathbb{N}$). However for the backward fractional difference equations (20)–(21) and (22)–(23)

$$({}_{\mathrm{C}}\overline{\nabla}^{\alpha}x)(1) = A(1)x(1) \quad \Leftrightarrow \quad (I - A(1))x(1) = x(0),$$
$$({}_{\mathrm{R\text{-}L}}\overline{\nabla}^{\alpha}x)(1) = A(1)x(1) \quad \Leftrightarrow \quad (I - A(1))x(1) = \alpha x(0),$$
$$({}_{\mathrm{C}}\widetilde{\nabla}^{\alpha}x)(2) = A(2)x(2) \quad \Leftrightarrow \quad (I - A(2))x(2) = x(1),$$
$$({}_{\mathrm{R\text{-}L}}\widetilde{\nabla}^{\alpha}x)(2) = A(2)x(2) \quad \Leftrightarrow \quad (I - A(2))x(2) = \alpha x(1).$$

As can be seen, $\mathrm{Im}(I - A(n))$ plays a role for the existence of solutions $x\colon \mathbb{N}_0 \to \mathbb{R}$. We now define initial value problems and remark on initial conditions before we show the unique existence of solutions to initial value problems under a sufficient condition on A.

Definition 4 *(Initial value problem)*
Let $x_0 \in \mathbb{R}^d$. Then $x\colon \mathbb{N}_0 \to \mathbb{R}$ with $x(0) = x_0$ is called *solution of the initial value problem* (18), (19), (20), (21), (22) or (23) *with initial value* x_0, if it satisfies the corresponding equation and in case of Eq. (22) it additionally satisfies $x(1) = x(0)$ and in case of (23) it additionally satisfies $(I - A(1))x(1) = x(0)$.

Remark 3 *(Initial value problems)*
Formally the initial value problems for (22)–(23) may be defined on $\mathbb{N}_1$ as a problem of finding, for a given $x_1 \in \mathbb{R}^d$, a sequence $x\colon \mathbb{N}_1 \to \mathbb{R}^d$ such that $x(1) = x_1$ and the corresponding equations are satisfied for all $n \in \mathbb{N}_2$. We choose the formulation of the initial value problem as in Definition 4 because this way all solutions of equations (22)–(23) are defined on the same set $\mathbb{N}_0$ and also in earlier papers on these equations (see [15, 22]) initial value problems are defined as in Definition 4.

In this paper we assume for the backward fractional difference equations (20)–(21) and (22)–(23) the condition

$$\det(I - A(n)) \neq 0 \quad \text{for each } n \in \mathbb{N}_0, \tag{24}$$

which is sufficient for unique existence of solutions.

Theorem 1 (Existence and uniqueness of solutions to initial value problems)
Let $A\colon \mathbb{N}_0 \to \mathbb{R}^{d\times d}$, $\alpha \in (0, 1)$ *and* $x_0 \in \mathbb{R}^d$.

(a) For each of the forward fractional difference equations (18)–(19) *there exists a unique solution* $x\colon \mathbb{N}_0 \to \mathbb{R}^d$ *with initial value* x_0, *denoted by* $\varphi_{\mathrm{C}}^{\Delta}(\cdot, x_0)$ *and* $\varphi_{\mathrm{R-L}}^{\Delta}(\cdot, x_0)$, *respectively.*

(b) Assume (24). *Then for each of the backward fractional difference equations* (20)–(21) *and* (22)–(23) *there exists a unique solution* $x\colon \mathbb{N}_0 \to \mathbb{R}^d$ *with initial value* x_0, *denoted by* $\varphi_{\mathrm{C}}^{\overline{\nabla}}(\cdot, x_0)$, $\varphi_{\mathrm{R-L}}^{\overline{\nabla}}(\cdot, x_0)$ *and* $\varphi_{\mathrm{C}}^{\widetilde{\nabla}}(\cdot, x_0)$, $\varphi_{\mathrm{R-L}}^{\widetilde{\nabla}}(\cdot, x_0)$, *respectively.*

Proof (a) Using (10) and (11) of Lemma 1, it follows that $x(n + 1)$ in Eqs. (18) and (19) is recursively defined from $x(0), \dots, x(n)$ for $n \in \mathbb{N}_0$.

(b) Under the assumption (24) and using (12)–(13), it follows that $x(n + 1)$ in Eqs. (20)–(21) is recursively defined from $x(0), \dots, x(n)$ for $n \in \mathbb{N}_0$. Similarly, using (14)–(15), $x(n + 1)$ in (22)–(23) is recursively defined from $x(1), \dots, x(n)$ for $n \in \mathbb{N}_1$ and defining $x(0)$ according to Definition 4 shows unique existence. □

4 Solution Representation with Volterra Convolution Sums

In this section we show that every solution of each of the forward equations (18)–(19) can be equivalently rewritten as a Volterra convolution equation

$$x(n+1) = \sum_{k=0}^{n} a(n-k)x(k) + g(k)$$

with appropriate sequences $a\colon \mathbb{N}_0 \to \mathbb{R}^{d\times d}$ and $g\colon \mathbb{N}_0 \to \mathbb{R}^d$. The backward equations (20)–(23) can be written as

$$(I - A(n))x(n+1) = \sum_{k=0}^{n} a(n-k)x(k) + g(k).$$

We use the convention $\sum\limits_{k=p}^{q} s(k) = 0$ for $q < p$.

Theorem 2 (Volterra sum representation of solutions)
Let $A\colon \mathbb{N}_0 \to \mathbb{R}^{d\times d}$, $\alpha \in (0,1)$, $x_0 \in \mathbb{R}^d$ and $x\colon \mathbb{N}_0 \to \mathbb{R}^d$ with $x(0) = x_0$. Then
(a) $\varphi_{\mathrm{C}}^{\Delta}(\cdot, x_0) = x$ is equivalent to each of the two statements (25) *or* (26)

$$x(n) = \sum_{k=0}^{n-1} (-1)^{n-1-k} \binom{-\alpha}{n-1-k} A(k)x(k) + x(0), \quad n \in \mathbb{N}_1, \tag{25}$$

$$\begin{aligned} x(n) = A(n-1)x(n-1) - \sum_{k=1}^{n-1} (-1)^{n-k} \binom{\alpha}{n-k} x(k) \\ + (-1)^{n-1} \binom{\alpha-1}{n-1} x(0), \quad n \in \mathbb{N}_1. \end{aligned} \tag{26}$$

(b) $\varphi_{\mathrm{R\text{-}L}}^{\Delta}(\cdot, x_0) = x$ is equivalent to each of the two statements (27) *or* (28)

$$x(n) = \sum_{k=0}^{n-1} (-1)^{n-1-k} \binom{-\alpha}{n-1-k} A(k)x(k) + (-1)^n \binom{-\alpha}{n} x(0), \quad n \in \mathbb{N}_1, \tag{27}$$

$$x(n) = A(n-1)x(n-1) - \sum_{k=0}^{n-1} (-1)^{n-k} \binom{\alpha}{n-k} x(k), \quad n \in \mathbb{N}_1. \tag{28}$$

We additionally assume (24). *Then also*
(c) $\varphi_{\mathrm{C}}^{\overline{\nabla}}(\cdot, x_0) = x$ is equivalent to each of the two statements (29) *or* (30)

$$x(n) = (I - A(n))^{-1} \left(\sum_{k=1}^{n-1} (-1)^{n-k} \binom{-\alpha}{n-k} A(k)x(k) + x(0) \right), \quad n \in \mathbb{N}_1, \quad (29)$$

$$x(n) = (I - A(n))^{-1} \cdot \left(- \sum_{k=1}^{n-1} (-1)^{n-k} \binom{\alpha}{n-k} x(k) + (-1)^{n-1} \binom{\alpha - 1}{n-1} x(0) \right), \quad n \in \mathbb{N}_1. \quad (30)$$

(d) $\varphi^{\overline{\nabla}}_{\mathrm{R-L}}(\cdot, x_0) = x$ *is equivalent to each of the two statements* (31) *or* (32)

$$x(n) = (I - A(n))^{-1} \cdot \left(\sum_{k=1}^{n-1} (-1)^{n-k} \binom{-\alpha}{n-k} A(k)x(k) + (-1)^n \binom{-\alpha}{n} x(0) \right), \quad n \in \mathbb{N}_1, \quad (31)$$

$$x(n) = -(I - A(n))^{-1} \sum_{k=0}^{n-1} (-1)^{n-k} \binom{\alpha}{n-k} x(k), \quad n \in \mathbb{N}_1. \quad (32)$$

(e) $\varphi^{\widetilde{\nabla}}_{\mathrm{C}}(\cdot, x_0) = x$ *is equivalent to* $x(1) = x(0)$ *and each of the two statements* (33) *or* (34)

$$x(n) = (I - A(n))^{-1} \left(x(1) + \sum_{k=2}^{n-1} (-1)^{n-k} \binom{-\alpha}{n-k} A(k)x(k) \right), \quad n \in \mathbb{N}_2, \quad (33)$$

$$x(n) = (I - A(n))^{-1} \cdot \left((-1)^{n-2} \binom{\alpha - 1}{n-2} x(1) - \sum_{k=2}^{n-1} (-1)^{n-k} \binom{\alpha}{n-k} x(k) \right), \quad n \in \mathbb{N}_2. \quad (34)$$

(f) $\varphi^{\widetilde{\nabla}}_{\mathrm{R-L}}(\cdot, x_0) = x$ *is equivalent to* $x(1) = (I - A(1))^{-1} x(0)$ *and each of the two statements* (35) *or* (36)

$$x(n) = (I - A(n))^{-1} \cdot \left((-1)^{n-1} \binom{-\alpha}{n-1} x(1) + \sum_{k=2}^{n-1} (-1)^{n-k} \binom{-\alpha}{n-k} A(k)x(k) \right), \quad n \in \mathbb{N}_2, \quad (35)$$

$$x(n) = -(I - A(n))^{-1}\left(\sum_{k=1}^{n-1}(-1)^{n-k}\binom{\alpha}{n-k}x(k)\right), \quad n \in \mathbb{N}_2. \tag{36}$$

Proof (a) Using (16) in Lemma 2, Eq. (25) is equivalent to (18), i.e. $\varphi_C^{\Delta}(\cdot, x_0) = x$. Assuming $\varphi_C^{\Delta}(\cdot, x_0) = x$, Eq. (18) follows and with (10) in Lemma 1 for $a = 0$ we get

$$\sum_{k=1}^{n+1}(-1)^{n-k+1}\binom{\alpha}{n-k+1}x(k) - (-1)^n\binom{\alpha-1}{n}x(0) = A(n)x(n), \qquad n \in \mathbb{N}_0,$$

and therefore

$$x(n+1) + \sum_{k=1}^{n}(-1)^{n-k+1}\binom{\alpha}{n-k+1}x(k) - (-1)^n\binom{\alpha-1}{n}x(0) = A(n)x(n)$$

or equivalently

$$x(n+1) = A(n)x(n) - \sum_{k=1}^{n}(-1)^{n-k+1}\binom{\alpha}{n-k+1}x(k) + (-1)^n\binom{\alpha-1}{n}x(0), \quad n \in \mathbb{N}_0,$$

which proves (26). Similarly (26) implies $\varphi_C^{\Delta}(\cdot, x_0) = x$.

(b)–(f) As in (a), the equivalences to (28), (30), (32), (34) and (36) follow with (11)–(15). Similarly as in (a), the equivalence to (27) is obtained from (17) and (20). The equivalence to (29) follows from [11, (Eq. (2.4)]. The equivalence to (31) is proved in [9, Eq. (3.4)]. Finally the equivalences to (33) and (35) are proved in [2, Eq. (5.9)] and [22, Eq. (2.6)]. □

Remark 4 *(Relation between $\varphi_C^{\overline{\nabla}}(\cdot, x_0)$ and $\varphi_C^{\widetilde{\nabla}}(\cdot, x_0)$ and between $\varphi_{R-L}^{\overline{\nabla}}(\cdot, x_0)$ and $\varphi_{R-L}^{\widetilde{\nabla}}(\cdot, x_0)$)*
Comparing (c) with (e) and (d) with (f) in Theorem 2, we observe the following relations.

(a) If $x\colon \mathbb{N}_0 \to \mathbb{R}^d$ satisfies (29) (or equivalently (30)), then

$$\widetilde{x}\colon \mathbb{N}_1 \to \mathbb{R}^d, \quad n \mapsto \widetilde{x}(n) := x(n-1)$$

satisfies (33) (or equivalently (34)) with A replaced by

$$\widetilde{A}\colon \mathbb{N}_2 \to \mathbb{R}^{d\times d}, \quad n \mapsto \widetilde{A}(n) := A(n-1).$$

Similarly, if $x\colon \mathbb{N}_0 \to \mathbb{R}^d$ satisfies (31) (or equivalently (32)), then $\widetilde{x}\colon \mathbb{N}_1 \to \mathbb{R}^d$, $\widetilde{x}(n) := x(n-1)$, satisfies (35) (or equivalently (36)) with A replaced by $\widetilde{A}\colon \mathbb{N}_2 \to \mathbb{R}^{d\times d}$, $\widetilde{A}(n) := A(n-1)$.

(b) Conversely, if $x\colon \mathbb{N}_1 \to \mathbb{R}^d$ satisfies (33) (or equivalently (34)) then

$$\overline{x}\colon \mathbb{N}_0 \to \mathbb{R}^d, \quad n \mapsto \overline{x}(n) := x(n+1)$$

satisfies (29) (or equivalently (30)) with A replaced by

$$\overline{A}\colon \mathbb{N}_1 \to \mathbb{R}^{d\times d}, \quad n \mapsto \overline{A}(n) := A(n+1).$$

Similarly, if $x\colon \mathbb{N}_1 \to \mathbb{R}^d$ satisfies (35) (or equivalently (36)), then $\overline{x}\colon \mathbb{N}_1 \to \mathbb{R}^d$, $\overline{x}(n) := x(n+1)$ satisfies (31) (or equivalently (32)) with A replaced by $\overline{A}\colon \mathbb{N}_1 \to \mathbb{R}^{d\times d}$, $\overline{A}(n) := A(n+1)$.

(c) From (a), (b) and Definition 4, it follows that for each $x_0 \in \mathbb{R}^d$

$$\varphi_{\mathrm{C}}^{\tilde{\nabla}}(n, x_0) = \varphi_{\mathrm{C}}^{\overline{\nabla}}(n-1, x_0), \qquad n \in \mathbb{N}_1,$$

and

$$\varphi_{\mathrm{R-L}}^{\tilde{\nabla}}(n, (I - A(1))x_0) = \varphi_{\mathrm{R-L}}^{\overline{\nabla}}(n-1, x_0), \qquad n \in \mathbb{N}_1.$$

5 Linear Time-Invariant Fractional Systems

In this section we consider linear time-invariant fractional systems (18)–(23) with constant linear part $A(n) = A \in \mathbb{R}^{d\times d}$ for $n \in \mathbb{N}_0$. We prove and cite explicit solution formulas. For another formulation of the stability problem, see [12, 24, 26, 29, 30].

Theorem 3 (Solution representation for linear time-invariant fractional systems) *Let $x_0 \in \mathbb{R}^d$, $A \in \mathbb{R}^{d\times d}$ and $\alpha \in (0, 1)$.*

(a) The solutions of (18)–(19) *with time-invariant A satisfy*

$$\varphi_{\mathrm{C}}^{\Delta}(n, x_0) = \sum_{k=0}^{n} A^k \binom{n-k+k\alpha}{n-k} x_0 = \sum_{k=0}^{n} A^k (-1)^{n-k} \binom{-k\alpha - 1}{n-k} x_0, \qquad n \in \mathbb{N}_0,$$

$$\begin{aligned}\varphi_{\mathrm{R-L}}^{\Delta}(n, x_0) &= \sum_{k=0}^{n} A^k \binom{n-k+(k+1)\alpha - 1}{n-k} x_0 \\ &= \sum_{k=0}^{n} A^k (-1)^{n-k} \binom{-(k+1)\alpha}{n-k} x_0, \qquad n \in \mathbb{N}_0.\end{aligned}$$

(b) If all eigenvalues of A lie inside the unit circle, then the solutions of (20)–(21) *and* (22)–(23) *with time-invariant A satisfy*

$$\varphi_{\mathrm{C}}^{\overline{\nabla}}(n, x_0) = \left(I + (-1)^{n-1} \sum_{k=1}^{\infty} A^k \binom{-k\alpha - 1}{n-1} \right) x_0, \qquad n \in \mathbb{N}_1, \tag{37}$$

$$\varphi_{\mathrm{R-L}}^{\overline{\nabla}}(n, x_0) = \sum_{k=0}^{\infty} A^k \binom{-k\alpha - \alpha}{n} (I - A)x_0, \qquad n \in \mathbb{N}_1, \tag{38}$$

and

$$\varphi_{\mathrm{C}}^{\widetilde{\nabla}}(n,x_0)=\left(I+(-1)^n\sum_{k=1}^{\infty}A^k\binom{-k\alpha-1}{n-2}\right)x_0,\qquad n\in\mathbb{N}_2,\tag{39}$$

$$\varphi_{\mathrm{R-L}}^{\widetilde{\nabla}}(n,x_0)=\sum_{k=0}^{\infty}A^k\binom{-k\alpha-\alpha}{n-1}(I-A)x_0,\qquad n\in\mathbb{N}_2.\tag{40}$$

Proof (a) The solution formulas for $\varphi_{\mathrm{C}}^{\Delta}(\cdot,x_0)$ and $\varphi_{\mathrm{R-L}}^{\Delta}(\cdot,x_0)$ are proved in [7, Remark 1].

(b) The solution representations for $\varphi_{\mathrm{R-L}}^{\overline{\nabla}}(\cdot,x_0)$ and $\varphi_{\mathrm{R-L}}^{\widetilde{\nabla}}(\cdot,x_0)$ follow from Remark 4(c) and [15, Theorem 18] or [8, Theorem 1], see also [10, Theorem 4.4] for an alternative proof using the Z-transform. Using Remark 4(c), the formula for $\varphi_{\mathrm{C}}^{\widetilde{\nabla}}(\cdot,x_0)$ follows from the formula (37) for $\varphi_{\mathrm{C}}^{\overline{\nabla}}(\cdot,x_0)$. To show (37) we follow the line of reasoning from Example 48 in [1] and show that the solution $\varphi_{\mathrm{C}}^{\overline{\nabla}}(\cdot,x_0)$ of the initial value problem for

$$({}_{\mathrm{C}}\overline{\nabla}^{\alpha}x)(n)=Ax(n),\qquad x\in\mathbb{N}_1,\tag{41}$$

under the assumption that all the eigenvalues of A lie inside the unit circle, is given by

$$\varphi_{\mathrm{C}}^{\overline{\nabla}}(0,x_0)=x_0$$

and

$$\varphi_{\mathrm{C}}^{\overline{\nabla}}(n,x_0)=(I-A)^{-1}\left(I+(-1)^{n-1}\sum_{k=1}^{\infty}A^k\binom{-k\alpha-1}{n-1}\right)x_0,\ n\in\mathbb{N}_1.\tag{42}$$

For each $x_0\in\mathbb{R}^d$ let us define a sequence $x=(x_m)_{m\in\mathbb{N}_0}$, of sequences $x_m:\mathbb{N}_1\to\mathbb{R}^d$ recursively by

$$\begin{aligned}x_0(n)&=x_0,\qquad n\in\mathbb{N}_1,\\ x_m(n)&=x_0+A\sum_{k=1}^{n}(-1)^{n-k}\binom{-\alpha}{n-k}x_{m-1}(k),\qquad m,n\in\mathbb{N}_1.\end{aligned}\tag{43}$$

In our further consideration we will use the following identities:

$$\sum_{k=1}^{n}(-1)^{n-k}\binom{-\alpha}{n-k}=(-1)^{n-1}\binom{-\alpha-1}{n-1},\qquad n\in\mathbb{N}_0,\tag{44}$$

and the Chu–Vandermonde identity

$$\sum_{k=0}^{n}\binom{s}{n-k}\binom{\nu}{k}=\binom{s+\nu}{n}, \qquad s,\nu\in\mathbb{R}, n\in\mathbb{N}_0, \tag{45}$$

(see [27, p. 165, (5.16)] and [27, Table 174]). For $m = 1$ we get

$$\begin{aligned}
x_1(n) &= x_0 + A\sum_{k=1}^{n}(-1)^{n-k}\binom{-\alpha}{n-k}x_0(k)\\
&= \left(I + A\sum_{k=1}^{n}(-1)^{n-k}\binom{-\alpha}{n-k}\right)x_0\\
&\overset{(44)}{=} \left(I + A(-1)^{n-1}\binom{-\alpha-1}{n-1}\right)x_0.
\end{aligned}$$

Next, we show by the induction that

$$x_m(n) = \left(I + (-1)^{n-1}\sum_{k=1}^{m}A^k\binom{-k\alpha-1}{n-1}\right)x_0. \tag{46}$$

As we have checked this formula is true for $m \in \{0, 1\}$. Assume that it is true for each $m \in \{0, 1, \dots, l\}$ for an $l \in \mathbb{N}_1$. For $m = l + 1$ we get

$$\begin{aligned}
x_{l+1}(n) &= x_0 + A\sum_{k=1}^{n}(-1)^{n-k}\binom{-\alpha}{n-k}x_l(k)\\
&= x_0 + A\sum_{k=1}^{n}(-1)^{n-k}\binom{-\alpha}{n-k}\left(\left(I + (-1)^{k-1}\sum_{j=1}^{l}A^j\binom{-j\alpha-1}{k-1}\right)x_0\right)\\
&= \left(I + A\sum_{k=1}^{n}(-1)^{n-k}\binom{-\alpha}{n-k}\right.\\
&\qquad\left. + A\sum_{k=1}^{n}(-1)^{n-k}\binom{-\alpha}{n-k}\left((-1)^{k-1}\sum_{j=1}^{l}A^j\binom{-j\alpha-1}{k-1}\right)\right)x_0\\
&\overset{(44)}{=} \left(I + A(-1)^{n-1}\binom{-\alpha-1}{n-1}\right.\\
&\qquad\left. + (-1)^{n-1}\sum_{j=1}^{l}A^{j+1}\left(\sum_{t=0}^{n-1}\binom{-\alpha}{n-1-t}\binom{-j\alpha-1}{t}\right)\right)x_0\\
&\overset{(45)}{=} \left(I + (-1)^{n-1}\sum_{k=1}^{l+1}A^k\binom{-k\alpha-1}{n-1}\right)x_0.
\end{aligned}$$

The proof of (46) for $m \in \mathbb{N}_0$ is completed.

We now show that for each $n \in \mathbb{N}_1$ the limit $\lim_{m\to\infty} x_m(n)$ exists. According to Lemma 5.6.10 in [19], the assumption that all the eigenvalues of A lie inside the unit circle is equivalent to the fact that there exists a matrix norm $\|\cdot\|_*$ such that $\|A\|_* < 1$. Therefore, according to the Cauchy-Hadamard theorem, to prove the existence of the limit $\lim_{m\to\infty} x_m(n)$ or equivalently to prove the convergence of the series $\sum_{k=1}^{\infty} A^k \binom{-k\alpha-1}{n-1}$, it is enough to show that

$$\lim_{k\to\infty} \left| \binom{-k\alpha - 1}{n-1} \right|^{\frac{1}{k}} = 1 \quad \text{for each } n \in \mathbb{N}_1. \tag{47}$$

Since

$$\binom{-k\alpha - 1}{n-1} = (-1)^{n-1} \binom{n-1+k\alpha}{n-1} = (-1)^{n-1} \frac{\Gamma(n+k\alpha)}{\Gamma(n)\Gamma(k\alpha+1)},$$

we have to show that

$$\lim_{k\to\infty} \left(\frac{\Gamma(n+k\alpha)}{\Gamma(k\alpha+1)} \right)^{\frac{1}{k}} = 1.$$

To compute the last limit we use the following well-known property of the Euler Gamma function [28, p. 415]

$$\Gamma(1+z) = z\Gamma(z),$$

from which it follows that

$$\Gamma(n+z) = (n-1+z)(n-2+z)\cdots z\Gamma(z), \qquad n \in \mathbb{N}_1.$$

Let us take $z = k\alpha$ in the last identity, then we have

$$\begin{aligned}
\lim_{k\to\infty} \left| \frac{\Gamma(n+k\alpha)}{\Gamma(k\alpha+1)} \right|^{\frac{1}{k}} &= \lim_{k\to\infty} \left| \frac{\Gamma(n-1+1+k\alpha)}{\Gamma(k\alpha+1)} \right|^{\frac{1}{k}} \\
&= \lim_{k\to\infty} \left| \frac{(n-1+k\alpha)(n-2+k\alpha)\cdots(k\alpha+1)\Gamma(k\alpha+1)}{\Gamma(k\alpha+1)} \right|^{\frac{1}{k}} \\
&= \lim_{k\to\infty} \left| (n-1+k\alpha)(n-2+k\alpha)\cdots(k\alpha+1) \right|^{\frac{1}{k}}.
\end{aligned}$$

Each of the factors in the last limit has the form $|k\alpha + b|$, where $b \in \{1, \ldots, n-1\}$, in particular, $b \geq 0$. Let us notice that

$$\lim_{k\to\infty} |k\alpha + b|^{\frac{1}{k}} = \lim_{k\to\infty} k^{\frac{1}{k}} \left| \alpha + \frac{b}{k} \right|^{\frac{1}{k}} = 1.$$

The proof of (47) is completed.

Observe that the sequence $x = (x(n))_{n\in\mathbb{N}_0}$ is given by

$$x(n) = \left(I + (-1)^{n-1} \sum_{k=1}^{\infty} A^k \binom{-k\alpha - 1}{n-1} \right) x_0$$

and satisfies the equation

$$x(n) = x_0 + A \sum_{k=1}^{n} (-1)^{n-k} \binom{-\alpha}{n-k} x(k), \qquad n \in \mathbb{N}_1. \tag{48}$$

In fact, we have

$$\begin{aligned} \lim_{m\to\infty} \sum_{k=1}^{n} (-1)^{n-k} \binom{-\alpha}{n-k} A x_m(k) &= \sum_{k=1}^{n} (-1)^{n-k} \binom{-\alpha}{n-k} A \lim_{m\to\infty} x_m(k) \\ &= \sum_{k=1}^{n} (-1)^{n-k} \binom{-\alpha}{n-k} A x(k). \end{aligned}$$

Using the last equality and passing to the limit for $m \to \infty$ in (43), we get (48). From (48) we have

$$x(n) = (I - A)^{-1} \left(x(0) + \sum_{k=1}^{n-1} (-1)^{n-k} \binom{-\alpha}{n-k} A x(k) \right)$$

and this is the Volterra convolution equation (25) which is equivalent to (41). This completes the proof of (42). □

Remark 4 leads us to the following useful analog observations for linear time-invariant systems.

Remark 5 *(Relation between solutions of backward time-invariant systems)* Assume that $I - A$ is invertible.

(a) If $x : \mathbb{N}_0 \to \mathbb{R}^d$ satisfies

$$x(n) = (I - A)^{-1} A \sum_{k=1}^{n-1} (-1)^{n-k} \binom{-\alpha}{n-k} x(k) + (I - A)^{-1} x(0), \quad n \in \mathbb{N}_1, \tag{49}$$

or equivalently

$$x(n) = -(I-A)^{-1}\sum_{k=1}^{n-1}(-1)^{n-k}\binom{\alpha}{n-k}x(k) + (I-A)^{-1}(-1)^{n-1}\binom{\alpha-1}{n-1}x(0), \quad n \in \mathbb{N}_1, \tag{50}$$

then $\widetilde{x}\colon \mathbb{N}_1 \to \mathbb{R}^d$, $\widetilde{x}(n) = x(n-1)$, satisfies

$$\widetilde{x}(n) = (I-A)^{-1}A\sum_{k=2}^{n-1}(-1)^{n-k}\binom{-\alpha}{n-k}\widetilde{x}(k) + (I-A)^{-1}\widetilde{x}(1), \quad n \in \mathbb{N}_2, \tag{51}$$

or equivalently

$$\widetilde{x}(n) = -(I-A)^{-1}\sum_{k=2}^{n-1}(-1)^{n-k}\binom{\alpha}{n-k}\widetilde{x}(k) + (I-A)^{-1}(-1)^{n-2}\binom{\alpha-1}{n-2}\widetilde{x}(1), \quad n \in \mathbb{N}_2. \tag{52}$$

Similarly if $x\colon \mathbb{N}_0 \to \mathbb{R}^d$ satisfies

$$x(n) = (I-A)^{-1}A\sum_{k=1}^{n-1}(-1)^{n-k}\binom{-\alpha}{n-k}x(k) + (I-A)^{-1}(-1)^{n}\binom{-\alpha}{n}x(0), \quad n \in \mathbb{N}_1 \tag{53}$$

or equivalently

$$x(n) = -(I-A)^{-1}\sum_{k=0}^{n-1}(-1)^{n-k}\binom{\alpha}{n-k}x(k), \quad n \in \mathbb{N}_1, \tag{54}$$

then $\widetilde{x}\colon \mathbb{N}_1 \to \mathbb{R}^d$, $\widetilde{x}(n) = x(n-1)$, satisfies

$$\widetilde{x}(n) = (I-A)^{-1}A\sum_{k=2}^{n-1}(-1)^{n-k}\binom{-\alpha}{n-k}\widetilde{x}(k) + (I-A)^{-1}(-1)^{n-1}\binom{-\alpha}{n-1}\widetilde{x}(1), \quad n \in \mathbb{N}_2, \tag{55}$$

or equivalently

$$\widetilde{x}(n) = -(I - A)^{-1} \sum_{k=1}^{n-1} (-1)^{n-k} \binom{\alpha}{n-k} \widetilde{x}(k), \quad n \in \mathbb{N}_2. \tag{56}$$

(b) Conversely, if $x: \mathbb{N}_1 \to \mathbb{R}^d$ satisfies (51) (or equivalently (52)), then $\overline{x}: \mathbb{N}_0 \to \mathbb{R}^d$, $\overline{x}(n) := x(n+1)$, satisfies (49) (or equivalently (50)). Similarly if $x: \mathbb{N}_1 \to \mathbb{R}^d$ satisfies (55) (or equivalently (56)) then $\overline{x}: \mathbb{N}_1 \to \mathbb{R}^d$, $\overline{x}(n) := x(n+1)$, satisfies (53) (or equivalently (54)).

(c) For $x_0 \in \mathbb{R}^d$

$$\varphi_{\mathrm{C}}^{\widetilde{\nabla}}(n, x_0) = \varphi_{\mathrm{C}}^{\overline{\nabla}}(n-1, x_0) \quad \text{and} \quad \varphi_{\mathrm{R-L}}^{\widetilde{\nabla}}(n, (I-A)x_0) = \varphi_{\mathrm{R-L}}^{\overline{\nabla}}(n-1, x_0), \qquad n \in \mathbb{N}_1.$$

In the next lemma we rewrite solutions of Volterra convolution equations as sums with recursively defined coefficients.

Lemma 3 (Volterra convolution representation)
Let $B, C \in \mathbb{R}^{d\times d}$, $a \in \mathbb{R}$, $c: \mathbb{N}_1 \to \mathbb{R}$, $g: \mathbb{N}_{a+1} \to \mathbb{R}$ and $x: \mathbb{N}_a \to \mathbb{R}^d$. If

$$x(n+a) = B \sum_{k=1}^{n-1} c(n-k)x(k+a) + Cg(n)x(a), \qquad n \in \mathbb{N}_1,$$

then

$$x(n+a) = \sum_{k=0}^{n-1} B^k C b(n,k) x(a), \qquad n \in \mathbb{N}_1, \tag{57}$$

where $b(n,k)$ for $n \in \mathbb{N}_1, k \in \{0, 1, \ldots, n-1\}$, are defined recursively by $b(n, 0) := g(n+a)$ and

$$b(n+1, k) := \sum_{j=k-1}^{n-1} c(n-j)b(j+1, k-1), \quad n \in \mathbb{N}_1, k \in \{1, 2, \ldots, n\}.$$

Proof We show the statement by induction over $n \in \mathbb{N}_1$. For $n = 1$ the statement is true. Suppose that (57) holds for $n \in \{1, \ldots, m\}$ for an $m \in \mathbb{N}_1$. Then

$$\begin{aligned}
x(m+1+a) &= B \sum_{k=1}^{m} c(m+1-k)x(k+a) + Cg(m+1+a)x(a) \\
&= B \sum_{k=1}^{m} c(m+1-k) \sum_{j=0}^{k-1} B^j C b(k,j) x(a) + Cg(m+1+a)x(a) \\
&= \left(\sum_{k=1}^{m} \sum_{j=0}^{k-1} c(m+1-k) B^{j+1} C b(k,j) + Cg(m+1+a) \right) x(a) \\
&= \left(\sum_{i=0}^{m-1} B^{i+1} C \sum_{j=i}^{m-1} c(m-j) b(j+1, i) + Cg(m+1+a) \right) x(a)
\end{aligned}$$

$$= \left(\sum_{k=1}^{m} B^k C \sum_{j=k-1}^{m-1} c(m-j) b(j+1, k-1) + C g(m+1+a) \right) x(a).$$

The last equality completes the proof. □

Applying the recursive representation of Volterra convolution equations in Lemma 3 to the time-invariant equations (49)–(56) of Remark 5, leads to the following result.

Theorem 4 (Recursive solution representation for backward time-invariant fractional systems)
Let $x_0 \in \mathbb{R}^d$, $A \in \mathbb{R}^{d\times d}$ and $\alpha \in (0, 1)$. Assume that $I - A$ is invertible.
(a) The solutions of (20) *with time-invariant linear part A satisfy*

$$\begin{aligned} \varphi_{\mathrm{C}}^{\nabla}(n, x_0) &= \sum_{k=0}^{n-1} (I-A)^{-k-1} A^k b_1(n,k) x_0 \\ &= \sum_{k=0}^{n-1} (-1)^k (I-A)^{-k-1} b_2(n,k) x_0, \qquad n \in \mathbb{N}_1, \end{aligned}$$

where $b_1(n, k)$ for $n \in \mathbb{N}_1$, $k \in \{0, 1, \ldots, n-1\}$, are defined recursively by $b_1(n, 0) := 1$ and

$$b_1(n+1, k) := \sum_{j=k-1}^{n-1} (-1)^{n-j} \binom{-\alpha}{n-j} b_1(j+1, k-1), \quad n \in \mathbb{N}_1, k \in \{1, 2, \ldots, n\},$$

and $b_2(n, k)$ for $n \in \mathbb{N}_1$, $k \in \{0, 1, \ldots, n-1\}$, are defined by $b_2(n, 0) := (-1)^{n-1} \binom{\alpha-1}{n-1}$ and

$$b_2(n+1, k) := \sum_{j=k-1}^{n-1} (-1)^{n-j} \binom{\alpha}{n-j} b_2(j+1, k-1), \quad n \in \mathbb{N}_1, k \in \{1, 2, \ldots, n\}.$$

(b) The solutions of (21) *with time-invariant linear part A satisfy*

$$\begin{aligned} \varphi_{\mathrm{R\text{-}L}}^{\nabla}(n, x_0) &= \sum_{k=0}^{n-1} (I-A)^{-k-1} A^k b_3(n,k) x_0 \\ &= \sum_{k=0}^{n-1} (I-A)^{-k-1} b_4(n,k) x_0, \qquad n \in \mathbb{N}_1, \end{aligned}$$

where $b_3(n,k)$ *for* $n \in \mathbb{N}_1$, $k \in \{0, 1, \ldots, n-1\}$, *are defined by* $b_3(n,0) := (-1)^n \binom{-\alpha}{n}$ *and*

$$b_3(n+1,k) := \sum_{j=k-1}^{n-1} (-1)^{n-j} \binom{-\alpha}{n-j} b_3(j+1, k-1), \quad n \in \mathbb{N}_1, k \in \{1, 2, \ldots, n\},$$

and $b_4(n,k)$ *for* $n \in \mathbb{N}_1$, $k \in \{0, 1, \ldots, n-1\}$, *are defined by* $b_4(n,0) := (-1)^n \binom{\alpha}{n}$ *and*

$$b_4(n+1,k) := \sum_{j=k-1}^{n-1} (-1)^{n-j} \binom{\alpha}{n-j} b_4(j+1, k-1), \quad n \in \mathbb{N}_1, k \in \{1, 2, \ldots, n\}.$$

(c) The solutions of (22) *with time-invariant linear part A satisfy*

$$\begin{aligned} \varphi_{\mathrm{C}}^{\widetilde{\nabla}}(n, x_0) &= \sum_{k=0}^{n-2} (I-A)^{-k-1} A^k b_1(n-1, k) x_0 \\ &= \sum_{k=0}^{n-2} (-1)^k (I-A)^{-k-1} b_2(n-1, k) x_0, \qquad n \in \mathbb{N}_2. \end{aligned}$$

(d) The solutions of (23) *with time-invariant linear part A satisfy*

$$\begin{aligned} \varphi_{\mathrm{R\text{-}L}}^{\widetilde{\nabla}}(n, x_0) &= \sum_{k=0}^{n-2} (I-A)^{-k-1} A^k b_3(n-1, k) x_0 \\ &= \sum_{k=0}^{n-2} (I-A)^{-k-1} b_4(n-1, k) x_0, \qquad n \in \mathbb{N}_2. \end{aligned}$$

Proof (a) Applying Lemma 3 with $B := (I-A)^{-1}A$, $C := (I-A)^{-1}$, $c(n) := (-1)^n \binom{-\alpha}{n}$ and $g(n) := 1$ for $n \in \mathbb{N}_1$ to (49) yields $\varphi_{\mathrm{C}}^{\overline{\nabla}}(n, x_0) = \sum_{k=0}^{n-1} (I-A)^{-k-1} A^k b_1(n,k) x_0$, and the second formula for $\varphi_{\mathrm{C}}^{\overline{\nabla}}(n, x_0)$ follows from Eq. (50) and again Lemma 3 with $B := -(I-A)^{-1}$, $C := (I-A)^{-1}$, $c(n) := (-1)^n \binom{\alpha}{n}$ and $g(n) := (-1)^{n-1} \binom{\alpha-1}{n-1}$ for $n \in \mathbb{N}_1$.

(b) As in (a), applying Lemma 3 to Eqs. (53) and (54) yields the result.

(c)–(d) This follows from Remark 5(c). □

We cite two results on asymptotic stability of (18) in the time-invariant case.

Theorem 5 (Characterization of asymptotic stability for time-invariant Caputo forward equations [3, Theorem 3.2])
Let $A \in \mathbb{R}^{d \times d}$, $\alpha \in (0, 1)$. *Then the following two statements are equivalent.*

(i) $({}_{\mathrm{C}}\overline{\Delta}^{\alpha}x)(n) = Ax(n)$ *is asymptotically stable, i.e.* $\lim_{n\to\infty}\varphi_{\mathrm{C}}^{\Delta}(n, x_0) = 0$ *for all* $x_0 \in \mathbb{R}^d$.

(ii) The isolated zeros, off the non-negative real axis, of $z \mapsto det(I - z^{-1}(1 - z^{-1})^{-\alpha}A)$ *lie inside the unit circle.*

To present another sufficient condition for asymptotic stability of (18) in the time-invariant case, let us denote

$$S_1^{\alpha} = \left\{ z \in \mathbb{C}\colon |z| < \left(2\cos\frac{|\arg z| - \pi}{2-\alpha}\right)^{\alpha} \text{ and } |\arg z| > \frac{\alpha\pi}{2} \right\}.$$

Theorem 6 (Asymptotic stability of time-invariant Caputo forward equations [13, Theorem 1.4])
Let $A \in \mathbb{R}^{d\times d}$.

(a) If $\lambda \in S_1^{\alpha}$ *for all eigenvalues* λ *of* A*, then* $({}_{\mathrm{C}}\overline{\Delta}^{\alpha}x)(n) = Ax(n)$ *is asymptotically stable. In this case, the solutions decay towards zero algebraically (and not exponentially), more precisely, for* $x_0 \in \mathbb{R}^d$

$$\|\varphi_{\mathrm{C}}^{\Delta}(n, x_0)\| = O(n^{-\alpha}) \quad as\ n \to \infty.$$

(b) If $\lambda \in \mathbb{C} \setminus \mathrm{cl}\, S_1^{\alpha}$ *for an eigenvalue* λ *of* A*, then* $({}_{\mathrm{C}}\overline{\Delta}^{\alpha}x)(n) = Ax(n)$ *is not stable and there exist* $x_0 \in \mathbb{R}^d$*,* $C > 0$ *and* $r > 1$ *with*

$$\|\varphi_{\mathrm{C}}^{\Delta}(n, x_0)\| \geq Cr^n, \qquad n \in \mathbb{N}_0.$$

The next result provides a sufficient condition for asymptotic stability of (19) in the time-invariant case.

Theorem 7 (Asymptotic stability of time-invariant Riemann-Liouville forward equations)
Let $A \in \mathbb{R}^{d\times d}$ *and* $\alpha \in (0, 1)$.

(a) If $\lambda \in S_1^{\alpha}$ *for all eigenvalues* λ *of* A*, then* $\varphi_{\mathrm{R-L}}^{\Delta}(\cdot, x_0) \in l^1$ *for each* $x_0 \in \mathbb{R}^d$*, hence* $({}_{\mathrm{R\text{-}L}}\overline{\Delta}^{\alpha}x)(n) = Ax(n)$ *is asymptotically stable.*

(b) If $\lambda \in \mathbb{C} \setminus \mathrm{cl}\, S_1^{\alpha}$ *for an eigenvalue* λ *of* A*, then* $({}_{\mathrm{R\text{-}L}}\overline{\Delta}^{\alpha}x)(n) = Ax(n)$ *is not stable and there exist* $x_0 \in \mathbb{R}^d$*,* $C > 0$ *and* $r > 1$ *with*

$$\|\varphi_{\mathrm{R-L}}^{\Delta}(n, x_0)\| \geq Cr^n, \qquad n \in \mathbb{N}_0. \tag{58}$$

Proof (a) The fact that the condition $\lambda \in S_1^{\alpha}$ for all eigenvalues λ of A implies stability of $({}_{\mathrm{R\text{-}L}}\overline{\Delta}^{\alpha}x)(n) = Ax(n)$ is proved in the first step of the proof of Theorem 1.4 in [13].

(b) Suppose that there exists an eigenvalue $\lambda \in \mathbb{C} \setminus \mathrm{cl}\, S_1^{\alpha}$ of A. As it has been shown in steps 1 and 2 of the proof of Theorem 1.4 in [13], this implies that the equation

$$\det\left(A - z(1 - z^{-1})^{\alpha}\right) = 0$$

has a solution $z_0 \in \mathbb{C}$ with $|z_0| > 1$. The Z-transform $y(z)$ of $\varphi^{\Delta}_{\text{R-L}}(n, x_0)$ is given by

$$y(z) = -\left(A - z(1 - z^{-1})^{\alpha}\right) z x_0,$$

and has a non-removable singularity at z_0. Hence the radius of convergence r of at least one of the coordinates $y_i(z)$ of $y(z)$ satisfies $r > 1$. Using the Cauchy-Hadamard theorem we get

$$r = \limsup_{n\to\infty} \sqrt[n]{|y_i(n)|} > 1$$

and consequently we get (58). □

To present stability results of the Riemann-Liouville backward equation (23) in the time-invariant case, let us denote

$$S_2^{\alpha} = \left\{ z \in \mathbb{C}\colon |z| > \left(2\cos\frac{|\arg z|}{\alpha}\right)^{\alpha} \text{ or } |\arg z| > \frac{\alpha\pi}{2} \right\}$$

and the interior of its complement in $\mathbb{C}$

$$U^{\alpha} = \left\{ z \in \mathbb{C}\colon |z| < \left(2\cos\frac{|\arg z| - \pi}{2-\alpha}\right)^{\alpha} \text{ and } |\arg z| < \frac{\alpha\pi}{2} \right\}.$$

Theorem 8 (Asymptotic stability of time-invariant Riemann-Liouville backward equations [15, Theorem 6])
Let $A \in \mathbb{R}^{d\times d}$, $\alpha \in (0,1)$. Assume that $I - A$ is invertible.

(a) If all eigenvalues of A lie in S_2^{α}, then $\varphi^{\tilde{\nabla}}_{\text{R-L}}(\cdot, x_0) \in l^1$, hence $({}_{\text{R-L}}\tilde{\nabla}^{\alpha} x)(n) = A(n)x(n)$ is asymptotically stable. Moreover, if all eigenvalues of $(I - A)^{-1}$ lie inside the open unit disc, then $\|\varphi^{\tilde{\nabla}}_{\text{R-L}}(n, x_0)\| = O(n^{-\alpha-1})$ as $n \to \infty$ for all $x_0 \in \mathbb{R}^d$.

(b) If there exists an eigenvalue λ of A such that $\lambda \in U^{\alpha}$, then $({}_{\text{R-L}}\tilde{\nabla}^{\alpha} x)(n) = A(n)x(n)$ is not stable and there exist $x_0 \in \mathbb{R}^d$, $C > 0$ and $r > 1$ with

$$\|\varphi^{\tilde{\nabla}}_{\text{R-L}}(n, x_0)\| \geq C r^n, \qquad n \in \mathbb{N}_0.$$

Theorem 8 does not answer the stability problem if some of the eigenvalues of A lie on the boundary of S_2^{α}. The following assertion from [15] demonstrates that all stability variants are possible in such a case.

Theorem 9 (Asymptotic behavior on the stability boundary of time-invariant Riemann-Liouville backward equations [15, Theorem 9])
Let $A \in \mathbb{R}^{d\times d}$, $\alpha \in (0,1)$. Assume that $I - A$ is invertible, that zero is an eigenvalue of A and that all nonzero eigenvalues of A belong to S_2^{α}. Denote by $r \in \mathbb{N}_1$ the maximal size of the Jordan blocks corresponding to the zero eigenvalue.

(a) If $r < \alpha^{-1}$, then $({}_{\text{R-L}}\tilde{\nabla}^{\alpha} x)(n) = A(n)x(n)$ is asymptotically stable and $\|\varphi^{\tilde{\nabla}}_{\text{R-L}}(n, x_0)\| = O(n^{r\alpha-1})$ as $n \to \infty$.

(b) If $r = \alpha^{-1}$, then $({}_{\text{R-L}}\tilde{\nabla}^{\alpha} x)(n) = A(n)x(n)$ is stable but not asymptotically stable.

(c) If $r > \alpha^{-1}$, *then* $({}_{\text{R-L}}\widetilde{\nabla}^{\alpha} x)(n) = A(n)x(n)$ *is not stable.*

Remark 6 Remark 5(c) implies that Theorems 8 and 9 are also valid for $({}_{\text{R-L}}\overline{\nabla}^{\alpha} x)(n) = A(n)x(n)$.

To discuss stability of the Caputo backward equation (52) in Remark 4 we rewrite it, using the fact [27] that $\binom{\alpha}{n-1} = \binom{\alpha-1}{n-1} + \binom{\alpha-1}{n-2}$, to get

$$\begin{aligned} x(n) = &-(I - A)^{-1} \sum_{k=1}^{n-1} (-1)^{n-k} \binom{\alpha}{n-k} x(k) \\ &+ (I - A)^{-1} (-1)^{n-1} \binom{\alpha - 1}{n - 1} x(1), \quad n \in \mathbb{N}_2. \end{aligned} \tag{59}$$

We will also use the following solution representation of inhomogeneous Volterra equations.

Theorem 10 (Inhomogeneous Volterra equation)
Let $C \in \mathbb{R}^{d\times d}$, $a \colon \mathbb{N}_1 \to \mathbb{R}$, $g \colon \mathbb{N}_2 \to \mathbb{R}^d$ *and* $R \colon \mathbb{N}_1 \to \mathbb{R}^{d\times d}$. *If*

$$R(n) = C \sum_{k=1}^{n-1} a(n-k) R(k), \qquad n \in \mathbb{N}_2$$

with initial condition $R(1) = I$, *then the unique solution* $x \colon \mathbb{N} \to \mathbb{R}^d$ *of the equation*

$$x(n) = C \sum_{k=1}^{n-1} a(n-k) x(k) + g(n), \qquad n \in \mathbb{N}_2$$

with the initial condition $x(1) = x_1 \in \mathbb{R}^d$, *is given by*

$$x(n) = R(n)x_1 + \sum_{k=1}^{n-1} R(k) g(n-k+1), \qquad n \in \mathbb{N}_1. \tag{60}$$

Proof For $n = 1$ the statement is true. Let $m \in \mathbb{N}_2$. Suppose that (60) holds for all $n \in \{1, \ldots, m\}$. Then it also holds for $m + 1$, since

$$\begin{aligned} x(m+1) &= C \sum_{k=1}^{m} a(m+1-k) x(k) + g(m+1) \\ &= C \sum_{k=1}^{m} a(m+1-k) \Big(R(k)x_1 + \sum_{j=1}^{k-1} R(j) g(k-j+1) \Big) + g(m+1) \end{aligned}$$

$$= C\sum_{k=1}^{m} a(m+1-k)R(k)x_1$$
$$+ C\sum_{k=1}^{m} a(m+1-k)\sum_{j=1}^{k-1} R(j)g(k-j+1) + g(m+1)$$
$$= R(m+1)x_1 + \sum_{k=1}^{m} R(k)g(m+1-k+1).$$

When we compare the Caputo backward equation (52) in its equivalent form (59) to the Riemann-Liouville backward equation (56) and use Theorem 10, we obtain the following relation between Caputo and Riemann-Liouville equations.

Lemma 4 (Relation between backward Caputo and Riemann Liouville equations) *Let $A \in \mathbb{R}^{d\times d}$, $\alpha \in (0, 1)$. If $X : \mathbb{N}_1 \to \mathbb{R}^{d\times d}$ is the solution of the Riemann-Liouville matrix equation*

$$X(n) = -(I-A)^{-1}\sum_{k=1}^{n-1}(-1)^{n-k}\binom{\alpha}{n-k}X(k), \qquad n \in \mathbb{N}_2, \tag{61}$$

with initial condition $X(1) = I$, then the solution $x : \mathbb{N} \to \mathbb{R}$ of the Caputo equation (59) *with initial condition $x(1) = x_1 \in \mathbb{R}^d$, is given by*

$$x(n) = X(n)x_1 + \sum_{k=1}^{n-1} X(k)g(n-k+1), \qquad n \in \mathbb{N}_2, \tag{62}$$

where

$$g(n) = (I-A)^{-1}(-1)^{n-1}\binom{\alpha-1}{n-1}x(1), \qquad n \in \mathbb{N}_2.$$

Using the representation (62) we show the following stability result for (22) in the time-invariant case.

Theorem 11 (Asymptotic stability of time-invariant Caputo backward equations) *Let $A \in \mathbb{R}^{d\times d}$. Assume that $I - A$ is invertible. If all eigenvalues of A lie in S_2^α, then*

$$({}_{\mathrm{C}}\widetilde{\nabla}^{\alpha}x)(n) = A(n)x(n)$$

is asymptotically stable and $\|\varphi_{\mathrm{C}}^{\widetilde{\nabla}}(n, x_0)\| = O(n^{-\alpha})$ as $n \to \infty$ for all $x_0 \in \mathbb{R}^d$.

Proof Suppose that all the eigenvalues of A lie in S_2^α and consider the sequence $X(n)$, given by (61) and $X(1) = I$. Since $\varphi_{\mathrm{R\text{-}L}}^{\widetilde{\nabla}}(n, x_0) = X(n)x_0$, by Theorem 8 we know that $X(n) \in l^1$ and therefore there exists a constant $C > 0$ such that

$$\|X(n)\| \le \frac{C}{n}, \qquad n \in \mathbb{N}_1. \tag{63}$$

It is also known [27] that for the binomial coefficients we have

$$\left|\binom{\alpha-1}{n}\right| \le \frac{C}{(n+1)^{\alpha}}, \qquad n \in \mathbb{N}_1, \tag{64}$$

for certain $C > 0$ (without loss of generality we may assume that the constants C are the same in the last two inequalities (63) and (64)). From (62) we have

$$\begin{aligned}\|\varphi_{\mathrm{C}}^{\widetilde{\nabla}}(n, x_0)\| &= \left\| X(n)x_0 + (I-A)^{-1}\sum_{k=1}^{n-1}(-1)^{n-k}X(k)\binom{\alpha-1}{n-k}x_0 \right\| \\ &\le C_1 \sum_{k=1}^{n} \|X(k)\| \left|\binom{\alpha-1}{n-k}\right| = C_1 \sum_{j=0}^{n-1} \|X(n-j)\| \left|\binom{\alpha-1}{j}\right|,\end{aligned}$$

where $C_1 = \|x_0\| \max\left\{1, \|(I-A)^{-1}\|\right\}$. Using (64) and dividing the sum into two parts, we get

$$\begin{aligned}\|\varphi_{\mathrm{C}}^{\widetilde{\nabla}}(n, x_0)\| &\le C_1 C \sum_{j=0}^{n-1} \frac{\|X(n-j)\|}{(j+1)^{\alpha}} \\ &= C_1 C \sum_{j=0}^{\lfloor n-1 \rfloor} \frac{\|X(n-j)\|}{(j+1)^{\alpha}} + C_1 C \sum_{j=\lfloor n-1 \rfloor+1}^{n-1} \frac{\|X(n-j)\|}{(j+1)^{\alpha}}.\end{aligned} \tag{65}$$

To estimate the first term we use (63) and the inequality

$$\sum_{i=1}^{l} \frac{1}{i^{\alpha}} \le \int_1^{l+1} x^{-\alpha} dx = \frac{(l+1)^{-\alpha+1}}{-\alpha+1} - \frac{1}{-\alpha+1}, \qquad l \in \mathbb{N}_1,$$

as follows

$$\begin{aligned}\sum_{j=0}^{\lfloor n-1 \rfloor} \frac{\|X(n-j)\|}{(j+1)^{\alpha}} &\le \frac{C}{n - \lfloor n-1 \rfloor} \sum_{j=0}^{\lfloor n-1 \rfloor} \frac{1}{(j+1)^{\alpha}} \\ &= \frac{C}{n - \lfloor n-1 \rfloor} \sum_{j=1}^{\lfloor n-1 \rfloor+1} \frac{1}{j^{\alpha}} \\ &\le \frac{C}{n - \lfloor n-1 \rfloor} \left(\frac{(\lfloor n-1 \rfloor + 2)^{-\alpha+1}}{-\alpha+1} - \frac{1}{-\alpha+1} \right).\end{aligned}$$

From the last inequality it is clear that

$$\sum_{j=0}^{\lfloor n-1 \rfloor} \frac{\|X(n-j)\|}{(j+1)^\alpha} \leq \frac{C_2}{n^\alpha}, \tag{66}$$

for certain $C_2 > 0$ and all $n \in \mathbb{N}_1$. To estimate the second term in (65) we proceed as follows

$$\sum_{j=\lfloor n-1 \rfloor+1}^{n-1} \frac{\|X(n-j)\|}{(j+1)^\alpha} \leq \frac{C_3}{n^\alpha} \sum_{j=\lfloor n-1 \rfloor+1}^{n-1} \|X(n-j)\| \leq \frac{C_4}{n^\alpha}, \tag{67}$$

for certain $C_3, C_4 > 0$ and all $n \in \mathbb{N}_1$. Applying (66) and (67) to (65), we obtain the statement of the theorem. □

6 Asymptotic Properties of Scalar Linear Fractional Equations

In this section we investigate one-dimensional linear fractional equations and discuss their asymptotic behavior in the time-invariant case in the first subsection. In the second subsection we study asymptotic behavior of time-varying backward equations.

6.1 *Scalar Time-Invariant Equations*

We consider one-dimensional systems (18)–(23) in the time-invariant case, i.e. we assume that $A(n) = \lambda \in \mathbb{R}$, $n \in \mathbb{N}_0$. For multi-dimensional time-invariant systems the picture of stability is not complete and the theorems from the previous section leave some cases unsolved. For one-dimensional equations the problem of stability and asymptotic stability is much more exhaustively described but even in this relatively simple situation it is not completely solved.

Theorem 12 (Asymptotic behavior of scalar time-invariant fractional equations) *Let $\lambda \in \mathbb{R}$ and $\alpha \in (0, 1)$.*

(a) $({}_{\text{C}}\overline{\Delta}^{\alpha} x)(n) = \lambda x(n)$ is asymptotically stable if and only if $\lambda \in (-2^\alpha, 0)$, and $({}_{\text{C}}\overline{\Delta}^{\alpha} x)(n) = \lambda x(n)$ is stable, but not asymptotically stable if $\lambda = 0$ or $\lambda = -2^\alpha$.
(b) $({}_{\text{R-L}}\overline{\Delta}^{\alpha} x)(n) = \lambda x(n)$ is asymptotically stable if and only if $\lambda \in (-2^\alpha, 0]$, and $({}_{\text{R-L}}\overline{\Delta}^{\alpha} x)(n) = \lambda x(n)$ is stable, but not asymptotically stable if $\lambda = -2^\alpha$.

(c) $({}_{\text{C}}\overline{\nabla}^{\alpha} x)(n) = \lambda x(n)$ with $\lambda \neq 1$ is asymptotically stable if $\lambda \in (-\infty, 0] \cup (2^\alpha, \infty)$.

(d) $({}_{\text{R-L}}\overline{\nabla}^{\alpha} x)(n) = \lambda x(n)$ with $\lambda \neq 1$ is asymptotically stable if and only if $\lambda \in (-\infty, 0] \cup (2^\alpha, \infty)$, $({}_{\text{R-L}}\overline{\nabla}^{\alpha} x)(n) = \lambda x(n)$ with $\lambda \neq 1$ is not stable if $\lambda \in (0, 2^\alpha)$.

(e) $({}_{\mathrm{C}}\widetilde{\nabla}^{\alpha}x)(n) = \lambda x(n)$ *with* $\lambda \neq 1$ *is asymptotically stable if* $\lambda \in (-\infty, 0] \cup (2^{\alpha}, \infty)$.

(f) $({}_{\mathrm{R\text{-}L}}\widetilde{\nabla}^{\alpha}x)(n) = \lambda x(n)$ *with* $\lambda \neq 1$ *is asymptotically stable if and only if* $\lambda \in (-\infty, 0] \cup (2^{\alpha}, \infty)$,
$({}_{\mathrm{R\text{-}L}}\overline{\nabla}^{\alpha}x)(n) = \lambda x(n)$ *with* $\lambda \neq 1$ *is not stable if* $\lambda \in (0, 2^{\alpha})$.

Proof (a) and (b) have been proved in [13] (see also [6, Example 29]). (c) follows from Theorem 11. (f) has been proved in [14]. (d) and (e) follow from Remark 5(c) together with (c) and (f), respectively. □

Theorem 12 leaves the following questions for (20)–(23) with $\lambda \neq 1$ open:

Scalar linear fractional difference equation	Open question
$({}_{\mathrm{C}}\overline{\nabla}^{\alpha}x)(n) = \lambda x(n)$ and $({}_{\mathrm{C}}\widetilde{\nabla}^{\alpha}x)(n) = \lambda x(n)$	asymptotic behavior for $\lambda \in (0, 2^{\alpha}]$
$({}_{\mathrm{R\text{-}L}}\overline{\nabla}^{\alpha}x)(n) = \lambda x(n)$ and $({}_{\mathrm{R\text{-}L}}\widetilde{\nabla}^{\alpha}x)(n) = \lambda x(n)$	stability for $\lambda = 2^{\alpha}$

For scalar time-invariant fractional equations, also some results about the convergence and divergence rates of solutions are known. These results are collected in the next two theorems.

Theorem 13 (Growth and decay rates for scalar linear time-invariant fractional equations)
Consider (20)–(23) *with linear part* $\lambda \in \mathbb{R}$. *Let* $\alpha \in (0, 1)$ *and* $x_0 \in \mathbb{R}$. *Then*

(a) $\lim_{n\to\infty} \varphi_{\mathrm{C}}^{\Delta}(n, x_0)n^{\alpha} = \frac{-x_0}{\lambda\Gamma(1-\alpha)}$ *if* $\lambda \in (-2\alpha, 0)$,

(b) $\lim_{n\to\infty} \varphi_{\mathrm{R\text{-}L}}^{\Delta}(n, x_0)n^{\alpha+1} = \frac{-x_0}{\lambda^2\Gamma(-\alpha)}$ *if* $\lambda \in (-2\alpha, 0)$,

(c) $\lim_{n\to\infty} \varphi_{\mathrm{C}}^{\nabla}(n, x_0)n^{\alpha} = \frac{-x_0}{\lambda\Gamma(1-\alpha)}$ *if* $\lambda \in (-\infty, 0) \cup (2, \infty)$,

(d) $\lim_{n\to\infty} \varphi_{\mathrm{R\text{-}L}}^{\widetilde{\nabla}}(n, x_0)n^{\alpha+1} = \frac{\alpha(1-\lambda)x_0}{\lambda^2\Gamma(1-\alpha)}$ *if* $\lambda \in (-\infty, 0) \cup (2, \infty)$, *and*
$\lim_{n\to\infty} \varphi_{\mathrm{R\text{-}L}}^{\widetilde{\nabla}}(n, x_0)n^{\alpha+1} = \frac{x_0}{\Gamma(\alpha)}$ *if* $\lambda = 0$,

(e) $\lim_{n\to\infty} \varphi_{\mathrm{C}}^{\overline{\nabla}}(n, x_0)n^{\alpha} = \frac{-x_0}{\lambda\Gamma(1-\alpha)}$ *if* $\lambda \in (-\infty, 0) \cup (2, \infty)$,

(f) $\lim_{n\to\infty} \varphi_{\mathrm{R\text{-}L}}^{\overline{\nabla}}(n, x_0)n^{\alpha+1} = \frac{\alpha(1-\lambda)^2x_0}{\lambda^2\Gamma(1-\alpha)}$ *if* $\lambda \in (-\infty, 0) \cup (2, \infty)$, *and*
$\lim_{n\to\infty} \varphi_{\mathrm{R\text{-}L}}^{\overline{\nabla}}(n, x_0)n^{\alpha+1} = \frac{x_0}{\Gamma(\alpha)}$ *if* $\lambda = 0$,

(g) *If* $0 < \lambda < 1$ *and* $x_0 > 0$, *then* $\varphi_{\mathrm{R\text{-}L}}^{\widetilde{\nabla}}(\cdot, x_0)$ *grows geometrically. More precisely,*

$$\frac{\lambda^{1/\alpha}x_0}{(1-\lambda^{1/\alpha})^n} < \varphi_{\mathrm{R\text{-}L}}^{\widetilde{\nabla}}(n, x_0) < \frac{x_0}{(1-\lambda^{1/\alpha})^n}, \qquad n \in \mathbb{N}_2.$$

In the proof of Theorem 13 we use the following Lemma from [5].

Lemma 5 ([5, Lemma 6])
Let $\alpha \in (0, 1)$ *and* $r, f : \mathbb{N}_1 \to \mathbb{R}$. *If*

$$\sup_{n\in\mathbb{N}_1}\left|\frac{r(n)}{n^{-\alpha-1}}\right| < \infty \quad \text{and} \quad \lim_{n\to\infty}\frac{f(n)}{n^{-\alpha}} =: d \text{ exists,} \tag{68}$$

then the convolution $r * f : \mathbb{N}_1 \to \mathbb{R}$, $(r * f)(n) = \sum_{i=0}^{n} r(n-i) f(i)$, *satisfies*

$$\lim_{n\to\infty} \frac{(r * f)(n)}{n^{-\alpha}} = d \sum_{i=1}^{\infty} r(i).$$

***Proof** (**of Theorem** 13)* (a) is proved in [5, Theorem 4].

(b) is shown in [13, Corollary 4.2].

(d) and (g) are proved in [14, Theorem 4.7].

(f) follows from (d) by Remark 5(c).

(e) is a consequence of (c) and Remark 5(c). It remains to prove (c).

(c) Let $x : \mathbb{N} \to \mathbb{R}$ be a solution of $({}_{\text{R-L}}\widetilde{\nabla}^{\alpha} x)(n) = \lambda x(n)$. Then by (36) in Lemma 2,

$$x(n) = -\frac{1}{1-\lambda} \sum_{k=1}^{n-1} (-1)^{n-k} \binom{\alpha}{n-k} x(k), \qquad n \in \mathbb{N}_2, \tag{69}$$

and by Theorem 8, if $\lambda \in (-\infty, 0] \cup (2^{\alpha}, \infty)$ then $x \in l^1(\mathbb{N}_1)$. We will calculate

$$S := \sum_{n=1}^{\infty} x(n).$$

Summing up the Eqs. (69) for n from 2 to ∞ we get

$$S - x(1) = -\frac{1}{1-\lambda} \sum_{n=2}^{\infty} \sum_{k=1}^{n-1} (-1)^{n-k} \binom{\alpha}{n-k} x(k). \tag{70}$$

Since the series $\sum_{i=1}^{\infty} (-1)^i \binom{\alpha}{i}$ and $\sum_{i=1}^{\infty} x(i)$ are absolutely convergent, their Cauchy product

$$\sum_{n=2}^{\infty} \sum_{k=1}^{n-1} (-1)^{n-k} \binom{\alpha}{n-k} x(k)$$

is also absolutely convergent and its sum is

$$\sum_{i=1}^{\infty} (-1)^i \binom{\alpha}{i} \sum_{i=1}^{\infty} x(i) = -S. \tag{71}$$

In the last step we use the well known formula

$$\sum_{i=0}^{\infty} \binom{\alpha}{i} w^i = (1+w)^{\alpha}, \qquad w \in [-1, 1],$$

with $w = -1$. Combining (70) with (71) we get

$$S = \frac{\lambda - 1}{\lambda} x(1). \tag{72}$$

From Lemma 4 we know that the solution $y(n)$, $n \in \mathbb{N}_1$, of the Caputo time-invariant one-dimensional equation (22) is given by

$$y(n) = (x * g)(n) = \sum_{j=1}^{n} x(n + 1 - j) g(j), \tag{73}$$

where $x(n)$, $n \in \mathbb{N}_1$, is the solution of (69) with $x(1) = 1$ and $g(n)$, $n \in \mathbb{N}_1$, is given by

$$g(n) = \begin{cases} y(1) & \text{for } n = 1, \\ (1 - \lambda)^{-1} (-1)^{n-1} \binom{\alpha - 1}{n-1} y(1) & \text{for } n \in \mathbb{N}_2. \end{cases}$$

Using formula (6) in [5]

$$\lim_{n \to \infty} (-1)^{n-1} \binom{\alpha - 1}{n - 1} n^{\alpha} = \frac{1}{\Gamma(1 - \alpha)},$$

we get

$$\lim_{n \to \infty} g(n) n^{\alpha} = \frac{y(1)}{(1 - \lambda) \Gamma(1 - \alpha)}.$$

Moreover, from Theorem 8 we know that the sequence $r(n) = x(n)$, $n \in \mathbb{N}_1$ satisfies condition (68) and therefore we may apply Lemma 5 to the sequences $r(n) = x(n)$ and $f(n) = g(n)$, $n \in \mathbb{N}_1$. This leads, in light of (72) and (73) to

$$\lim_{n \to \infty} \frac{y(n)}{n^{-\alpha}} = -\frac{x(1)}{\lambda \Gamma(1 - \alpha)}.$$

The last equality completes the proof of (c). □

6.2 Scalar Time-Varying Backward Equations

In this subsection we consider one dimensional time-varying fractional backward equations (20)–(23) with linear part $\lambda \colon \mathbb{N}_0 \to \mathbb{R}$, i.e.

$$({}_{\text{C}}\overline{\nabla}^{\alpha} x)(n) = \lambda(n) x(n) \quad \text{and} \quad ({}_{\text{R-L}}\overline{\nabla}^{\alpha} x)(n) = \lambda(n) x(n), \quad n \in \mathbb{N}_1, \tag{74}$$

$$({}_{\text{C}}\widetilde{\nabla}^{\alpha} x)(n) = \lambda(n) x(n) \quad \text{and} \quad ({}_{\text{R-L}}\widetilde{\nabla}^{\alpha} x)(n) = \lambda(n) x(n), \quad n \in \mathbb{N}_2. \tag{75}$$

We first provide conditions under which an order relation between two linear parts and initial conditions implies an order of the two corresponding solutions.

Theorem 14 (Comparison theorem for scalar backward equations)
Let $x_1, x_2 \in \mathbb{R}$, $\lambda_1, \lambda_2 \colon \mathbb{N}_0 \to \mathbb{R}$, $\alpha \in (0,1)$, and assume the order relations

$$x_1 \geq x_2 > 0 \quad \text{and} \quad \lambda_1(n) \geq \lambda_2(n) \quad \text{for each } n \in \mathbb{N}_0.$$

For $\widehat{\nabla}^\alpha \in \{{}_{\mathrm{C}}\overline{\nabla}^\alpha, {}_{\mathrm{R\text{-}L}}\overline{\nabla}^\alpha, {}_{\mathrm{C}}\widetilde{\nabla}^\alpha, {}_{\mathrm{R\text{-}L}}\widetilde{\nabla}^\alpha\}$ let $\varphi_1^{\widehat{\nabla}}(\cdot, x_1)$ and $\varphi_2^{\widehat{\nabla}}(\cdot, x_2)$ denote the solutions of

$$(\widehat{\nabla}^\alpha x)(n) = \lambda_1(n)x(n) \quad \text{and} \quad (\widehat{\nabla}^\alpha x)(n) = \lambda_2(n)x(n)$$

with initial condition x_1 and x_2, respectively.

If either $\lambda_1(n) < 1$ for each $n \in \mathbb{N}_0$, or $\lambda_2(n) > 1$ for each $n \in \mathbb{N}_0$, then

$$\varphi_1^{\widehat{\nabla}}(n, x_1) \geq \varphi_2^{\widehat{\nabla}}(n, x_2), \qquad n \in \mathbb{N}_0. \tag{76}$$

Proof Under the assumption that $\lambda_1(n) < 1$ for each $n \in \mathbb{N}_0$, the statement has been proved for $\widehat{\nabla}^\alpha = {}_{\mathrm{C}}\overline{\nabla}^\alpha$ in [21, Theorem 2.4] and for $\widehat{\nabla}^\alpha = {}_{\mathrm{R\text{-}L}}\overline{\nabla}^\alpha$ in [20, Theorem 2.5]. Remark 4(c) implies the statement also for $\widehat{\nabla}^\alpha = {}_{\mathrm{C}}\widetilde{\nabla}^\alpha$ and $\widehat{\nabla}^\alpha = {}_{\mathrm{R\text{-}L}}\widetilde{\nabla}^\alpha$.

Assume now that $\lambda_2(n) > 1$ for each $n \in \mathbb{N}_0$. We consider first the case $\widehat{\nabla}^\alpha = {}_{\mathrm{R\text{-}L}}\overline{\nabla}^\alpha$. For $n = 1$ the statement is true, since a direct calculation shows that for $i = 1, 2$

$$\varphi_i^{\widehat{\nabla}}(n, x_i) = \frac{\alpha x_i}{-\lambda_i(1) + 1}.$$

Suppose that (76) holds for $n \in \{0, 1, \ldots, m\}$ for an $m \in \mathbb{N}_1$, then according to (32) we have

$$\begin{aligned}
(1 - \lambda_1(m+1))\varphi_1^{\widehat{\nabla}}(m+1, x_1) &= -\sum_{k=0}^{m} (-1)^{m+1-k} \binom{\alpha}{m+1-k} \varphi_1^{\widehat{\nabla}}(k, x_1) \\
&\geq -\sum_{k=0}^{m} (-1)^{m+1-k} \binom{\alpha}{m+1-k} \varphi_2^{\widehat{\nabla}}(k, x_1) \\
&= (1 - \lambda_2(m+1))\varphi_2^{\widehat{\nabla}}(m+1, x_1),
\end{aligned}$$

i.e. (76) holds for $n = m + 1$ and the proof is completed. In the same way, using the representation (30), the statement follows for $\widehat{\nabla}^\alpha = {}_{\mathrm{C}}\overline{\nabla}^\alpha$. Finally using Remark 4(c) again, the statement follows also for $\widehat{\nabla}^\alpha = {}_{\mathrm{C}}\widetilde{\nabla}^\alpha$ and $\widehat{\nabla}^\alpha = {}_{\mathrm{R\text{-}L}}\widetilde{\nabla}^\alpha$. □

The next theorem presents sufficient conditions for asymptotic stability of the Eqs. (74) and (75).

Theorem 15 (Asymptotic stability of scalar backward equations)
Let $\lambda \colon \mathbb{N}_0 \to \mathbb{R}$, $\alpha \in (0,1)$.

(a) If $\sup_{n \in \mathbb{N}_0} \lambda(n) < 0$ then

$$({}_{\mathrm{C}}\overline{\nabla}^\alpha x)(n) = \lambda(n)x(n) \quad \text{and} \quad ({}_{\mathrm{C}}\widetilde{\nabla}^\alpha x)(n) = \lambda(n)x(n)$$

are asymptotically stable.

(b) *If* $\inf_{n\in\mathbb{N}_0} |1-\lambda(n)| \geq 1$ *then*

$$({}_{\text{R-L}}\overline{\nabla}^{\alpha} x)(n) = \lambda(n)x(n) \quad \textit{and} \quad ({}_{\text{R-L}}\widetilde{\nabla}^{\alpha} x)(n) = \lambda(n)x(n)$$

are asymptotically stable.

Proof (a) The asymptotic stability of $({}_{\text{C}}\overline{\nabla}^{\alpha} x)(n) = \lambda(n)x(n)$ was proved in [21, Theorem B and D]. Remark 4(c) implies the asymptotic stability of $({}_{\text{C}}\widetilde{\nabla}^{\alpha} x)(n) = \lambda(n)x(n)$.

(b) [20, Theorem $\widehat{\text{B}}$] implies the asymptotic stability of $({}_{\text{R-L}}\overline{\nabla}^{\alpha} x)(n) = \lambda(n)x(n)$. Applying again Remark 4(c) completes the proof. □

Finally we present a result about divergence of solutions of (74) and (75).

Theorem 16 (Divergence of solutions of scalar backward equations)
Let $\lambda\colon \mathbb{N}_0 \to \mathbb{R}$, $\alpha \in (0,1)$. *For* $\widehat{\nabla}^{\alpha} \in \{{}_{\text{C}}\overline{\nabla}^{\alpha}, {}_{\text{R-L}}\overline{\nabla}^{\alpha}, {}_{\text{C}}\widetilde{\nabla}^{\alpha}, {}_{\text{R-L}}\widetilde{\nabla}^{\alpha}\}$ *and* $x_0 \in \mathbb{R}$, *let* $\varphi^{\widehat{\nabla}}(\cdot, x_0)$ *denote the solution of*

$$(\widehat{\nabla}^{\alpha} x)(n) = \lambda(n)x(n)$$

with initial condition x_0. *If there exists a* $\lambda_0 > 0$ *such that*

$$1 > \lambda(n) \geq \lambda_0 > 0, \qquad n \in \mathbb{N}_0,$$

then for each $x_0 \in \mathbb{R}$

$$\lim_{n\to\infty} |\varphi^{\widehat{\nabla}}(n, x_0)| = \infty.$$

Proof For $\widehat{\nabla}^{\alpha} \in \{{}_{\text{C}}\overline{\nabla}^{\alpha}, {}_{\text{R-L}}\overline{\nabla}^{\alpha}\}$ the result is proved in [21, Theorems A and C] and [20, Theorems A and $\widehat{\text{A}}$], respectively. Using Remark 4(c), we get the conclusion also for $\widehat{\nabla}^{\alpha} \in \{{}_{\text{C}}\widetilde{\nabla}^{\alpha}, {}_{\text{R-L}}\widetilde{\nabla}^{\alpha}\}$. □

7 Separation of Solutions

The next theorem contains the main result of this paragraph.

Theorem 17 (Separation of solutions of Caputo equations)
Let $\alpha \in (0,1)$, $A\colon \mathbb{N}_0 \to \mathbb{R}^{d\times d}$ *with* $\sup_{n\in\mathbb{N}_0} \|A(n)\| < \infty$, $\lambda \in \mathbb{R}$ *with* $\lambda > \frac{\alpha}{1-\alpha}$, $x, y \in \mathbb{R}^d$ *with* $x \neq y$, *and* $x_0 \in \mathbb{R}^d \setminus \{0\}$.

(a) Forward equation $({}_{\text{C}}\overline{\Delta}^{\alpha} x)(n) = A(n)x(n)$*:*

$$\limsup_{n\to\infty} n^{\lambda} \|\varphi_{\text{C}}^{\Delta}(n,x) - \varphi_{\text{C}}^{\Delta}(n,y)\| = \infty \quad \textit{and} \quad \limsup_{n\to\infty} \frac{1}{n} \ln \|\varphi_{\text{C}}^{\Delta}(n,x_0)\| = \infty.$$

(b) Backward equations $({}_{\text{C}}\overline{\nabla}^{\alpha} x)(n) = A(n)x(n)$ and $({}_{\text{C}}\widetilde{\nabla}^{\alpha} x)(n) = A(n)x(n)$*:*

$$\limsup_{n\to\infty} n^{\lambda}\|\varphi_{\mathrm{c}}^{\overline{\nabla}}\Delta(n,x)-\varphi_{\mathrm{c}}^{\overline{\nabla}}(n,y)\|=\infty \quad \textit{and} \quad \limsup_{n\to\infty}\frac{1}{n}\ln\|\varphi_{\mathrm{c}}^{\overline{\nabla}}(n,x_0)\|=\infty,$$

if $(I-A(n))^{-1}$ *exists for each* $n\in\mathbb{N}_0$*. The same holds for* $\varphi_{\mathrm{c}}^{\widetilde{\nabla}}$.

In the proof of this theorem we use the following fact.

Lemma 6 ([5, Lemma 1])
Let $\alpha>0$ *and the sequence* $(u_{-\alpha}(k))_{k\in\mathbb{N}_0}$ *be defined by*

$$u_{-\alpha}(k)=(-1)^k\binom{-\alpha}{k},\qquad k\in\mathbb{N}_0. \tag{77}$$

Then the following statements hold:

(a) $u_{-\alpha}(k)>0$ *for* $k\in\mathbb{N}_0$.
(b) If $0<\alpha<1$*, then* $(u_{-\alpha}(k))_{k\in\mathbb{N}_0}$ *is a decreasing sequence.*
(c) $\sum_{k=0}^{n}u_{-\alpha}(k)=u_{-\alpha-1}(n)$ *for* $n\in\mathbb{N}_0$.
(d) There exist $\overline{m},\overline{M}>0$ *such that*

$$\frac{\overline{m}}{n^{1-\alpha}}<u_{-\alpha}(n)<\frac{\overline{M}}{n^{1-\alpha}},\qquad n\in\mathbb{N}_1.$$

Proof (of Theorem 17) (a) This is proved in [5, Theorem 5].

(b) Assume that $I-A(n)$ is invertible for each $n\in\mathbb{N}_0$. We show the claim for $({}_{\mathrm{c}}\overline{\nabla}^{\alpha}x)(n)=A(n)x(n)$. The result for the solution $\varphi_{\mathrm{c}}^{\widetilde{\nabla}}$ of $({}_{\mathrm{c}}\widetilde{\nabla}^{\alpha}x)(n)=A(n)x(n)$ follows then by Remark 4(c). To this end let $x,y\in\mathbb{R}^d$ with $x\neq y$ and $\lambda>\frac{\alpha}{1-\alpha}$. Suppose the contrary, i.e. there exists $K\in\mathbb{R}$ such that

$$\limsup_{n\to\infty} n^{\lambda}\|\varphi_{\mathrm{c}}^{\overline{\nabla}}(n,x)-\varphi_{\mathrm{c}}^{\overline{\nabla}}(n,y)\|<K,$$

which implies that

$$\lim_{n\to\infty}\|\varphi_{\mathrm{c}}^{\overline{\nabla}}(n,x)-\varphi_{\mathrm{c}}^{\overline{\nabla}}(n,y)\|=0 \tag{78}$$

and therefore by the boundedness of A that

$$\lim_{n\to\infty}\|(I-A(n))\varphi_{\mathrm{c}}^{\overline{\nabla}}(n,x)-\varphi_{\mathrm{c}}^{\overline{\nabla}}(n,y)\|=0.$$

Let us denote

$$L:=\sup_{n\in\mathbb{N}_0}\|\varphi_{\mathrm{c}}^{\overline{\nabla}}(n,x)-\varphi_{\mathrm{c}}^{\overline{\nabla}}(n,y)\|<\infty. \tag{79}$$

Furthermore, there exists $N\in\mathbb{N}_0$ such that

$$\|\varphi_C^{\overline{\nabla}}(n,x)-\varphi_C^{\overline{\nabla}}(n,y)\| \le Kn^{-\lambda}, \qquad n\ge N. \tag{80}$$

Considering the Caputo equation in the form given by (29), we have

$$\begin{aligned}
&(I-A(n))(\varphi_C^{\overline{\nabla}}(n,x)-\varphi_C^{\overline{\nabla}}(n,y))\\
&=x-y+\sum_{k=1}^{n} u_{-\alpha}(n-k)A(k)(\varphi_C^{\overline{\nabla}}(k,x)-\varphi_C^{\overline{\nabla}}(k,y))\\
&=x-y+\sum_{k=1}^{n} B(n,k)(\varphi_C^{\overline{\nabla}}(k,x)-\varphi_C^{\overline{\nabla}}(k,y)),
\end{aligned}$$

where

$$B(n,k):=u_{-\alpha}(n-k)A(k),$$

with $u_{-\alpha}(\cdot)$ given by (77). Thus,

$$\|x-y\|\le \|(I-A(n))(\varphi_C^{\overline{\nabla}}(n,x)-\varphi_C^{\overline{\nabla}}(n,y))\|+\left\|\sum_{k=1}^{n} B(n,k)(\varphi_C^{\overline{\nabla}}(k,x)-\varphi_C^{\overline{\nabla}}(k,y))\right\|.$$

Letting $n\to\infty$ and using (78), we obtain that

$$\limsup_{n\to\infty}\left\|\sum_{k=1}^{n} B(n,k)(\varphi_C^{\overline{\nabla}}(k,x)-\varphi_C^{\overline{\nabla}}(k,y))\right\|>0. \tag{81}$$

Since $\lambda>\frac{\alpha}{1-\alpha}$, there exists $\delta\in(\frac{\alpha}{\lambda},1-\alpha)$. Thus, to get a contradiction to inequality (81), it is sufficient to show that

$$\limsup_{n\to\infty}\sum_{k=1}^{\lceil n^\delta\rceil-1} B(n,k)(\varphi_C^{\overline{\nabla}}(k,x)-\varphi_C^{\overline{\nabla}}(k,y))=0 \tag{82}$$

and

$$\limsup_{n\to\infty}\sum_{k=\lceil n^\delta\rceil}^{n} B(n,k)(\varphi_C^{\overline{\nabla}}(k,x)-\varphi_C^{\overline{\nabla}}(k,y))=0. \tag{83}$$

By definition of $B(n,k)$ and non-negativity of the sequence $(u_{-\alpha}(n))$ by Lemma 6(a), we have

$$\left\|\sum_{k=1}^{\lceil n^\delta\rceil-1} B(n,k)(\varphi_C^{\overline{\nabla}}(k,x)-\varphi_C^{\overline{\nabla}}(k,y))\right\|$$

$$\leq \sum_{k=1}^{\lceil n^\delta \rceil - 1} \|B(n,k)\| \|(\varphi_c^{\overline{\nabla}}(k,x) - \varphi_c^{\overline{\nabla}}(k,y))\|$$
$$\leq \sum_{k=1}^{\lceil n^\delta \rceil - 1} M u_{-\alpha}(n-k) \|(\varphi_c^{\overline{\nabla}}(k,x) - \varphi_c^{\overline{\nabla}}(k,y))\|$$
$$\leq ML \sum_{k=1}^{\lceil n^\delta \rceil - 1} u_{-\alpha}(n-k),$$

where we used (79) to obtain the last inequality. By Lemma 6(b), the sequence $(u_{-\alpha}(n))$ is decreasing. Thus,

$$\Big\| \sum_{k=1}^{\lceil n^\delta \rceil - 1} B(n,k)(\varphi_c^{\overline{\nabla}}(k,x) - \varphi_c^{\overline{\nabla}}(k,y)) \Big\| \leq ML \lceil n^\delta \rceil u_{-\alpha}(n - \lceil n^\delta \rceil).$$

Using Lemma 6(d), we obtain that

$$\Big\| \sum_{k=1}^{\lceil n^\delta \rceil - 1} B(n,k)(\varphi_c^{\overline{\nabla}}(k,x) - \varphi_c^{\overline{\nabla}}(k,y)) \Big\| \leq ML(n^\delta + 1) \frac{\overline{M}}{(n - n^\delta)^{1-\alpha}},$$

which, together with the fact that $\delta < 1 - \alpha$, proves (82). To conclude the proof we show (83). For this purpose, we use the estimate

$$\Big\| \sum_{k=\lceil n^\delta \rceil}^{n} B(n,k)(\varphi_c^{\overline{\nabla}}(k,x) - \varphi_c^{\overline{\nabla}}(k,y)) \Big\|$$
$$\leq \sum_{k=\lceil n^\delta \rceil}^{n} \|B(n,k)\| \|(\varphi_c^{\overline{\nabla}}(k,x) - \varphi_c^{\overline{\nabla}}(k,y))\|$$
$$\leq M \sum_{k=\lceil n^\delta \rceil}^{n} u_{-\alpha}(n-k) \|(\varphi_c^{\overline{\nabla}}(k,x) - \varphi_c^{\overline{\nabla}}(k,y))\|.$$

Let $n \in \mathbb{N}$ such that $n^\delta \geq N$. Using (80), we obtain that

$$\Big\| \sum_{k=\lceil n^\delta \rceil}^{n} B(n,k)(\varphi_c^{\overline{\nabla}}(k,x) - \varphi_c^{\overline{\nabla}}(k,y)) \Big\| \leq MK \lceil n^\delta \rceil^{-\lambda} \sum_{k=\lceil n^\delta \rceil}^{n} u_{-\alpha}(n-k).$$

By Lemma 6(a) and the fact that

$$\sum_{k=1}^{n} (-1)^k \binom{\alpha}{k} = (-1)^n \binom{\alpha - 1}{n}, \qquad n \in \mathbb{N}_0,$$

we have

$$\sum_{k=\lceil n^{\delta}\rceil}^{n} u_{-\alpha}(n-k) \leq \sum_{k=1}^{n} u_{-\alpha}(n-k) = u_{-(\alpha+1)}(n).$$

Thus,

$$\left\| \sum_{k=\lceil n^{\delta}\rceil}^{n} B(n,k)(\varphi_{\mathrm{C}}^{\overline{\nabla}}(k,x) - \varphi_{\mathrm{C}}^{\overline{\nabla}}(k,y)) \right\| \leq MK\lceil n^{\delta}\rceil^{-\lambda} u_{-(\alpha+1)}(n).$$

In light of Lemma 6(d) for $\alpha + 1$, we have

$$\left\| \sum_{k=\lceil n^{\delta}\rceil}^{n} B(n,k)(\varphi_{\mathrm{C}}^{\overline{\nabla}}(k,x) - \varphi_{\mathrm{C}}^{\overline{\nabla}}(k,y)) \right\| \leq MKn^{-\delta\lambda}\frac{\overline{M}}{n^{-\alpha}}.$$

Note that $\delta\lambda > \alpha$, (83) is proved and the proof is complete. □

Our hypothesis is that a similar result holds true for the Riemann-Liouville forward equation $({}_{\text{R-L}}\overline{\Delta}^{\alpha}x)(n) = A(n)x(n)$ and backward equations $({}_{\text{R-L}}\overline{\nabla}^{\alpha}x)(n) = A(n)x(n)$ and $({}_{\text{R-L}}\widetilde{\nabla}^{\alpha}x)(n) = A(n)x(n)$ but we cannot provide a proof.

8 Conclusions

In this work we considered the asymptotic properties of six types of linear fractional equations in discrete time described by the equations (18)–(23). The first two are the Caputo and Riemann-Liouville forward equations in which the difference operator is defined as the composition of the classical forward difference with the fractional order sum. In the case of the Caputo equation, the order of these operators is such that the difference operator acts first and the fractional sum operator acts second and in the case of the Riemann-Liouville equation, the order of these operators is reversed. The next four equations, i.e. (20)–(23) are the Caputo and Riemann-Liouville backwards equation, in which the difference operator is defined as composing the classical backward difference with the sum of the fractional order. Additionally, in the case of these equations, we distinguish between two sum definitions, which include and do not include the initial condition.

For each of the equations under consideration we have given a precise formulation of the initial value problem (see Definition 4 and Remark 3) and discussed the existence and uniqueness of solutions to initial value problems (see Theorem 1). In Theorem 2 we show that each of the equations considered can be represented in two different ways as a convolution-type Volterra equation. These preparations play a key role in obtaining the further results of our work. One of them is included in Remark 5, which shows that the solutions of the Caputo backwards equations defined with a sum that takes into account the initial conditions and a sum that does not take it into account are closely related and in particular, have the same asymptotic properties. The same is true for the backwards Riemann-Liouville equations.

In Theorem 3, we present explicit formulas for solutions to stationary equations. It should be noted that the first two points of this theorem provide the formula for the solution in the form of a polynomial of the variable A, where A is the coefficient of the equation under consideration without additional assumptions about the matrix A. The remaining points present the formula for the solution in the form of a series and require an additional assumption about the matrix A: that it has all eigenvalues inside the unit circle, although a solution also exists when A has eigenvalues outside the unit circle. The formulas for solutions of the backward equations in the general case are an open problem. A step towards its solution may be Theorem 4, where such formulas are given in the form of polynomials of the variable $(I - A)^{-1}$, unfortunately the coefficients of these polynomials are given in recursive form and therefore these formulas cannot be considered as satisfactory, explicit formulas.

Theorems 5, 6, 7, 8, 9 and 11 provide sufficient conditions for the stability and instability of the equations (18)–(23). They have the following form: if the eigenvalues of the matrix A belong to an open set S, then the equation is asymptotically stable (the form of the set S depends on the equation) and if at least one eigenvalue of A belongs to the set $\mathbb{C} \setminus \mathrm{cl}\, S$, the equation is unstable. The problem of the asymptotic behavior of these equations when certain eigenvalues of the matrix A lie on the boundary of the stability region S remains an open problem. Moreover, these theorems say that in the case of stable Caputo equations, the rate of decay to zero as $n \to \infty$ is not greater than $n^{-\alpha}$ and in the case of the Riemann-Liouville equations, not greater than $n^{-\alpha-1}$, and that the rate of growth to infinity in the case of unstable equations is not less than r^n, $r > 1$. The problem of giving the exact growth and decay rates is also an open problem. In a special case when the system is one-dimensional, the stability problem is completely solved and its solution is given by Theorem 12. However even in the one-dimensional case, the problem of the exact rate of convergence to zero is a problem that is not completely solved. Its solution for some subsets of the stability set is given by Theorem 13. Finally, Theorems 14, 15 and 16 provide some conditions that are sufficient for the asymptotic stability and instability of one-dimensional equations with variable coefficients.

Theorem 17 is a complement of the picture of rate of convergence of solutions. It says that for the considered Caputo time-varying equations, this rate is not faster than $n^{-\lambda}$ with a certain $\lambda > 0$. Our hypothesis is that a similar result holds true for the Riemann-Liouville equations but we cannot provide a proof.

Acknowledgements The research of A. Czornik was supported by the Polish National Agency for Academic Exchange (NAWA), during the implementation of the project PPN/BEK/2020/1/00188 within the Bekker NAWA Programme. The research of the A. Babiarz was funded by the National Science Centre in Poland granted according to decision DEC-2017/01/X/ST7/00313. The research of the second author was done when he visited the Vietnam Institute for Advanced Study in Mathematics (VIASM). He would like to thank the VIASM for its hospitality and financial support. The last author was partially supported by an Alexander von Humboldt Polish Honorary Research Fellowship.

References

1. Abdeljawad, T.: On delta and nabla Caputo fractional differences and dual identities. Discrete Dyn. Nat. Soc. **2013**, 12 pages (2013). Article ID 406910
2. Abdeljawad, T., Atici, F.M.: On the definitions of nabla fractional operators. Abstr. Appl. Anal. **2012** (2012). Article ID 406757
3. Abu-Saris, R., Al-Mdallal, Q.: On the asymptotic stability of linear system of fractional-order difference equations. Fract. Calc. Appl. Anal. **16**(3), 613–629 (2013)
4. Anastassiou, G.A.: Intelligent mathematics: computational analysis. Intelligent Systems Reference Library, vol. 5. Springer (2011)
5. Anh, P., Babiarz, A., Czornik, A., Niezabitowski, M., Siegmund, S.: Asymptotic properties of discrete linear fractional equations. Bull. Pol. Acad. Sci. Tech. Sci. **67**(4), 749–759 (2019)
6. Anh, P.T., Babiarz, A., Czornik, A., Kitzing, K., Niezabitowski, M., Siegmund, S., Trostorff, S., Tuan, H.T.: A Hilbert space approach to fractional difference equations. In: Bohner, M., Siegmund, S., Šimon Hilscher, R., Stehlík, P. (eds.) Difference Equations and Discrete Dynamical Systems with Applications, pp. 115–131. Springer (2020)
7. Anh, P.T., Babiarz, A., Czornik, A., Niezabitowski, M., Siegmund, S.: Variation of constant formulas for fractional difference equations. Arch. Control Sci. **28**(4), 617–633 (2018)
8. Anh, P.T., Babiarz, A., Czornik, A., Niezabitowski, M., Siegmund, S.: Some results on linear nabla Riemann-Liouville fractional difference equations. Math. Methods Appl. Sci. **43**(13), 7815–7824 (2020)
9. Atıcı, F., Eloe, P.: Initial value problems in discrete fractional calculus. Proc. Am. Math. Soc. **137**(3), 981–989 (2009)
10. Atıcı, F.M., Eloe, P.W.: Linear systems of fractional nabla difference equations. Rocky Mt. J. Math. **41**(2), 353–370 (2011)
11. Atıcı, F.M., Eloe, P.W.: Gronwall's inequality on discrete fractional calculus. Comput. Math. Appl. **64**(10), 3193–3200 (2012)
12. Busłowicz, M., Ruszewski, A.: Necessary and sufficient conditions for stability of fractional discrete-time linear state-space systems. Bull. Pol. Acad. Sci. Tech. Sci. **61**(4), 779–786 (2013)
13. Čermák, J., Győri, I., Nechvátal, L.: On explicit stability conditions for a linear fractional difference system. Fract. Calc. Appl. Anal. **18**(3), 651–672 (2015)
14. Čermák, J., Kisela, T., Nechvátal, L.: Stability and asymptotic properties of a linear fractional difference equation. Adv. Differ. Equ. **2012**(1), 122 (2012)
15. Čermák, J., Kisela, T., Nechvátal, L.: Stability regions for linear fractional differential systems and their discretizations. Appl. Math. Comput. **219**(12), 7012–7022 (2013)
16. Diaz, J., Osler, T.: Differences of fractional order. Math. Comput. **28**(125), 185–202 (1974)
17. Goodrich, C., Peterson, A.C.: Discrete Fractional Calculus. Springer (2015)
18. Gray, H.L., fan Zhang, N.: On a new definition of the fractional difference. Math. Comput. **50**(182), 513–529 (1988)
19. Horn, R.A., Johnson, C.R.: Matrix Analysis. Cambridge University Press (2012)
20. Jia, B., Erbe, L., Peterson, A.: Comparison theorems and asymptotic behavior of solutions of discrete fractional equations. Electron. J. Qual. Theory Differ. Equ. **89**, 1–18 (2015)
21. Jia, B., Erbe, L., Peterson, A.: Comparison theorems and asymptotic behavior of solutions of Caputo fractional equations. Int. J. Differ. Equ. **11**, 163–178 (2016)
22. Jonnalagadda, J.M.: Analysis of nonlinear fractional nabla difference equations. Int. J. Anal. Appl. **7**(1), 79–95 (2015)
23. Kaczorek, T.: Selected Problems of Fractional Systems Theory, vol. 411. Springer (2011)
24. Mozyrska, D., Wyrwas, M.: Explicit criteria for stability of two – dimensional fractional difference systems (2016)
25. Nechvátal, L.: On asymptotics of discrete Mittag-Leffler function. Mathematica Bohemica **139**(4), 667–675 (2014)
26. Ostalczyk, P.: Equivalent descriptions of a discrete-time fractional-order linear system and its stability domains. Int. J. Appl. Math. Comput. Sci. **22**(3), 533–538 (2012)

27. Patashnik, O., Knuth, D., Graham, R.L.: Concrete Mathematics: A Foundation for Computer Science. Pearson Education (1994)
28. Shen, J., Tang, T., Wang, L.L.: Spectral Methods: Algorithms, Analysis and Applications, vol. 41. Springer (2011)
29. Stanisławski, R., Latawiec, K.J.: Stability analysis for discrete-time fractional-order LTI state-space systems. Part i: new necessary and sufficient conditions for the asymptotic stability. Bull. Pol. Acad. Sci. Tech. Sci. **61**(2), 353–361 (2013)
30. Stanisławski, R., Latawiec, K.J.: Stability analysis for discrete-time fractional-order LTI state-space systems. Part ii: new stability criterion for FD-based systems. Bull. Pol. Acad. Sci. Tech. Sci. **61**(2), 363–370 (2013)

Variable-, Fractional-Order Linear System State-Space Description Transformation

Piotr Ostalczyk

Abstract In the paper linear time-variant SISO systems described by difference equations with variable-, fractional-order Grünwald–Letnikov backward differences are analysed. The state-space-like form (similar to an observability matrix) is proposed. The solution of the linear time-variant fractional order SISO system is derived. For a special case: time invariant SISO system, a similarity-like transformation to diagonal form is defined.

Keywords Fractional systems · Time-varying systems · Discrete event modelling and simulation

1 Introduction

There is a wide class of dynamic systems in which the time-variance of parameters is noticeable. One can mention devices with observable lowering mass of the fuel (for instance rockets). Here the parameter changes in time due to the ageing of system elements or devices, or the influence of external conditions (for instance temperature) to the mathematical model, should be taken into account.

Fractional calculus [9, 12, 26, 29] has for years been considered a very efficient mathematical tool for dynamic system modelling and control strategies. It encompasses many fields of science ranging from mathematics and physics, technics, biology, medicine to economics. Though all physical systems and signals in their nature are continuous in time, now they are treated as discrete ones. Hence, the derivatives [14, 15], in their mathematical models are approximated by differences [27]. In this paper the generalization to the variable-, fractional orders is proposed [16, 17, 28, 30]. The proposed VFOBDs are much more versatile in comparison with the classical integer order backward differences or even constant fractional ones [7, 19]. Guided by practical requirements in time-signals processing, only the backward difference

P. Ostalczyk (✉)
Faculty of Electrical Engineering, Institute of Control and Industrial Electronics, Warsaw University of Technology, ul. Koszykowa 75, 00-662 Warsaw, Poland
e-mail: piotr.ostalczyk@ee.pw.edu.pl

P. Kulczycki et al. (eds.), *Fractional Dynamical Systems: Methods, Algorithms and Applications*, Studies in Systems, Decision and Control 402,
https://doi.org/10.1007/978-3-030-89972-1_6

will be considered. On the base of VFOBD one builds variable, fractional-order difference equations (VFODE). Here, one focuses on the variable, fractional-order Grünwald–Letnikov backward difference (GL-VFOBD).

In the chapter a novelty vector-matrix description of the VFODS will be applied. It is based on the square upper triangular matrices [6] of growing dimensions due to the consecutive steps of a transient process [19, 22, 24]. For constant fractional-orders the matrices are upper triangular band ones [11]. The discrete independent variable in the VFOBD is real. In the description of the discrete dynamical systems, the variable is called a discrete-time. Though in time domain it has similar properties to the classical description based on the one-sided $\mathcal{Z}$ transform. This means that there is an analogue to the polynomial numerator and denominator matrix in the Matrix Fraction Description (MFD) [3, 8, 10, 13]. These analogs will be further called MFD-like. Related to it, there is a discrete transfer function called discrete transfer-function-like. There is also an analogous description with the state-space one [5] called a state-space-like. The mentioned forms enable easy description of fundamental configurations of the subsystems.

There are analogous applications of the VFOBD, as for so called classical difference equations. Here, one can mention the main ones: discrete-time control algorithms [20, 25] described by the VFOBD, especially in VFO PID (VFOPID) controllers [4, 20, 21], dynamic SISO and MIMO system identification [1, 20, 31] and analysis [21].

The chapter is divided into several sections. First some basic notions related to the variable-, fractional order Grünvald–Letnikov backward difference are given. In Sect. 3 the matrix notation of the VFOBD is analysed. The main results are presented in Sects. 4 and 5 where the state-space-like SISO linear time-variant system is proposed and similarity transformation for its time-invariant case is given.

2 Mathematical Preliminaries

In this chapter a common notation will be used: $\mathbb{N}$ is the set of natural numbers (including zero). Its elements will be in general denoted by letters i, j, k, m, n. The elements of the set of real $\mathbb{R}$ in general will be denoted by Greek letters ν, μ.

First, the basic definitions of mathematical notions used in the paper will be presented. Square matrices of dimensions $(k+1) \times (k+1)$ will be denoted by bold uppercases $\mathbf{A}$, $\mathbf{B}$, $\mathbf{C}$, $\mathbf{T}_k$ whereas column vectors of dimensions $(k+1) \times 1$ by lower cases $\mathbf{a}$, $\mathbf{x}$, $\mathbf{u}_k$. Especially, $\mathbf{1}_k$ will denote a unit matrix of dimensions $(k+1)\times(k+1)$.

Definition 1 For two discrete variables $k, l \in \mathbb{N}$ and a bounded order function $\nu(k) \in \mathbb{R}$ a kernel function of two discrete variables is defined as

$$a^{[\nu(l)]}(k) = \begin{cases} 1 & \text{for } k = 0 \\ \frac{\nu(l)(\nu(l)-1)\cdots(\nu(l)-k+1)}{(-1)^k k!} & \text{for } k \in \mathbb{N} \end{cases}. \quad (1)$$

The function defined above will be called an "oblivion" or "decay" function for $\nu(k) \in \mathbb{R}_+$ and a "collection" one for $\nu(k) \in \mathbb{R}_-$. In order to standardize the notation, the order function $\nu(k)$ will be limited to positive values only. For a variable-, fractional-order the notation $-\nu(k) < 0$ will be used.

Definition 2 The VFOBD with defined by an order function ν, with values $\nu(k) \in [0, 1]$, is defined as a finite sum, provided that the series is convergent

$$\begin{aligned} {}_{k_0}\Delta_k^{\nu(k)} f(k) &= \sum_{i=0}^{k-k_0} a^{\nu(k)}(i) f(k-i) \\ &= \left[1 \ a^{\nu(k)}(1) \ a^{\nu(k)}(2) \ \cdots \ a^{\nu(k)}(k-k_0)\right] \begin{bmatrix} f(k) \\ f(k-1) \\ \vdots \\ f(k_0) \end{bmatrix}. \end{aligned} \quad (2)$$

Remark 1 There are different definitions of the VFOBD [17, 32, 34]. The main dissimilarity is in the form of oblivion or collection function (2).

Definition 3 The variable-, fractional-order backward sum (VFOBS) with an order function $-\nu$, with values $\nu(k) \in [0, 1]$, is defined as a finite sum, provided that the series is convergent

$$\begin{aligned} {}_{k_0}\Delta_k^{-\nu(k)} f(k) &= \sum_{i=0}^{k-k_0} a^{-\nu(i)}(i) f(k-i) \\ &= \left[1 \ a^{-\nu(1)}(1) \ a^{-\nu(2)}(2) \ \cdots \ a^{-\nu(k-k_0)}(k-k_0)\right] \begin{bmatrix} f(k) \\ f(k-1) \\ \vdots \\ f(k_0) \end{bmatrix}, \end{aligned} \quad (3)$$

or similarly

$$
\begin{aligned}
{}_{k_0}\Delta_k^{-\nu(k)} f(k) &= \sum_{i=k_0}^{k} a^{-\nu(k-i)}(k-i) f(i) \\
&= \left[a^{-\nu(k-k_0)}(k-k_0)\ a^{\nu(k-k_0-1)}(k-k_0+1) \cdots \times \right. \\
&\times \cdots a^{-\nu(k-k_0)}(1)\ 1\Big] \begin{bmatrix} f(k_0) \\ f(k_0+1) f(k_0+2) \\ \vdots \\ f(k) \end{bmatrix}.
\end{aligned} \tag{4}
$$

The considered VFOBDs are valid also for $k-1, k-2, \ldots, k_0$. Hence, there are $k-k_0+1$ equalities which may be expressed as one equality

$$
{}_{k_0}^{GL}\Delta_k^{[\nu(k)]}\mathbf{f}(k) = {}_{k_0}\mathbf{A}_k^{[\nu(k)]}\mathbf{f}(k)\,, \tag{5}
$$

where

$$
\mathbf{f}(k) = \begin{bmatrix} f(k) \\ f(k-1) \\ \vdots \\ f(k_0+1) \\ f(k_0) \end{bmatrix}, \quad {}_{k_0}^{GL}\Delta_k^{[\nu(k)]}\mathbf{f}(k) = \begin{bmatrix} {}_{k_0}^{GL}\Delta_k^{[\nu(k)]} f(k) \\ {}_{k_0}^{GL}\Delta_{k-1}^{[\nu(k)]} f(k-1) \\ \vdots \\ {}_{k_0}^{GL}\Delta_{k_0+1}^{[\nu(k)]} f(k_0+1) \\ {}_{k_0}^{GL}\Delta_{k_0}^{[\nu(k)]} f(k_0) \end{bmatrix} \tag{6}
$$

$$
{}_{k_0}\mathbf{A}_k^{[\nu(k)]} = \begin{bmatrix} 1 & a^{[\nu(k)]}(1) & a^{[\nu(k)]}(2) & \cdots & a^{[\nu(k)]}(k-k_0) \\ 0 & 1 & a^{[\nu(k-1)]}(1) & \cdots & a^{[\nu(k-1)]}(k-k_0-1) \\ \vdots & \vdots & \vdots & & \vdots \\ 0 & 0 & 1 & \cdots & a^{[\nu(1)]}(1) \\ 0 & 0 & 0 & \cdots & 1 \end{bmatrix}. \tag{7}
$$

2.1 Selected Properties of a Matrix ${}_{k_0}\mathbf{A}_k^{[\nu(k)]}$

The set of matrices ${}_{k_0}\mathbf{A}_k^{[\nu(k)]}$ is not a group [33].

For $\nu_1(k), \nu_2(k) \in \mathbb{R}$

$$
{}_{k_0}\mathbf{A}_k^{[\nu_1(k)]} + {}_{k_0}\mathbf{A}_k^{[\nu_2(k)]} = {}_{k_0}\mathbf{A}_k^{[\nu_2(k)]} + {}_{k_0}\mathbf{A}_k^{[\nu_1(k)]}, \tag{8}
$$

$$
{}_{k_0}\mathbf{A}_k^{[\nu_1(k)]}\ {}_{k_0}\mathbf{A}_k^{[\nu_2(k)]} \neq {}_{k_0}\mathbf{A}_k^{[\nu_1(k)+\nu_2(k)]} \tag{9}
$$

Proposition 1 *For* $\nu_1(k) \in \mathbb{R}$ *and* $\nu_2 = \text{const} \in \mathbb{R}$ *a following equality holds*

$$ {}_{k_0}\mathbf{A}_k^{[\nu_1(k)]}\,{}_{k_0}\mathbf{A}_k^{[\nu_2]} = {}_{k_0}\mathbf{A}_k^{[\nu_1(k)+\nu_2]}. \tag{10} $$

Proof Without lost of generality there will be assumed $k_0 = 0$. Then, for $k = 1$ is

$$ {}_0\mathbf{A}_0^{[\nu_1(1)]}\,{}_0\mathbf{A}_0^{[\nu_2]} = {}_0\mathbf{A}_0^{[\nu_1(1)+\nu_2]} = [1]. \tag{11} $$

For $k = 2$

$$ \begin{aligned} {}_0\mathbf{A}_0^{[\nu_1(2)]}\,{}_0\mathbf{A}_0^{[\nu_2]} &= \begin{bmatrix} 1 & -\nu_1(1) \\ 0 & 1 \end{bmatrix} \begin{bmatrix} 1 & -\nu_2 \\ 0 & 1 \end{bmatrix} \\ = \begin{bmatrix} 1 & -(\nu(1)+\nu_2) \\ 0 & 1 \end{bmatrix} &= {}_0\mathbf{A}_0^{[\nu_1(0)+\nu_2]} = [1]. \end{aligned} \tag{12} $$

For $k > 2$ one can realize that matrix (7) can be expressed as a block matrix

$$ \begin{aligned} {}_0\mathbf{A}_k^{[\nu_1(k)]}\,{}_0\mathbf{A}_k^{[\nu_2]} &= \begin{bmatrix} {}_k\mathbf{A}_{k-1}^{[\nu_1(k)]} & \mathbf{a}_{k-1}^{[\nu_1(k)]} \\ \mathbf{0}_{1,k-1} & 1 \end{bmatrix} \begin{bmatrix} {}_k\mathbf{A}_{k-1}^{[\nu_2]} & \mathbf{a}_{k-1}^{[\nu_2]} \\ \mathbf{0}_{1,k-1} & 1 \end{bmatrix} \\ & \begin{bmatrix} {}_k\mathbf{A}_{k-1}^{[\nu_1(k)+\nu_2]} & {}_k\mathbf{A}_{k-1}^{[\nu_1(k)]}\mathbf{a}_{k-1}^{[\nu_2]} + \mathbf{a}_{k-1}^{[\nu_1(k)]} \\ \mathbf{0}_{1,k-1} & 1 \end{bmatrix}, \end{aligned} \tag{13} $$

where

$$ \mathbf{a}_{k-1}^{[\nu_1(k)]} = \begin{bmatrix} a^{[\nu_1(k)]}(k) \\ a^{[\nu_1(k-1)]}(k-1) \\ \vdots \\ a^{[\nu_1(1)]}(1) \\ 1 \end{bmatrix}, \quad \mathbf{a}_{k-1}^{[\nu_2]} = \begin{bmatrix} a^{[\nu_2]}(k) \\ a^{[\nu_2]}(k-1) \\ \vdots \\ a^{[\nu_2]}(1) \\ 1 \end{bmatrix}. \tag{14} $$

Performing calculations step by step from bottom rows of submatrix ${}_k\mathbf{A}_{k-1}^{[\nu_1(k)]}\mathbf{a}_{k-1}^{[\nu_2]} + \mathbf{a}_{k-1}^{[\nu_1(k)]}$ gives

$$ {}_k\mathbf{A}_{k-1}^{[\nu_1(k)]}\mathbf{a}_{k-1}^{[\nu_2]} + \mathbf{a}_{k-1}^{[\nu_1(k)]} = \begin{bmatrix} a^{[\nu_1(k)+]nu_2]}(k) \\ a^{[\nu_1(k-1)+\nu_2]}(k-1) \\ \vdots \\ a^{[\nu_1(1)+\nu_2]}(1) \end{bmatrix}. \tag{15} $$

Remark 2 For $\nu_1(k), \nu_2(k) = \text{const} \in \mathbb{R}$ inequality (9) becomes equality.

Next, there is

$$ {}_{k_0}\mathbf{A}_k^{[\nu_1(k)]}\,{}_{k_0}\mathbf{A}_k^{[\nu_2(k)]} \neq {}_{k_0}\mathbf{A}_k^{[\nu_2(k)]}\,{}_{k_0}\mathbf{A}_k^{[\nu_1(k)]}. \tag{16} $$

Remark 3 For $\nu_1(k)$, $\nu_2(k) = \text{const} \in \mathbb{R}$ inequality presented above becomes equality.

For $\nu(k) = 0$

$$ {}_{k_0}\mathbf{A}_k^{[0]} = \mathbf{1}_{k+1}. \tag{17} $$

For any $\nu(k) \in \mathbb{R}$ matrix (7)

$$ \det\left[{}_{k_0}\mathbf{A}_k^{[\nu(k)]}\right] \neq 0. \tag{18} $$

Hence, matrix (7) is always nonsingular.

3 VFO Linear Time-Variant SISO Systems

Following the classical form of the linear time-variant difference equation with the backward differences one defines the variable-, fractional-order linear time-variant difference equation

$$ \sum_{i=0}^{n} a_i(k)\,{}_{k_0}^{GL}\Delta_k^{[\nu_i(k)]} y(k) = \sum_{i=0}^{m} b_i(k)\,{}_{k_0}^{GL}\Delta_k^{[\mu_i(k)]} u(k), \tag{19} $$

with

$$ \begin{aligned} m &\leqslant n, \\ \nu_n(k) > \nu_{n-1}(k) > \cdots \nu_1(k) > \nu_0(k) &= 0, \\ \mu_n(k) > \mu_{n-1}(k) > \cdots \mu_1(k) &> \mu_0(k), \\ a_i(k), b_j(k) \in \mathbb{R} \text{ for } i = 0, \ldots, n \;\; and \;\; j &= 0, \ldots, m \\ a_n &= 1, \end{aligned} \tag{20} $$

or in an equivalent matrix-vector form

$$ \sum_{i=0}^{n} \mathbf{S}_i(k)\,{}_{k_0}\mathbf{A}_k^{[\nu_i(k)]}\mathbf{y}(k) = \sum_{i=0}^{m} \mathbf{T}_i(k)\,{}_{k_0}\mathbf{A}_k^{[\mu_i(k)]}\mathbf{u}(k), \tag{21} $$

where

$$\mathbf{S}_i(k) = \begin{bmatrix} a_i(k) & 0 & \cdots & 0 & 0 \\ 0 & a_i(k-1) & \cdots & 0 & 0 \\ 0 & 0 & \cdots & 0 & 0 \\ \vdots & \vdots & & \vdots & \vdots \\ 0 & 0 & \cdots & a_i(1) & 0 \\ 0 & 0 & \cdots & 0 & a_i(0) \end{bmatrix}, \tag{22}$$

$$\mathbf{T}_i(k) = \begin{bmatrix} b_i(k) & 0 & \cdots & 0 & 0 \\ 0 & b_i(k-1) & \cdots & 0 & 0 \\ 0 & 0 & \cdots & 0 & 0 \\ \vdots & \vdots & & \vdots & \vdots \\ 0 & 0 & \cdots & b_i(1) & 0 \\ 0 & 0 & \cdots & 0 & b_i(0) \end{bmatrix}. \tag{23}$$

Investigating the BDs one must take into account the following assumptions:

- the input signal: $u(k) = 0$ for $k < k_0$,
- the initial conditions and input signal vectors, respectively

$$\mathbf{y}_{k_0-1} = \begin{bmatrix} y_{k_0-1} \\ y_{k_0-2} \\ \vdots \\ y_{k_0} \end{bmatrix}, \ \mathbf{u}(k) = \begin{bmatrix} u(k) \\ u(k-1) \\ \vdots \\ u(k_0) \end{bmatrix}, \tag{24}$$

- inequality preserving the maximal order:

$$1 + \sum_{i=0}^{n-1} a_i(k) \neq 0 \text{ for } k \in \mathbb{N}. \tag{25}$$

3.1 Matrix Fraction Description Like of VFODS

Now one defines the matrices

$${}_{k_0}\mathbf{D}_k^n = \sum_{i=0}^{n} \mathbf{S}_i(k)_{k_0}\mathbf{A}_k^{[\nu_i(k)]}, \tag{26}$$

$${}_{k_0}\mathbf{N}_k^m = \sum_{i=0}^{m} \mathbf{T}_i(k)_{k_0}\mathbf{A}_k^{[\mu_i(k)]}. \tag{27}$$

Under assumption (25) matrix (26) is invertible for all $k \geqslant k_0$

$$\det\left[{}_{k_0}\mathbf{D}_k^n\right] \neq 0. \tag{28}$$

Matrix (27) is usually singlular

$$\det\left[{}_{k_0}\mathbf{N}_k^m\right] = 0. \tag{29}$$

This is related to the physical realizability of a dynamical system which describes Eq. (27). An exception occurs when $m = n$. Hence, one can write

$${}_{k_0}\mathbf{D}_k^n\mathbf{y}(k) = {}_{k_0}\mathbf{N}_k^m\mathbf{u}(k) \tag{30}$$

and

$$\mathbf{y}(k) = \left[{}_{k_0}\mathbf{D}_k^n\right]^{-1}\left[{}_{k_0}\mathbf{N}_k^m\right]\mathbf{u}(k). \tag{31}$$

The equation given above is similar to the Matrix Fraction Description (MFD) [13] where matrices ${}_{k_0}\mathbf{D}_k^n$ and ${}_{k_0}\mathbf{N}_k^n$ play roles of a left denominator and nominator, respectively. One should emphasise that the description (20) is in the discrete-time domain whereas the classical MFD is in the complex domain [3, 10].

MDF is a non-unique description of a dynamical system. For instance, matrices

$$\mathbf{D'}_k^n = \mathbf{E}_{k+1}\left({}_{k_0}\mathbf{D}_k^n\right) \tag{32}$$

and

$$\mathbf{N'}_k^m = \mathbf{E}_{k+1}\left({}_{k_0}\mathbf{N}_k^m\right), \tag{33}$$

where $\mathbf{E}_{k+1}$ is square non-singular matrix in upper triangular form lead to the same Eq. (30).

3.2 Transfer-Function-Like Description of the VFODS

A product of matrices in (30) defines a new one

$${}_{k_0}\mathbf{G}_k = \left[{}_{k_0}\mathbf{D}_k^n\right]^{-1}\left[{}_{k_0}\mathbf{N}_k^m\right], \tag{34}$$

which will be further named as a transfer-function-like matrix though it describes SISO VFODS. The method enables simple description of fundamental connections of the considered systems. The following signals and transfer-function-like of subsystems denote:

- $\mathbf{u}(k)$—the main input signal;
- $\mathbf{y}(k)$—the main output signal;
- $\mathbf{u}_i(k)$—the input of subsystem $i = 1, 2$;
- $\mathbf{y}_i(k)$—the output of subsystem $i = 1, 2$;

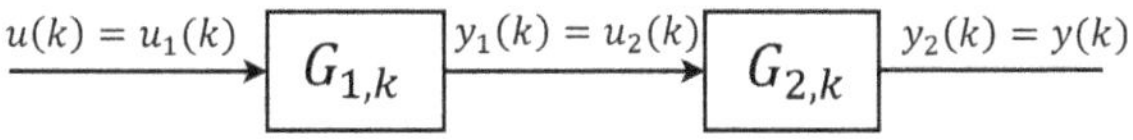

Fig. 1 Series connection of two transfer-functions-like

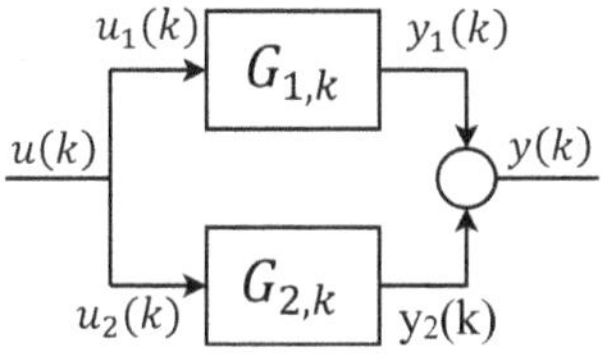

Fig. 2 Parallel connection of two transfer-functions-like

- $\mathbf{e}(k)$—the closed-loop system error signal;
- $\mathbf{G}_{i,k}(k) = \left[{}_{k_0}\mathbf{D}^{n}_{i,k}\right]^{-1}\left[{}_{k_0}\mathbf{N}^{m}_{i,k}\right]$—the transfer-function-like and MFD-like of subsystem $i = 1, 2$.

First a series connection of two SISO VFODSs will be considered. The connection is depicted in Fig. 1. There are the following relations

$$\begin{array}{r} \mathbf{u}(k) = \mathbf{u}_1(k), \\ \mathbf{y}_1(k) = \mathbf{u}_2(k), \\ \mathbf{y}(k) = \mathbf{y}_1(k), \\ \mathbf{y}_i(k) = \mathbf{G}_{i,k}\mathbf{u}_i(k) = \left[{}_{k_0}\mathbf{D}^{n_i}_{k}\right]^{-1}\left[{}_{k_0}\mathbf{N}^{m_i}_{k}\right]\mathbf{u}_i(k) \text{ for } i = 1, 2 \end{array}. \tag{35}$$

Combining Eqs. (35) one obtains

$$\begin{aligned} \mathbf{y}(\mathbf{k}) &= \mathbf{G}_{2,k}\mathbf{G}_{1,k}\mathbf{u}(\mathbf{k}) \\ &= \left[{}_{k_0}\mathbf{D}^{n_2}_{2,k}\right]^{-1}\left[{}_{k_0}\mathbf{N}^{m_2}_{2,k}\right]\left[{}_{k_0}\mathbf{D}^{n_1}_{1,k}\right]^{-1}\left[{}_{k_0}\mathbf{N}^{m_1}_{1,k}\right]\mathbf{u}(\mathbf{k}). \end{aligned} \tag{36}$$

Next, one considers a parallel connection of two subsystems presented in Fig. 2.
Appropriate relations are as follows

$$\begin{array}{r} \mathbf{u}_1(k) = \mathbf{u}_2(k) = \mathbf{u}(k), \\ \mathbf{y}(k) = \mathbf{y}_1(k) + \mathbf{y}_2(k), \\ \mathbf{y}_i(k) = \mathbf{G}_{i,k}\mathbf{u}_i(k) = \left[{}_{k_0}\mathbf{D}^{n_i}_{k}\right]^{-1}\left[{}_{k_0}\mathbf{N}^{m_i}_{k}\right]\mathbf{u}_i(k) \text{ for } i = 1, 2. \end{array} \tag{37}$$

From the above equations one gets

$$\begin{aligned} \mathbf{y}(\mathbf{k}) &= \left[\mathbf{G}_{1,k} + \mathbf{G}_{2,k}\right]\mathbf{u}(\mathbf{k}) \\ &= \left\{\left[{}_{k_0}\mathbf{D}^{n_1}_{1,k}\right]^{-1}{}_{k_0}\mathbf{N}^{m_1}_{1,k} + \left[{}_{k_0}\mathbf{D}^{n_2}_{2,k}\right]^{-1}{}_{k_0}\mathbf{N}^{m_2}_{2,k}\right\}\mathbf{u}(\mathbf{k}) \\ &= \left[{}_{k_0}\mathbf{D}^{n_1}_{1,k}\right]^{-1}\left\{{}_{k_0}\mathbf{N}^{m_1}_{1,k} + {}_{k_0}\mathbf{D}^{n_1}_{1,k}\left[{}_{k_0}\mathbf{D}^{n_2}_{2,k}\right]^{-1}{}_{k_0}\mathbf{N}^{m_2}_{2,k}\right\}\mathbf{u}(\mathbf{k}). \end{aligned} \tag{38}$$

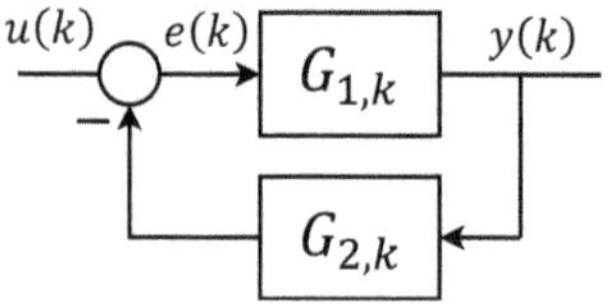

Fig. 3 Closed-loop connection of two transfer-functions-like

Finally the closed-loop system given in Fig. 3 is considered. The relations between the signals are given below

$$\begin{aligned} \mathbf{e}(k) &= \mathbf{u}(k) - \mathbf{G}_{2,k}\mathbf{y}(k) \\ \mathbf{y}(k) &= \mathbf{G}_{1,k}\mathbf{e}(k). \end{aligned} \tag{39}$$

From (39) one gets

$$\left[\mathrm{I}_k + \mathbf{G}_{1,k}\mathbf{G}_{2,k}\right]\mathbf{y}(k) = \mathbf{G}_{1,k}\mathbf{u}(k). \tag{40}$$

By assumption (29) the transfer-function-like matrices $\mathbf{G}_{i,k}$ $(i = 1, 2)$ are mostly singular. Hence,

$$\det\left[\mathrm{I}_k + \mathbf{G}_{1,k}\mathbf{G}_{2,k}\right] \neq 0. \tag{41}$$

In the case of the above mentioned exception, when the condition (29) is not satisfied (41) is still valid because of the form of a matrix product (34) being an upper triangular matrix with positive elements on the main diagonal. This permits us to write

$$\mathbf{y}(k) = \left[\mathrm{I}_k + \mathbf{G}_{1,k}\mathbf{G}_{2,k}\right]^{-1}\mathbf{G}_{1,k}\mathbf{u}(k), \tag{42}$$

or

$$\begin{aligned} \mathbf{y}(k) &= \left\{\mathrm{I}_k + \left[{}_{k_0}\mathbf{D}^{n_1}_{1,k}\right]^{-1}{}_{k_0}\mathbf{N}^{m_1}_{1,k}\left[{}_{k_0}\mathbf{D}^{n_2}_{2,k}\right]^{-1}{}_{k_0}\mathbf{N}^{m_2}_{2,k}\right\}^{-1} \times \\ &\qquad \times \left[{}_{k_0}\mathbf{D}^{n_1}_{1,k}\right]^{-1}{}_{k_0}\mathbf{N}^{m_1}_{1,k}\mathbf{u}(k) \\ &= \left\{{}_{k_0}\mathbf{D}^{n_1}_{1,k} + {}_{k_0}\mathbf{N}^{m_1}_{1,k}\left[{}_{k_0}\mathbf{D}^{n_2}_{2,k}\right]^{-1}{}_{k_0}\mathbf{N}^{m_2}_{2,k}\right\}^{-1}{}_{k_0}\mathbf{N}^{m_1}_{1,k}\mathbf{u}(k). \end{aligned} \tag{43}$$

The transfer-matrix-like and MFD-like forms and elementary connections of dynamical subsystems described by them reveal similar transformations leading to similar forms to the classical linear SISO continuous and discrete-time systems.

3.3 Solution of the VFODE

VFODE solution is of the form

$$\left[\sum_{i=0}^{n} \mathbf{S}_i(k)_{k_0}\mathbf{A}_k^{[\nu_i(k)]} \; \sum_{i=0}^{n} \mathbf{S}_i(k)_{-\infty}\mathbf{A}_{k_0-1}^{[\nu_i(k)]}\right]\begin{bmatrix}\mathbf{y}(k)\\ \mathbf{y}_{k_0-1}\end{bmatrix} = \sum_{i=0}^{m} \mathbf{T}_i(k)_{k_0}\mathbf{A}_k^{[\mu_i(k)]}\mathbf{u}(k), \tag{44}$$

or

$$\sum_{i=0}^{n} \mathbf{S}_i(k)_{k_0}\mathbf{A}_k^{[\nu_i(k)]}\mathbf{y}(k) = \sum_{i=0}^{m} \mathbf{T}_i(k)_{k_0}\mathbf{A}_k^{[\mu_i(k)]}\mathbf{u}(k) - \sum_{i=0}^{n} \mathbf{S}_i(k)\;_{-\infty}\mathbf{A}_{k_0-1}^{[\nu_i(k)]}\mathbf{y}_{k_0-1}. \tag{45}$$

An initial conditions term will be further denoted as

$$_{-\infty}\mathbf{I}_{k_0-1} = \sum_{i=0}^{n} \mathbf{S}_i(k)\;_{-\infty}\mathbf{A}_{k_0-1}^{[\nu_i(k)]} = \mathbf{S}_i(k)\begin{bmatrix} a^{[\nu(k)]}(k-k_0+1) & a^{[\nu(k)]}(k-k_0+2) & \cdots \\ a^{[\nu(k)]}(k-k_0) & a^{[\nu(k)]}(k-k_0+1) & \cdots \\ \vdots & \vdots & \\ a^{[\nu(k)]}(2) & a^{[\nu(k)]}(3) & \cdots \\ a^{[\nu(k)]}(1) & a^{[\nu(k)]}(2) & \cdots \end{bmatrix}. \tag{46}$$

3.4 Non-commensurate and Commensurate VFODEs

VFODE with parameters (20) will be further called non-commensurate ones whereas those satisfying conditions

$$\nu_i(k) = i\,[\nu(k)] \text{ for } i = 0, 1, \ldots, n, \tag{47}$$

$$\mu_i(k) = i\,[\mu(k)] \text{ for } i = 0, 1, \ldots, m,$$

$$\text{where } 0 < \nu(k) \leqslant 1,\, 0 < \mu(k) \leqslant 1, \tag{48}$$

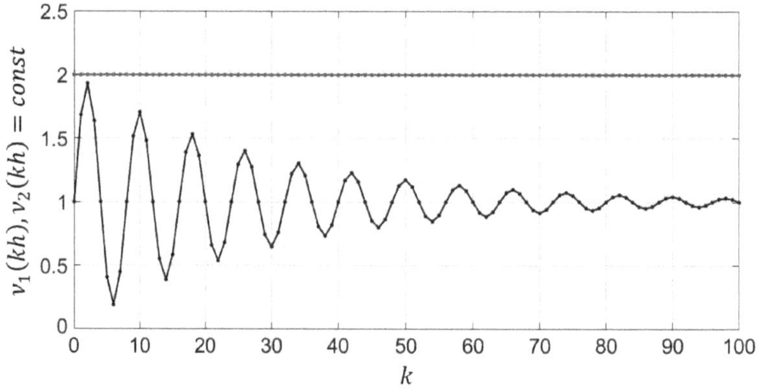

Fig. 4 Plots of the order functions $\nu_2(k)$ (in blue), $\nu_1(k)$ (in black)

commensurate ones [2]. The following example deals with the defined system.

Example 1 Consider commensurate VFODE (8) where $n = 2, m = 1$. Hence,

$$\begin{aligned}\left[{}_{k_0}\mathbf{A}_k^{[\nu_2(k)]} + a_1(k)_{k_0}\mathbf{A}_k^{[\nu_1(k)]} + a_0(k)\mathbf{1}_k\right]\mathbf{y}(k)\\ = \left[b_1(k)_{k_0}\mathbf{A}_k^{[\mu_i(k)]} + b_0(k)\mathbf{1}_k\right]\mathbf{u}(k).\end{aligned} \tag{49}$$

One takes $a_1 = 0.7, a_0 = 1, b_1 = 0.1, b_0 = 0.5$ and the following orders functions

$$\begin{aligned}\nu_2(k) &= 2 = \text{const},\\ \nu_1(k) &= 1 + e^{-0.025(k-1)}\sin\left(\frac{(k-1)\pi}{4}\right),\\ \mu_1(k) &= 1.\end{aligned} \tag{50}$$

There are assumed $k_0 = 0$, zero initial conditions and an input function in the form of a shifted discrete step function $u(k) = \mathbf{1}(k-1)$. In Figs. 4 and 5 there are plotted order functions $\nu_2(k)$ (in blue), $\nu_1(k)$ (in black) and solution (21) (in black), respectively.

Example 2 Consider commensurate VFODE (21) where $n = 2, m = 1$ in which one takes $a_0 = 1, b_1 = 0.1, b_0 = 0.5$ and

$$\begin{aligned}a_1(k) &= 0.7 - 1.4e^{-0.002(k-1)}\cos\left(\frac{(k-1)\pi}{150}\right),\\ \nu_2(k) &= 2 = \text{const},\\ \nu_1(k) &= 1 + e^{-0.025(k-1)}\sin\left(\frac{(k-1)\pi}{4}\right),\\ \mu_1(k) &= 1.\end{aligned}$$

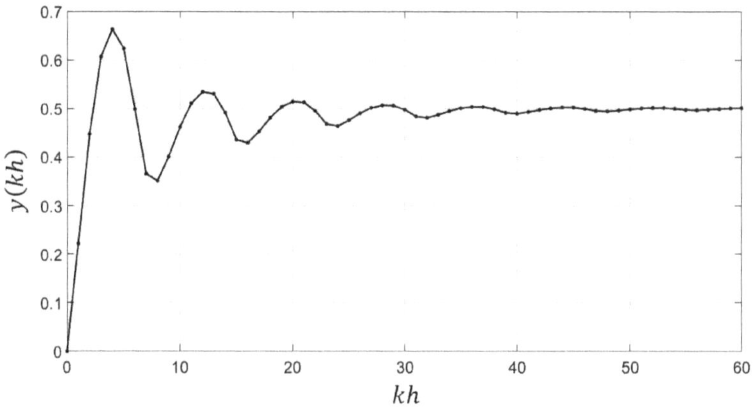

Fig. 5 Plot of the solution of (21)

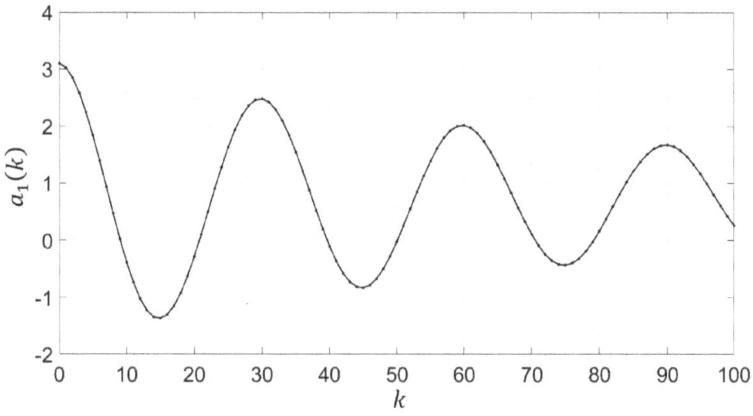

Fig. 6 Plot of the coefficient $a_1(k)$

Similarly to Example 1 $k_0 = 0$, zero initial conditions are also assumed and an input function in the form of a shifted discrete step function $u(k) = \mathbf{1}(k-1)$. In Figs. 6 and 7 there are plotted coefficient $a_1(k)$ and solution of (21), respectively.

The solution looks very strange. Though the system is linear, the plot suggests the non-linear behaviour of the considered system.

4 VFO Linear Time-Variant State-Space-Like Description

In this section an equivalent to the VFODE of the considered linear time-variant SISO system is presented. It will be called a state-space-like (VFOSS-L) one. Consider the linear time-variant VFODE (19) with the following statement. For a known input

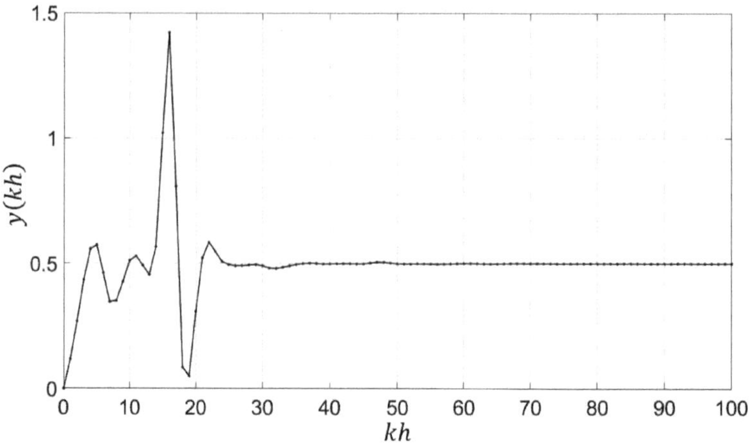

Fig. 7 Plot of the solution of (21)

signal $\mathbf{u}(k)$ and known right hand side of the considered equation one can denote

$$\mathbf{u}'(k) = \sum_{i=0}^{m} \mathbf{T}_i(k)_{k_0}\mathbf{A}_k^{[\mu_i(k)]}\mathbf{u}(k). \tag{51}$$

Then, new vector variables are defined

$$\begin{aligned} \mathbf{y}(k) &= \mathbf{x}_1(k), \\ {}_{k_0}\mathbf{A}_k^{[\nu_1(k)]}\mathbf{x}_1(k) &= \mathbf{x}_2(k), \\ {}_{k_0}\mathbf{A}_k^{[\nu_2(k)]}\left[{}_{k_0}\mathbf{A}^{[\nu_1(k)]}\right]^{-1}\mathbf{x}_2(k) &= \mathbf{x}_3(k), \\ &\vdots \\ {}_{k_0}\mathbf{A}_k^{[\nu_{n-1}(k)]}\left[{}_{k_0}\mathbf{A}^{[\nu_{n-2}(k)]}\right]^{-1}\mathbf{x}_{n-1}(k) &= \mathbf{x}_n(k) \end{aligned} \tag{52}$$

$${}_{k_0}\mathbf{A}_k^{[\nu_n(k)]}\left[{}_{k_0}\mathbf{A}^{[\nu_{n-1}(k)]}\right]^{-1}\mathbf{x}_n(k) = -\sum_{i=0}^{n}\mathbf{R}_{i-1}(k)\mathbf{x}_i(k) + \mathbf{u}'(k)$$

Defining $(n(k-k_0+1)) \times 1$ vectors

$$\tilde{\mathbf{x}}(k) = \begin{bmatrix} \mathbf{x}_1(k) \\ \mathbf{x}_2(k) \\ \vdots \\ \mathbf{x}_n(k) \end{bmatrix}, \quad \tilde{\mathbf{u}}(k) = \mathbf{u}(k), \tag{53}$$

and matrices $n(k-k_0+1) \times n(k-k_0+1)$, $n(k-k_0+1) \times k-k_0+1$ and $k-k_0+1 \times n(k-k_0+1)$

$$\tilde{\mathbf{A}}_k = \begin{bmatrix} \mathbf{0}_k & \mathbf{1}_k & \mathbf{0}_k & \cdots & \mathbf{0}_k & \mathbf{0}_k \\ \mathbf{0}_k & \mathbf{0}_k & \mathbf{1}_k & \cdots & \mathbf{0}_k & \mathbf{0}_k \\ \vdots & \vdots & \vdots & & \vdots & \vdots \\ \mathbf{0}_k & \mathbf{0}_k & \mathbf{0}_k & \cdots & \mathbf{0}_k & \mathbf{1}_k \\ -\mathbf{R}_0(k) & -\mathbf{R}_1(k) & -\mathbf{R}_2(k) & \cdots & -\mathbf{R}_{n-2}(k) & -\mathbf{R}_n(k) \end{bmatrix}, \tag{54}$$

$$\tilde{\mathbf{B}}_k = \begin{bmatrix} \mathbf{0}_k \\ \mathbf{0}_k \\ \vdots \\ \mathbf{1}_k \end{bmatrix}, \quad \tilde{\mathbf{C}}_k = \begin{bmatrix} \mathbf{1}(k) \; \mathbf{0}(k) \cdots \mathbf{0}_k \end{bmatrix}, \tag{55}$$

one gets state-space-like equations

$$\tilde{\mathbf{F}}_k \tilde{\mathbf{x}}(k) = \tilde{\mathbf{A}}_k \tilde{\mathbf{x}}(k) + \tilde{\mathbf{B}}_k \tilde{\mathbf{u}}(k) \tag{56}$$

$$\mathbf{y}(k) = \tilde{\mathbf{C}}_k \tilde{\mathbf{x}}(k), \tag{57}$$

where

$$\tilde{\mathbf{F}}_k = \begin{bmatrix} {}_{k_0}\mathbf{A}_k^{[\nu_1(k)]} & \mathbf{0}_k & \mathbf{0}_k \cdots \\ \mathbf{0}_k & {}_{k_0}\mathbf{A}_k^{[\nu_2(k)]} \left[{}_{k_0}\mathbf{A}_k^{[\nu_1(k)]} \right]^{-1} & \mathbf{0}_k \cdots \\ \vdots & \vdots & \vdots \\ \mathbf{0}_k & \mathbf{0}_k & \mathbf{0}_k \cdots \\ \mathbf{0}_k & \mathbf{0}_k & \mathbf{0}_k \cdots \end{bmatrix} \tag{58}$$

$$\begin{matrix} \cdots & \mathbf{0}_k & \mathbf{0}_k \\ \cdots & \mathbf{0}_k & \mathbf{0}_k \\ & \vdots & \vdots \\ \cdots & {}_{k_0}\mathbf{A}_k^{[\nu_{n-2}(k)]}\left[{}_{k_0}\mathbf{A}_k^{[\nu_{n-1}(k)]}\right]^{-1} & \mathbf{0}_k \\ \cdots & -\mathbf{0}_k & {}_{k_0}\mathbf{A}_k^{[\nu_n(k)]}\left[{}_{k_0}\mathbf{A}_k^{[\nu_{n-1}(k)]}\right]^{-1} \end{matrix}\Bigg].$$

Remark 4 For non-commensurate system (56)–(57) term $\tilde{\mathbf{F}}_k\tilde{\mathbf{x}}(k)$ takes the form

$$\tilde{\mathbf{F}}_k\tilde{\mathbf{x}}(k) = \begin{bmatrix} {}_{k_0}\mathbf{A}_k^{[\nu(k)]} & \mathbf{0}_k & \cdots & \mathbf{0}_k \\ \mathbf{0}_k & {}_{k_0}\mathbf{A}_k^{[\nu(k)]} & \cdots & \mathbf{0}_k \\ \vdots & \vdots & & \vdots \\ \mathbf{0}_k & \mathbf{0}_k & \cdots & {}_{k_0}\mathbf{A}_k^{[\nu(k)]} \end{bmatrix} \begin{bmatrix} \mathbf{x}_1(k) \\ \mathbf{x}_2(k) \\ \vdots \\ \mathbf{x}_n(k) \end{bmatrix} = \begin{bmatrix} {}_{k_0}\mathbf{A}_k^{[\nu(k)]}\mathbf{x}_1(k) \\ {}_{k_0}\mathbf{A}_k^{[\nu(k)]}\mathbf{x}_2(k) \\ \vdots \\ {}_{k_0}\mathbf{A}_k^{[\nu(k)]}\mathbf{x}_n(k) \end{bmatrix}. \tag{59}$$

Matrices in (56)–(57) play roles of: a system matrix, input distribution and output gain, respectively. The considered state-space-like equations are similar to the so called observability form [10].

5 VFO Linear Time-Invariant Commensurate State-Space-Like State Vectors Transformation

Further simplifications of the system described by (19) (and its equivalence (21)) will now be imposed. Aside from the commensurity (expressed by conditions (19) and (20) a time-invariance is assumed: $a_i(k) = a_i = \text{const}$, $b_i(k) = b_i = \text{constant}$. Then, matrices (11)–(12) take the forms

$$\mathbf{S}_i(k) = a_i\mathbf{1}_k \text{ for } i = 0, 1, \ldots, n, \tag{60}$$

$$\mathbf{T}_i(k) = b_i\mathbf{1}_k \text{ for } i = 0, 1, \ldots, m. \tag{61}$$

Under such assumptions the final form of the considered equation is as follows

$$\sum_{i=0}^{n} a_i\, {}^{GL}_{k_0}\Delta_k^{[\nu^i(k)]} y(k) = \sum_{i=0}^{m} b_i\, {}^{GL}_{k_0}\Delta_k^{[\mu^i(k)]} u(k) \tag{62}$$

or equivalently

$$\sum_{i=0}^{n} a_i \left[{}_{k_0}\mathbf{A}_k^{[\nu(k)]}\right]^i y(k) = \sum_{i=0}^{m} b_i \left[{}_{k_0}\mathbf{A}_k^{[\nu(k)]}\right]^i u(k) \tag{63}$$

Then the state-space-like form simplifies to

$$\begin{bmatrix} {}_{k_0}\mathbf{A}_k^{[\nu(k)]}\mathbf{x}_1(k) \\ {}_{k_0}\mathbf{A}_k^{[\nu(k)]}\mathbf{x}_2(k) \\ \vdots \\ {}_{k_0}\mathbf{A}_k^{[\nu(k)]}\mathbf{x}_n(k) \end{bmatrix} = \begin{bmatrix} \mathbf{0}_k \\ \mathbf{0}_k \\ \vdots \\ \mathbf{0}_k \\ \mathbf{1}_k \end{bmatrix} \mathbf{u}(k) + \begin{bmatrix} \mathbf{0}_k & \mathbf{1}_k & \cdots & \mathbf{0}_k & \mathbf{0}_k \\ \mathbf{0}_k & \mathbf{0}_k & \cdots & \mathbf{0}_k & \mathbf{0}_k \\ \vdots & \vdots & & \vdots & \vdots \\ \mathbf{0}_k & \mathbf{0}_k & \cdots & \mathbf{0}_k & \mathbf{1}_k \\ -a_0\mathbf{1}_k & -a_1\mathbf{1}_k & \cdots & -a_{n-2}\mathbf{1}_k & -a_{n-1}\mathbf{1}_k \end{bmatrix} \begin{bmatrix} \mathbf{x}_1(k) \\ \mathbf{x}_2(k) \\ \vdots \\ \mathbf{x}_{n-1}(k) \\ \mathbf{x}_n(k) \end{bmatrix}$$

$$= \mathbf{b}\mathbf{u}(k) + \mathbf{A}_k\mathbf{x}(k) \tag{64}$$

$$\mathbf{y}(k) = \begin{bmatrix} b_0\mathbf{1}_k \; b_1\mathbf{1}_k \; \cdots \; b_m\mathbf{1}_k \; \mathbf{0}_k \; \cdots \; \mathbf{0}_k \end{bmatrix} \begin{bmatrix} \mathbf{x}_1(k) \\ \mathbf{x}_2(k) \\ \vdots \\ \mathbf{x}_m(k) \\ \mathbf{x}_{m+1}(k) \\ \vdots \\ \mathbf{x}_n(k) \end{bmatrix} = \mathbf{c}_k\mathbf{x}(k). \tag{65}$$

Equations (64) and (65) are a system state-space-like description. To the classical state-space description a characteristic polynomial is related. Treating ${}_{k_0}\mathbf{A}_k^{[\nu(k)]}$ in (63) as an independent scalar variable w one gets a characteristic polynomial

$$\sum_{i=0}^{n} a_i w^i = 0, \; a_n = 1. \tag{66}$$

Now it is assumed that

$$\sum_{i=0}^{n} a_i w^i = \prod_{i=1}^{n} (w - p_i), \; p_i \neq p_j \text{ for } i \neq j \; p_i \in \mathbb{R}. \tag{67}$$

Remark 5 Values w_i will be called poles-like of the VFOS.

Proposition 2 *Given state-space-like description* (64) *and* (65). *If invertible matrices exist*

$$\tilde{\mathbf{G}}_k = \begin{bmatrix} \mathbf{0}_k & \mathbf{0}_k & \cdots & \mathbf{0}_k & \mathbf{1}_k \\ \mathbf{0}_k & \mathbf{0}_k & \cdots & \mathbf{1}_k & \mathbf{G}_{1,k} \\ \vdots & \vdots & & \vdots & \vdots \\ \mathbf{0}_k & \mathbf{1}_k & \cdots & \mathbf{G}_{n-3} & \mathbf{G}_{n-2,k} \\ \mathbf{1}_k & \mathbf{G}_{1,k} & \cdots & \mathbf{G}_{n-2,k} & \mathbf{G}_{n-1,k} \end{bmatrix} \tag{68}$$

with

$$\mathbf{G}_{i,k} = -\sum_{j=0}^{i-1} a_{n-j}\mathbf{G}_{i-1-j,k},\ \mathbf{G}_{0,k} = \mathbf{1}_k,\ i = 1, \ldots, nm \tag{69}$$

and

$$\tilde{\tilde{\mathbf{G}}}_k = \begin{bmatrix} -a_1\mathbf{1}_k & -a_2\mathbf{1}_k & \cdots & a_{n-1}\mathbf{1}_k & \mathbf{1}_k \\ -a_2\mathbf{1}_k & -a_3\mathbf{1}_k & \cdots & \mathbf{1}_k & \mathbf{0}_k \\ \vdots & \vdots & & \vdots & \vdots \\ -a_{n-1}\mathbf{1}_k & \mathbf{1}_k & \cdots & \mathbf{0}_k & \mathbf{0}_k \\ \mathbf{1}_k & \mathbf{0}_k & \cdots & \mathbf{0}_k & \mathbf{0}_k \end{bmatrix} \tag{70}$$

then, there exists a linear similarity transformation matrix $\mathbf{M}_k$

$$\mathbf{M}_k = \tilde{\mathbf{G}}_k\tilde{\tilde{\mathbf{G}}}_k \tag{71}$$

converting the state and input matrices in (64)–(65) *to the forms*

$$\tilde{\mathbf{A}}_k = \begin{bmatrix} p_1\mathbf{1}_k & \mathbf{0}_k & \cdots & \mathbf{0}_k & \mathbf{0}_k \\ \mathbf{0}_k & p_2\mathbf{1}_k & \cdots & \mathbf{0}_k & \mathbf{0}_k \\ \vdots & \vdots & & \vdots & \vdots \\ \mathbf{0}_k & \mathbf{0}_k & \cdots & p_{n-1}\mathbf{1}_k & \mathbf{1}_k \\ \mathbf{0}_k & \mathbf{0}_k & \cdots & \mathbf{0}_k & p_n\mathbf{1}_k \end{bmatrix},\ \tilde{\mathbf{B}}_k = \begin{bmatrix} \mathbf{1}_k \\ \mathbf{1}_k \\ \vdots \\ \mathbf{1}_k \\ \mathbf{1}_k \end{bmatrix}. \tag{72}$$

Proof It is easy to check by direct calculation that

$$\tilde{\mathbf{A}}_k\tilde{\mathbf{G}}_k\tilde{\tilde{\mathbf{G}}}_k = \tilde{\mathbf{G}}_k\tilde{\tilde{\mathbf{G}}}_k\tilde{\mathbf{A}}_k \tag{73}$$

is satisfied. Then,

$$\tilde{\mathbf{A}}_k = \tilde{\mathbf{G}}_k\tilde{\tilde{\mathbf{G}}}_k\mathbf{A}_k\left[\tilde{\mathbf{G}}_k\tilde{\tilde{\mathbf{G}}}_k\right]^{-1} = \mathbf{M}_k\mathbf{A}_k\left[\mathbf{M}_k\right]^{-1}, \tag{74}$$

$$\tilde{\mathbf{b}}_k = \tilde{\mathbf{G}}_k \tilde{\tilde{\mathbf{G}}}_k \mathbf{b}_k = \mathbf{M}_k \tilde{\mathbf{B}}_k. \tag{75}$$

The state-space-like equations obtained after the transformation are similar to the so called diagonal (or Jordan) form [10].

6 VFO Linear Time-Invariant Commensurate State-Space-Like Iterative Form

Consider Eq. (63). Following the procedure performed on its left-hand side leading to the characteristic polynomial (67) one obtains a similar form of the right-hand side. For $m = n - 1$ one writes

$$\sum_{i=0}^{n-1} b_i w^i = 0 \tag{76}$$

and next presents in an equivalent form

$$\sum_{i=0}^{n-1} b_i w^i = \prod_{i=1}^{n-1} (w - z_i), \; z_i \in \mathbb{R}, \tag{77}$$

where z_i will be called zeros-like of the state-space-like description. In the considered case one admits multiple poles- and zeros-like. The vector variables are defined as follows

$$\begin{aligned}
{}_{k_0}\mathbf{A}_k^{[\nu(k)]}\mathbf{x}_1(k) &= p_1\mathbf{x}_1(k) + \mathbf{u}(k),\\
{}_{k_0}\mathbf{A}_k^{[\nu(k)]}\mathbf{x}_2(k) &= (p_1 - z_1)\mathbf{x}_1(k) + s_2\mathbf{x}_2(k) + \mathbf{u}(k),\\
&\vdots\\
{}_{k_0}\mathbf{A}_k^{[\nu(k)]}\mathbf{x}_{n-1}(k) &= (p_1 - z_1)\mathbf{x}_1(k) +\\
+(p_2 - z_2)\mathbf{x}_2(k) + \cdots &+ (p_{n-1} - z_{n-1})\mathbf{x}_{n-1}(k) +\\
&+ p_n\mathbf{x}_n(k) + \mathbf{u}(k)
\end{aligned} \tag{78}$$

$$\mathbf{y}(k) = \mathbf{x}_n(k). \tag{79}$$

In the vector-matrix form, equations given above are as follows

$$\hat{\mathbf{F}}\hat{\mathbf{x}}(k) = \hat{\mathbf{A}}_k\hat{\mathbf{x}}(k) + \hat{\mathbf{B}}_k\mathbf{u}(k) \tag{80}$$

$$\mathbf{y}(k) = \hat{\mathbf{C}}_k\hat{\mathbf{x}}(k), \tag{81}$$

where

$$\hat{\mathbf{F}}_k = \tilde{\mathbf{F}}_k \tag{82}$$

$$\hat{\mathbf{A}}_k = \begin{bmatrix} p_1\mathbf{1}_k & \mathbf{0}_k & \mathbf{0}_k & \cdots \\ (p_1 - z_1)\mathbf{1}_k & p_2\mathbf{1}_k & \mathbf{0}_k & \cdots \\ (p_1 - z_1)\mathbf{1}_k & (p_2 - z_2)\mathbf{1}_k & p_3\mathbf{1}_k & \cdots \\ \vdots & \vdots & \vdots & \\ (p_1 - z_1)\mathbf{1}_k & (p_2 - z_2)\mathbf{1}_k & (p_3 - z_3)\mathbf{1}_k & \cdots \\ (p_1 - z_1)\mathbf{1}_k & (p_2 - z_2)\mathbf{1}_k & (p_3 - z_3)\mathbf{1}_k & \cdots \end{bmatrix} \tag{83}$$

$$\begin{matrix} \ldots & \mathbf{0}_k & \mathbf{0}_k & \mathbf{0}_k \\ \ldots & \mathbf{0}_k & \mathbf{0}_k & \mathbf{0}_k \\ \ldots & \mathbf{0}_k & \mathbf{0}_k & \mathbf{0}_k \\ & \vdots & \vdots & \\ \ldots & (p_{n-2} - z_{n-2})\mathbf{1}_k & p_{n-1}\mathbf{1}_k & \mathbf{0}_k \\ \ldots & (p_{n-2} - z_{n-2})\mathbf{1}_k & (p_{n-1} - z_{n-1})\mathbf{1}_k & p_n\mathbf{1}_k \end{matrix}\Bigg]$$

$$\hat{\mathbf{C}}_k = \begin{bmatrix} \mathbf{0}_k & \mathbf{0}_k & \cdots & \mathbf{0}_k & b_{n-1}\mathbf{1}_k \end{bmatrix}, \quad \hat{\mathbf{B}}_k = \begin{bmatrix} \mathbf{1}_k \\ \mathbf{1}_k \\ \vdots \\ \mathbf{1}_k \\ \mathbf{1}_k \end{bmatrix}, \tag{84}$$

Proposition 3 *One assumes distinct poles-like of the VFODS e.g.* $p_i \neq p_j$ *for* $i \neq j$. *Then, there exists a non-singular state-vector transformation matrix* $\mathbf{T}_k$ *such that*

$$\hat{\mathbf{A}}_k = [\mathbf{T}_k]^{-1}\tilde{\mathbf{A}}_k\mathbf{T}_k \tag{85}$$

with

$$\mathbf{T}_k = \begin{bmatrix} \mathbf{T}_{1,1,k} & \mathbf{0}_k & \cdots & \mathbf{0}_k & \mathbf{0}_k \\ \mathbf{T}_{2,1,k} & \mathbf{T}_{2,2,k} & \cdots & \mathbf{0}_k & \mathbf{0}_k \\ \vdots & \vdots & & \vdots & \vdots \\ \mathbf{T}_{n-1,1,k} & \mathbf{T}_{n-1,2,k} & \cdots & p_{n-1}\mathbf{T}_{n-1,n-1,k} & \mathbf{0}_k \\ \mathbf{T}_{n,1,k} & \mathbf{T}_{n,2,k} & \cdots & \mathbf{T}_{n,n-1,k} & p_n\mathbf{T}_{n,n,k} \end{bmatrix} \tag{86}$$

$$\mathbf{T}_{i,i,k} = \mathbf{1}_k \text{ for } i = 1, 2, \ldots, k, \tag{87}$$

$$\mathbf{T}_{i,i-1,k} = \frac{p_{i-1} - z_{i-1}}{p_{i-1} - p_i}\mathbf{1}_k \text{ for } i = 2, 3, \ldots, k, \tag{88}$$

and in general

$$\mathbf{T}_{i,j,k} = \frac{1}{p_j - p_i}\sum_{l=j}^{i-1}(p_l - z_l)\,\mathbf{T_{l,j,k}} \text{ for } i > j. \tag{89}$$

Proof Equation (85) can be immediately written in the form

$$\mathbf{T}_k\hat{\mathbf{A}}_k = \tilde{\mathbf{A}}_k\mathbf{T}_k. \tag{90}$$

Matrices $\tilde{\mathbf{A}}_k$ and $\hat{\mathbf{A}}_k$ are in lower triangular block form, so one concludes that the transformation matrix $\hat{\mathbf{T}}_k$ is in lower triangular block one. Hence, comparing appropriate matrices on both sides of Eq. (90) one gets (86).

7 Final Conclusions

In the paper a novelty description of the VFODE is investigated. The description in the time-domain with matrices of linearly growing dimensions has similar properties to the classical description based on the one-sided Laplace transform. The state-space like form reveals similar transformation to the well known ones used in the continuous SISO systems forms.

There are several open problems which are mentioned briefly below.

- Defined matrices (26) and (27) play similar role to the polynomial matrices in MFD. Here, a problem of the greatest common left (or right) divisor arises. For two matrices ${}_{k_0}\mathbf{D}_k^n$ and ${}_{k_0}\mathbf{N}_k^m$ an open problem is to find a matrix $\mathbf{E}_{k+1} \neq \mathbf{I}_{k+1}$ satisfying (31) and (32).
- Assumption of the invertibility of matrices (43) is related to the system controllability. The mentioned notion should be explored.
- One may expect that the VFO-PID time-variant controller in the closed-loop control system [18, 23] can essentially improve its transient properties.
- The analysis of the VFO time-variant SISO systems can be extended to MIMO systems with known difficulties.

Acknowledgements This work was supported by the National Science Centre Poland Grant Number 2016/23/B/ST7/03686.

References

1. Baleanu, D., Diethelm, K., Trujillo, J.J.: Fractional Calculus: Models and Numerical Methods. Series on Complexity, Nonlinearity and Chaos. World Scientific, Singapore (2012)
2. Bąkała, M., Nowakowski, J., Ostalczyk, P.: Commensurate and non-commensurate fractional-order discrete model of an independent wheel electrical drive of the autonomous platform. In: Proceedings of the International Conference on Nonlinear Dynamics and Complexity, Poland, NDC 17 (2017)
3. Calier, F.M., Desoer, C.A.: Multivariable Feedback Systems. Springer, New York (1982)
4. Dabiri, A., Moghaddam, B.P., Machado, J.T.: Optimal variable-order fractional PID controllers for dynamical systems. J. Comput. Appl. Math. **339**, 40–48 (2018)
5. Dorčák, L., Petráš, I., Koštial, I., Trepák, J.: Fractional-order state space models. In: International Carpathian Control Conference ICCC', Czech Republic, Malenovice, pp. 193–198 (2002)
6. Fiedler, J.: Special Matrices and Their Applications in Numerical Mathematics. Martinus Nijhoff Publishers. Kluwer Academic Publishers Group, Dordrecht (1986)
7. Goodrich, C., Peterson, A.C.: Discrete Fractional Calculus. Springer International Publishing, Switzerland (2015)
8. Kaczorek, T.: Linear Control Systems. Wiley. Research Studies Press LTD., New York (1993)
9. Kaczorek, T.: Selected Problems of Fractional System Theory. Springer, Heidelberg (2011)
10. Kailath, T.: Linear Systems. Prentice-Hall, Inc., Englewood Cliffs (1980)
11. Kilac, E., Stanica, P.: The inverse of banded matrices. J. Comput. Appl. Math. **237**, 126–135
12. Kilbas, A.A., Srivastawa, H.M., Trujilo, J.J.: Theory and Applications of Fractional Differential Equations. North-Holland Mathematics Studies, Amsterdam (2006)
13. Kučera, V.: Discrete Linear Control: the Polynomial Equation Approach. Wiley, London (1979)
14. Malesza, W., Macias, M., Sierociuk, D.: Matrix approach and analog modeling for solving fractional variable order differential equations. In: Latawiec, K.J., et al. (eds.) Advances in Modeling and Control of Non-integer Order Systems. Lecture Notes in Electrical Engineering, vol. 320, pp. 71–79. Springer, Berlin (2015)
15. Malesza, W., Macias, M., Sierociuk, D.: Analytical solution of fractional variable order differential equations. J. Comput. Appl. Math. **348**, 214–236 (2019)
16. Mozyrska, D., Ostalczyk, P.: Variable-, fractional-order oscillation element. In: Babiarz, A., Czornik, A., Klamka, J., Niezabitowski, M. (eds.) Theory and Applications of Non-integer Order Systems. Lecture Notes in Electrical Engineering, vol. 407, pp. 65–76. Springer, Heidelberg (2016)
17. Mozyrska, D., Ostalczyk, P.: Variable-, fractional-order discrete-time integrator. Complexity, Hindawi Wiley **2017**, Article ID 3452409, 11 pp
18. Ostalczyk, P.: Variable-, fractional-order discrete PID controllers. In: 17th International Conference on Methods and Models in Automation and Robotics, Miedzyzdroje, Poland (2012)
19. Ostalczyk, P.: Discrete Fractional Calculus: Applications in Control and Image Processing (Series in Computer Vision Book 4). World Scientific Publishing, Singapore (2016)
20. Ostalczyk, P.: Variable-, fractional-order linear filter with orders depending on the signal values. In: RRNR 11th International Conference on Non-integer Order Calculus and Its Applications, Czstochowa, Poland (2019)
21. Ostalczyk, P.: Analysis of a closed-loop system with DC micro-motor electrical drive with propeller and variable- fractional order PID controller. In: Proceedings of the International Carpathian Control Conference, Krakow, Poland (2019)
22. Ostalczyk, P.: WHO child growth standards modelling by variable-, fractional-order difference equation. In: Kumar, D., Singh, J. (eds.) Fractional Calculus in Medical and Health Science. CRC Press, Boca Raton (2020)
23. Ostalczyk, P., Duch, P.: Closed loop system synthesis with the variable-, fractional order PID controller. In: 17th International Conference on Methods and Models in Automation and Robotics, Miedzyzdroje, Poland (2012)

24. Ostalczyk, P., Mozyrska, D.: Variable-fractional-order linear time-invariant system description and response. In: Proceedings of the International Conference on Fractional Differentiation and Its Applications, Novi Sad, Serbia, pp. 799–807 (2016)
25. Ostalczyk, P., Rybicki, T.: Variable-fractional-order dead-beat control of an electromagnetic servo. J. Vib. Control **14**(9–10), 1457–1471 (2008)
26. Oustaloup, A.: Diversity and Non-integer Differentiation for System Dynamics. Wiley, Hoboken (2014)
27. Pawluszewicz, E.: Constrained controllability of the -difference fractional control systems with Caputo type operator. Discret. Dyn. Nat. Soc. **2015**, Article ID 638420, 7 pp
28. Pawluszewicz, E., Koszewnik, A., Burzynski, P.: On Grünwald-Letnikov fractional operator with measurable order on continuous-discrete-time scale. Acta mechanica et automatica **14**(3), 161–165 (2020)
29. Podlubny, I.: Fractional Differential Equations. Academic, New York (1999)
30. Sierociuk, D., Malesza, W.: Fractional variable order discrete-time systems, their solutions and properties. Int. J. Syst. Sci. **48**(14), 3098–3105 (2017)
31. Sierociuk, D., Twardy, M.: Duality of variable fractional order difference operators and its application to identification. Bull. Pol. Acad. Sci.: Tech. Sci. **62**(4), 809–815 (2014)
32. Sierociuk, D., Malesza, W., Macias, M.: On the recursive fractional variable-order derivative: equivalent switching strategy, duality, and analog modelling. Circuit Syst. Signal Process. **34**(4), 1077–1113 (2015)
33. Underwood, R.G.: Fundamentals of Modern Algebra: a Global Perspective. World Scientific Publishing Co Pte Ltd, Singapore (2016)
34. Valerio, D., da Costa, J.S.: Variable-order fractional derivatives and their numerical approximations. Signal Process. **91**(3), 470–483 (2011)

Balanced Truncation Model Reduction in Approximation of Nabla Difference-Based Discrete-Time Fractional-Order Systems

Rafał Stanisławski, Marek Rydel, and Krzysztof J. Latawiec

Abstract This chapter surveys new results in applying Balanced Truncation (BT) model reduction methods to approximate LTI discrete-time noncommensurate fractional-order systems. For the fractional-order difference, the nabla-difference-based fractional-order difference under Grünwald–Letnikov definition is used. The method employs the Fourier-based decomposition of the fractional-order system and the BT model order reduction method. The presented algorithm's main advantage is the specific representation of the fractional-order system enabling a simple, analytical formula for the determination of the Gramians. This contributes to a significant improvement of the computational efficiency of the BT reduction method. The simulation experiments confirm the introduced method's effectiveness in terms of high modeling accuracy and low computational cost.

1 Introduction

It is well known that fractional-order systems are much more challenging in the implementation process than their classical, integer-order counterparts. This is because the continuous-time fractional-order derivatives and discrete-time fractional-order differences cannot be implemented in practice in the original form due to the infinite complexity of its equations. Therefore to implement the fractional-order system, we have to use finite-length implementations/approximations of the fractional-order derivatives/differences applied in the system. Regardless of the continuous and discrete-time systems, the approximation/implementation of the fractional-order

R. Stanisławski (✉) · M. Rydel · K. J. Latawiec
Department of Electrical, Control and Computer Engineering, Opole University of Technology, ul. Prószkowska 76, 45-758 Opole, Poland
e-mail: r.stanislawski@po.edu.pl

M. Rydel
e-mail: m.rydel@po.edu.pl

K. J. Latawiec
e-mail: k.latawiec@po.edu.pl

P. Kulczycki et al. (eds.), *Fractional Dynamical Systems: Methods, Algorithms and Applications*, Studies in Systems, Decision and Control 402,
https://doi.org/10.1007/978-3-030-89972-1_7

system can be realized in two ways: (a) substitution of the fractional-order derivatives/differences used in the system by their finite-length approximations and (b) approximation of the fractional-order system as a whole.

In the first case, various integer-order approximations of fractional order derivatives are the most often used. In the continuous-time case, the Oustaloup approximation is effectively used, which models a fractional-order derivative in a given frequency range [3, 29]. This approach is computationally simple and can lead to effective approximation but stability issues may occur for higher model lengths [4]. In the discrete-time case, the use of finite-length implementations of the Grünwald–Letnikov difference is popular [44]. This FIR-like approach is conceptually simple but in order to obtain satisfactory accuracy, we have to use very high implementation lengths (usually in the thousands). Therefore in this area we can find various more effective approaches, including discrete-time versions of the Oustaloup methods [4], the Laguerre-filter based approximators [22, 44], algorithms based on continuous fraction expansion [26, 27, 44] and many other approaches [8, 11, 25]. Note that, irrespectively of the approximation method used for fractional-order derivative/difference, these ways lead to a high length of model for a whole fractional-order system due to the model length rising multiplicatively with the length of fractional-order difference approximators and the fractional-order system's length.

The second way of implementing the fractional-order system is modeling the system as a whole using integer-order approximation. In this way, in the discrete-time case, which is exploited hereinafter, we can find the methods based on fractional-order generalizations of the Orthonormal Basis Functions-based models [42], using model-order reduction techniques [12, 18, 33, 35, 38, 45] and other approaches [6, 23]. Note that the use of these methods depends on a specific definition of fractional-order difference used in the system. In Refs. [18, 45] it is shown that using model order reduction techniques can lead to an effective approximation of the fractional-order systems by a relatively simple integer-order model. In particular, in Ref. [45] it is presented that a specific representation of the forward shifted fractional-order difference-based commensurate system by use of the Fourier model, leads to a significant simplification of the BT reduction method, through analytically-driven solutions of Lyapunov equations. As a result, we obtain a simple, so called FIRBT-based integer-order model of the fractional-order system.

In this paper we demonstrate that using the BT order reduction method applied to the Fourier-based model presented in Ref. [45] can also be effectively used for approximation of the nabla difference-based fractional-order noncommensurate state space system.

This chapter is organized as follows: Having introduced the problem of implementation of the nabla difference-based discrete-time fractional-order systems in Sect. 1, the representation of the nabla difference-based noncommensurate fractional-order state space system is presented in Sect. 2. Two different stability analysis methods for this kind of system, based on f-poles of the characteristic pseudopolynomial and so called Mikhailov curve, are recalled in Sect. 3. The main result in terms of a new approximation/implementation method for discrete-time nabla difference-based state space system is presented in Sect. 4. In particular, in this section we offer a specific form of FIR-based state space representation for the system, with the fractional properties modeled in a specific FIR filter and analytical implementation of the BT square root model order reduction algorithm to obtain a low order and accurate approximation of the system. Additionally, in Sect. 4, we offer an algorithm to generate an integer-order approximation of the considered system. An analysis of the proposed methodology's efficiency is presented in Sect. 5 and conclusions in Sect. 6 complete the paper.

2 System Representation

In this paper we consider the nabla-difference based noncommensurate fractional-order state space LTI MIMO system defined in the uniform time scale related to the continuous/discrete unification theory $\mathbb{T} = h\mathbb{Z} = \{0, h, 2h, 3h, \ldots\}$, with $h \in \mathbb{R}_+$ [15, 28]. The system is described by equations

$$\begin{aligned} \nabla^{\alpha} x(t) &= Ax(t) + Bu(t) \\ y(t) &= Cx(t) + Du(t) \end{aligned} \tag{1}$$

where $t \in \mathbb{T}$, $x(t) \in \mathbb{R}^n$, $u(t) \in \mathbb{R}^{n_u}$, $y(t) \in \mathbb{R}^{n_y}$ are the state, input and output vectors, respectively, and $A \in \mathbb{R}^{n\times n}$, $B \in \mathbb{R}^{n\times n_u}$, $C \in \mathbb{R}^{n_y\times n}$ and $D \in \mathbb{R}^{n_y\times n_u}$ are the system matrices respectively. Since the system (1) is noncommensurate order, the $\nabla^{\alpha} x(t)$ denotes the fractional-order difference vector

$$\nabla^{\alpha} x(t) = [\nabla^{\alpha_1} x_1(t),\ \nabla^{\alpha_2} x_2(t),\ \ldots,\ \nabla^{\alpha_n} x_n(t)]^T \tag{2}$$

where $x(t) = [x_1(t), \ldots, x_n(t)]^T$ and ∇^{α_i}, $i = 1, \ldots, n$, are the nabla fractional differences of order $\alpha_i \in (0, 2)$. In this paper, to describe the fractional-order difference we use the well known Grünwald–Letnikov definition

$$\nabla^{\alpha_i} x_i(t) = h^{-\alpha_i} \sum_{j=0}^{t/h} (-1)^j \binom{\alpha_i}{j} x_i(t - jh) \tag{3}$$

with $i = 1, \ldots, n$ and $\binom{\alpha_i}{j}$ being the Newton binomial. Note that under some conditions, the Grünwald–Letnikov difference can be equivalent to the Riemann–Liouville and Caputo ones [1, 37].

The $\mathscr{Z}$-transform of the state space system of Eq. (1) is as follows

$$\begin{aligned} w(z)X(z) &= AX(z) + BU(z) \\ Y(z) &= CX(z) + DU(z) \end{aligned} \tag{4}$$

where $U(z)$, $X(z)$ and $Y(z)$ are the $\mathscr{Z}$-transforms of the inputs u_k, states x_k and inputs y_k, respectively. The $w(z)$ denotes the $\mathscr{Z}$-transform of the nabla fractional difference vector

$$w(z) = diag\{w^{\alpha_1}(z), w^{\alpha_2}(z), \ldots, w^{\alpha_n}(z)\} \tag{5}$$

with $w^{\alpha_i}(z)$, $i = 1, \ldots, n$, being the $\mathscr{Z}$-transform of the nabla fractional-order difference (3)

$$w^{\alpha_i}(z) = h^{-\alpha_i} \sum_{j=0}^{\infty} (-1)^j \binom{\alpha_i}{j} z^{-j} = h^{-\alpha_i} \left(1 - z^{-1}\right)^{\alpha_i} \tag{6}$$

A characteristic pseudo-polynomial of the system (1) can be presented in the form

$$p(z) = \det\{w(z) - A\} = a_{\underline{n}} w^{\beta_{\underline{n}}}(z) + a_{\underline{n}-1} w^{\beta_{\underline{n}-1}}(z) + \cdots + a_1 w^{\beta_1}(z) + a_0 \tag{7}$$

where $\underline{n} = \sum_{i=1}^{n} \binom{n}{i}$ and β_j, $j = 1, \ldots, \underline{n}$, are sums of all i-combinations of the set $\{\alpha_1, \ldots, \alpha_n\}$ for all $i = 1, \ldots, n$.

In a specific SISO case, the system (4) can be presented in the scalar transfer function form as

$$G(z) = C(w(z) - A)^{-1} B \tag{8}$$

$$= \frac{b_{\underline{n}} w^{\beta_{\underline{n}}}(z) + b_{\underline{n}-1} w^{\beta_{\underline{n}-1}}(z) + \cdots + b_1 w^{\beta_1}(z) + b_0}{a_n w^{\beta_{\underline{n}}}(z) + a_{\underline{n}-1} w^{\beta_{\underline{n}-1}}(z) + \cdots + a_1 w^{\beta_1}(z) + a_0} \tag{9}$$

where β_i, $i = 1, \ldots, \underline{n}$, are as in Eq. (7).

The simple and numerically effective implementation of the system of Eq. (1) is the main topic of the paper.

Note that the cases of implementation/approximation methods considered in this chapter can be only used for asymptotically stable systems. Therefore we have to analyze the system's stability before its implementation. The methods for stability analysis are presented in the next subsection.

3 Stability of the Nabla Difference-Based Fractional-Order System

This section presents two alternative methods for commensurate and noncommensurate nabla difference-based fractional-order systems. The first method is analytically-driven and is based on f-poles of the characteristic pseudopolynomial of the system. The second, a more general method, is based on the so-called Mikhailov curve for the fractional-order system.

3.1 Commensurate-Order Case

The system of Eq. (1), and consequently of Eq. (4), can be considered in a commensurate-order form, with $\alpha_i = \alpha \ \forall\, i = 1, \ldots, n$ and $\underline{n} = n$. In this case the characteristic pseudopolynomial is

$$p(z) = a_n w^{\alpha n}(z) + a_{n-1} w^{\alpha(n-1)}(z) + \cdots + a_0 \tag{10}$$

$$= a_n\big(w^{\alpha}(z) - \lambda_1^f\big)\big(w^{\alpha}(z) - \lambda_2^f\big)\ldots\big(w^{\alpha}(z) - \lambda_n^f\big) \tag{11}$$

$$= a_n \prod_{j=1}^{n} p_j(z) \tag{12}$$

where $p_j(z) = (w^{\alpha}(z) - \lambda_j^f)$, $j = 1, \ldots, n$, and $\lambda_j^f \in \mathbb{C}$, $j = 1, \ldots, n$, are the zeros of the characteristic pseudopolynomial, which can be called pseudo-poles or, as in Refs. [40, 41], f-poles of the system (4). The simple analytically driven stability test for the nabla-based commensurate fractional-order system has been introduced in Ref. [39].

Theorem 1 *The fractional commensurate-order system, with characteristic pseudopolynomial as in Eq.* (11), *with* $\alpha_i = \alpha \in (0, 2)$, *is asymptotically stable if and only if*

$$(i)\ \ \varphi_i^f \in \Big(\alpha\frac{\pi}{2}, 2\pi - \alpha\frac{\pi}{2}\Big)\ \ or \tag{13}$$

$$(ii)\ \ \varphi_i^f \in \Big[-\alpha\frac{\pi}{2}, \alpha\frac{\pi}{2}\Big]\ \ and\ |\lambda_i^f| > |w_i|, \tag{14}$$

where $|w_i| = \Big(\frac{2}{T}\Big|\Big(\sin\frac{\pi}{2} - \frac{\varphi_i^f}{\alpha}\Big)\Big|\Big)^{\alpha}$, *and* φ_i^f *and* $|\lambda_i^f|$, $i = 1, \ldots, n$, *are the argument and modulus, respectively, of the ith* f*-pole of the system.*

Theorem 1 is originally driven for the SISO case, where the system is presented in the transfer function form of Eq. (9). However, taking into account that the stability of the commensurate-order case of the system (1) is based on the same characteristic pseudopolynomial (11), the commensurate-order case of the system (1) can also be considered by this theorem. Note that in order to use Theorem 1, we have to employ the canonical form of the characteristic equation as in Eq. (11), which is possible to do for commensurate-order systems only.

On the other hand, consider the system with rational orders $\alpha_i \in \mathbb{Q}$, $i = 0, \ldots, n$. In this case, the orders of the characteristic pseudopolynomial (7) can be presented in the form of

$$\beta_i = \frac{l_{\beta i}}{d_{\beta i}} = \frac{k_i}{d_{\alpha 0}} = k_i \alpha_0 \quad \forall \ \ i = 1, \ldots, \underline{n} \tag{15}$$

where $l_{\beta i}, d_{\beta i}, k_i, d_{\alpha 0} \in \mathscr{Z}_{+}$. The fractional-order α_0 can be calculated on the basis of the common divisor as $\alpha_0 = \frac{1}{\prod_{j=1}^{\underline{n}} d_{\beta j}}$. The characteristic pseudopolynomial (7) can be presented in form of

$$p(z) = a_{\underline{n}} w^{\alpha_0 k_{\underline{n}}}(z) + a_{\underline{n}-1} w^{\alpha_0 k_{\underline{n}-1}}(z) + \cdots + a_1 w^{\alpha_0 k_1}(z) + a_0 \tag{16}$$

where $w^{\alpha_0 k_j}(z) = (1 - z^{-1})^{\alpha_0 k_j}$, $j = 1, \ldots, \underline{n}$. Finally, the characteristic pseudopolynomial (16) can be presented in the canonical form as follows

$$p(z) = a_{\underline{n}} \big(w^{\alpha_0}(z) - \zeta_1\big)\big(w^{\alpha_0}(z) - \zeta_2\big) \ldots \big(w^{\alpha_0}(z) - \zeta_{k_{\underline{n}}}\big) \tag{17}$$

Therefore every system with rational orders can be considered an 'implicit' commensurate-order system, and we can still use Theorem 1 for the stability analysis. In this case $\alpha \leftarrow \alpha_0$, $n \leftarrow k_{\underline{n}}$ and $\lambda_i^f \leftarrow \zeta_i$, $i = 1, \ldots, k_{\underline{n}}$. However note that for 'implicit' commensurate-order systems we have $k_n >> n$. So stability analysis of this class of system may be much more challenging than 'explicit' commensurate-order ones. This issue has been presented in the Examples of stability analysis subsection. This issue is avoided in another stability analysis method which is a topic of the next subsection.

3.2 General Case

In this section, we present a simple stability test using an extension of the Mikhailov criterion. The results presented in this section are based on Ref. [43]. The stability test is based on the so-called Mikhailov curve, which is defined as

Definition 1 ([43]) Consider the fractional-order system presented in Eq. (1) with the nabla fractional-order difference as in Eq. (3). Then the Mikhailov curve for the system is defined as follows

$$p(e^{i\varphi}) = a_{\underline{n}} w^{\beta_{\underline{n}}}(e^{i\varphi}) + a_{\underline{n}-1} w^{\beta_{\underline{n}-1}}(e^{i\varphi}) + \cdots + a_1 w^{\beta_1}(e^{i\varphi}) + a_0 \tag{18}$$

for $\varphi \in [0, \pi]$, where $w^{\beta_j}(e^{i\varphi}) = h^{-\alpha_j}(1 - e^{-i\varphi})^{\beta_j}$, $j = 1, \ldots, \underline{n}$, and a_l, $l = 0, \ldots, \underline{n}$, are as in Eq. (7).

On the basis of the above definition an extended Mikhailov stability criterion is as follows

Theorem 2 ([43]) *Consider the discrete-time fractional-order system* (1) *with the nabla fractional-order difference defined in Eqs.* (2) *and* (3). *The system is asymptotically stable if and only if its Mikhailov curve* $p(e^{i\varphi})$, $\varphi \in [0, \pi]$, *satisfies the following two conditions*

(1) $p(e^{i\varphi}) \neq 0 \; \forall \, \varphi \in [0, \pi]$,
(2) $\Delta \arg p(e^{i\varphi})|_0^\pi = 0$.

where $\Delta \arg p_j(e^{i\varphi})|_0^\pi = \arg p_j(e^{i\pi}) - \arg p_j(e^{i0})$ *is a change in the argument of the function* $p_j(e^{i\varphi})$ *for* φ *ranging from 0 to* π.

Note that the above Theorem 2 can be used both for commensurate and noncommensurate order cases and also when the system (1) considers non-rational orders $\exists i : \alpha_i \in \mathbb{R}\backslash\mathbb{Q}, i = 1, \ldots, n$.

3.3 Stability Analysis Examples

Example 1 Consider the discrete-time nabla difference-based fractional-order state space system of commensurate-order $\alpha_1 = \alpha_2 = \alpha_3 = \alpha = 0.7$ with the two different sampling periods $h_1 = 0.2$ and $h_2 = 0.15$ and the parameters as follows

$$\left[\begin{array}{c|c} A & B \\ \hline C & D \end{array}\right] = \left[\begin{array}{ccc|c} -1 & -0.2 & -1 & 1 \\ 1 & 0 & 0 & 0 \\ 0 & 1 & 0 & 0 \\ \hline 0 & 0 & 1 & 0 \end{array}\right]$$

The characteristic pseudopolynomial for the system is $p(z) = w^{2.7}(z) + w^{1.8}(z) + 0.25w^{0.9}(z) + 1$ and the f-poles of the system are $\lambda_1^f = -1.3801040$, $\lambda_{2,3}^f = 0.190052 \pm 0.829737$. The stability analysis results by the use of Theorem 1 are presented in Table 1.

It can be seen from Table 1 that for $h = 0.15$ the modulus criterion *(ii)* of Theorem 1 is not satisfied and the system is unstable. Alternatively, for $h = 0.2$ the criterion *(ii)* of Theorem 1 is fulfilled and the considered system is asymptotically stable.

These results are confirmed by the use of the Mikhailov stability criterion of Theorem 2. The Mikhailov curves for both sampling periods are presented in Fig. 1.

It can be seen from Fig. 1 that for $h = 0.2$ we have $\Delta \arg p_j(e^{i\varphi})|_0^\pi = 0$ and the asymptotic stability condition of Theorem 2 is fulfilled. For $h = 0.15$ we have $\Delta \arg p_j(e^{i\varphi})|_0^\pi = -2\pi$ and taking into account Theorem 2 the considered system is unstable.

Table 1 Stability analysis for the system of Example 1

i	λ_i^f	$\lvert\lambda_i^f\rvert$	φ_i^f	$\lvert w_i\rvert$ for $h = 0.15$	$\lvert w_i\rvert$ for $h = 0.20$
1	1.3801040	1.3801040	3.141592	0	0
2	0.190052 + 0.829737	0.851224	1.345629	1.006960	0.777262
3	0.190052 − 0.829737	0.851224	−1.345629	1.006960	0.777262
				System unstable	**System stable**

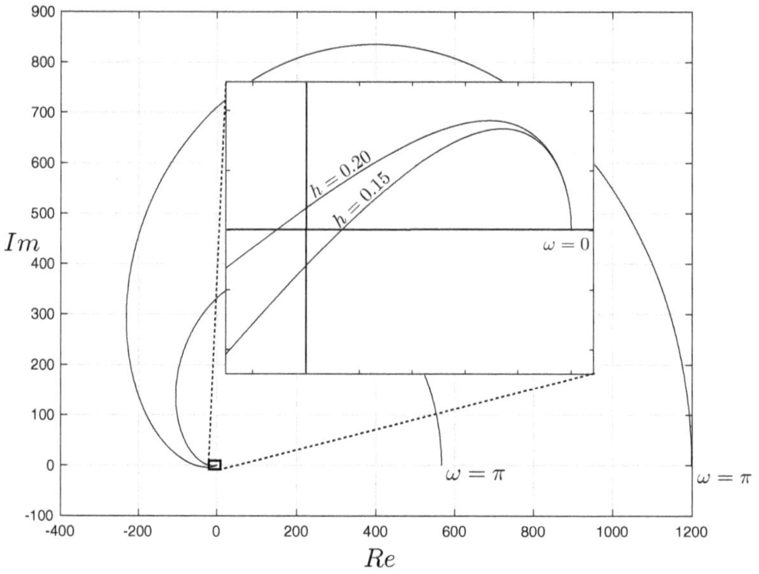

Fig. 1 Mikhailov curve for the system of Example 1

Example 2 Consider the nabla difference-based discrete-time noncommensurate fractional-order state space system with non-rational orders $\alpha_1 = \frac{\sqrt{3}}{2}, \alpha_2 = \frac{\sqrt{2}}{2}, \alpha_3 = 0.7$, the sampling period $h = 1$, and the parameters as follows

$$\left[\begin{array}{c|c} A & B \\ \hline C & D \end{array}\right] = \left[\begin{array}{ccc|c} -0.1 & -0.04 & -0.03 & 1 \\ 1 & 0 & 0 & 0 \\ 0 & 1 & 0 & 0 \\ \hline 0 & 0 & 1 & 0 \end{array}\right]$$

The characteristic pseudopolynomial for the system is $p(z) = w^{\left(\frac{\sqrt{3}}{2}+\frac{\sqrt{2}}{2}+0.7\right)}(z) + 0.1w^{\left(\frac{\sqrt{2}}{2}+0.7\right)}(z) + 0.04w^{0.7}(z) + 0.03$. In the considered system case we cannot obtain f-poles nor use Theorem 1. The only method to use for a simple stability test of this system is to employ the Mikhailov stability criterion of Theorem 2. The Mikhailov curve for the considered system is plotted in Fig. 2.

It can be seen from Fig. 2 that we have $\Delta \arg p_j(e^{i\varphi})|_0^\pi = 0$, so the system is asymptotically stable.

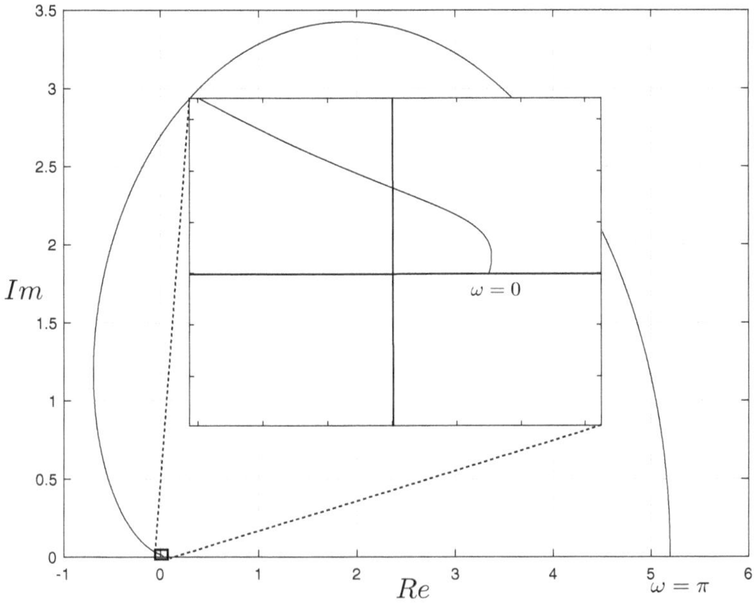

Fig. 2 Mikhailov curve for the system of Example 2

4 System Implementation

In order to approximate the system of Eq. (1) we have to firstly obtain time-domain responses of the system. We present the Laurent expansion method for the considered system and the FIR-based model of the system. This model has been used to calculate controllability and observability Gramians so that the fast implementation of the BT order reduction method can be achieved. As a result, we obtain a low, integer-order model of the system.

Calculation of the time response of the state equation of the system (1) can be realized by a formula resulting from the combination of Eqs. (1)–(3)

$$\begin{bmatrix} h^{-\alpha_1} \sum_{j=0}^{t/h} (-1)^j \binom{\alpha_1}{j} x_1(t-jh) \\ h^{-\alpha_2} \sum_{j=0}^{t/h} (-1)^j \binom{\alpha_2}{j} x_2(t-jh) \\ \vdots \\ h^{-\alpha_n} \sum_{j=0}^{t/h} (-1)^j \binom{\alpha_n}{j} x_n(t-jh) \end{bmatrix} = Ax(t) + Bu(t) \tag{19}$$

$$\left[diag\left(h^{-\alpha_1},\, h^{-\alpha_2},\, \ldots,\, h^{\alpha_n}\right) - A\right]x(t) = Bu(t) - \begin{bmatrix} h^{-\alpha_1} \sum_{j=1}^{t/h} (-1)^j \binom{\alpha_1}{j} x_1(t-jh) \\ h^{-\alpha_2} \sum_{j=1}^{t/h} (-1)^j \binom{\alpha_2}{j} x_2(t-jh) \\ \vdots \\ h^{-\alpha_n} \sum_{j=1}^{t/h} (-1)^j \binom{\alpha_n}{j} x_n(t-jh) \end{bmatrix}$$

Finally, the state equation can be presented in the form of

$$x(t) = \underline{A}^{-1}Bu(t) + \underline{A}^{-1}\sum_{j=1}^{t/h} R_j x(t-jh) \tag{20}$$

where $\underline{A} = \left[diag\left(h^{-\alpha_1},\, h^{-\alpha_2},\, \ldots,\, h^{\alpha_n}\right) - A\right]$, $R_j = diag\big(h^{-\alpha_1}(-1)^{j+1}\binom{\alpha_1}{j},\, h^{-\alpha_2}(-1)^{j+1}\binom{\alpha_2}{j},\, \ldots,\, h^{-\alpha_n}(-1)^{j+1}\binom{\alpha_n}{j}\big)^T$, $j = 1, \ldots, t/h$. Similar results in calculation of the time-domain characteristics of the nabla difference-based fractional-order systems have been presented in Ref. [17]. In that paper it has also been presented that the time responses of the nabla difference-based systems are similar to those based on the forward shifted delta difference, which are also often used in modeling of fractional-order systems.

On the basis of the time-domain results we can easily calculate the Laurent expansion and FIR model for the considered system.

4.1 FIR-Based Model

It is well known that discrete-time dynamical system $G(z)$ can be described by the Laurent expansion as $G(z) = \sum_{i=0}^{\infty} \eta_i z^{-i}$, where η_i, $i = 0, 1, \ldots$ are the model parameters in terms of the system impulse response coefficients and z is the $\mathscr{Z}$-transform operator. On the basis of the results above we can propose following

Theorem 3 *Consider the noncommensurate fractional-order state space system of Eq.* (1). *The system can be described by formula*

$$y(t) = \sum_{i=1}^{\infty} \eta_i u(t-ih) \tag{21}$$

where $\eta_i = C\Phi_i + D$, *with* Φ_i, $i = 1, 2, \ldots$ *calculated in a recursive way as*

$$\Phi_i = \underline{A}^{-1}Bu(t) + \underline{A}^{-1}\sum_{j=1}^{j} R_j \Phi_{i-j} \tag{22}$$

where $R_j = diag\left(h^{-\alpha_1}(-1)^{j+1}\binom{\alpha_1}{j},\ h^{-\alpha_2}(-1)^{j+1}\binom{\alpha_2}{j},\ \ldots,\ h^{-\alpha_n}(-1)^{j+1}\binom{\alpha_n}{j}\right)$ *and* $\Phi_0 = x(0)$.

Proof It is well known that the system of Eq. (1) can be described by the Laurent expansion as in Eq. (21). Then immediately from Eq. (20) we obtain an iterative way of calculation of Φ_i, $i = 1, 2, \ldots$ of the system as

$$\Phi_1 = x(h) = \underline{A}^{-1}Bu(h) + \underline{A}^{-1}\begin{bmatrix} -\binom{\alpha_1}{1} & 0 & \ldots & 0 \\ 0 & -\binom{\alpha_2}{1} & \ldots & 0 \\ \\ 0 & 0 & \ldots & -\binom{\alpha_n}{1} \end{bmatrix}\Phi_0$$

$$\Phi_2 = x(2h) = \underline{A}^{-1}Bu(2h) + \underline{A}^{-1}\begin{bmatrix} -\binom{\alpha_1}{1} & 0 & \ldots & 0 \\ 0 & -\binom{\alpha_2}{1} & \ldots & 0 \\ \\ 0 & 0 & \ldots & -\binom{\alpha_n}{1} \end{bmatrix}\Phi_1$$

$$+A^{-1}\begin{bmatrix} -\binom{\alpha_1}{2} & 0 & \ldots & 0 \\ 0 & -\binom{\alpha_2}{2} & \ldots & 0 \\ \\ 0 & 0 & \ldots & -\binom{\alpha_n}{2} \end{bmatrix}\Phi_0 \qquad (23)$$

$$\vdots$$

$$\Phi_i = x(ih) = \underline{A}^{-1}Bu(ih) + \underline{A}^{-1}\sum_{j=1}^{i}\begin{bmatrix} -\binom{\alpha_j}{1} & 0 & \ldots & 0 \\ 0 & -\binom{\alpha_j}{1} & \ldots & 0 \\ \\ 0 & 0 & \ldots & -\binom{\alpha_n}{j} \end{bmatrix}\Phi_{i-j}$$

Now, taking into account the output equation of Eq. (1) we immediately arrive at Theorem 3.

Note again, that to describe the system on the basis of the Laurent expansion of Theorem 3 we have to use an infinite sum in Eq. (21). Therefore in practical applications we have to use the finite length model in terms of the FIR model. Taking into account Lemma 1 of Ref. [45] we can obtain FIR-based state space model for MIMO nabla difference-based fractional-order system with matrices

$$A_{FIR} = \begin{bmatrix} 0 & 0 & \ldots & 0 & 0 \\ I_{n_u \times n_u} & 0 & \ldots & 0 & 0 \\ 0 & I_{n_u \times n_u} & \ldots & 0 & 0 \\ \vdots & \vdots & \ddots & \vdots & \vdots \\ 0 & 0 & \ldots & I_{n_u \times n_u} & 0 \end{bmatrix} \qquad (24)$$

$$B_{FIR} = \begin{bmatrix} I_{n_u \times n_u} & 0_{n_u \times n_u(L-1)} \end{bmatrix}^T \qquad (25)$$

$$C_{FIR} = \begin{bmatrix} \eta_1 & \dots & \eta_L \end{bmatrix} \tag{26}$$

$$D_{FIR} = \eta_0 \tag{27}$$

where $A_{FIR} \in \mathbb{R}^{Ln_u \times Ln_u}$, $B_{FIR} \in \mathbb{R}^{Ln_u \times n_u}$, $C_{FIR} \in \mathbb{R}^{n_y \times Ln_u}$ and $D_{FIR} \in \mathbb{R}^{n_y \times n_u}$, with η_i, $i = 0, \dots, L$ as in Theorem 3.

Note that the implementation of the system by use of the FIR-based model of Eqs. (24)–(27) leads to high model lengths. Usually, to obtain satisfactory accuracy, we have to use $L \geq 5000$. However Ref. [45] has shown that the model can easily be reduced using the BT reduction method, which is presented in the next subsection.

4.2 Balanced Truncation Model Order Reduction Method

Consider a discrete-time state space dynamical system of the FIR-based integer-order state space model with matrices as in Eqs. (24)–(27). Model order reduction then leads to approximation of this system by a state space model of a lower order

$$\begin{aligned} x_r(t+h) &= A_r x_r(t) + B_r u(t) \\ y_r(t) &= C_r x_r(t) + D_r u(t) \end{aligned} \tag{28}$$

with $A_r \in \mathbb{R}^{k \times k}$, $B_r \in \mathbb{R}^{k \times n_u}$. $C_r \in \mathbb{R}^{n_y \times k}$ and $D_r \in \mathbb{R}^{n_y \times n_u}$, with $k << L$.

Note that order reduction of the system is not a unique operation and there are various techniques for model order reduction involving Krylov projection-based methods [2, 7, 10, 12] and Singular Value Decomposition-based (SVD-based) methods [2, 14, 21, 31]. One of the most popular algorithms among SVD-based methods is the Balanced Truncation approximation [20, 24, 32, 36]. This method can lead to good modeling performance with an 'a priori' given approximation error.

Moreover there exist methods that extend the BT by introducing weighting functions. They can be implemented in the form of direct input/output weighting functions [9, 16, 33, 34, 48, 49] or by Gramians calculation in a restricted frequency or time intervals [5, 13, 19, 46, 51]. These methods can lead to higher model accuracy in selected frequency ranges.

The BT method is based on so-called Balanced Realization and determination on this basis a dominant part of the model. The Balanced Realization is reached through the controllability and observability Gramians, W_c and W_o respectively, which are usually obtained as the solutions of Lyapunov equations. Taking into account that we have to apply the reduction method to the system of Eqs. (24)–(27), the Lyapunov equations are

$$A_{FIR} W_c A_{FIR}^T - W_c + B_{FIR} B_{FIR}^T = 0 \tag{29}$$

$$A_{FIR}^T W_o A_{FIR} - W_o + C_{FIR}^T C_{FIR} = 0 \tag{30}$$

where A_{FIR}, B_{FIR} and C_{FIR} are the system matrices of Eqs. (24)–(27). Note that the solution of the Lyapunov equation is a computationally complex process and is the most time-consuming element of the whole BT method.

The balanced form of the model is obtained by the use of linear transformation involving controllability and observability Gramians as follows

$$TW_cT^T = \left(T^T\right)^{-1} W_o T^{-1} = \Sigma = diag(\sigma_i) \tag{31}$$

where σ_i, $i = 1, \ldots, L$, are called the Hankel Singular Values of the system.

Finally, we obtain the balanced form of the system in state space form with matrices $\{\overline{A} = TA_{FIR}T^{-1}, \overline{B} = TB_{FIR}, \overline{C} = C_{FIR}T^{-1}, D_{FIR}\}$. This balanced model can be easily divided to two parts in terms of the 'dominant' (order k) and 'weak' (order $Ln_u - k$) as

$$\begin{aligned} \begin{bmatrix} \overline{x}_1(t+1) \\ \overline{x}_2(t+1) \end{bmatrix} &= \begin{bmatrix} \overline{A}_{11} & \overline{A}_{12} \\ \overline{A}_{21} & \overline{A}_{22} \end{bmatrix} \begin{bmatrix} \overline{x}_1(t) \\ \overline{x}_2(t) \end{bmatrix} + \begin{bmatrix} \overline{B}_1 \\ \overline{B}_2 \end{bmatrix} u(t) \\ \tilde{y} &= \begin{bmatrix} \overline{C}_1 & \overline{C}_2 \end{bmatrix} \begin{bmatrix} \overline{x}_1(t) \\ \overline{x}_2(t) \end{bmatrix} + D_{FIR}\, u(t) \end{aligned} \tag{32}$$

The reduced model constitutes the 'dominant' part of the model Eq. (32), so $A_r = \overline{A}_{11}$ $B_r = \overline{B}_1$ $C_r = \overline{C}_1$.

In several papers we can find various methods for selection of the transformation matrix T (see e.g. [2, 14, 20, 36, 47]). However, the selection of the whole matrix may be a time-consuming process. Therefore, in practical applications, we often use the rectangular truncation matrix $T \in \mathbb{R}^{k \times Ln_u}$ and its right inverse $T^\# \in \mathbb{R}^{k \times Ln_u}$ such that $TT^\# = I$. In this case, the reduced model is finally obtained as

$$A_r = TA_{FIR}T^\# \quad B_r = TB_{FIR} \quad C_r = C_{FIR}T^\# \tag{33}$$

Determining of the matrices T and $T^\#$ may be realized in various ways. The computationally simplest and most effective method is the square root (SR) algorithm [20]. This algorithm can be effective but for the specific system may generate numerical issues. The more robust but also computationally more involving, is the balancing-free square root algorithm [2, 47, 48]. Both algorithms are based on the Cholesky factorizations of controllability and observability Gramians, which can be obtained through square-root solvers for discrete-time Lyapunov equations [30] or approximate solutions $W_c \approx \hat{W}_c = \hat{S}^T\hat{S}$ and $W_o \approx \hat{W}_o = \hat{R}^T\hat{R}$ (see e.g. alternating directions implicit iteration methods [31, 50]). Note that the Cholesky factorization selection for the Gramians is the most time-consuming operation for the presented algorithms.

On the other hand, in Ref. [45] it has been presented that time-consuming solution of the Lyapunov equations for the system in the form of (28), can be substituted by the simple, analytical formula of calculation Cholesky factorizations of controllability and observability grammarians as

$$S = I_{Ln_u \times Ln_u} \tag{34}$$

$$R = \begin{bmatrix} \eta_L & 0 & 0 & \dots 0 \\ \eta_{L-1} & \eta_L & 0 & \dots 0 \\ \eta_{L-2} & \eta_{L-1} & \eta_L & \dots 0 \\ \vdots & \vdots & \vdots & \ddots \vdots \\ \eta_1 & \eta_2 & \eta_3 & \dots \eta_L \end{bmatrix} \tag{35}$$

where η_i, $i = 1, \dots, L$, are as in Eq. (26). This leads to significant reduction of computational complexity of the model order reduction methods.

Finally, we can present an implementation/approximation procedure, which is presented in the next subsection.

4.3 Implementation Algorithm

The selection procedure of simple, approximation/implementation of the fractional-order system needs to specify two values in terms of L and k, describing the length of FIR model used in the modeling procedure and the order of the final (integer-order) model of the system.

(1) Calculation of the impulse response η_i, $i = 0, \dots, L$, of the nabla difference-based fractional-order system (1) as in Theorem 3.
(2) Selection of Cholesky factorization of the observability Gramian as in Eq. (35).
(3) Calculation of Hankel Singular Values and unitary matrices containing left- and right-singular vectors by the use of Singular Value Decomposition method

$$[U, \Sigma, V] = SVD(SR^T) = SVD(R^T) \tag{36}$$

where $\Sigma = diag(\sigma_1, \dots, \sigma_L)$ contains singular values such that $\sigma_1 \geq \sigma_2 \geq \dots \geq \sigma_L$ and U and V are matrices containing left- and right-singular vectors, respectively. Note that $U \Sigma V^T = R^T$.
(4) Selection of the submatrices $U_1 = U(1 : L, 1 : k)$, $\Sigma_1 = \Sigma(1 : k, 1 : k)$, and $V_1 = V(1 : L, 1 : k)$, with k being the order of the final model.
(5) Calculation of the balanced matrices T and $T^\#$ by use of the following formulas

$$T = \Sigma_1^{-0.5} V_1^T R \tag{37}$$

$$T^\# = U_1 \Sigma_1^{-0.5} \tag{38}$$

(6) Calculation of the final state space model of Eq. (28) where the matrices A_r, B_r, C_r are as

$$A_r = TA_{FIR}T^{\#} \quad (39)$$
$$B_r = TB_{FIR} \quad (40)$$
$$C_r = C_{FIR}T^{\#} \quad (41)$$
$$D_r = D_{FIR} \quad (42)$$

Note that steps (3) and (4) can be substituted by the compact SVD decomposition method, where we immediately obtain U_1, Σ_1, and V_1 matrices of respective orders. The compact SVD method is computationally simpler than classical SVD but in this case, we cannot determine the approximation error on the basis of the well-known inequality

$$||G_{A_{FIR},B_{FIR},C_{FIR},D_{FIR}}(z) - \hat{G}_{A_r,B_r,C_r,D_r}(z)||_{\mathscr{H}_\infty} \leq 2 \sum_{j=k+1}^{L} \sigma_i \quad (43)$$

where $G_{A_{FIR},B_{FIR},C_{FIR},D_{FIR}}(z)$ and $G_{A_r,B_r,C_r,D_r}(z)$ are the systems of Eqs. (23)–(26) and (39)–(41), respectively, $||.||_{\mathscr{H}_\infty}$ denotes the $\mathscr{H}_\infty$ norm.

Remark 1 It is important to note that the square root model order reduction algorithm can only be implemented in asymptotically stable systems. Moreover, the FIR-based model can be effectively used in the case of asymptotically stable processes. Therefore in the modeling process, we have to firstly check the stability of the modeled system.

Remark 2 In the modeling process we have to determine two parameters in terms of L and k. The parameter L is the length of the FIR model used in the order reduction process. The value of this parameter has to be selected with regard to the frequency adequacy range of the final model.

5 Simulation Examples

This section presents the continuation of Examples 1 and 2 shown in the stability analysis section.

Example 3 Consider the fractional-order state space system of Example 1. It has been already presented that the system is asymptotically stable for $h = 0.2$ and unstable for $h = 0.15$. Taking into account that the proposed methodology is dedicated to asymptotically stable systems, we use the example system with sampling period $h = 0.2$ only. We use the introduced methodology to model the system with $L = 10000$ and two various $k = 3$ and $k = 5$. By use of the procedure presented in the Sect. 4.3 we obtain the models of the system

- $k = 3$:

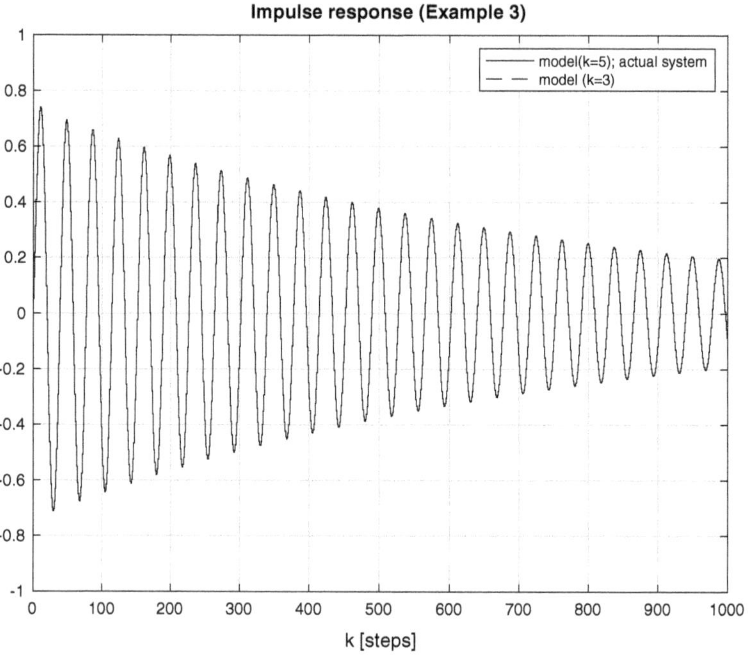

Fig. 3 Impulse responses of the models versus actual system (Example 3)

$$\left[\begin{array}{c|c} A_r & B_r \\ \hline C_r & D_r \end{array}\right] = \left[\begin{array}{ccc|c} 0.98494251 & -0.16629185 & -0.00126454 & 0.27250687 \\ 0.16629185 & 0.98448385 & -0.00180687 & -0.27149270 \\ -0.00126454 & 0.00180687 & 0.76080780 & 0.15738987 \\ \hline 0.27250687 & 0.27149270 & 0.15738986 & 0.01029865 \end{array}\right]$$

- $k = 5$:

$$\left[\begin{array}{c|c} A_r & B_r \\ \hline C_r & D_r \end{array}\right] =$$

$$\left[\begin{array}{ccccc|c} 0.98494251 & -0.16629185 & -0.00126454 & -1.39700e-4 & 4.94920e-7 & 0.27250687 \\ 0.16629185 & 0.98448385 & -0.00180687 & -1.19845e-4 & 7.02990e-7 & -0.27149270 \\ -0.00126454 & 0.00180687 & 0.76080780 & -0.06017513 & 2.17290e-4 & 0.15738986 \\ -1.39700e-4 & 1.98455e-4 & -0.06017513 & 0.64614239 & 0.00241121 & 0.01732902 \\ 4.94920e-7 & -7.02990e-7 & 2.17290e-4 & 0.00241121 & 0.99989794 & 6.13878e-5 \\ \hline 0.27250687 & -0.27149270 & 0.15738986 & 0.01732902 & 6.13878e-5 & 0.010298654 \end{array}\right]$$

The impulse responses of the actual system and both models shown above are presented in Fig. 3.

It can be seen from Fig. 3 that impulse responses of the system and both models (for $k = 3$ and 5) are indistinguishable from each other. The $\mathscr{H}_2$ norm of the error of the impulse responses for the particular models are $||G(q)\delta(t) - G_r(q)\delta(t)||_{\mathscr{H}_2} =$

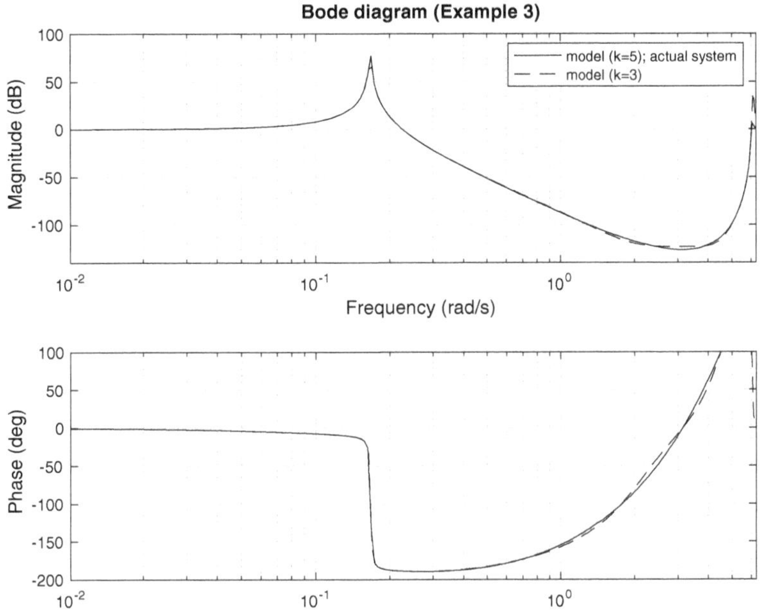

Fig. 4 Bode plots of the models versus actual system (Example 3)

$1.14028e-5$ for $k=3$ and $||G(q)\delta(t)-G_r(q)\delta(t)||_{\mathscr{H}_2}=2.03849e-7$ for $k=5$, respectively. Therefore, both models are very effective approximations of the system in the time domain of this example. Additionally the bode plots of the considered models versus actual system are presented in Fig. 4.

Figure 4 presents that the bode plots for the model with $k=5$ are indistinguishable from the actual system. Even the responses for the model with $k=3$ are hardly distinguishable from the actual system. This fact confirms that both considered models are very good approximations of the actual system. This example shows that the proposed methodology can be very useful in modeling discrete-time nabla-based fractional-order systems.

Example 4 Consider the fractional-order state space system of Example 2. It has been already presented that the system is asymptotically stable. We use the introduced methodology to model the system with $L=1000$ and two various $k=3$ and $k=5$. By use of the procedure presented in the Sect. 4.3 we obtain the models of the system

- $k=3$:

$$\left[\begin{array}{c|c} A_r & B_r \\ \hline C_r & D_r \end{array}\right] = \left[\begin{array}{ccc|c} 0.93767647 & -0.19946797 & -0.02074329 & 1.90067015 \\ 0.19946797 & 0.91782930 & -0.04565874 & -1.46625112 \\ -0.02074329 & 0.04565874 & 0.89327213 & 0.51078622 \\ \hline 1.900670150 & 1.46625116 & 0.51078623 & 0.85470085 \end{array}\right]$$

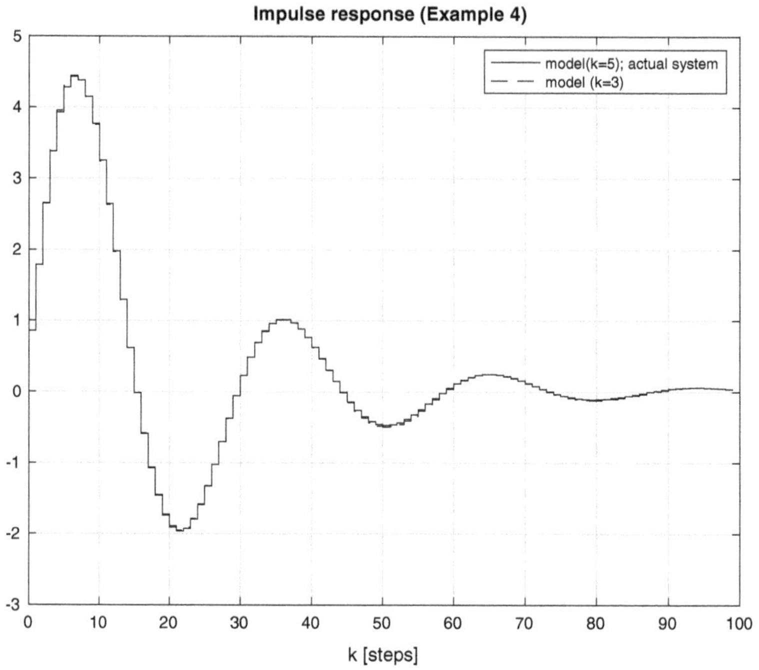

Fig. 5 Impulse responses of the models versus actual system (Example 4)

- $k = 5$:

$$\left[\begin{array}{c|c} A_r & B_r \\ \hline C_r & D_r \end{array}\right] =$$

$$\left[\begin{array}{ccccc|c} 0.93767647 & -0.19946797 & -0.02074329 & 0.00892238 & -0.00481044 & 1.90067014 \\ 0.19946797 & 0.91782930 & -0.04565874 & 0.01749585 & -0.00918664 & -1.46625116 \\ -0.02074329 & 0.04565874 & 0.89327213 & 0.07473478 & -0.04693619 & 0.51078622 \\ 0.00892238 & -0.01749585 & 0.07473478 & 0.90602487 & 0.08318155 & -0.20791583 \\ -0.00481044 & 0.00918664 & -0.04693619 & 0.08318155 & 0.89010439 & 0.11067545 \\ \hline 1.90067014 & 1.46625116 & 0.51078622 & -0.20791583 & 0.11067545 & 0.85470085 \end{array}\right]$$

The impulse responses of the actual system and both models shown above are presented in Fig. 5.

It can be seen from Fig. 5 that impulse responses of the system and both models (for $k = 3$ and 5) are indistinguishable from each other. The $\mathscr{H}_2$ norm of the error of models impulse responses is $||G(q)\delta(t) - G_r(q)\delta(t)||_{\mathscr{H}_2} = 0.01657186$ for model of order $k = 3$ and $||G(q)\delta(t) - G_r(q)\delta(t)||_{\mathscr{H}_2} = 1.6905232e - 4$ for $k = 5$. So we can see in the considered example, that the integer-order model with $k = 3$ provides a good approximate of a noncommensurate fractional-order system. The integer order model with $k = 5$ produces excellent results in this example. The time-domain results are confirmed by bode plots of the considered models versus actual system presented in Fig. 6.

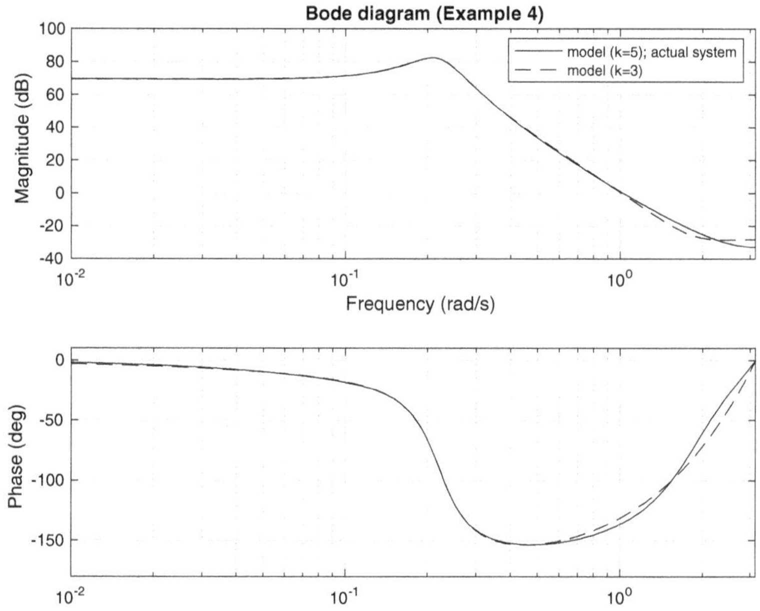

Fig. 6 Bode plots of the models versus actual system (Example 4)

It can be seen from Fig. 6 that again, the bode plots for the model with $k = 5$ are indistinguishable from the actual system. Also, the simpler model with $k = 3$ generates satisfactory results in the frequency domain. In this specific example, we can see that the introduced methodology can be effectively used in approximating fractional-order systems by simple integer-order models. In this case, we have a fast-dynamic system with a relatively short impulse response. This kind of system can be well approximated by relatively simple ordinary models.

6 Conclusions

This paper has produced new results in the application of Balanced Truncation model reduction methods for approximation of LTI discrete-time noncommensurate fractional-order systems. As a fractional-order difference, the nabla difference under the Grünwald–Letnikov definition. The methods employ the FIR-based decomposition of fractional-order systems. The classical square-root algorithm has been used to obtain the integer order model of the system. The main advantage of the presented algorithms is the specific representation of the fractional-order system enabling simple, analytical formulas for determination of the Gramians. This contributes to significant improvement of the computational efficiency of the reduction methods. In this paper we also recall some simple stability tests designed for commensurate and noncommensurate-order systems. The simulation experiments confirm the effectiveness of the introduced methods both in terms of high modeling accuracy and low computational cost.

References

1. Abdeljawad, T.: On Riemann and Caputo fractional differences. Comput. Math. Appl. **62**(3), 1602–1611 (2011)
2. Antoulas, A.: Approximation of Large-Scale Dynamical System. Society for Industrial and Applied Mathematics, Philadelphia (2005)
3. Baranowski, J., Bauer, W., Zagorowska, M., Dziwinski, T., Piatek, P.: Time-domain Oustaloup approximation. In: 20th International Conference on Methods and Models in Automation and Robotics (MMAR), pp. 116–120 (2015). https://doi.org/10.1109/MMAR.2015.7283857
4. Baranowski, J., Bauer, W., Zagorowska, M.: Stability properties of discrete time-domain Oustaloup approximation. Theoretical Developments and Applications of Non-integer Order Systems. Lecture Notes in Electrical Engineering. Springer, Berlin (2016)
5. Benner, P., Kürschner, P., Saak, J.: Frequency-limited balanced truncation with low-rank approximations. SIAM J. Sci. Comput. **38**(1), A471–A499 (2016). https://doi.org/10.1137/15M1030911
6. Bishehniasar, M., Salahshour, S., Ahmadian, A., Ismail, F., Baleanu, D.: An accurate approximate-analytical technique for solving time-fractional partial differential equations. Complexity **2017** (2017). Article ID 8718209, 12 pp
7. Boley, D.: Krylov space methods on state-space control models. Circuits Syst. Signal Process. **13**(6), 733–758 (1994)
8. Ditzian, Z.: Fractional derivatives and best approximation. Acta Math. Hung. **81**(4), 323–348 (1998)
9. Enns, D.: Model reduction with balanced realizations: an error bound and frequency weighted generalization. In: 23rd IEEE Conference on Decision and Control, pp. 127–132 (1984)
10. Freund, R.W.: Model reduction methods based on Krylov subspaces. Acta Numer. **12**, 267–319 (2003). https://doi.org/10.1017/S0962492902000120
11. Gao, Z., Liao, X.: Rational approximation for fractional-order system by particle swarm optimization. Nonlinear Dyn. **67**(2), 1387–1395 (2012)
12. Garrappa, R., Maione, G.: Model order reduction on Krylov subspaces for fractional linear systems. IFAC Proc. Vol. **46**(1), 143–148 (2013)
13. Gawronski, W., Juang, J.: Model reduction in limited time and frequency intervals. Int. J. Syst. Sci. **21**(2), 349–376 (1990)
14. Glover, K.: All optimal Hankel-norm approximations of linear multivariable systems and their $l,^{\infty}$ error bounds. Int. J. Control **39**(6), 1115–1193 (1984)
15. Hilger, S.: Analysis on measure chains—a unified approach to continuous and discrete calculus. Result Math. **18**(1), 18–56 (1990)
16. Imran, M., Ghafoor, A., Sreeram, V.: A frequency weighted model order reduction technique and error bounds. Automatica **50**(12), 3304–3309 (2014). https://doi.org/10.1016/j.automatica.2014.10.062, http://www.sciencedirect.com/science/article/pii/S0005109814004439
17. Kaczorek, T., Ostalczyk, P.: Responses comparison of the two discrete-time linear fractional state-space models. Fract. Calc. Appl. Anal. **19**(4), 789–805 (2016)
18. Krajewski, W., Viaro, U.: A method for the integer-order approximation of fractional-order systems. J. Frankl. Inst. **351**(1), 555–564 (2014)
19. Kürschner, P.: Balanced truncation model order reduction in limited time intervals for large systems. Adv. Comput. Math. **44**(6), 1821–1844 (2018). https://doi.org/10.1007/s10444-018-9608-6

20. Laub, A., Heath, M., Paige, C., Ward, R.: Computation of system balancing transformations and other applications of simultaneous diagonalization algorithms. IEEE Trans. Autom. Control **AC–32**(2), 115–122 (1987)
21. Liu, Y., Anderson, B.: Singular perturbation approximation of balanced system. In: 28th IEEE Conference on Decision and Control, vol. 2, pp. 1355–1360 (1989)
22. Maione, G.: On the Laguerre rational approximation to fractional discrete derivative and integral operators. IEEE Trans. Autom. Control **58**(6), 1579–1585 (2013)
23. Mansouri, R., Bettayeb, M., Djennoune, S.: Comparison between two approximation methods of state space fractional systems. Signal Process. **91**(3), 461–469 (2011)
24. Moore, B.: Principal component analysis in linear systems: controllability, observability and model reduction. IEEE Trans. Autom. Control **AC–26**(1), 17–32 (1981)
25. Odibat, Z.: Approximations of fractional integrals and Caputo fractional derivatives. Appl. Math. Comput. **176**(2), 527–533 (2006)
26. Oprzedkiewcz, K., Mitkowski, W., Gawin, E.: The PLC Implementation of Fractional-Order Operator Using CFE Approximation. Springer, Berlin (2017)
27. Oprzedkiewicz, K., Stanisławski, R., Gawin, E., Mitkowski, W.: A new algorithm for a CFE-approximated solution of a discrete-time non integer-order state equation. Bull. Pol. Acad. Sci. Tech. Sci. **65**(4), 429–437 (2017)
28. Ortigueira, M.D., Machado, J.T.: New discrete-time fractional derivatives based on the bilinear transformation: definitions and properties. J. Adv. Res. **25**, 1–10 (2020)
29. Oustaloup, A., Levron, F., Nanot, F.: Frequency band complex non integer differentiator: characterization and synthesis. IEEE Trans. Circuits Syst. I: Fundam. Theory Appl. **47**(1), 25–40 (2000)
30. Penzl, T.: Numerical solution of generalized Lyapunov equations. Adv. Comput. Math. **8**(1), 33–48 (1998). https://doi.org/10.1023/A:1018979826766
31. Penzl, T.: Algorithms for model reduction of large dynamical systems. Linear Algebra Appl. **415**, 322–343 (2006)
32. Pernebo, L., Silverman, L.: Model reduction via balanced state space representations. IEEE Trans. Autom. Control **27**(2), 382–387 (1982). https://doi.org/10.1109/TAC.1982.1102945
33. Rydel, M.: New integer-order approximations of discrete-time non-commensurate fractional-order systems using the cross Gramian. Adv. Comput. Math. **45**(2), 631–653 (2019). https://doi.org/10.1007/s10444-018-9633-5
34. Rydel, M., Stanisławski, R.: A new frequency weighted Fourier-based method for model order reduction. Automatica **88**, 107–112 (2018)
35. Rydel, M., Stanisławski, R.: Computation of controllability and observability Gramians in modeling of discrete-time noncommensurate fractional-order systems. Asian J. Control **22**(3), 1052–1064 (2020). https://doi.org/10.1002/asjc.2060
36. Safonov, M.G., Chiang, R.Y.: A Schur method for balanced-truncation model reduction. IEEE Trans. Autom. Control **34**(7), 729–733 (1989)
37. Scherer, R., Kalla, S.L., Tang, Y., Huang, J.: The Grünwald-Letnikov method for fractional differential equations. Comput. Math. Appl. **62**(3), 902–917 (2011)
38. Shen, J., Lam, J.: H_∞ model reduction for positive fractional order systems. Asian J. Control **16**(2), 441–450 (2014)
39. Stanisławski, R.: New results in stability analysis for LTI SISO systems modeled by GL-discretized fractional-order transfer functions. Fract. Calc. Appl. Anal. **20**(1), 243–259 (2017). https://doi.org/10.1515/fca-2017-0013
40. Stanisławski, R., Latawiec, K.J.: Stability analysis for discrete-time fractional-order LTI state-space systems. Part I: new necessary and sufficient conditions for asymptotic stability. Bull. Pol. Acad. Sci. Tech. Sci. **61**(2), 353–361 (2013)
41. Stanisławski, R., Latawiec, K.J.: Stability analysis for discrete-time fractional-order LTI state-space systems. Part II: new stability criterion for FD-based systems. Bull. Pol. Acad. Sci. Tech. Sci. **61**(2), 362–370 (2013)
42. Stanisławski, R., Latawiec, K.J.: Fractional-order discrete-time Laguerre filters – a new tool for modeling and stability analysis of fractional-order LTI SISO systems. Discret. Dyn. Nat. Soc. **2016**, 1–9 (2016). Article ID: 9590687

43. Stanisławski, R., Latawiec, K.J.: A modified Mikhailov stability criterion for a class of discrete-time noncommensurate fractional-order systems. Commun. Nonlinear Sci. Numer. Simul. **96**, 105,697 (2021). https://doi.org/10.1016/j.cnsns.2021.105697, https://www.sciencedirect.com/science/article/pii/S1007570421000083
44. Stanisławski, R., Latawiec, K.J., Łukaniszyn, M.: A comparative analysis of Laguerre-based approximators to the Grünwald-Letnikov fractional-order difference. Math. Probl. Eng. **2015**, 1–10 (2015). Article ID: 512104
45. Stanisławski, R., Rydel, M., Latawiec, K.J.: Modeling of discrete-time fractional-order state space systems using the balanced truncation method. J. Frankl Inst. **354**(7), 3008–3020 (2017)
46. Toor, H.I., Imran, M., Ghafoor, A., Kumar, D., Sreeram, V., Rauf, A.: Frequency limited model reduction techniques for discrete-time systems. IEEE Trans. Circuits Syst. II: Express Briefs **67**(2), 345–349 (2020). https://doi.org/10.1109/TCSII.2019.2909122
47. Varga, A.: Balancing-free square-root algorithm for computing singular perturbation approximations. In: 30th IEEE Conference on Decision and Control, vol. 2, pp. 1062–1065 (1991)
48. Varga, A., Anderson, B.D.: Accuracy-enhancing methods for balancing-related frequency-weighted model and controller reduction. Automatica **39**(5), 919–927 (2003). https://doi.org/10.1016/S0005-1098(03)00030-X
49. Wang, G., Sreeram, V., Liu, W.Q.: A new frequency-weighted balanced truncation method and an error bound. IEEE Trans. Autom. Control **44**(9), 1734–1737 (1999)
50. Zhou, Y., Sorensen, D.C.: Approximate implicit subspace iteration with alternating directions for LTI system model reduction. Numer. Linear Algebra Appl. **15**(9), 873–886 (2008). https://doi.org/10.1002/nla.602
51. Zulfiqar, U., Imran, M., Ghafoor, A., Liaquat, M.: A new frequency-limited interval Gramians-based model order reduction technique. IEEE Trans. Circuits Syst. II: Express Briefs **64**(6), 680–684 (2017)

State Feedback Law for Discrete-Time Fractional Order Nonlinear Systems

Ewa Pawłuszewicz, Andrzej Koszewnik, and Piotr Burzynski

The analysis of output regulation for fractional nonlinear vibration control system of a smart plate designed by using the optimal LQR controller versus a closed-loop system design based on an integer order model is presented. It is shown that if the output of a linearized fractional system can be regulated to a finite number of steps around a constant output then the output of the nonlinear fractional system can also be regulated in the same number of steps around this output. The result is valid for systems with the Caputo, Riemann–Liouville-types h-difference operators. The obtained simulation and experimental results in both time and frequency domains show that the closed-loop system design based on nonlinear fractional order control plant model, similar to a closed-loop system with an integer order model, effectively dampens the vibration of the chosen mechanical structure by using piezo-element works as sensors and actuators.

1 Introduction

In engineering practice most often one knows the inputs and measurements of the investigated process. Also a relationship between input and output variables is often required. So there is a question about possible systems that provide a good description

E. Pawłuszewicz (✉) · A. Koszewnik · P. Burzynski
Department of Robotics and Mechatronics, Bialystok University of Technology, 45C Bialystok, Wiejska, Poland
e-mail: e.pawluszewicz@pb.edu.pl

A. Koszewnik
e-mail: a.koszewnik@pb.edu.pl

P. Burzynski
e-mail: p.burzynski@doktoranci.pb.edu.pl

P. Kulczycki et al. (eds.), *Fractional Dynamical Systems: Methods, Algorithms and Applications*, Studies in Systems, Decision and Control 402,
https://doi.org/10.1007/978-3-030-89972-1_8

of the observed system's input–output behavior. This is the crucial idea of the realization problem. Realization of an input–output map that describes a system's behavior means finding a dynamical state-space form of the system with input and output, that is able to reproduce, when initialized at some state for the given input–output behavior. In [1] a theoretical approach to this problem (see for example in [2–5]) for both continuous-time and discrete-time linear systems is generalized and extended to any time domain. An extension to a more general case of a differential/difference order, i.e. in systems defined by fractional order operators, has been presented in [5, 6]. The term *fractional* basically implies all non-integer numbers. There are many processes that can be more accurately modelled using fractional differ-integrals operators, see for example [7–11]. The rapid development of computer techniques has caused parallel investigations in this field, among others, combinatorics tools and difference equations. This is the reason that in modeling of the real phenomena, the generalizations of n-th order differences to their fractional forms and the state-space equations of control systems in discrete-time are used, see for example [8, 12, 13] and its applications [14–16].

In engineering applications, especially in automatic control and automatization, realization given in state-space form should be stable. The stability conditions for a linear open-loop, fractional difference system have been presented in [17]. However in automatic control engineers work with closed-loop systems. This leads to construction of proper feedback and in effect to the problem of stabilization. Conditions for stabilization of h-difference control systems have been given in [18–20]. Stability of the closed-loop system given by a fractional transfer function has been discussed in [21].

Our goal is to study the stabilization conditions for the given fractional-order h-difference nonlinear system. For this purpose, the tools often used are derived from differential geometry, [22, 23]. Such attempt is not possible for systems defined by fractional differences or derivatives operators. So, we adopt another approach to nonlinear difference systems with fractional operators. Similarly as in [24], by applying the Inverse Function Theorem and Implicit Function Theorem, the problem of stabilization in finite number of steps of a nonlinear system is studied. The idea of such approach comes from [24, 25]. The obtained results are similar to the classical ones but it should be emphasized that the final result is definitely influenced by the fractional dynamics of the system that can also be seen on examples of real systems in the second part of the paper. For this purpose a harvesting system has been consider. These systems, especially with piezo-elements are nonlinear ones. Special attention has been put on the SFSF (Simply-supported Free-Simply-supported-Free) rectangular, aluminum plate with piezo-sensors and piezo-actuators. This system has been described in details in [26]. Here only a short description of this structure is given. From the nature of piezo materials it is known that the behavior of these piezo-elements caused nonlinearity between output displacement and voltage to its electrodes [27–29]. This effect leads to difficulties in the process of designing a vibration control system and appearance in some cases of the Spillover effect, especially in the low frequency range. In order to counteract such phenomena an appropriate law of the (vibration) control should be designed. This process is much easier for a linearized system. However in practice, such an approach in some cases, can leads to obtaining two different measurement signals with the same constant output power of both linear and non-linear systems. This is due to the different dynamics description of the considered mechanical structure.

The paper is organized as follows: At the beginning, the definitions of Caputo-, Riemann–Liouville- and Grünwald-Letnikov-type of fractional order h-difference operators are presented. Properties of nonlinear, discrete-time fractional order systems with these operators are recalled. Next the theoretical stabilization problem is discussed. In order to illustrate the idea of the obtained results, a SFSF rectangular aluminum plate with piezo-sensors and piezo-actuators is consider. Simulation and experimental results indicate that the behavior of the vibration control system designed for linear and nonlinear order models are very close. As a result, both control systems with an optimal controller can ensure proper damping of structure vibration especially in the range of low frequencies.

2 Fractional h-Difference Operators

Let α be any number and s any integer. Then:

$$\binom{\alpha}{s} = \begin{cases} 0 & for\ s < 0 \\ 1 & for\ s = 0 \\ \frac{\alpha(\alpha-1)\dots(\alpha-s+1)}{s!} & for\ s > 0 \end{cases} \tag{1}$$

denotes the classical binomial coefficient. The family of binomial functions φ_u parametrised by $\mu > 0$ will be denoted as:

$$\varphi_\mu(n) = \begin{cases} \binom{n+\mu-1}{n} & for\ n \in N_0 \\ 0 & for\ n < N_0 \end{cases} \tag{2}$$

Recall that the discrete two-parameter Mittag–Leffler function is defined as, [17]:

$$E_{(\alpha,\beta)}(\lambda, n) := \sum_{k=0}^{\infty} \lambda^k \varphi_{k\alpha+\beta}(n-k) = \sum_{k=0}^{\infty} A^k \binom{n-k+k\alpha+\beta-1}{n-k}. \tag{3}$$

If $\alpha = \beta$ then $E_{(\alpha,\alpha)}(A, n) = \sum_{k=0}^{\infty} A^k \binom{n-(k+1)(\alpha-1)}{n-k}$. Note that $E_{(0,1)}(A, n) = \sum_{k=0}^{\infty} A^k$ is well known geometrical series.

Let h be a positive real number. For any real a let $(hN)_a = \{a, a+h, a+2h, \dots\}$. Consider a function $x : (hN)_a \to R$. Recall that if $\overline{x}(s) := x(a+sh)$ and "$*$"

denotes the convolution operator, then $(\varphi_\mu * \overline{x})(n) := \sum_{s=0}^{n} \binom{n-s+\mu-1}{n-s} \overline{x}(s)$. The fractional h-sum of order $\alpha > 0$ for a function $x : (hN)_a \rightarrow R$ is defined as [17]:

$$\left({}_a\Delta_h^{-\alpha} x\right)(t) := h^\alpha (\varphi_\alpha * \overline{x})(n), \tag{4}$$

where $t = a + (\alpha + n)h$ for any natural n. Let $\alpha \in (0, 1]$ and $\Delta_h x = \frac{x(t+h)-x(t)}{h}$ denotes the classical forward h-difference operator . The Caputo-type h- difference operator ${}_a\Delta_{h,*}^\alpha$ of order α or a function $x : (hN)_a \rightarrow R$ is defined as, [30]:

$$\left({}_a\Delta_{h,*}^\alpha x\right)(t) := \left({}_a\Delta_h^{-(1-\alpha)} (\Delta_h x)\right)(t) \tag{5}$$

for any $t \in (hN)_{a+(1-\alpha)h}$. The Riemann–Liouville-type fractional h-difference operator ${}_a\Delta_h^\alpha$ of order $\alpha \in (0, 1]$ for a function $x : (hN)_a \rightarrow R$ is defined as, [30, 31]:

$$\left({}_a\Delta_h^\alpha x\right)(t) := \left(\Delta_h \left({}_a\Delta_h^{-(1-\alpha)} x\right)\right)(t), \tag{6}$$

where $t \in (hN)_{a+(1-\alpha)h}$. The last operator we are going to consider is the Grünwald–Letnikov-type fractional h-difference operator ${}_a\tilde{\Delta}_h^\alpha x$ of a real order α defined for a function $x : (hN)_a \rightarrow R$ as, [30]:

$$\left({}_a\tilde{\Delta}_h^\alpha x\right)(t) := \sum_{s=0}^{\frac{t-a}{h}} a_s^{(\alpha)} x(t - sh), \tag{7}$$

where $a_s^{(\alpha)} = (-1)^s \binom{\alpha}{s} \frac{1}{h^\alpha}$. If $a = (\alpha - 1)h$, then

$$\left({}_a\tilde{\Delta}_h^\alpha y\right)(t + h) := \left({}_a\Delta_h^\alpha x\right)(t), \tag{8}$$

where $x(t) = y(t - a)$ for $t \in (hN)_a$, [30]. Also, for $\alpha \in (0, 1]$, it holds that

$$\left({}_a\Delta_{h,*}^\alpha x\right)(t) := \left({}_a\Delta_h^\alpha x\right)(t) - \frac{x(a)}{h^\alpha} \binom{\frac{t-a}{h}}{-\alpha} \tag{9}$$

for $t \in (hN)_{a+(1-\alpha)h}$.

Taking into account relations (8) and (9), one can use the common symbol defined by its values:

$$\left({}_a\gamma_h^\alpha x\right)(t) = \begin{cases} \left({}_a\Delta_{h,*}^\alpha x\right)(t) \quad or \ \left({}_a\Delta_h^\alpha x\right)(t) \ for \ a = (\alpha - 1)h \\ \left({}_a\tilde{\Delta}_h^\alpha x\right)(t+h) \qquad\qquad for \quad a = 0 \end{cases}.$$

It is known that the single-sided Z-transform of a sequence $\{y(n)\}_{n\in N_0}$ is a complex function *Y(z)* given by $Y(z) := Z[y](z) = \sum_{k=0}^{\infty} \frac{y(k)}{z^k}$, where z is a complex variable for which series $\sum_{k=0}^{\infty} \frac{y(k)}{z^k}$ converges absolutely.

Proposition 3 [17]: *Let $a \in R$ and $\alpha \in (0, 1]$ Define $y(n) := \left({}_a\gamma_h^\alpha x\right)(t)$ where $t \in (hN)_{a+(1-\alpha)h}$ and $t = a + (1-\alpha)h + nh$ Then :*

$$Z\left[\left({}_a\gamma_h^\alpha x\right)(t)\right](z) = z\left(\frac{hz}{z-1}\right)^{-\alpha}(X(z) - x(a)), \tag{10}$$

where $X(z) = Z\left[\tilde{x}\right](z)$, $\tilde{x}(n) := x(a + nh)$ and *$\beta = \alpha$ for the Riemann–Liouville- or Grünwald–Letnikov-type $h-$ difference operators and $\beta = 1$ for the Caputo-type $h-$ difference operator, and $\beta = \alpha$ for the Riemann–Liouville- or Caputo-type operators and for $\alpha = 0$ the Grünwald–Letnikov-type operator.*

3 Linear Fractional Order Systems

Recall that a continuously differentiable function $F : R^p \to R^m$ is called a higher order function if $F(0) = 0$ and $\frac{\partial F}{\partial x}\Big|_{x=0} = 0$ where $x = (x_1, ..., x_p) \in R^p$, [24]. The class of the higher order functions will be denoted by $H_{p,m}$. Note that if $A \in R^{m \times m}$ is a stationary matrix and F is the higher order function then AF also is the higher order function and $AF(x) = AF(x)$. If $F_1, F_2 \in H_{p,p}$ then also $F_1 + F_2 \in H_{p,p}$ and $F_1F_2 \in H_{p,p}$. Also, if $F \in H_{p,m}$ is such function that $F(0) = 0$ and a map $g : R^p \to R^p$ is continuously differentiable, then the composition $F \circ g \in H_{p,m}$.

Lemma 4 [24] **:** *Suppose that $A \in R^{m \times m}$ is a nonsingular matrix (x, y) belongs to an open subset V_0 of R^{pxm} containing the origin and $f \in H_{p+m,p}$. If $z = Ax + By + f(x, y) : V_0 \to R^m$ then locally $x = A^{-1}(z - By) + \overline{g}(y, z)$ where $\overline{g} \in H_{p+m,p}$.*

Let us consider the following common form control of nonlinear fractional order control system

$$({}_a\gamma^\alpha x)(nh) = F(x(nh + a), u(nh)) \tag{11a}$$

$$y(nh) = H(x(nh + a), u(nh)) \tag{11b}$$

with the initial condition $x(a) = x_0$, where $n \in N_0$, $x : (hN)_a \to R^p$ denotes the state vector $y : (hN)_0 \to R^r$, is the output vector, the value $u(nh)$ of control u are elements of an arbitrary set $\Omega \subseteq R^m$, functions F, H are smooth and $F(0,0) = 0$, $H(0) = 0$. The set Ω is called the control space and satisfies the following property: $\Omega \subseteq R^m$ is such that $\Omega \subseteq \mathrm{int}\overline{\Omega}$ and any two points in the same connected component of Ω can be jointed by a smooth curve lying in Ω, except for end points. Equation (11a) defines dynamics of control system (11). Equation (11b) defines the output of this system. Using Taylor expansion of functions F and H the system (11) can be rewritten in the form

$$({}_a\gamma^\alpha x)(nh) = A(x(nh+a) + Bu(nh) + f(x(nh+a), u(nh)) \tag{12a}$$

$$y(nh) = Cx(nh+a) + Du(nh) + g(x(nh+a), u(nh)), \tag{12b}$$

where $A = \frac{\partial F}{\partial x}(0,0)$, $B = \frac{\partial F}{\partial u}(0,0)$, $C = \frac{\partial H}{\partial x}(0,0)$ and $D = \frac{\partial H}{\partial u}(0,0)$. Additionally $f : R^p \times \Omega \to R^p$,$g : R^p \to R^r$ and $f \in H_{p+m,p}$, $g \in H_{p,r}$. The local properties of the system (12) are based on properties on the linearized system

$$({}_a\gamma^\alpha x)(t) = Ax(nh+a) + Bu(nh) \tag{13a}$$

$$y(nh) = Cx(nh+a) + Du(nh). \tag{13b}$$

Proposition 4 [25]: *Let $A \in R^{p\times p}$, $B \in R^{m\times p}$, $\alpha \in (0,1]$,$a = (\alpha - 1)h$ and $u : (hN)_0 \to \Omega$ be the fixed control. Then system (11) with initial condition $x(a) = x_0$ has the unique solution given by.*

$$x(nh+a) = E_{(\alpha,\beta)}(Ah^\alpha, n)x_0 + \left(E^\rho_{(\alpha,\alpha)}(Ah^\alpha, \cdot)\right) * B\overline{u})(n) + F_n(x_0, U),$$

where $\overline{u}(n) = \alpha u(nh)$, $\beta = 1$ for the Caputo–type operator, for $\beta = \alpha$ the Riemman-Liouville- and Grünwald-Letnikov-type operators and $F_n(x_0, U) = \left(E^\rho_{(\alpha,\alpha)}(Ah^\alpha, \cdot) * \overline{f}\right)(n)$ with $\overline{f}(n) := h^\alpha f(x(nh+a), \bar{u}(n) := u(nh))$, $E^\rho_{(\alpha,\alpha)}(Ah^\alpha, n) = E^\rho_{(\alpha,\alpha)}(Ah^\alpha, n-1)$ and $F_n(\cdot, U) \in H_{p,p}$.

Let $J_0(m)$ denote the set of all sequences $U = (u_0, u_1, ...)$ where $u_n := u(nh) \in \Omega$, $n \in N_0$. Let $\xi(\cdot, x_0, U)$ be defined by its values $\xi(nh+a, x_0, U) = x(nh+a)$ and denotes the state forward trajectory of system (11) i.e. a solution of system's dynamic (11a) which is uniquely defined by initial state x_0 and control sequence $U \in J_0(m)$. The reachable set from the given initial state x_0 in q steps, denoted as $R^q(x_0)$ is the set of all states to which the given system can be steered from x_0 in q steps by control $u \in J_0(m)$. Additionally,$R^0(x_0) := \{x_0\}$. Then, the set $R^q(x_0) := \bigcup_{q\in N_0} R^q(x_0)$ is the set of all states reachable from x_0.

System (11) is locally controllable in q steps from x_0 if there exists a neighborhood $V \subset R^n$ of x_0 such that $V \subset R^q(x_0)$. System (11) is (globally) controllable from x_0 in q steps if $R^q(x_0) = R^p$.

Proposition 5 [25, 32]: *System (13) is controllable in q steps if and only if the rank of controllability matrix* $Q_q = [B, E_{(\alpha,\alpha)}(Ah^\alpha, 1)B, ..., E_{(\alpha,\alpha)}(Ah^\alpha, q-1)B]$ *is full.*

From the Rank Matrix Theorem it follows that any state $x_0 \in R^p$ can be steered to a final state $x_f \in R^n$ in no more than p steps.

Theorem 5 [25, 32]: *System (11) is locally controllable from the origin in q steps if its linear approximation (13) is globally controllable from the given initial state* x_0 *in q steps.*

Let us consider a nonlinear autonomous fractional order system of the form

$$({}_a\gamma^\alpha x)(nh) = F(x(nh + a)) \tag{14}$$

with the initial condition $x(a) = x_0$ where $n \in N_0$, $x : (hN)_a \to R^p$. The equilibrium cannot necessarily be the zero state. In the case of a system with the Caputo-type h-difference operator, the equilibrium is any constant state x^{eq} such that $F(x^{eq}) = 0$ independently on a number of steps. For the Riemann–Liouville-type h-difference operator the equilibrium is any constant state x^{eq} such that $F(x^{eq}) = \varphi_{1-\alpha}(p+1)h^\alpha x^{eq}$ in p steps, [20]. Without loss of generality we assume that $x^{eq} = 0$. The equilibrium point $x^{eq} = 0$ of (14) (or equivalently, the system (14)) is.

(i) *stable* if, for each $\varepsilon > 0$ there exists $\delta = \delta(\varepsilon) > 0$ such that $\|x_0\| < \delta$ implies $\|\overline{x}(n)\| < \varepsilon$ for all $n \in N_0$;
(ii) *asymptotically stable* if it is stable here exists $\delta > 0$ such that $\|x_0\| < \delta$ implies $\lim_{n\to\infty} \overline{x}(n) = 0$.

The fractional system (14) is called *stable/ asymptotically stable* if their equilibrium points $x^{eq} = 0$ is stable/asymptotically stable.

Using Taylor expansion of the dynamics of system (14), described by function F, this system can be rewritten in the form

$$\left({}_a\gamma^\alpha x\right)(nh) = Ax(nh + a) + \overline{f}(x(nh + a)) \tag{15}$$

with $A = \frac{\partial F}{\partial x}(0, 0)$ and $\tilde{f} : R^p \to R^p, \tilde{f} \in H_p$. Then the linearisation of (14) is of the form

$$\left({}_a\gamma^\alpha x\right)(nh) = Ax(nh + a). \tag{16}$$

Proposition 6 [17]: *Let P denotes the set of all roots of the equation* $\det(I - \frac{h^{\alpha}}{z}\left(\frac{z}{z-1}\right)^{\alpha} A) = 0$. *If all elements from P are strictly inside the unit circle, then the system (14) is asymptotically stable.*

System (13) is called stabilizable if there exists the linear state-feedback controller with the control gain $K \in R^{m \times p}$ i.e. with $u(nh) = Kx(nh+a)$, such that the closed loop system

$$({}_{a}\gamma^{\alpha} x)(nh) = (A + BK)x(nh + a) \tag{17}$$

is asymptotically stable.

Proposition 7 *If linear system (13) is controllable in p steps then nonlinear system (15) is stabilizable in p steps by a static feedback law $u(nh) = Ku(x(nh + a)) + \eta(x(nh + a))$, $n \in N_0$* with $\eta(\cdot) \in H_{p,p}$.

Proof From definitions of stability it follows that final state $x_f = 0$ of the system (13) can be reached in at most p steps from a state $x(t_0) = x_0$ using uniquely determined control sequence $U_p = (u_0, u_1, ..., u_{p-1})$ with elements.

$$U_i = -Q_i^{-1} E_{(\alpha,\alpha)}(h^{\alpha} A^{\alpha}, i) + \eta(x_0, 0), \; i = 0, 1, ..., p - 1,$$

where $\eta(...) \in H_{p,p}$. Let x_1 be the state reached by the system after application of the control

$$u_0 := Kx(t_0) + \eta(x(t_0)). \tag{18}$$

The equation

$$u_1 = -Q_i^{-1} E_{(\alpha,\beta)}(hA^{\alpha}, 1)x_0 - F_n(x_0, u_1) \tag{19}$$

with $F_n \in H_{p,p}$ has at least one solution $u_1 := u_1(x_0)$, depending on $x_0 = x(t_0)$, of the form $u_1(x_0) = Kx(h + a) + \eta(x(h + a))$. Then, the remaining control sequence $(u_1(x_0), r(u_1(x_0)))$ with the first element of control sequence U_{p-1} denoted as $u_1(x_0)$ and next elements—as $r_1(u_1(x_0))$, transfers state x_1 to equilibrium $x_f = 0$ in $p-1$ steps. By uniqueness it follows that $U_{p-2} = (u_1(x_0), r(u_1(x_0)), 0)$ and from assumption and (18) it holds that

$$u_2(x_0) := u_1(x_1) = Kx(2h + a) + \eta(x(2h + a)). \tag{20}$$

Continuing the same reasoning successively for the next elements of the given control sequence one can conclude that feedback law of the form $u(t) = Kx(nh + a) + \eta(x(nh + a))$ stabilizes the given system in at least p steps.

Consider nonlinear system (11) and stabilizing feedback law of the form

$$u(nh) := Kx(nh + a) + \nu \tag{21}$$

with constant function $\nu(nh) = \nu$. Applying this feedback to (11) one obtains

$$({}_a\gamma^\alpha x)(nh) = (A + BK)x(nh + a) + B\nu + f(x(nh + a), Kx(nh + a) + \nu) \tag{22a}$$

$$y(nh) = (C + DK)x(nh + a) + Dv + g(x(nh + a)), Kx(nh + a) + v), \tag{22b}$$

where $A + BK$ is the stable matrix. Let

$$x^{eq} := (I - A - BK)^{-1}Bv + \xi(x^{eq}, Kx^{eq} + \nu) \tag{23}$$

and consider the coordinate transformation $z = x - x^{eq}$ such that x^{eq} is transformed into the origin in z-coordinates. Then

$$\begin{aligned}({}_a\gamma^\alpha z)(nh) = ({}_a\gamma^\alpha x)(nh) - \left({}_a\gamma^\alpha x^{eq}\right)(nh) = (A + BK)(z(nh + a) +\\ f(z(nh + a) + x^{eq}, Kz(nh + a) + Kx^{eq} + \nu) - \left({}_a\gamma^\alpha x^{eq}\right)(nh)\end{aligned} \tag{24}$$

where $({}_a\gamma^\alpha x^{eq})(nh) = 0$ for the Caputo-type h-difference operator and $({}_a\gamma^\alpha x^{eq})(nh) = \varphi_{L-\alpha}(p + 1)h^\alpha x^{eq}$ for the Riemann–Liouville-type h-difference operator. With $F = \frac{\partial f}{\partial x}$ taking into account (24) one obtains

$$({}_a\gamma^\alpha z)(nh) = \overline{A}(v) + o(z), \tag{25}$$

where $\overline{A}(v)$ is a constant matrix for any given constant function $v(nh) = v$. Since $f \in H_{p+m,p}$, then $\frac{\partial f}{\partial x}(0, 0) = 0$ and $\overline{A}(v)$ is a stable matrix for enough small v. This means that $\lim_{n\to\infty} \overline{x}(n) = x^{eq}$, $\lim_{n\to\infty} z(nh + a) = 0$, so $\lim_{n\to\infty} \overline{x}(nh + a) = x^{eq}$. In this way we obtain the following.

Proposition 8 *Under the coordinate transformation* $z = x - \overline{x}$ *such that* $\overline{x}$ *is transformed into origin in* z*-coordinates the system (12) a constant input produces in a finite number of steps, a constant output of this system.*

Proposition 9 *If the output of a linearized system can be regulated a finite number of steps around a constant*

$$r = C(I - \overline{A})^{-1}Bv + c\eta(v) + h(v) \tag{26}$$

and $h(v) := h(I - \overline{A})^{-1}Bv + \eta(v)$, *in a neighborhood of state* $x = 0$ *then the output of the nonlinear system also can be regulated in the same number of steps around this* r *in a neighborhood of* $x = 0$ *for both Caputo and Riemann–Liouville types h-difference operators.*

Remark Constant r can be directly determined on the basis of system equations and control law. Output regulation results look similar to those in discrete-time case, but their assumptions are different and strictly connected with fractional order operators. These results need to tested on real systems.

4 Smart Plate as a Control Plant of a Non-Collocated Control System

In this section the smart aluminum plate with two piezo-stripes sensors and two piezo-actuators (Fig. 1) is chosen as a control plant to check:

1. the influence of fractional order of the system onto regulation of output around a some constant r;
2. the active vibration control system by using an optimal output controller.

Parameters of the described structure with piezo-elements attached to both its surfaces are collected in Table 1. It can be seen (Fig. 1) that the piezo-elements

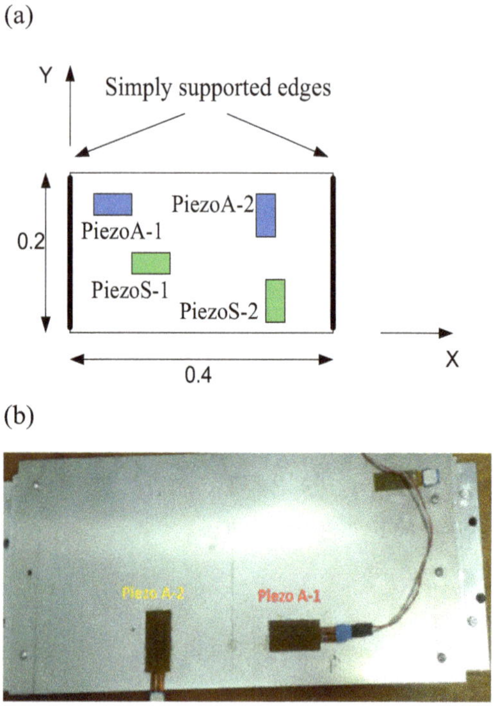

Fig. 1 The host structure with attached piezo-sensors and piezo-actuators **a** scheme diagram, **b** photo of the real structure

Table 1 Parameters of the plate and piezo-elements works as actuator and sensor

Parameter	Plate		Piezo-element	Actuator/sensor QP20N/QP10N
Length [m]	L	0.4	l_p/ l_{peh}	0.05
Width [m]	W	0.2	w_p/w_{peh}	0.025
Thickness [m]	h_{plate}	0.0002	h_a/h_{peh}	0.000782/0.000381
Young module [GPa]	E_b	70	E_p	0.18
Density [kg/m^3]	ρ_{plate}	2720	ρ_p	7200
Strain constant of piezo [pm/V]	-	-	d_{33}	– 125
Piezoelectric stress/charge constant [pC/m^2]	-	-	e_{31}	-

(actuators and sensors) are oriented on these surfaces at quasi-optimal locations in two perpendicular directions X and Y, respectively. The process of determining these locations is described in detail in [26]. As a result, it allows for a low energy vibration control system to obtain a non-collocated system.

Due to the existence of the nonlinearity effect between output displacement and voltage applied to its electrodes, the piezoelectric actuators used to control vibration of the host structure are nonlinear elements. As a result, the hysteresis effect in the PCA (piezo ceramic actuator) appears. The hysteresis component $z(t)$ of this piezo is expressed in the form:

$$\dot{z}(t) = \left\{A - \left[\beta \mathrm{sgn}(\dot{u}(t)z(t)) + \gamma\right]|z(t)|^n\right\}\dot{u}(t), \tag{27}$$

where $u(t)$ denotes the voltage applied to the piezo-actuators, A, β, γ, n are parameters of the Bouc-Wen hysteresis [33]. Taking into account the Bouc-Wen model of the piezo structure, the hysteresis component and the output displacement have been linearized with respect to a working point corresponding to voltage U = 80 V applied to the piezo. The linearized the output displacement from the piezo is expressed in the form:

$$x(t) = k_v u(t) + x_0, \tag{28}$$

where k_v denotes a constant representing the ratio between the output displacement and applied voltage, x_0—the initial displacement without applied voltage. The linearized form of the output displacement from the piezo simplifies the identification procedure of the structure on the lab stand. In this way, the mathematical model of the smart structure has been obtained. To this aim, the excitation signal in the form of a chirp signal $u(t) = 5sin(\omega t)$ in the frequency from the interval (10 Hz; 250 Hz) containing the first five natural frequencies has been firstly generated from the Agilent signal generator and next applied to both piezo-actuators A-1 and A-2 by the bipolar amplifier. At the same time, the vibration of the plate has been measured by two piezo-sensors:

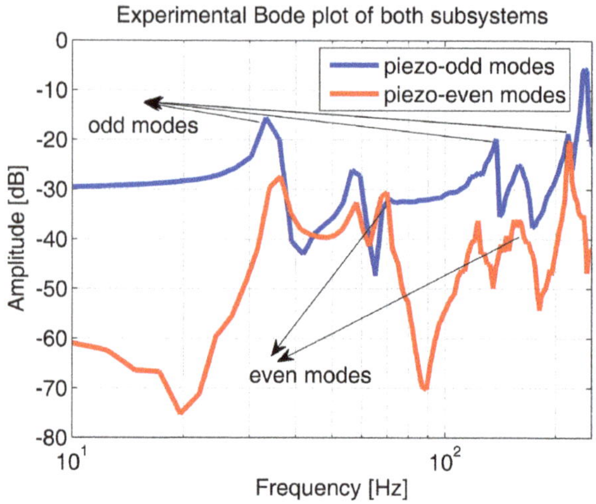

Fig. 2 Experimental amplitude plot of both systems: odd modes and even modes [26]

- by the piezo-sensor S-1 oriented parallel to X axis to measure amplitude of vibrations of the plate for odd modes
- by the piezo-sensor S-2 oriented parallel to Y axis to obtain the amplitude plot of the even modes.

The results of this experiment are shown in Fig. 2. Moreover, the obtained identification results have shown that the given TITO (Two-Input-Two-Output) system could be decoupled onto two SISO subsystem. It allows us to obtain the results that first subsystem contains three main frequency resonances: 34.8 Hz; 137.6 Hz; 218.9 Hz and three anti-resonance frequencies: 41.9 Hz; 140.5 Hz; 222.6 Hz, that it can be called *odd modes* and that the second one contains only two main resonance frequencies: 69.2 Hz; 163.1 Hz and two anti-resonance frequencies: 89.2 Hz; 181.6 Hz that it can be called "even modes". Finally the form of these subsystems are given in Eq. (30) and Eq. (31), respectively.

$$G_{odd}(s) = \frac{Y_{S-1}(s)}{U(s)} = 0.0168\frac{\left(s^2 + 11.23s + 6.482e^4\right)\left(s^2 + 46.47s + 9.353e^5\right)\left(s^2 + 62.66s + 2.685e^6\right)}{\left(s^2 + 7.641s + 4.885e^4\right)\left(s^2 + 36.7s + 8.443e^5\right)\left(s^2 + 50.08s + 2.574e^6\right)} \quad (29)$$

$$G_{even}(s) = \frac{Y_{S-2}(s)}{U(s)} = -0.00142\frac{\left(s^2 + 6.32s + 3.138e^5\right)\left(s^2 + 25.98s + 130e^6\right)}{\left(s^2 + 22.84s + 1.887e^5\right)\left(s^2 + 60.88s + 1.049e^6\right)}. \quad (30)$$

Summarizing from Fig. 2 the proper piezo-sensor locations effect on measurement of vibration of the plate can be seen. Evidently it is shown in the case of the second resonance frequency ($f_2 = 69$ Hz) where amplitude of vibration is measured by only the piezo-sensor S-2.

5 The Influence of Fractional Order on Vibration Behavior of the Smart Plate

In this section dynamical behavior of the smart plate for fractional order system has been analyzed due to different values of the non-integer order. It is known that each term of (30) describes vibration phenomena of the plate especially in indicated resonance or anti-resonance frequencies. In order to check how changing of a fractional order influences the structure's dynamic, simulations and experiments have been performed in two steps. In the first one, only the highest order of particular integer order model (second order) describing single resonance of smart structure has been replaced by fractional order close to two. As a result, it leads to determination of four different fractional orders models with orders from interval [1.97;2.01], see Fig. 3.

As can be seen in Fig. 3, even a slight change in the highest of the non-integral exponents significantly affects the dynamics of the mechanical structure. It is especially evident in the Bode diagram, where a small change mainly leads to a shift of resonance anti-resonance frequencies to frequencies versus frequency resonance of the integer order model. In the case of an increase of this order from 1.97 to 1.99 one can also observe increasing accuracy between frequencies of non-integer and integer order models and a light decreasing of the damping coefficient but without its exact curve fitting. Further increasing this parameter leads to instability of the system and its omission when designing the optimal control system. As a result, the highest fractional order of any consideration ensuring its stability should not be greater than 1.99.

In the next stage, the damping parameters of the system under consideration are analyzed according to the model of the highest fractional order equal to 1.99. To this aim, the middle part of the second-order model of each model describing single frequency resonance or frequency antiresonance, in the model given by (30) has also been replaced by the fractional order from interval [0.85; 1.05]. The results presented in Fig. 4 show that this change has a significant impact on the value of the damping coefficient. As a result, reducing this parameter to a value of 0.86 leads to an increase in the plate vibration amplitude as well as an improvement in the accuracy of the nodal plots between the integral and fractional models. Taking into account the above analysis, the estimated model (30) for designing the optimal control system can be rewritten in the form:

$$G_{odd\_fo}(s) = 0.0168\frac{\left(s^{1.99}+11.23s^{0.86}+6.482e^{4}\right)\left(s^{1.99}+46.47s^{0.86}+9.353e^{5}\right)\left(s^{1.99}+62.66s^{0.86}+2.685e^{6}\right)}{\left(s^{1.99}+7.641s^{0.86}+4.885e^{\partial}\right)\left(s^{1.99}+36.7s^{0.86}+8.443e^{5}\right)\left(s^{1.99}+50.08s^{0.86}+2.574e^{6}\right)} \tag{31}$$

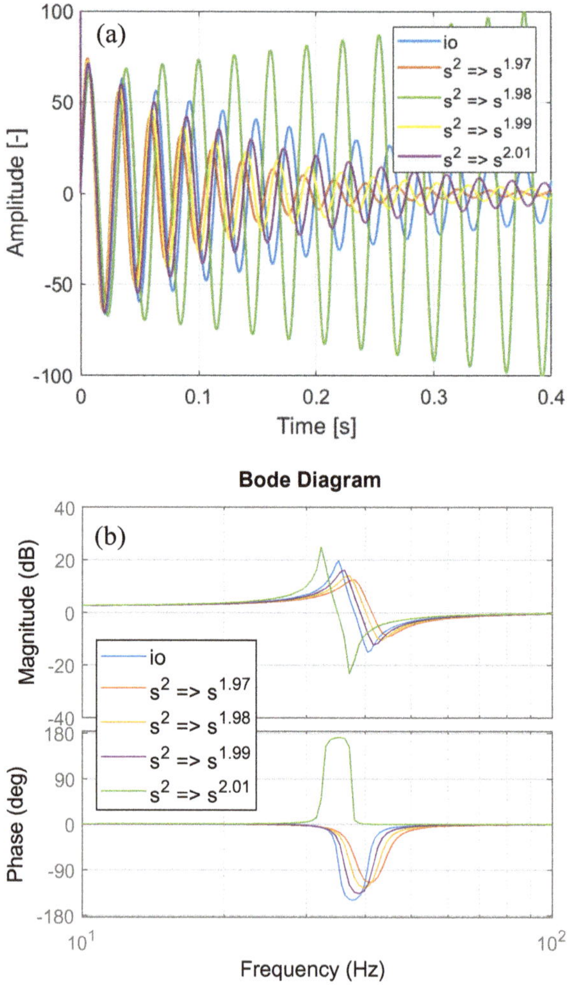

Fig. 3 The comparison behavior of the integer order system and fractional order systems **a** in the time domain (impulse responses), **b** in the frequency domain (Bode plots)

6 Oustaloup Approximation

The determined fractional order transfer function given by (31) requires proper approximation for further analysis. An important problem during this step is correctly assigning approximation parameters to obtain high accuracy between both non-integer and approximated models. The best known technique of approximation in frequency domain is the one presented by Oustaloup [34]. This is caused by fact that for fractional order systems, the Bode magnitude plot can be drawn exactly and its parameters can be applied to approximation calculations. Taking into account Oustaloup's method, the Oustaloup filter the approximation is designed for frequency ω form interval [0.001 rad/s;1000 rad/s] and approximation's order N is equal to 5.

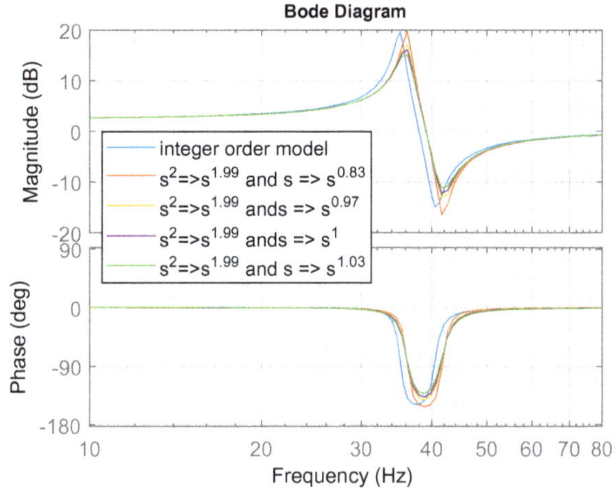

Fig. 4 The comparison of Bode plots of the integer order model and fractional order model by changing the order from 0.86 to 1.03

Further analysis of the resonance and anti-resonance frequencies of the approximated 240-order model presented in Fig. 5 show high accuracy between both models. This is especially evident in the vicinity of frequency resonances, where the amplitudes of the approximate model are close to the total model amplitudes. In the case of antiresonances less accuracy can be observed. However from the control point of view, it should not have a significant influence on the design of the optimal controller. Finally, it can be said that the process is proceeding well.

The aforementioned order of the approximated model from a strategy control point of view on the lab stand using real processors is too high and requires significantly reduction. This process is performed using the balance method. But first this process requires expressing of the approximated model in the state space form. This has led

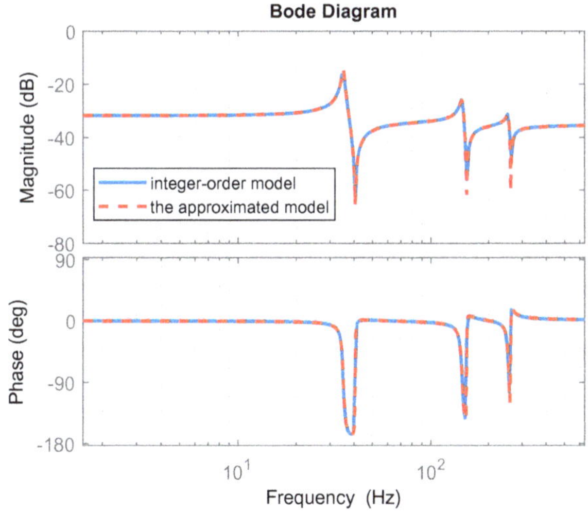

Fig. 5 The comparison of the approximated model (240th order) with integer order model (6th order)

to a cancellation of these states that from a control point of view can be omitted. As a result, the reduced 18th order model has been obtained. Next, the reduction process of the model's order has been repeated for cancellation of these resonances and anti-resonances, which frequencies have been very close to each other as it is shown in Fig. 6.

Finally, taking into account theory of flexible structure, the reduced order model can be described in the state space form as

$$\dot{\mathbf{x}}_o = \mathbf{A}_o x_o + \mathbf{B}_o u_o$$

Fig. 6 Root locus plot **a** the approximated 18th order model, **b** the integer 6th order model and the reduced approximate 6th order model

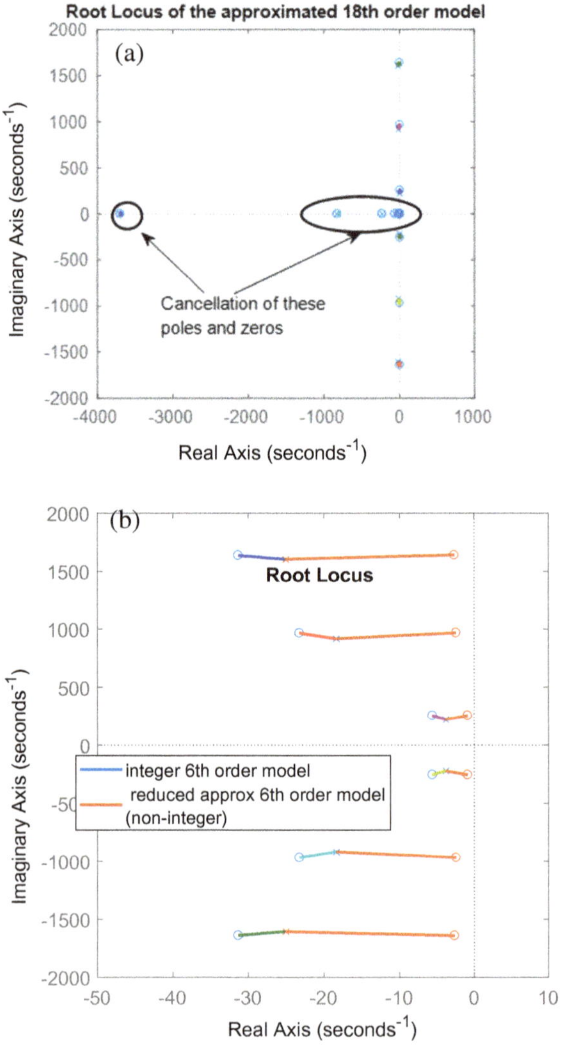

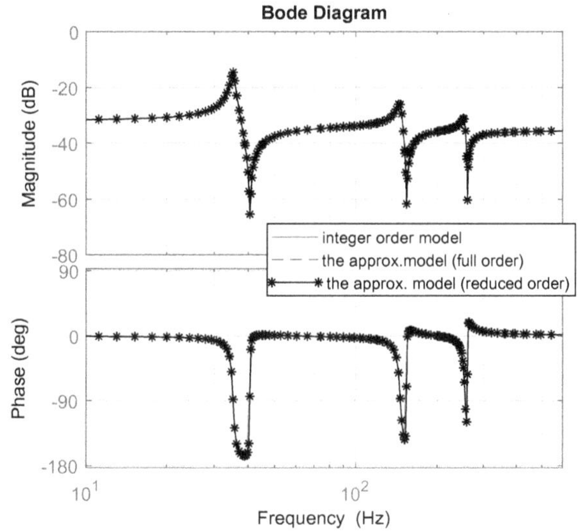

Fig. 7 The comparison of Bode plot of the estimated model (integer order model), the approximated full order model (240th order) and the reduced approximate order model (18th order)

$$y = \mathbf{C}_o x_o + \mathbf{D}_o u_o, \tag{32}$$

where:

$$A_o = \begin{bmatrix} -0.003\ amp;\ 0.221\ amp;\ 0\ amp;\ 0\ amp;\ 0\ amp;\ 0 \\ -0.221 & -0.003 & 0.085 & -0.026 & 0.068 & -0.041 \\ 0 & 0 & -0.018 & 0.918 & 0 & 0 \\ 0 & 0 & -0.918 & -0.018 & 0.082 & -0.050 \\ 0 & 0 & 0 & 0 & -0.025 & 1.604 \\ 0 & 0 & 0 & 0 & -1.604 & -0.025 \end{bmatrix}, B_o = \begin{bmatrix} 0 \\ 1.072 \\ 0 \\ 1.296 \\ 0 \\ 1.159 \end{bmatrix},$$
$$C_o = \begin{bmatrix} 1.152 & -0.092 & 1.335 & -0.413 & 1.067 & -0.650 \end{bmatrix}, D_o = 0.0168$$

and with the state vector $\mathbf{x}_0$ and the control vector $\mathbf{u}_0$ of the odd subsystem G_{odd_fo}. One can note that system (32) is the 6th order system. Taking into account the results of order truncation shown in Figs. 6 and 7 it can be concluded that this process has been properly performed.

7 Designing of Optimal LQR Controller

Firstly, the linear square regulator (LQR) for the reduced order model has been designed. This has been done in order to describe odd modes shapes of a smart plate in the selected interval [10 Hz; 250 Hz]. For this goal the performance index

$$J_1 = \frac{1}{2}\int_0^{\infty} \left(\mathbf{Y}^{\mathrm{T}}\mathbf{Q}_{odd}\mathbf{Y} + \mathbf{U}^{\mathrm{T}}\mathbf{R}_{odd}\mathbf{U}\right)dt \tag{33}$$

with matrices $\mathbf{Q}_{odd} = \mathrm{diag}(300) * I(6 \times 6)$ diag $= 300$ and $R_{odd} = 1$ ensuring proper damping of structure vibration has been minimized by correctly chosen values of input and control weighting matrices $\mathbf{Q}_{odd}$ and $\mathbf{R}_{odd,}$espectively, [35]. The high level damping and very short period of the vibrations have been chosen as criteria for designing weighting matrix $\mathbf{Q}_{odd}$ of this controller. The values of the feedback gains of the optimal local controller have been chosen during computer simulation, by comparison of the time plots (response to step, response to impulse and response to sinusoidal force excitation) and Bode plots. Minimization of the quadratic cost function (33) for the assumed weighting matrices leads to the determination of optimal feedback gain

$$\mathbf{K}_{odd} = \left[-0.043\ 13.820\ 0.235\ 11.418\ 0.721\ 8.083\right] \tag{34}$$

that allows to dampen vibrations of the plate in the selected frequency range.

The computer simulations of vibration the control system designed for the fractional order model have been performed with using Matlab software. Obtained results as well as in time and in frequency domains have shown that the best damping parameters are achieved for the closed-loop system with such a designed controller. Taking into account the step response of the open-loop system and the closed-loop system, respectively (Fig. 8), different time scale and amplitude in both plots can be noted. In this case the answer of the closed-loop system is four times faster than the answer of the open-loop system. The similar behavior of the closed-loop system can be seen in Figs. 9 and 10 where both systems have been excited by the Dirac impulse and harmonically signal with frequency equal to the first natural frequency respectively.

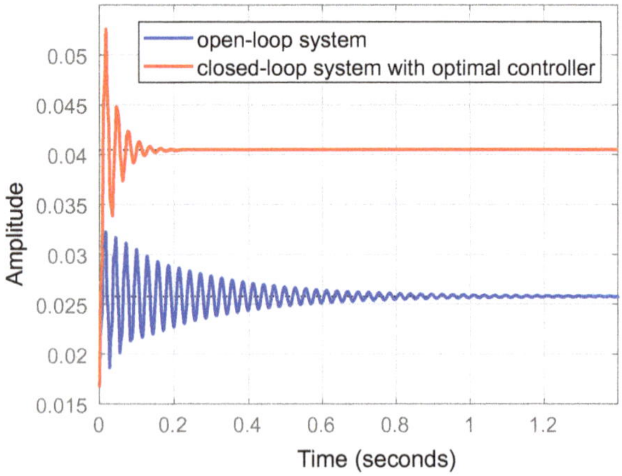

Fig. 8 The comparison of the step response of the open-loop system and the closed-loop system with LQR controller

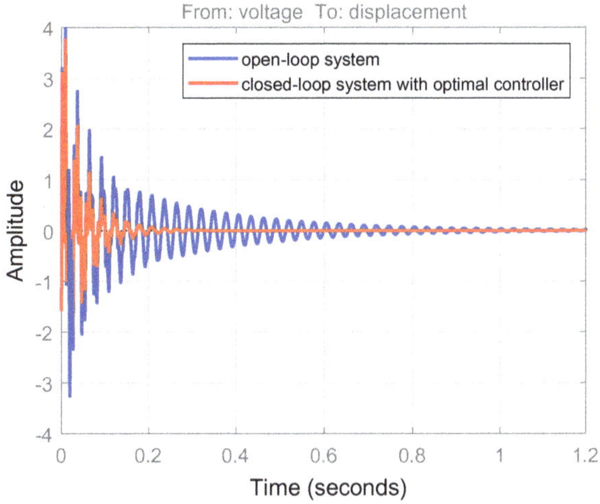

Fig. 9 The comparison of the impulse response of the open-loop system and the closed-loop system with LQR controller

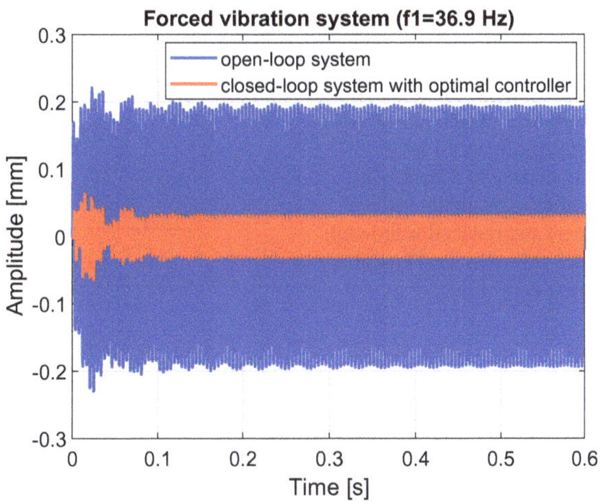

Fig. 10 The comparison of the harmonic response of the open-loop system and the closed-loop system with LQR controller

In the next step, the given system has been checked in the frequency domain. For this goal the Bode plots of the open-loop system and the closed-loop system with linear quadratic controller have been realized. The obtained results are shown in Fig. 11. It can be noted that the closed-loop system is strongly damped, especially in the case of the first resonance. Then, the amplitude of the vibration is about 10 dB lower than the amplitude of the open-loop system. Similar results have been obtained for two other natural frequencies in the selected frequency range where the amplitude vibrations of the closed-loop model have been decreased slightly versus the amplitude of the open-loop system.

In the last step, responses of both the closed-loop system and open-loop system in the frequency domain are compared. Taking into account Fig. 12 it can be noted

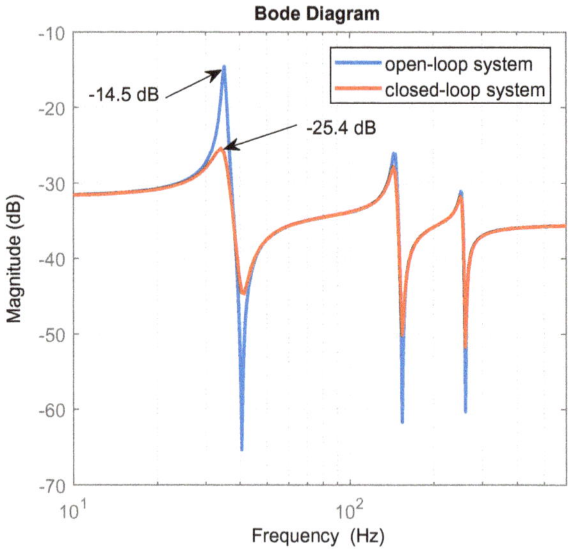

Fig. 11 The comparison of the amplitude plot of the open-loop system and the closed-loop system with LQR controller

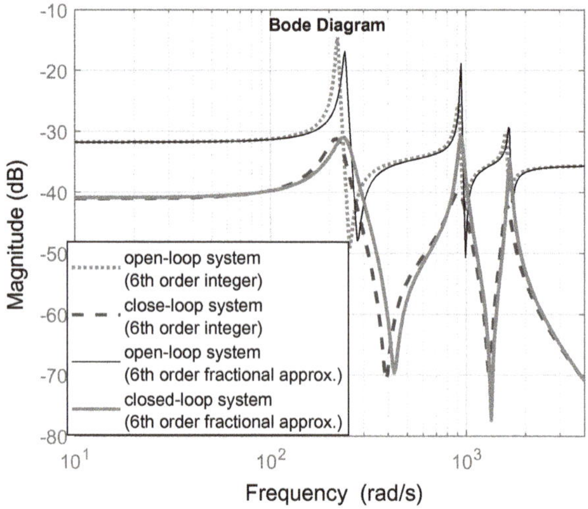

Fig. 12 The comparison of the Bode plot of open-loop system and closed loop system designed based on integer order model (6th order) and reduced fractional order approximated model (6th order)

that the dynamics of the fractional order closed-loop system is close to dynamics of the integer order closed-loop system. The exception is the region of the first mode shape, where values of particular resonance and anti-resonance frequencies of both fractional order and integer order models are divergent. Besides this phenomena, it can be observed that the fractional order closed-loop system also properly dampnes vibration of the plate in the selected frequency from the interval [10 Hz; 250 Hz]. As a result, it can be concluded that the desired dynamics of the system described by fractional order and also by the proper choice of parameters of the controller, lightly influence the value of constant r as the output of the linear system.

8 Experimental Setup

The designed active vibration control system of the plate has been also experimentally investigated on the lab stand to verify the simulation results. To this aim, the aluminum plate, being a host structure with additional piezo-stripes (actuators QP20N and sensors QP10N) attached to both has been used. Apart from this, the lab stand has been additionally equipped into the piezo amplifier developed by the Piezomechanik company, using it to drive the piezo-actuators and the piezo-charge amplifier 5018A1000, developed by the Kistler company to measure longitudinal direction vibration of the plate. In order to control vibration of the plate this stand has also been equipped with DSP processor 1005. Experimental verification of this control system has also been carried out in time and frequency domains. In both cases, the vibration control system using cards A/D 2103 and D/A 2002 have been created in Simulink software, shown in Fig. 13 and next, have been implemented to the DSP processor.

The experimental test in the time domain has been performed in the following form. As a excitation signal a chirp signal *u(t)* has been chosen, with amplitude 5 V and frequency linearly increasing up to 250 Hz that has been generated from the Agilent signal generator. The amplified periodic signal *u(t)* has been applied to both piezo-patch actuators (A-1 and A-2) located on the upper surface of the plate. At the same time, the vibration of the plate has been measured by the piezo-sensor (S-1). As a result, the recorded measurement signal from piezo S-1 by switching the off/on control is shown in Fig. 14.

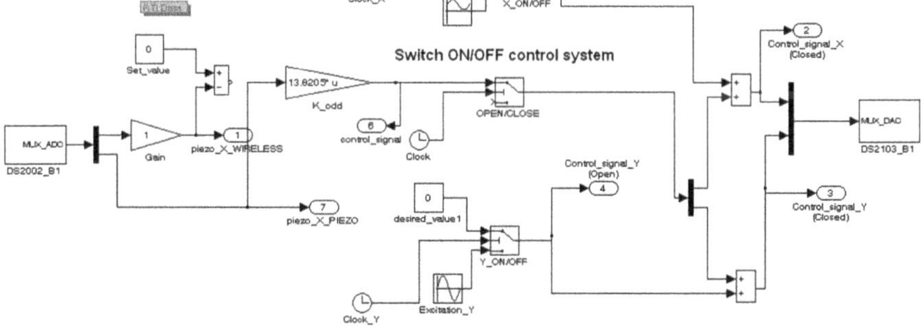

Fig. 13 The optimal control system with LQR controller implemented to the real time processor DSP 1005

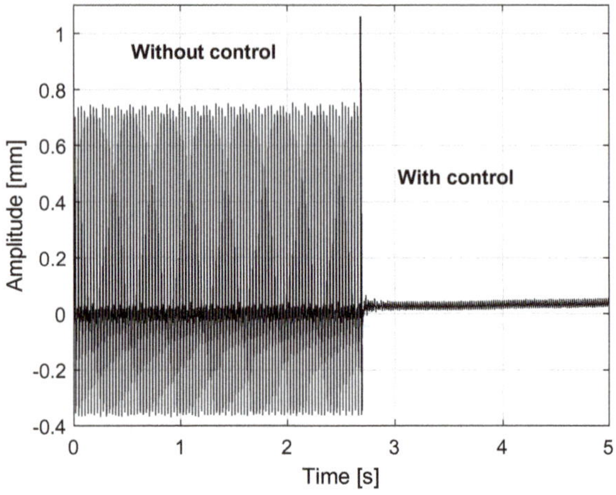

Fig. 14 The recorded measurement signal by switching the optimal control system off/on

In the next step, a test has been carried out on the frequency domain for both open-loop and closed-loop systems. To this aim the Dynamical Signal Analyzer HP35670A has been used to generate the chirp signal (excitation signal) and also to measure amplitude vibration of the plate from the piezo-sensor S-1. The recorded signals from two channels allows us to plot the frequency responses of this structure shown in Fig. 15.

Taking into account Fig. 15 it can be noted that the generated frequency response of the closed-loop system properly verified the simulations results (Fig. 10). In both cases the magnitude of the closed-loop system with LQR controller designed for the fractional order model of control plant given by (32) versus the magnitude of the open-loop system, especially for the first odd mode, is reduced by about 10–12 dB. Finally it can be concluded that this control system is correct.

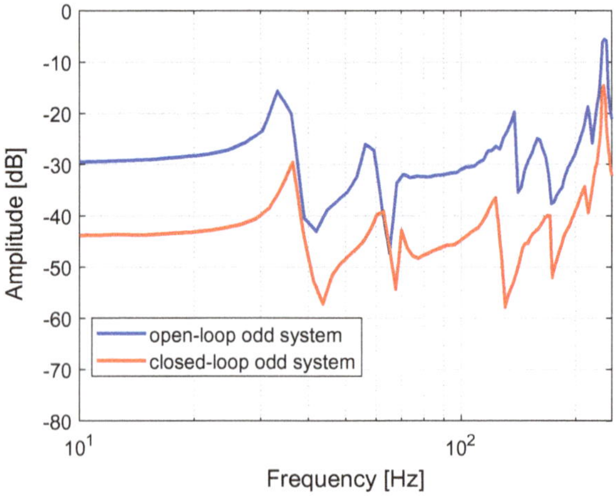

Fig. 15 The comparison of frequency responses of the open-loop system and closed-loop system with LQR controller

9 Summary and conclusions

The paper presents analysis of output regulation of a fractional vibration control system of the smart plate designed by using the optimal LQR controller versus a closed-loop system design based on an integer order model. To this aim, the definitions of Caputo-, Riemann–Liouville- and Grünwald-Letnikov-type fractional order h-difference operators and properties of nonlinear, discrete-time fractional order systems with these operators and relationship between them have been recalled. It has been shown that the if the generalized fractional order linear approximation of the control system is controllable in a finite number of steps steps then the nonlinear system is stabilized in the same number of steps by a static feedback. Also if the output of the linearized the system can be regulated, a finite number of steps around a constant output in the neighborhood of a zero state, then the output of the nonlinear system can also be regulated in the same number of steps around this output in this state for both Caputo and Riemann–Liouville types h-difference operators. Next the active vibration control system for the nonlinear, discrete-time fractional order model of the rectangular plate is designed. In order to do this, analysis of the fractional order model is performed. The obtained results shown in Figs. 3 and 4 indicate that the best vibration performances of this model are achieved in the model given by (32) due to the fact that the vibration amplitudes of the fractional order model are very close to the amplitudes of an integer order model.

In the next step, process approximation of a fractional order model is performed based on the model given by (32). It allows us to obtain the reduced approximated order model and design an optimal controller. The obtained simulation and experimental results of the closed-loop for the chosen parameters of a feedback gain vector and the reduced order model given by (34) in time and frequency domain (see Figs. 8, 9, 10, 11 and 15) show that such a designed control system ensures proper vibration damping of the structure in the selected frequency interval (10 Hz; 50 Hz). Next, vibration amplitude of the closed-loop system versus amplitude of the open-loop system has been reduced about 1–12 dB. A similar effect has been achieved during analysis of the fractional order model also. Finally amplitudes of the open-loop and the closed-loop systems of the fractional order model are also very close to the amplitudes of the integer order model. As a result, it can be concluded that the output of the nonlinear system can be regulated in the same number of steps around this r as in the case of output for the linear system.

Acknowledgements The work of A. Koszewnik and E. Pawluszewicz is supported by the *University Grant* no WZ/WM-IIM/1/2019 of Faculty of Mechanical Engineering, Bialystok University of Technology. The work of P. Burzynski is supported by *University Grant* no WI/WM-IIM/7/2020 of Faculty of Mechanical Engineering, Bialystok University of Technology.

References

1. Bartosiewicz, Z., Pawluszewicz, E.: Realizations of linear control systems on time scales. Control Cybern. **35**(4), (2006)
2. Zabczyk, J.: Mathematical Control Theory. Birkhäuser, Basel (2008)
3. Sontag, E.D.: Mathematical Control Theory. Springer (1998)
4. Halás, M., Kotta, Ü.: A transfer function approach to the realization problem of nonlinear systems. Int. J. Control **85**(3), 320–331 (2012)
5. Bartosiewicz, Z., Pawluszewicz, E.: Realizations of nonlinear control systems on time scales. IEEE Trans. Autom. Control **53**(2), (2008)
6. Pawluszewicz, E.: Fractional vector-order h-realization of the impulse response function. Acta Mechanica et Automatica **14**(2), 108–113 (2020)
7. Teplajkov, A.: Fractional order Modelling and Control of Dynamic systems. Publisher Springer (2017)
8. Bastos, N.R.O., Ferreira, R.A.C., Torres, D.F.M.: Necessary optimality conditions for fractional difference problems of the calculus of variations. Discrete Contin. Dynam. Syst. **29**(2), 417–437 (2011)
9. Ferreira, R.A.C., Torres, D.F.M.: Fractional h-difference equations arising from the calculus of variations. Appl. Anal. Discret. Math. **5**(1), 110–121 (2011)
10. Oprzędkiewicz, K.: Fractional order, discrete model of heat transfer process using time and spatial Grünwald-Letnikov operator. Bull. Pol. Acad. Sci.: Tech. Sci. **69**(1), No. e135843 (2021)
11. Kopka, R.: Changes in derivative orders for fractional models of supercapacitors as a function of operating temperature. IEEE Access **7**(8684234), 47674–47681 (2019)
12. Das, S.: Functional Fractional Calculus for System Identification and Controls. Springer (2008)
13. Ambroziak, L., Lewon, D., Pawluszewicz, E.: The use of fractional order operators in modeling of RC-electrical systems. Control. Cybern. **45**(3), 275–288 (2016)
14. Koszewnik, A., Nartowicz, T., Pawluszewicz, E.: Fractional order controller to control pump in FESTO MPS® PA Compact Workstation. In: Proceedings of the International Carpathian Control Conference, pp. 364–367 (2016)
15. Oprzędkiewicz, K., Gawin, E.: A noninteger order, state space model for one dimensional heat transfer process. Arch. Control Sci. **26**(2), 261–275 (2016)
16. Sierociuk, D., Dzieliński, A., Sarwas, G., Petras, I., Podlubny, I., Skovranek, T.: Modelling heat transfer in heterogenous media using fractional calculus. Philos. Trans. R. Soc. A-Math., Phys. Eng. Sci. 20120146, 371 (2013)
17. Mozyrska, D., Wyrwas, M.: The Z-transform method and delta type fractional difference operators. Discret. Dyn. Nat. Soc. 852734 (2015)
18. Mozyrska, D., Wyrwas, M., Pawluszewicz, E.: Stabilization of linear multi-parameter fractional difference control systems. In: Proceedings of the International Conference on Methods and Models in Automation and Robotics, Międzyzdroje, Poland, pp. 315–319 (2015)
19. Mozyrska, D., Wyrwas, M.: Stability by linear approximation and the relation between the stability of difference and differentional fractional systems. Math. Methods Appl. Sci. (2016)
20. Wyrwas, M., Pawluszewicz, E., Girejko, E.: Stability of nonlinear H-difference systems with n fractional orders. Kybernetika **51**(1), 112–136 (2015)
21. Koszewnik, A., Pawluszewicz, E., Ostaszewski, M.: Experimental studies of the fractional PID and TID controllers for industrial process. Int. J. Control Autom. Syst. accepted for publication in 2021, doi.org/https://doi.org/10.1007/s12555-020-0123-4
22. Isidori, A.: Nonlinear Control Systems. Springer (1996)
23. Marino, R., Tomei, P.: Nonlinear Control Design Geometric, Adaptive and Robust. Prentice Hall (1995)
24. Chen, L., Narendra, K.: Identification and Control of a Nonlinear Discrete-Time System Based on its Linearization: a Unified Framework. IEEE Trans. Neural Netw. **15**(3), 663–673 (2004)
25. Mozyrska, D., Wyrwas, M., Pawluszewicz, E.: Local observability and controllability of nonlinear discrete-time fractional order systems based on their linearization. Int. J. Syst. Sci. **48**(4), 788–794 (2017)

26. Koszewnik, A.: The Design Vibration Control System for Aluminum Plate with piezo-stripes based on residues analysis of mode. Eur. Phys. J. Plus **133**, 1–15 (2018)
27. Litak, G., Ambrożkiewicz, B., Wolszczak, P.: Dynamics of a nonlinear energy harvester with subharmonic responses. J. Phys: Conf. Ser. **1736**, 1–12 (2021)
28. Litak, G., Margielewicz, J., Gaska, D., Wolszczak, P., Zhou, S.: Multiple Solutions of the Tristable Energy Harvester. Energies **14**(5), 1–17 (2021)
29. Czaban, J., Litak, G., Ambrożkiewicz, B., Gardynski, L., Staczek, P., Wolszczak, P.: Impact-based piezoelectric energy harvesting system excited from diesel engine suspension. Appl. Comput. Sci. **16**(3), 16–29 (2020)
30. Mozyrska, D., Girejko, E.: Overview of the fractional h-differences operator. In: Almeida, A., Castro, L., Speck, F.O. (eds.) Advances in Harmonic Analysis and Operator Theory. Operator Theory: Advances and Applications, p. 229. Birkhäuser, Basel (2013)
31. Mozyrska, D., Girejko, E., Wyrwas, M.: Comparison of h-difference fractional operators. in: Theory and Application of Non-integer Order systems, LNEE 257, pp.191–197. Springer (2013)
32. MozyrskaD., Pawluszewicz E., Local controllability of nonlinear discrete-time fractional order systems, Bull Pol Acad Sci. Tech Sci **61**(1), 251–156 (2013)
33. Zhu, W., Rui, X.-T.: Hysteresis modeling and displacement control of piezoelectric actuators with the frequency-dependent behavior using a generalized Bouc-Wen model. Precis. Eng. **43**, 299–307 (2016)
34. Petras, I.: Fractional-Order Nonlinear Systems: Modeling, Analysis and Simulation, Nonlinear Physical Science. Springer (2011)
35. Koszewnik, A.: The optimal vibration control of the plate structure by using piezo-actuators. In: Proceedings of International Carpathian Control Conference, ICCC (2015), DOI:https://doi.org/10.1109/CarpathianCC.2016. 7501123

Some Specific Properties of Positive Standard and Fractional Interval Systems

Tadeusz Kaczorek and Łukasz Sajewski

Abstract Some specific properties of positive standard and fractional linear systems with interval state matrices are analyzed. The stability, positivity and transfer matrices of positive different orders fractional continuous-time linear systems are considered. New necessary and sufficient conditions for the asymptotic stability of positive different orders fractional linear systems are established. It is shown that the transfer matrices of positive asymptotically stable different orders fractional linear systems have only nonnegative coefficients. New conditions for the interval stability of positive standard and fractional linear systems are established. It is shown that the adjoint matrix of a singular Metzler matrix with zero sum of entries of each row (column) has all equal entries.

1 Introduction

A dynamical system is called fractional if it is described by a fractional order differential or difference equation. The fundamentals of fractional calculus and fractional systems have been given in monographs [16, 25, 29–32].

In positive systems inputs, state variables and outputs take only nonnegative values for any nonnegative inputs and nonnegative initial conditions. Examples of positive systems are industrial processes involving chemical reactors, heat exchangers and distillation columns, storage systems, compartmental systems and water and atmospheric pollution models. A variety of models that have positive behavior can be found in engineering, management science, economics, social sciences, biology and medicine, etc. An overview of state of the art positive systems theory is given in the monographs [2, 7, 13]. Positivity and its applications in linear and nonlinear systems

T. Kaczorek · Ł. Sajewski (✉)
Faculty of Electrical Engineering, Białystok University of Technology Bialystok, Wiejska 45D, 15-351 Białystok, Poland
e-mail: l.sajewski@pb.edu.pl

T. Kaczorek
e-mail: t.kaczorek@pb.edu.pl

P. Kulczycki et al. (eds.), *Fractional Dynamical Systems: Methods, Algorithms and Applications*, Studies in Systems, Decision and Control 402,
https://doi.org/10.1007/978-3-030-89972-1_9

has been considered in many papers [1, 9, 10, 14, 15, 27, 37]. The stability, robust stability, practical stability, absolute stability and super stability of positive linear and nonlinear standard and fractional systems have been considered in [4–6, 8–10, 12, 17, 22, 24, 26, 33–36]. Stability of interval positive systems has been considered in [18–21]. The positive fractional linear systems of different orders have been introduced in [14, 15] and stability and stabilization of positive fractional systems consisting of n subsystems with different fractional orders have been investigated in [5, 36].

The existence and determination of the set of Metzler matrices for given, stable polynomials have been considered in [11] and dynamical properties of Metzler systems in [28].

The chapter is organized as follows. In Sect. 2 the basic definitions and theorems concerning the standard and fractional positive linear systems are recalled. The new stability conditions and transfer matrices with nonnegative coefficients for different orders fractional positive linear systems are given in Sects. 3 and 4 respectively. The stability of fractional positive systems with interval state matrices is addressed in Sect. 5 and the interval stability of positive standard linear systems is investigated in Sect. 6. In Sect. 7 it is shown that the adjoint matrix of a singular Metzler matrix with zero sum of entries of each row (column) has all equal entries. Concluding remarks are given in Sect. 8.

The following notation will be used: $\Re$—the set of real numbers, $\Re^{n\times m}$—the set of $n \times m$ real matrices, $\Re_+^{n\times m}$—the set of $n \times m$ real matrices with nonnegative entries and $\Re_+^n = \Re_+^{n\times 1}$, M_n—the set of $n \times n$ Metzler matrices (real matrices with nonnegative off-diagonal entries), I_n- the $n \times n$ identity matrix.

2 Standard and Fractional Linear Systems

Consider the continuous-time linear system

$$\dot{x}(t) = Ax(t) + Bu(t), \tag{1a}$$

$$y(t) = Cx(t), \tag{1b}$$

where $x(t) \in \Re^n$, $u(t) \in \Re^m$, $y(t) \in \Re^p$ are the state, input and output vectors and $A \in \Re^{n\times n}$, $B \in \Re^{n\times m}$, $C \in \Re^{p\times n}$.

Definition 1 [7, 13] The system (1a), (1b) is called (internally) positive if $x(t) \in \Re_+^n$ and $y(t) \in \Re_+^p$, $t \geq 0$ for any initial conditions $x(0) \in \Re_+^n$ and all inputs $u(t) \in \Re_+^m$, $t \geq 0$.

A real matrix $A = [a_{ij}] \in \Re^{n\times n}$ is called a Metzler matrix if its off-diagonal entries are nonnegative, i.e. $a_{ij} \geq 0$ for $i \neq j$. The set of $n \times n$ Metzler matrices will be denoted by M_n.

Theorem 1 [7, 13] The system (1a), (1b) is positive if and only if

$$A \in M_n, \quad B \in \Re_+^{n\times m}, \quad C \in \Re_+^{p\times n}. \tag{2}$$

Definition 2 [7, 13] The positive system (1a), (1b) with $u(t) = 0$ is called asymptotically stable (the matrix $A \in M_n$ is Hurwitz) if

$$\lim_{t\to\infty} x(t) = 0. \tag{3}$$

Theorem 2 [7, 13] The positive system (1a), (1b) is asymptotically stable if and only if one of the equivalent conditions is satisfied:

(1) All coefficient of the characteristic polynomial

$$\det[I_n s - A] = s^n + a_{n-1}s^{n-1} + \cdots + a_1 s + a_0 \tag{4}$$

are positive, i.e. $a_k > 0$ for $k = 0, 1, ..., n-1$.

(2) All principal minors $\overline{M}_i$, $i = 1, ..., n$ of the matrix $-A$ are positive, i.e.

$$\overline{M}_1 = |-a_{11}| > 0,\ \overline{M}_2 = \begin{vmatrix} -a_{11} & -a_{12} \\ -a_{21} & -a_{22} \end{vmatrix} > 0\overline{M}_2 = \begin{vmatrix} -a_{11} & -a_{12} \\ -a_{21} & -a_{22} \end{vmatrix} > 0, ..., \overline{M}_n = \det[-A] > 0 \tag{5}$$

(3) There exists strictly positive vector $\lambda^T = [\lambda_1 \cdots \lambda_n]^T$, $\lambda_k > 0$, $k = 1, ..., n$ such that

$$A\lambda < 0 \text{ or } A^T\lambda < 0 \tag{6}$$

Consider the fractional continuous-time linear system.

$$\frac{d^\alpha x(t)}{dt^\alpha} = Ax(t) + Bu(t), 0 < \alpha < 1 \tag{7a}$$

$$y(t) = Cx(t) + Du(t) \tag{7b}$$

where $x(t) \in \Re^n$, $u(t) \in \Re^m$, $y(t) \in \Re^p$ are the state, input and output vectors and $A \in \Re^{n\times n}$, $B \in \Re^{n\times m}$, $C \in \Re^{p\times n}$, $D \in \Re^{p\times m}$. The fractional derivative of α order is defined be [32]

$$_0D_t^\alpha f(t) = \frac{d^\alpha f(t)}{dt^\alpha} = \frac{1}{\Gamma(1-\alpha)}\int_0^t \frac{\dot{f}(\tau)}{(t-\tau)^\alpha} d\tau, 0 < \alpha < 1 \tag{8}$$

where $\dot{f}(\tau) = \frac{df(\tau)}{d\tau}$ and $\Gamma(x) = \int_0^\infty t^{x-1}e^{-t}dt$, $\mathrm{Re}(x) > 0$ is the Euler gamma function.

Definition 3 [16] The fractional system (7a), (7b) is called (internally) positive if $x(t) \in \Re_+^n$ and $y(t) \in \Re_+^p$, $t \geq 0$ for any initial conditions $x(0) \in \Re_+^n$ and all inputs $u(t) \in \Re_+^m$, $t \geq 0$.

Theorem 3 [16] The fractional system (7a), (7b) is positive if and only if.

$$A \in M_n,\ B \in \Re_+^{n\times m},\ C \in \Re_+^{p\times n},\ D \in \Re_+^{p\times m} \tag{9}$$

Definition 4 [16] The fractional positive system (7a), (7b) for $u(t) = 0$ is called asymptotically stable if.

$$\lim_{t\to\infty} x(t) = 0 \text{ for all } x(0) \in \Re_+^n \tag{10}$$

Theorem 4 [16] The fractional positive system (7a), (7b) for $u(t) = 0$ is asymptotically stable if and only if one of the following equivalent conditions is satisfied:

(1) The eigenvalues λ_i, $i = 1, \ldots, n$ of the matrix $A \in M_n$ satisfy the condition

$$\mathrm{Re}\lambda_i < 0,\ i = 1, \ldots, n \tag{11}$$

(2) All coefficients of the characteristic polynomial

$$\det[I_n\gamma - A] = \gamma^n + a_{n-1}\gamma^{n-1} + \ldots + a_1\gamma + a_0 \tag{12}$$

are positive, i.e. $a_k > 0$ for $k = 0, 1, \ldots, n-1$ and $\gamma = s^\alpha$.

(3) There exists strictly positive vector $\lambda^T = [\lambda_1 \cdots \lambda_n]^T$, $\lambda_i > 0$, $i = 1, \ldots, n$ such that

$$A\lambda < 0 \text{ or } \lambda^T A < 0 \tag{13}$$

Theorem 5 The fractional positive system (7a), (7b) is asymptotically stable if the sum of entries of each column (row) of the matrix A is negative.

Proof Using (13) we obtain

$$A\lambda = \begin{bmatrix} a_{11} & \ldots & a_{1n} \\ \vdots & \ldots & \vdots \\ a_{n1} & \ldots & a_{nn} \end{bmatrix} \begin{bmatrix} \lambda_1 \\ \vdots \\ \lambda_n \end{bmatrix} = \begin{bmatrix} a_{11} \\ \vdots \\ a_{n1} \end{bmatrix} \lambda_1 + \cdots + \begin{bmatrix} a_{1n} \\ \vdots \\ a_{nn} \end{bmatrix} \lambda_n < \begin{bmatrix} 0 \\ \vdots \\ 0 \end{bmatrix} \tag{14}$$

and the sum of entries of each column of the matrix A is negative since $\lambda_k > 0$, $k = 1, \ldots, n$. The proof for rows is similar. □

The transfer matrix of the system (7a), (7b) is given by.

$$T(\gamma) = C[I_n\gamma - A]^{-1}B + D \tag{15}$$

where $\gamma = s^\alpha$.

Now consider the fractional linear system with two different fractional orders.

$$\begin{bmatrix} \frac{d^\alpha x_1(t)}{dt^\alpha} \\ \frac{d^\beta x_2(t)}{dt^\beta} \end{bmatrix} = \begin{bmatrix} A_{11} & A_{12} \\ A_{21} & A_{22} \end{bmatrix}\begin{bmatrix} x_1(t) \\ x_2(t) \end{bmatrix} + \begin{bmatrix} B_1 \\ B_2 \end{bmatrix}u(t) \tag{16a}$$

$$y(t) = [\,C_1 \; C_2\,]\begin{bmatrix} x_1(t) \\ x_2(t) \end{bmatrix} \tag{16b}$$

where $0 < \alpha, \beta < 1, x_1(t) \in \Re^{n_1}$ and $x_2(t) \in \Re^{n_2}$ are the state vectors, $A_{ij} \in \Re^{n_i \times n_j}$, $B_i \in \Re^{n_i \times m}$, $C_i \in \Re^{p \times n_i}$; $i, j = 1{,}2$; $u(t) \in \Re^m$ is the input vector and $y(t) \in \Re^p$ is the output vector. Initial conditions for (16a) have the form.

$$x_1(0) = x_{10}, x_2(0) = x_{20} \text{ and } x_0 = \begin{bmatrix} x_{10} \\ x_{20} \end{bmatrix} \tag{17}$$

Theorem 6 The solution of the Eq. (16a) for $0 < \alpha < 1;\; 0 < \beta < 1$ with initial conditions (17) has the form.

$$x(t) = \begin{bmatrix} x_1(t) \\ x_2(t) \end{bmatrix} = \Phi_0(t)x_0 + \int_0^t M(t-\tau)u(\tau)d\tau \tag{18}$$

where

$$\begin{aligned} M(t) &= \Phi_1(t)B_{10} + \Phi_2(t)B_{01} \\ &= \begin{bmatrix} \Phi_{11}^1(t) & \Phi_{12}^1(t) \\ \Phi_{21}^1(t) & \Phi_{22}^1(t) \end{bmatrix}\begin{bmatrix} B_1 \\ 0 \end{bmatrix} + \begin{bmatrix} \Phi_{11}^2(t) & \Phi_{12}^2(t) \\ \Phi_{21}^2(t) & \Phi_{22}^2(t) \end{bmatrix}\begin{bmatrix} 0 \\ B_2 \end{bmatrix} \\ &= \begin{bmatrix} \Phi_{11}^1(t)B_1 + \Phi_{12}^2(t)B_2 \\ \Phi_{21}^1(t)B_1 + \Phi_{22}^2(t)B_2 \end{bmatrix} = \begin{bmatrix} \Phi_{11}^1(t) & \Phi_{12}^2(t) \\ \Phi_{21}^1(t) & \Phi_{22}^2(t) \end{bmatrix}\begin{bmatrix} B_1 \\ B_2 \end{bmatrix} \end{aligned} \tag{19a}$$

and.

$$\Phi_0(t) = \sum_{k=0}^{\infty}\sum_{l=0}^{\infty} T_{kl}\frac{t^{k\alpha + l\beta}}{\Gamma(k\alpha + l\beta + 1)} \tag{19b}$$

$$\Phi_1(t) = \sum_{k=0}^{\infty}\sum_{l=0}^{\infty} T_{kl} \frac{t^{(k+1)\alpha + l\beta - 1}}{\Gamma[(k+1)\alpha + l\beta]} \tag{19c}$$

$$\Phi_2(t) = \sum_{k=0}^{\infty}\sum_{l=0}^{\infty} T_{kl} \frac{t^{k\alpha + (l+1)\beta - 1}}{\Gamma[k\alpha + (l+1)\beta]} \tag{19d}$$

$$T_{kl} = \begin{cases} I_n \text{ for } k = l = 0 \\ \begin{bmatrix} A_{11} & A_{12} \\ 0 & 0 \end{bmatrix} \text{ for } k = 1, l = 0 \\ \begin{bmatrix} 0 & 0 \\ A_{21} & A_{22} \end{bmatrix} \text{ for } k = 0, l = 1 \\ T_{10}T_{k-1,l} + T_{01}T_{k,l-1} \text{ for } k + l > 1 \end{cases} \tag{19e}$$

Proof is given in [14, 15].

Definition 5 The fractional system (16a), (16b) is called positive if $x_1(t) \in \Re_+^{n_1}$ and $x_2(t) \in \Re_+^{n_2}$, $t \geq 0$ for any initial conditions $x_{10} \in \Re_+^{n_1}$, $x_{20} \in \Re_+^{n_2}$ and all input vectors $u \in \Re_+^m$ and outputs $y \in \Re_+^p$, $t \geq 0$.

Theorem 7 [14, 15] The fractional system (16a), (16b) for $0 < \alpha < 1;\ 0 < \beta < 1$ is positive if and only if.

$$\tilde{A} = \begin{bmatrix} A_{11} & A_{12} \\ A_{21} & A_{22} \end{bmatrix} \in M_N,\ \tilde{B} = \begin{bmatrix} B_1 \\ B_2 \end{bmatrix} \in \Re_+^{N\times m},\ \tilde{C} = [\,C_1\ C_2\,] \in \Re_+^{p\times n} (N = n_1 + n_2) \tag{20}$$

Definition 6 [16] The fractional system (16a), (16b) for $u(t) = 0$ is called asymptotically stable if.

$$\lim_{t\to\infty} \begin{bmatrix} x_1(t) \\ x_2(t) \end{bmatrix} = \begin{bmatrix} 0 \\ 0 \end{bmatrix} \text{ for all } x_{10} \in \Re_+^{n_1}, x_{20} \in \Re_+^{n_2} \tag{21}$$

Theorem 8 The positive fractional system (16a), (16b) is asymptotically stable if and only if one of the following equivalent conditions is satisfied:

(1) The eigenvalues $\tilde{\lambda}_i$, $i = 1, \ldots, n$ of the matrix $\tilde{A} \in M_n$ satisfy the condition

$$\mathrm{Re}\tilde{\lambda}_i < 0, i = 1, \ldots, n \tag{22}$$

(2) All coefficients of the characteristic polynomial

$$\det\left[\begin{bmatrix} I_{n_1}\gamma & 0 \\ 0 & I_{n_2}\delta \end{bmatrix} - \tilde{A}\right] = \gamma^n\delta^n + \tilde{a}_{n,n-1}\gamma^n\delta^{n-1} + \tilde{a}_{n-1,n}\gamma^{n-1}\delta^n + \ldots + \tilde{a}_{1,1}\gamma\delta + \tilde{a}_{1,0}\gamma + \tilde{a}_{0,1}\delta + \tilde{a}_{00} \tag{23}$$

are positive, i.e. $\tilde{a}_{i,j} > 0$ for $i = 0, 1, ..., n_1 - 1$; $j = 0, 1, ..., n_2 - 1$ and $\gamma = s^{\alpha}$, $\delta = s^{\beta}$.

Proof is similar to the proof of Theorem 4, which is given in [16].
The transfer matrix of the system (16a), (16b) is given by.

$$T(s^{\alpha}, s^{\beta}) = \tilde{C}\left[\begin{bmatrix} I_{n_1}s^{\alpha} & 0 \\ 0 & I_{n_2}s^{\beta} \end{bmatrix} - \tilde{A}\right]^{-1}\tilde{B} \tag{24}$$

3 New Stability Conditions for Different Orders Fractional Positive Linear Systems

In this section new necessary and sufficient condition for the asymptotic stability of different orders fractional positive systems will be established.

Theorem 9 The positive different orders fractional linear system (16a), (16b) is asymptotically stable if and only if there exists strictly positive vector $\lambda = [\lambda_1 \cdots \lambda_n]$ with all positive $\lambda_k > 0$, $k = 1, ..., n$ such that.

$$\tilde{A}\lambda < 0 \text{ or } \lambda^T\tilde{A} < 0 \tag{25}$$

Proof Using the fractional integration of (16a) on the interval of $(0, +\infty)$ for $\tilde{B} = 0$ we obtain

$$\begin{aligned} x_1(\infty) - x_1(0) &= A_{11}\int_0^{\infty} x_1(\tau)d\tau + A_{12}\int_0^{\infty} x_2(\tau)d\tau, \\ x_2(\infty) - x_2(0) &= A_{21}\int_0^{\infty} x_1(\tau)d\tau + A_{22}\int_0^{\infty} x_2(\tau)d\tau. \end{aligned} \tag{26}$$

If the system is asymptotically stable then $x_1(\infty) = 0$, $x_2(\infty) = 0$ and arbitrary $x_1(0) \in \Re_+^{n_1}$ and $x_2(0) \in \Re_+^{n_2}$ we obtain.

$$[A_{11}\ A_{12}]\lambda < 0 \text{ and } [A_{21}\ A_{22}]\lambda < 0 \tag{27}$$

since for positive system $\int_0^{\infty} x_k(\tau)d\tau > 0$ for $k = 1,2$.

Therefore, the positive fractional different orders system is asymptotically stable if and only if the conditions (25) are satisfied. □

Example 1 Check the asymptotic stability of the positive fractional linear system (16a), (16b) with the matrix.

$$\tilde{A} = \begin{bmatrix} A_{11} & A_{12} \\ A_{21} & A_{22} \end{bmatrix} = \begin{bmatrix} -2 & 1 & 0 & 0.5 \\ 1.5 & -3 & 1 & 0.2 \\ 1 & 2 & -4 & 1 \\ 0 & 1 & 2 & -4 \end{bmatrix}, \quad n_1 = n_2 = 2 \tag{28}$$

Applying the condition (25) to the matrix (28) for $\lambda = [\,1 \;\; 1 \;\; 1 \;\; 0.8\,]^T$ we obtain.

$$\begin{bmatrix} -2 & 1 & 0 & 0.5 \\ 1.5 & -3 & 1 & 0.2 \end{bmatrix} \begin{bmatrix} 1 \\ 1 \\ 1 \\ 0.8 \end{bmatrix} = \begin{bmatrix} -0.6 \\ -0.34 \end{bmatrix} < 0 \text{ and } \begin{bmatrix} 1 & 2 & -4 & 1 \\ 0 & 1 & 2 & -4 \end{bmatrix} \begin{bmatrix} 1 \\ 1 \\ 1 \\ 0.8 \end{bmatrix} = \begin{bmatrix} -0.2 \\ -0.2 \end{bmatrix} < 0 \tag{29}$$

Therefore, the positive fractional system with the matrix (28) is asymptotically stable.

The same result we obtain using condition (2) of Theorem 8 to the matrix (28) for $\gamma = s^\alpha$ and $\delta = s^\beta$, where

$$\begin{aligned} \det\left[\begin{bmatrix} I_2\gamma & 0 \\ 0 & I_2\delta \end{bmatrix} - \tilde{A}\right] &= \gamma^2\delta^2 + 8\gamma^2\delta + 14\gamma^2 + 5\gamma\delta^2 \\ &+ 37.8\gamma\delta + 58.4\gamma + 4.5\delta^2 + 298.5\delta + 27.9. \end{aligned} \tag{30}$$

All coefficients of the characteristic polynomial are positive and the positive fractional system with the matrix (28) is asymptotically stable.

4 Transfer Matrices with Nonnegative Coefficients

In this section it will be shown that the transfer matrices of positive asymptotically stable different orders fractional linear systems have only nonnegative coefficients.

Theorem 10 The transfer matrix (24) of the positive asymptotically stable different fractional orders linear system (16a), (16b) has positive coefficients of the denominator and nonnegative coefficients of the numerators.

Proof Firstly applying the method by induction we shall show that the rational matrix in $\gamma = s^\alpha$ and $\delta = s^\beta$

$$\begin{bmatrix} I_{n_1}\gamma - A_{11} & -A_{12} \\ -A_{21} & I_{n_2}\delta - A_{22} \end{bmatrix}^{-1} \tag{31}$$

has positive coefficients. For $n_1 = n_2 = 1$ we obtain

$$\begin{bmatrix} \gamma - a_{11} & -a_{12} \\ -a_{21} & \delta - a_{22} \end{bmatrix}^{-1} = \frac{1}{\gamma\delta + a_{10}\gamma + a_{01}\delta + a_{00}} \begin{bmatrix} \gamma - a_{22} & a_{12} \\ a_{21} & \delta - a_{11} \end{bmatrix} \tag{32}$$

where $a_{10} = -a_{22} > 0,\ a_{01} = -a_{11} > 0,\ a_{00} = a_{11}a_{22} - a_{12}a_{21}$.

Assuming that the hypothesis is valid for the matrix $[I_{n-1}z - \tilde{A}_{n-1}]^{-1}$ for $n-1 > 2$ and $z = \begin{bmatrix} \gamma & 0 \\ 0 & \delta \end{bmatrix}$, we shall show that it is also true for the matrix $[I_n z - \tilde{A}_n]^{-1}$ with size n. It is easy to check that the inverse of the matrix

$$[I_n z - \tilde{A}_n] = \begin{bmatrix} I_{n-1}z - \tilde{A}_{n-1} & u_n \\ v_n & z + a_{nn} \end{bmatrix},\ u_n = -\begin{bmatrix} a_{1n} \\ \vdots \\ a_{n-1,n} \end{bmatrix},\ v_n = -[\,a_{n1}\ \dots\ a_{n,n-1}\,] \tag{33}$$

has the form

$$^{-1} = \begin{bmatrix} [I_{n-1}z - \tilde{A}_{n-1}]^{-1} + \frac{[I_{n-1}z-\tilde{A}_{n-1}]^{-1}u_n v_n [I_{n-1}z-\tilde{A}_{n-1}]^{-1}}{a_n} & -\frac{[I_{n-1}z-\tilde{A}_{n-1}]^{-1}u_n}{a_n} \\ -\frac{v_n[I_{n-1}z-\tilde{A}_{n-1}]^{-1}}{a_n} & \frac{1}{a_n} \end{bmatrix}, \tag{34a}$$

where

$$a_n = (z + a_{nn}) - v_n[I_{n-1}z - \tilde{A}_{n-1}]^{-1}u_n \tag{34b}$$

By assumption, the matrix $[I_{n-1}z - \tilde{A}_{n-1}]^{-1}$ has all positive coefficients and the rational function (34b) has positive coefficients. Taking into account that u_n and v_n have nonnegative entries, we conclude that $-\frac{[I_{n-1}z-\tilde{A}_{n-1}]^{-1}u_n}{a_n}$ and $-\frac{v_n[I_{n-1}z-\tilde{A}_{n-1}]^{-1}}{a_n}$ are column and row rational vectors with positive coefficients. By the same arguments the matrix

$$\frac{[I_{n-1}z - \tilde{A}_{n-1}]^{-1}u_n v_n[I_{n-1}z - \tilde{A}_{n-1}]^{-1}}{a_n} \tag{35}$$

also has all rational entries in z with positive coefficients.

Note that theorem is valid for the transfer matrix (24) if and only if the matrices $\tilde{B} \in \Re_+^{n\times m}$, $\tilde{C} \in \Re_+^{p\times n}$. This accomplishes the proof. □

5 Stability of Fractional Interval Positive Linear Systems

Consider the interval fractional positive linear continuous-time system.

$$\frac{d^\alpha x(t)}{dt^\alpha} = Ax(t), 0 < \alpha < 1 \tag{36}$$

where $x(t) \in \Re^n$ is the state vector and the matrix $A \in M_n$ is defined by.

$$\underline{A} \leq A \leq \overline{A} \text{ or } equivalently\ A \in [\underline{A}, \overline{A}] \tag{37}$$

The interval asymptotic stability of linear systems has been investigated in [19].

Definition 7 The interval fractional positive system (36) is called asymptotically stable if the system is asymptotically stable for all matrices $A \in M_n$ satisfying the condition (37).

By condition (3) of Theorem 4 the fractional positive system (36) is asymptotically stable if there exists strictly positive vector $\lambda > 0$ such that the condition (13) is satisfied.

For two fractional positive linear systems.

$$\frac{d^\alpha x_1(t)}{dt^\alpha} = \underline{A}x_1(t), \underline{A} \in M_n, 0 < \alpha < 1 \tag{38a}$$

and

$$\frac{d^\alpha x_2(t)}{dt^\alpha} = \overline{A}x_2(t), \overline{A} \in M_n, 0 < \alpha < 1 \tag{38b}$$

there exists a strictly positive vector $\lambda \in \Re_+^n$ such that.

$$\underline{A}\lambda < 0 \text{ and } \overline{A}\lambda < 0 \tag{39}$$

if and only if the systems (38a), (38b) are asymptotically stable.

Theorem 11 If the matrices $\underline{A}$ and $\overline{A}$ of fractional positive systems (38a), (38b) are asymptotically stable then their convex linear combination.

$$A = (1-q)\underline{A} + q\overline{A} \text{ for } 0 \leq q \leq 1 \tag{40}$$

is also asymptotically stable.

Proof By condition (3) of Theorem 4 if the fractional positive linear systems (38a), (38b) are asymptotically stable then there exists strictly positive vector $\lambda \in \Re_+^n$ such that.

$$\underline{A}\lambda < 0 \text{ and } \overline{A}\lambda < 0 \tag{41}$$

Using (40) and (41) we obtain

$$A\lambda = [(1-q)\underline{A} + q\overline{A}]\lambda = (1-q)\underline{A}\lambda + q\overline{A}\lambda < 0 \tag{42}$$

for $0 \le q \le 1$. Therefore, if the positive linear systems (38a), (38b) are asymptotically stable then their convex linear combination (40) is also asymptotically stable. □

Theorem 12 The interval positive system (36) is asymptotically stable if and only if the positive linear systems (38a), (38b) are asymptotically stable.

Proof By condition (3) of Theorem 4 if the matrices $\underline{A} \in M_n$, $\overline{A} \in M_n$ are asymptotically stable, then there exists a strictly positive vector $\lambda \in \Re_+^n$ such that (13) holds. The convex linear combination (40) satisfies the condition $A\lambda < 0$ if and only if (41) holds. Therefore, the interval system (36) is asymptotically stable if and only if the positive linear systems (38a), (38b) are asymptotically stable. □

Example 2 Consider the fractional interval positive linear continuous-time system (36) with the matrices.

$$\underline{A} = \begin{bmatrix} -3 & 2 \\ 2 & -4 \end{bmatrix}, \quad \overline{A} = \begin{bmatrix} -4 & 3 \\ 2 & -5 \end{bmatrix} \tag{43}$$

Using the condition (3) of Theorem 4 we choose $\lambda = [1 \quad 1]^T$ and we obtain

$$\underline{A}\lambda = \begin{bmatrix} -3 & 2 \\ 2 & -4 \end{bmatrix}\begin{bmatrix} 1 \\ 1 \end{bmatrix} = \begin{bmatrix} -1 \\ -2 \end{bmatrix} < 0 \tag{44a}$$

and

$$\overline{A}\lambda = \begin{bmatrix} -4 & 3 \\ 2 & -5 \end{bmatrix}\begin{bmatrix} 1 \\ 1 \end{bmatrix} = \begin{bmatrix} -1 \\ -3 \end{bmatrix} < 0. \tag{44b}$$

Therefore, the matrices (43) are Hurwitz.

6 Interval Stability of Positive Linear Systems

Consider the positive autonomous linear continuous-time asymptotically stable system

$$\dot{x}(t) = Ax(t), \tag{45}$$

where $x(t) \in \Re_+^n$ is the state vector and $A = [a_{ij}] \in M_n$.

Theorem 13 If the positive system (45) is asymptotically stable then the positive system with the matrix

$$A_d = DA, D = \mathrm{diag}[d_1, ..., d_n] \tag{46}$$

is also asymptotically stable if and only if all diagonal entries of D are positive, $d_k > 0$ for $k = 1, ..., n$.

Proof By assumption, the matrix A is Hurwitz and by Theorem 2 it satisfies the condition

$$A\lambda < 0, \lambda = [\lambda_1, ..., \lambda_n]^{\mathrm{T}}, \lambda_k > 0, k = 1, ..., n \tag{47a}$$

or equivalently.

$$A\lambda = \begin{bmatrix} A_1 \\ \vdots \\ A_n \end{bmatrix} \lambda = -\begin{bmatrix} c_1 \\ \vdots \\ c_n \end{bmatrix} \tag{47b}$$

for $c = [c_1, ..., c_n]^{\mathrm{T}}, c_k > 0, k = 1, ..., n$ and A_k is the kth row of A. Premultiplying (47b) by the matrix D we obtain

$$DA\lambda = \mathrm{diag}[d_1, ..., d_n] \begin{bmatrix} A_1 \\ \vdots \\ A_n \end{bmatrix} \lambda = \begin{bmatrix} A_1\lambda d_1 \\ \vdots \\ A_n\lambda d_n \end{bmatrix} = -\begin{bmatrix} c_1 \\ \vdots \\ c_n \end{bmatrix} = -Dc \tag{48a}$$

and

$$A\overline{\lambda} < 0, \overline{\lambda} = [\overline{\lambda}_1, ..., \overline{\lambda}_n]^{\mathrm{T}} = [\lambda d_1, ..., \lambda d_n]^{\mathrm{T}}, \overline{\lambda}_k > 0, k = 1, ..., n \tag{48b}$$

where $\overline{c} = Dc$ with all positive entries $\overline{c}_k > 0$, $k = 1, ..., n$.

Therefore, by Theorem 2 the positive system with the matrix (46) is asymptotically stable if and only if all diagonal entries of the matrix D are positive.

Example 3 Premultiplying the Metzler matrix

$$A = \begin{bmatrix} -3 & 1 & 1 \\ 1 & -4 & 2 \\ 1 & 2 & -4 \end{bmatrix} \quad (49)$$

by the diagonal matrix we obtain

$$DA\lambda = \begin{bmatrix} -3d_1 & d_1 & d_1 \\ d_2 & -4d_2 & 2d_2 \\ d_3 & 2d_3 & -4d_3 \end{bmatrix} \begin{bmatrix} \lambda_1 \\ \lambda_2 \\ \lambda_3 \end{bmatrix} = \begin{bmatrix} (-3\lambda_1 + \lambda_2 + \lambda_3)d_1 \\ (-4\lambda_2 + \lambda_1 + \lambda_3)d_2 \\ (-4\lambda_3 + \lambda_1 + 2\lambda_2)d_3 \end{bmatrix}. \quad (50)$$

From (50) it follows that $DA\lambda < 0$ if and only if $A\lambda < 0$ since by assumption $d_k > 0$, $k = 1, 2, 3$. Let

$$\begin{aligned} \underline{A} &= D_1 A \in M_n, \\ \overline{A} &= D_2 A \in M_n \end{aligned} \quad (51a)$$

where

$$D_1 = \text{diag}[d_{11}, ..., d_{1n}], D_2 = \text{diag}[d_{21}, ..., d_{2n}], d_{1k} < d_{2k}, k = 1, ..., n. \quad (51b)$$

From Theorem 12 it follows that $\underline{A}$ and $\overline{A}$ are Hurwitz Metzler matrices. Therefore by Theorem 12 we have the following important Theorem.

Theorem 14 The positive interval system with (51a) is asymptotically stable if and only if the system (45) is asymptotically stable.

Now let us consider more general case in which the diagonal matrix D will be substituted by the matrix $P \in \Re_+^{n \times n}$ with the same sum of elements of each row (column) satisfying the equality.

$$PA = AP \quad (52)$$

Using the Kronecker product of two matrices $P = [p_{ij}]$ and $A = [a_{ij}]$ defined by [23]

$$P \otimes A = \begin{bmatrix} p_{11}A & p_{12}A & ... & p_{1n}A \\ p_{21}A & p_{22}A & ... & p_{2n}A \\ ... & ... & ... & ... \\ p_{n1}A & p_{n2}A & ... & p_{nn}A \end{bmatrix} \quad (53)$$

we may write (52) in the form

$$\hat{A}x = 0, \tag{54}$$

where $\hat{A} = A \otimes I_n - I_n \otimes A^{\mathrm{T}}$, $x = [x_1, ..., x_n]^{\mathrm{T}} \in \Re^{n^2}$ and $x_k \in \Re^n$ is the k-th row of the matrix P. The Eq. (54) has nonzero solution if and only if [23]

$$\det \hat{A} = 0. \tag{55}$$

In this case the Eq. (54) has many solutions. Therefore, the following theorem has been proved.

Theorem 15 There exist many matrices P satisfying (52) if and only if (55) holds.

Example 4 Find the matrix

$$P = \begin{bmatrix} p_1 & p_2 \\ p_3 & p_4 \end{bmatrix} \tag{56}$$

satisfying the equality (52) for the given matrix

$$A = \begin{bmatrix} -2 & 1 \\ 2 & -3 \end{bmatrix}. \tag{57}$$

In this case the Eq. (54) has the form

$$\hat{A}x(t) = \begin{bmatrix} 0 & -2 & 1 & 0 \\ -1 & 1 & 0 & 1 \\ 2 & 0 & -1 & -2 \\ 0 & 2 & -1 & 0 \end{bmatrix} \begin{bmatrix} p_1 \\ p_2 \\ p_3 \\ p_4 \end{bmatrix} = \begin{bmatrix} 0 \\ 0 \\ 0 \\ 0 \end{bmatrix} \tag{58}$$

and the condition (55) is satisfied since rank $\hat{A} = 2$. From the first two linearly independent rows of (58) we have

$$\begin{bmatrix} 1 & 0 \\ 0 & 1 \end{bmatrix} \begin{bmatrix} p_3 \\ p_4 \end{bmatrix} = \begin{bmatrix} 0 & 2 \\ 1 & -1 \end{bmatrix} \begin{bmatrix} p_1 \\ p_2 \end{bmatrix} \tag{59}$$

and $p_3 = 2p_2$, $p_4 = p_1 - p_2$ for any nonnegative p_1 and p_2. If we choose $p_1 = 3$, $p_2 = 2$ then we obtain

$$P = \begin{bmatrix} 3 & 2 \\ 4 & 1 \end{bmatrix} \tag{60a}$$

and for $p_1 = 2$, $p_2 = 1$ we have.

$$P = \begin{bmatrix} 2 & 1 \\ 2 & 1 \end{bmatrix} \tag{60b}$$

Theorem 16 If λ_i, $i = 1, ..., n$ are the eigenvalues of the matrix $A \in \Re^{n\times n}$ and μ_j, $j = 1, ..., m$ are the eigenvalues of the matrix $B \in \Re^{n\times n}$ then $\lambda_i + \mu_j$ are the eigenvalues of the matrix $A \otimes I_n + I_n \otimes B$.

Proof is given in [3].

In the particular case for $B = -A$ we obtain the following conclusions.

Conclusion 1 If λ_i, $i = 1, ..., n$ are the eigenvalues of the Hurwitz Metzler matrix $A \in M_n$ then the eigenvalues of the matrix $\hat{A} = A \otimes I_n - I_n \otimes A^{\mathrm{T}}$ are $\lambda_i - \lambda_j$ for $i, j = 1, ..., n$.

Conclusion 2 The matrix $\hat{A}$ is singular, that is $\det \hat{A} = 0$.

Example 5 The eigenvalues of the Hurwitz Metzler matrix (57) are $\lambda_1 = -1$, $\lambda_2 = -4$ since

$$\det[I_2\lambda - A] = \begin{vmatrix} \lambda + 2 & -1 \\ -2 & \lambda + 3 \end{vmatrix} = \lambda^2 + 5\lambda + 4. \tag{61}$$

In this case the matrix

$$\hat{A} = A \otimes I_2 - I_2 \otimes A^{\mathrm{T}} = \begin{bmatrix} 0 & -2 & 1 & 0 \\ -1 & 1 & 0 & 1 \\ 2 & 0 & -1 & -2 \\ 0 & 2 & -1 & 0 \end{bmatrix} \tag{62}$$

has the characteristic polynomial

$$\det[I_4\lambda - \hat{A}] = \begin{vmatrix} \lambda & 2 & -1 & 0 \\ 1 & \lambda - 1 & 0 & -1 \\ -2 & 0 & \lambda + 1 & 2 \\ 0 & -2 & 1 & \lambda \end{vmatrix} = \lambda^2(\lambda + 3)(\lambda - 3) \tag{63}$$

and according to Conclusion 1, zeros of the matrix (62) are the following: $\hat{\lambda}_1 = \lambda_1 - \lambda_1 = 0$, $\hat{\lambda}_2 = \lambda_1 - \lambda_2 = 3$, $\hat{\lambda}_3 = \lambda_2 - \lambda_1 = -3$, $\hat{\lambda}_4 = \lambda_2 - \lambda_2 = 0$.

From the above considerations follows the Theorem.

Theorem 17 If A is a Hurwitz Metzler matrix then the matrix $\hat{A}$ contains a nonsingular submatrix $A_r \in M_r$, $r = n(n - 1)$, $B_r \in \Re_+^n$ such that

$$[\,A_r \;\; B_r\,]\overline{x} = 0, \tag{64}$$

where $\overline{x} = [\overline{x}_r^T, \; \overline{x}_n^T]^T$, $\overline{x}_r \in \Re_+^r$, $\overline{x}_n \in \Re_+^n$ and

$$\overline{x}_r = -A_r^{-1} B_r \overline{x}_n. \tag{65}$$

Proof From (64) we have

$$A_r \overline{x}_r = -B_r \overline{x}_n \tag{66}$$

and

$$\overline{x}_r = -A_r^{-1} B_r \overline{x}_n \tag{67}$$

since $-A_r^{-1} \in \Re_+^{r \times r}$ and $B_r \in \Re_+^{r \times n}$. □

The desired matrix P can be computed by the use of the following procedure.

Procedure 1.

Step 1. Knowing the matrix A compute the matrix $\hat{A}$ and find its rank r ($r = rank\ \hat{A}$).

Step 2. Choose the nonsingular matrix $A_r \in M_r$ and $B_r \in \Re_+^{r \times n}$. of the matrix $\hat{A}$.

Step 3. Choose the components of the vector $\overline{x}_n$ (entries of the matrix P corresponding to the matrix B_r).

Step 4. Using (67) compute the vector $\overline{x}_r$ for the given A_r and B_r.

Step 5. Knowing $\overline{x}_n$ and $\overline{x}_r$ find the matrix P.

Example 6 Compute the matrix P satisfying (52) for the matrix

$$A = \begin{bmatrix} -3 & 1 \\ 2 & -4 \end{bmatrix}. \tag{68}$$

Using Procedure 1 and (68) we obtain.

Step 1. The matrix $\hat{A}$ has the form

$$\hat{A} = A \otimes I_2 - I_2 \otimes A^{\mathrm{T}} = \begin{bmatrix} 0 & -2 & 1 & 0 \\ -1 & 1 & 0 & 1 \\ 2 & 0 & -1 & -2 \\ 0 & 2 & -1 & 0 \end{bmatrix} \tag{69}$$

and its rank is $r = 2$.

Step 2. We choose the matrices A_r and B_r of $\hat{A}$ in the forms

$$A_r = \begin{bmatrix} -2 & 1 \\ 0 & -1 \end{bmatrix},$$

$$B_r = \begin{bmatrix} 0 & 0 \\ 2 & -2 \end{bmatrix}. \tag{70}$$

Step 3. In this case we choose

$$\overline{x}_n = \begin{bmatrix} p_1 \\ p_4 \end{bmatrix}, \ \overline{x}_r = \begin{bmatrix} p_2 \\ p_3 \end{bmatrix}. \tag{71}$$

Step 4. Using (65) and (70) we obtain

$$\begin{bmatrix} p_2 \\ p_3 \end{bmatrix} = -A_r^{-1} B_r \overline{x}_{n-r} = -\begin{bmatrix} -2 & 1 \\ 0 & -1 \end{bmatrix}^{-1} \begin{bmatrix} 0 & 0 \\ 2 & 2 \end{bmatrix} \begin{bmatrix} p_1 \\ p_4 \end{bmatrix} = \begin{bmatrix} 1 & -1 \\ 2 & -2 \end{bmatrix} \begin{bmatrix} p_1 \\ p_4 \end{bmatrix} \tag{72}$$

and for $[p_1 \ \ p_4]^T = [2 \ 1]$ we obtain

$$\begin{bmatrix} p_2 \\ p_3 \end{bmatrix} = \begin{bmatrix} 1 & -1 \\ 2 & -2 \end{bmatrix} \begin{bmatrix} 2 \\ 1 \end{bmatrix} = \begin{bmatrix} 1 \\ 2 \end{bmatrix}. \tag{73}$$

Step 5. In this case the desired matrix P has the form

$$P = \begin{bmatrix} p_1 & p_2 \\ p_3 & p_4 \end{bmatrix} = \begin{bmatrix} 2 & 1 \\ 2 & 1 \end{bmatrix}. \tag{74}$$

Let

$$\underline{A} = P_1 A,$$

$$\overline{A} = P_2 A, \tag{75}$$

where $P_1 \in \Re_+^{n\times n}$, $P_2 \in \Re_+^{n\times n}$ satisfying the condition

$$P_2 > P_1. \tag{76}$$

Note that $\underline{A}$ and $\overline{A}$ are Hurwitz Metzler matrices and we have the following important Theorem.

Theorem 18 The positive interval system with the matrices (75) is asymptotically stable if and only if the system (45) is asymptotically stable.

Example 7 Continuation of Example 6.

For the positive system with the state matrix (68) the entries of the matrix P are related by (72). For $p_{11} = 0.2$, $p_{14} = 0.1$ we obtain

$$P_1 = \begin{bmatrix} 0.2 & 0.1 \\ 0.2 & 0.1 \end{bmatrix} \tag{77}$$

and for $p_{21} = 20$, $p_{24} = 10$ we have

$$P_2 = \begin{bmatrix} 20 & 10 \\ 20 & 10 \end{bmatrix}. \tag{78}$$

In this case

$$\underline{A} = P_1 A = \begin{bmatrix} -0.4 & -0.2 \\ -0.4 & -0.2 \end{bmatrix},$$
$$\overline{A} = P_2 A = \begin{bmatrix} -40 & -20 \\ -40 & -20 \end{bmatrix}. \tag{79}$$

Therefore by Theorem 18, the positive interval system with the matrices (79) is not asymptotically stable for all matrices A satisfying $\underline{A} \leq A \leq \overline{A}$. Note that the system is asymptotically stable for all diagonal matrix P with positive diagonal entries.

7 Adjoint Matrix of the Singular Metzler Matrix

In this section some properties of singular Metzler matrices with the sum of entries of each row (column) equal to zero will be considered.

Theorem 19 If the Metzler matrix $A = [a_{ij}] \in M_n$ satisfies the assumption

$$\sum_{j=1}^{n} a_{ij} = 0,$$
$$i = 1, ..., n,$$
$$\sum_{i=1}^{n} a_{ij} = 0,$$
$$j = 1, ..., n,$$
$$\begin{cases} a_{ij} < 0 & \text{for} \quad i = j \\ a_{ij} \geq 0 & \text{for} \quad i \neq j \end{cases} \tag{80}$$

then all entries d_{ij} of its adjoint matrix $A_{ad} = [d_{ij}]$ have the same value i.e. $d_{ij} = c$ for $i, j = 1, ..., n$.

Proof From assumption (80) we have

$$A\mathbf{1}_n = 0 \tag{81}$$

where $\mathbf{1}_n = [1, ..., 1]^T \in \Re^n$. Taking into account the first $n - 1$ rows of the matrix A we obtain.

$$\begin{bmatrix} a_{11} & a_{12} & \dots & a_{1,n-1} \\ a_{21} & a_{22} & \dots & a_{2,n-1} \\ \dots & \dots & \dots & \dots \\ a_{n-1,1} & a_{n-1,2} & \dots & a_{n-1,n-1} \end{bmatrix} \begin{bmatrix} 1 \\ 1 \\ \dots \\ 1 \end{bmatrix} = - \begin{bmatrix} a_{1n} \\ a_{2n} \\ \dots \\ a_{n-1,n} \end{bmatrix} \tag{82}$$

Using the well-known Cramer formula to (82) we obtain.

$$d_{11} = d_{1j} = c \text{ for } j = 2, ..., n \tag{83}$$

In a similar way, taking into account $n - 1$ for any others rows of the matrix A, we may prove the thesis in general case. □

Example 8 The matrix

$$A = \begin{bmatrix} -3 & 2 & 1 \\ 1 & -4 & 3 \\ 2 & 2 & -4 \end{bmatrix} \tag{84}$$

satisfies the assumption of the Theorem 19 since the sum of all entries of each row (column) is equal zero. Taking into account the first two rows of (84) we obtain

$$\begin{bmatrix} -3 & 2 & 1 \\ 1 & -4 & 3 \end{bmatrix} \begin{bmatrix} 1 \\ 1 \\ 1 \end{bmatrix} = \begin{bmatrix} 0 \\ 0 \end{bmatrix} \tag{85a}$$

or equivalently

$$\begin{bmatrix} -3 & 2 \\ 1 & -4 \end{bmatrix} \begin{bmatrix} 1 \\ 1 \end{bmatrix} = - \begin{bmatrix} 1 \\ 3 \end{bmatrix}. \tag{85b}$$

From (85b) we have

$$\det \begin{bmatrix} -3 & 2 \\ 1 & -4 \end{bmatrix} = \det \begin{bmatrix} -1 & 2 \\ -3 & -4 \end{bmatrix} = \det \begin{bmatrix} -3 & -1 \\ 1 & -3 \end{bmatrix} = 10. \tag{86}$$

Taking into account the first and third rows of (84) we obtain

$$\begin{bmatrix} -3 & 2 \\ 2 & 2 \end{bmatrix}\begin{bmatrix} 1 \\ 1 \end{bmatrix} = \begin{bmatrix} -1 \\ 4 \end{bmatrix} \tag{87}$$

and

$$\det\begin{bmatrix} -3 & 2 \\ 2 & 2 \end{bmatrix} = \det\begin{bmatrix} -1 & 2 \\ 4 & 2 \end{bmatrix} = \det\begin{bmatrix} -3 & -1 \\ 2 & 4 \end{bmatrix} = 10. \tag{88}$$

Taking into account the second and third rows of (84) we obtain

$$\begin{bmatrix} 1 & -4 \\ 2 & 2 \end{bmatrix}\begin{bmatrix} 1 \\ 1 \end{bmatrix} = \begin{bmatrix} -3 \\ 4 \end{bmatrix} \tag{89}$$

and

$$\det\begin{bmatrix} 1 & -4 \\ 2 & 2 \end{bmatrix} = \det\begin{bmatrix} -3 & -4 \\ 4 & 2 \end{bmatrix} = \det\begin{bmatrix} 1 & -3 \\ 2 & 4 \end{bmatrix} = 10. \tag{90}$$

This confirms the Theorem 19. Consider the state matrix

$$\begin{aligned} A &= [a_{ij}] \in \Re_+^{n\times n}, \\ &\sum_{j=1}^{n} a_{ij} = 1, \\ &\sum_{i=1}^{n} a_{ij} = 1, \end{aligned} \tag{91}$$

of positive discrete-time linear system with the sum of each row (column) equal 1. Note that the matrix

$$\overline{A} = A - I_n = [\overline{a}_{ij}] \in \Re^{n\times n} \tag{92}$$

satisfies the assumptions of Theorem 19. Therefore for the state matrix (92) of the positive discrete-time linear system, we have the following Theorem.

Theorem 20 All entries of the adjoint matrix of the matrix (92) have the same value.

Example 9 The sum of entries of each row (column) of the matrix

$$A = \begin{bmatrix} 0.5 & 0.25 & 0.25 \\ 0 & 0.7 & 0.3 \\ 0.5 & 0.05 & 0.45 \end{bmatrix} \tag{93}$$

is equal 1 and the sum of entries of each row (column) of the matrix

$$\overline{A} = A - I_n = \begin{bmatrix} -0.5 & 0.25 & 0.25 \\ 0 & -0.3 & 0.3 \\ 0.5 & 0.05 & -0.55 \end{bmatrix} \tag{94}$$

is equal to zero. Therefore by Theorem 20, all entries of the adjoint matrix have the same value equal to $c = 0.15$, i.e.

$$\overline{A}_{ad} = \begin{bmatrix} 0.15 & 0.15 & 0.15 \\ 0.15 & 0.15 & 0.15 \\ 0.15 & 0.15 & 0.15 \end{bmatrix}. \tag{95}$$

8 Concluding Remarks

Some specific properties of positive standard and fractional linear systems with interval state matrices have been considered. The stability, positivity and transfer matrices of positive different orders fractional continuous-time linear systems have been presented. New necessary and sufficient conditions for the asymptotic stability of positive different orders fractional linear systems have been established. It has been shown that the transfer matrices of positive asymptotically stable different orders fractional linear systems have only nonnegative coefficients. New conditions for the interval stability of positive standard and fractional linear systems have been established. At the end, it has been shown that the adjoint matrix of a singular Metzler matrix with zero sum of entries of each row (column) has all equal entries.

Acknowledgements This work was supported by National Science Centre in Poland under work No. 2017/27/B/ST7/02443.

References

1. Benvenuti, L., Farina, L.: A tutorial on the positive realization problem. IEEE Trans. Autom. Control **49**(5), 651–664 (2004)
2. Berman, A., Plemmons R.J.: Nonnegative matrices in the mathematical sciences. SIAM (1994)
3. Białas, S.: Matrices. Publisher AGH, Kraków (2006).(in Polish)
4. Busłowicz, M.: Robust stability of positive discrete-time interval systems with time-delays. Bull. Pol. Acad.: Tech. **52**(2), 99–102 (2004)
5. Busłowicz, M.: Stability analysis of continuous-time linear systems consisting of n subsystems with different fractional orders. Bull. Pol. Acad.: Tech. **60**(2), 279–284 (2012)
6. Busłowicz, M., Kaczorek, T.: Simple conditions for practical stability of positive fractional discrete-time linear systems. Int. J. Appl. Math. Comput. Sci. **19**(2), 263–69 (2009)

7. Farina, L., Rinaldi, S.: Positive Linear Systems. Theory and Applications. Wiley, New York (2000)
8. Kaczorek, T.: Absolute stability of a class of fractional positive nonlinear systems. Int. J. Appl. Math. Comput. Sci. **29**(1), 93–98 (2019)
9. Kaczorek, T.: Analysis of positivity and stability of discrete-time and continuous-time nonlinear systems. Comput. Probl. Electr. Eng. **5**(1), 11–16 (2015)
10. Kaczorek, T.: Analysis of positivity and stability of fractional discrete-time nonlinear systems. Bull. Pol. Acad.: Tech. **64**(3), 491–494 (2016)
11. Kaczorek, T.: Existence and determination of the set of Metzler matrices for given stable polynomials. Int. J. Appl. Math. Comput. Sci. **22**(2), 389–399 (2012)
12. Kaczorek, T.: Linear Control Systems: Analysis of Multivariable Systems. Wiley, New York (1992)
13. Kaczorek, T.: Positive 1D and 2D Systems. Springer, London (2002)
14. Kaczorek, T.: Positive linear systems with different fractional orders. Bull. Pol. Acad.: Tech. **58**(3), 453–458 (2010)
15. Kaczorek, T.: Positive linear systems consisting of n subsystems with different fractional orders. IEEE Trans. Circuits Syst. **58**(7), 1203–1210 (2011)
16. Kaczorek, T.: Selected Problems of Fractional Systems Theory. Springer, Berlin (2011)
17. Kaczorek, T.: Stability of fractional positive nonlinear systems. Arch. Control Sci. **25**(4), 491–496 (2015)
18. Kaczorek, T.: Stability of interval positive continuous-time linear systems. Bull. Pol. Acad.: Tech. **66**(1), 31–35 (2018)
19. Kaczorek, T.: Stability of interval positive fractional continuous time linear systems. In: European Conference on Electrical Engineering and Computer Science (EECS), Bern, pp. 314–317 (2017)
20. Kaczorek, T.: Stability of interval positive fractional discrete–time linear systems. Int. J. Appl. Math. Comput. Sci. **28**(3), 451–456 (2018)
21. Kaczorek, T.: Stability of interval positive of integer and fractional orders continuous-time linear systems. In: 19th International Conference Computational Problems of Electrical Engineering, Banska Stiavnica, pp. 1–6 (2018)
22. Kaczorek, T.: Superstabilization of positive linear electrical circuit by state-feedbacks. Bull. Pol. Acad.: Tech. **65**(5), 703–708 (2017)
23. Kaczorek, T.: Vector and matrices. PWN, Warsaw (2008).(in Polish)
24. Kaczorek, T., Borawski, K.: Stability of positive nonlinear systems. In: 22nd International Conference on Methods and Models in Automation and Robotics, Międzyzdroje, Poland (2017)
25. Kaczorek, T., Rogowski, K.: Fractional Linear Systems and Electrical Circuits. Springer, Cham (2015)
26. Kharitonov, V.L.: Asymptotic stability of an equilibrium position of a family of systems of differential equations. Differentsialnye urawnienia **14**, 2086–2288 (1978)
27. Li, TT., Tung, SL., Juang, YT.: Stabilization of positive continuous-time interval systems. In: Zhu M. (eds.) Electrical Engineering and Control. Lecture Notes in Electrical Engineering, vol 98, pp. 335–342. Springer, Berlin (2011).
28. Mitkowski, W.: Dynamical properties of Metzler systems. Bull. Pol. Acad.: Tech. **56**(4), 309–312 (2008)
29. Oldham, K.B., Spanier, J.: The Fractional Calculus. Academic Press, New York (1974)
30. Ostalczyk, P.: Discrete Fractional Calculus: Selected Applications in Control and Image Processing. Series in Computer Vision, vol. 4 (2016)
31. Ostalczyk, P.: Epitome of the fractional calculus: Theory and its applications in automatics, Wydawnictwo Politechniki Łódzkiej, Łódź (2008)
32. Podlubny, I.: Fractional Differential Equations. Academic Press, San Diego (1999)
33. Radwan, A.G., Soliman, A.M., Elwakil, A.S., Sedeek, A.: On the stability of linear systems with fractional-order elements. Chaos. Solitones Fractals **40**(5), 2317–2328 (2009)
34. Ruszewski, A.: Stability of discrete-time fractional linear systems with delays. Arch. Control Sci. **29**(3), 549–567 (2019)

35. Sajewski, Ł.: Decentralized stabilization of descriptor fractional positive continuous-time linear systems with delays. In: 22nd International Conference on Methods and Models in Automation and Robotics, Międzyzdroje, Poland, pp. 482–487 (2017)
36. Sajewski, Ł.: Stabilization of positive descriptor fractional discrete-time linear systems with two different fractional orders by decentralized controller, Bull. Pol. Acad.: Tech. **65**(5), 709–714 (2017)
37. Shu, Z., Lam, J., Gao, H., Du, B., Wu, L.: Positive Observers and Dynamic Output-Feedback Controllers for Interval Positive Linear Systems. IEEE Trans. Circuits Syst. I, Regul. **55**(10), 3209–3222 (2008)

Stability and Controllability

Global Stability of Nonlinear Fractional Dynamical Systems

Tadeusz Kaczorek

Abstract New sufficient conditions for the global stability of different classes of nonlinear fractional feedback systems are presented. The linear parts of the systems are positive systems with interval state matrices. The nonlinear parts are described by static nonlinear characteristics located in the first and third quarter of the plane. The feedbacks are described in general case by matrices with positive entries. The sufficient conditions for the global stability are given for the following classes of the nonlinear systems: Positive interval continuous-time feedback nonlinear systems; Fractional positive interval continuous-time feedback nonlinear systems; Positive interval discrete-time feedback nonlinear systems; Descriptor nonlinear feedback discrete-time systems and Positive nonlinear electrical circuits. Procedures are given for calculations of gain matrices of the characteristics of nonlinear elements of the systems. The effectiveness of the procedures are demonstrated on numerical examples of nonlinear systems.

1 Introduction

In positive systems inputs, state variables and outputs take only nonnegative values for any nonnegative inputs and nonnegative initial conditions [1, 4, 13]. Examples of positive systems are industrial processes involving chemical reactors, heat exchangers and distillation columns, storage systems, compartmental systems and water and atmospheric pollution models. A variety of models having positive behavior can be found in engineering, management science, economics, social sciences, biology and medicine, etc. An overview of state of the art positive systems theory is given in the monographs [1, 4, 13, 22, 29, 30].

The mathematical fundamentals of fractional calculus are given in the monographs [13, 22, 29, 30]. Positive fractional linear systems have been investigated in [3, 5, 7,

T. Kaczorek (✉)
Faculty of Electrical Engineering, Białystok University of Technology Bialystok, Wiejska 45D, 15-351 Białystok, Poland
e-mail: t.kaczorek@pb.edu.pl

P. Kulczycki et al. (eds.), *Fractional Dynamical Systems: Methods, Algorithms and Applications*, Studies in Systems, Decision and Control 402,
https://doi.org/10.1007/978-3-030-89972-1_10

9, 14–19, 22–25, 29, 33–35]. Positive linear systems with different fractional orders have been addressed in [14, 15, 35]. Descriptor positive systems have been analyzed in [2, 11, 12] and their stabilization in [34, 35]. Linear positive electrical circuits have been addressed in [22]. The superstabilization of positive linear electrical circuits by state feedbacks have been analyzed in [20]. The global stability of nonlinear systems with negative feedbacks and positive asymptotically stable linear parts has been investigated in [6, 8]. The global stability of nonlinear standard and fractional positive feedback systems has been considered in [8–10, 21, 26, 32].

In this paper the global stability of nonlinear fractional orders feedback multi-input multi-output systems with interval matrices of positive linear parts will be addressed.

In this chapter new sufficient conditions for the global stability of different classes of nonlinear fractional feedback systems are presented. The linear parts of the systems are positive systems with interval state matrices. The nonlinear parts are described by static nonlinear characteristics located in the first and third quarter of the plane. The feedbacks are described in general case by matrices with positive entries. The chapter is organized as follows. In Sect. 2 the basic definitions and theorems concerning positive fractional linear continuous-time systems are recalled. The basic definitions and theorems concerning fractional continuous-time linear systems are given in Sect. 3 and for fractional discrete-time systems in Sect. 4. Descriptor discrete-time positive linear systems are introduced in Sect. 5. The stability of fractional positive interval linear systems is analyzed in Sect. 6. New sufficient conditions for the global stability of the feedback nonlinear systems with positive linear parts are established in Sect. 7. In Sect. 8 the global stability of fractional feedback systems are analyzed. The new sufficient conditions for global stability of nonlinear feedback discrete-time systems are given in Sect. 9 and for the descriptor nonlinear feedback systems in Sect. 10. The new stability conditions are applied to positive nonlinear systems in Sect. 11. The summary of the results is given in Sect. 12.

The following notation will be used: $\Re$—the set of real numbers, $\Re^{n\times m}$—the set of $n \times m$ real matrices, $\Re_+^{n\times m}$—the set of $n \times m$ real matrices with nonnegative entries and $\Re_+^n = \Re_+^{n\times 1}$, M_n—the set of $n \times n$ Metzler matrices (real matrices with nonnegative off-diagonal entries), I_n—the $n \times n$ identity matrix.

2 Positive Continuous-Time Linear Systems

Consider the continuous-time linear system

$$\dot{x} = Ax + Bu, \tag{1a}$$

$$y = Cx, \tag{1b}$$

where $x = x(t) \in \Re^n$, $u = u(t) \in \Re^m$, $y = y(t) \in \Re^p$ are the state, input and output vectors and $A \in \Re^{n\times n}$, $B \in \Re^{n\times m}$, $C \in \Re^{p\times n}$.

Definition 1 [4, 13, 17, 22] The continuous-time linear system (1a), (1b) is called (internally) positive if $x(t) \in \Re_+^n$, $y(t) \in \Re_+^p$, $t \geq 0$ for any initial conditions $x(0) \in \Re_+^n$ and all inputs $u(t) \in \Re_+^m$, $t \geq 0$.

Theorem 1 [4, 13, 17, 22] The continuous-time linear system (1a), (1b) is positive if and only if

$$A \in M_n,\ B \in \Re_+^{n\times m},\ C \in \Re_+^{p\times n}. \tag{2}$$

Definition 2 [4, 13, 17, 22] The positive continuous-time system (1a), (1b) for $u(t) = 0$ is called asymptotically stable if

$$\lim_{t\to\infty} x(t) = 0 \text{ for any } x(0) \in \Re_+^n. \tag{3}$$

Theorem 2 [17, 22] The positive continuous-time linear system (1a), (1b) for $u(t) = 0$ is asymptotically stable if and only if one of the following equivalent conditions is satisfied:

(1) All coefficient of the characteristic polynomial

$$p_n(s) = \det[I_n s - A] = s^n + a_{n-1}s^{n-1} + \cdots + a_1 s + a_0 \tag{4}$$

are positive, i.e. $a_i > 0$ for $i = 0, 1, \ldots, n-1$.

(2) There exists strictly positive vector $\lambda^T = [\lambda_1 \cdots \lambda_n]^T$, $\lambda_k > 0$, $k = 1, ..., n$ such that

$$A\lambda < 0 \text{ or } \lambda^T A < 0. \tag{5}$$

If the matrix A is nonsingular then we can choose $\lambda = A^{-1}c$, where $c \in \Re^n$ is strictly positive.

Theorem 3 The positive system (1a), (1b) is asymptotically stable if the sum of entries of each column (row) of the matrix A is negative.

Proof Using (5) we obtain

$$A\lambda = \begin{bmatrix} a_{11} & \ldots & a_{1n} \\ \vdots & \ldots & \vdots \\ a_{n1} & \ldots & a_{nn} \end{bmatrix} \begin{bmatrix} \lambda_1 \\ \vdots \\ \lambda_n \end{bmatrix} = \begin{bmatrix} a_{11} \\ \vdots \\ a_{1n} \end{bmatrix} \lambda_1 + \cdots + \begin{bmatrix} a_{n1} \\ \vdots \\ a_{nn} \end{bmatrix} \lambda_n < \begin{bmatrix} 0 \\ \vdots \\ 0 \end{bmatrix} \tag{6}$$

and the sum of entries of each column of the matrix A is negative since $\lambda_k > 0$, $k = 1, \ldots, n$. The proof for rows is similar. □

Consider the positive asymptotically stable system (1a), (1b).

Lemma 1 If the matrix

$$A_n = \begin{bmatrix} -a_{11} & a_{12} & \dots & a_{1n} \\ a_{21} & -a_{22} & \dots & a_{2n} \\ \vdots & \vdots & \ddots & \vdots \\ a_{n1} & a_{n2} & \dots & -a_{nn} \end{bmatrix} \in M_n,\ a_{ij} > 0,\ i, j = 1, \dots, n \tag{7}$$

is asymptotically stable (Hurwitz) then the rational matrix in s

$$P_{A_n}(s) = [I_n s - A_n]^{-1} \in \Re^{n \times n}(s) \tag{8}$$

has positive coefficients.

Proof is given in [12].

Theorem 4 If the matrix $A \in M_n$ is Hurwitz and $B \in \Re_+^{n \times m}$, $C \in \Re_+^{p \times n}$, $D \in \Re_+^{p \times m}$ then all coefficients of the transfer matrix

$$T(s) = C[I_n s - A]^{-1} B + D \tag{9}$$

of the positive system (1) are positive.

Proof By Lemma 1 if the matrix $A \in M_n$ is Hurwitz then the inverse matrix (8) has all nonnegative coefficients. Therefore, if $B \in \Re_+^{n \times m}$, $C \in \Re_+^{p \times n}$ and $D \in \Re_+^{p \times m}$ then all coefficients of the matrix (9) are positive. □

3 Fractional Positive Continuous-Time Linear Systems

The following Caputo definition of the fractional derivative of α order will be used [17, 22, 30, 31]

$${}_0D_t^\alpha f(t) = \frac{d^\alpha f(t)}{dt^\alpha} = \frac{1}{\Gamma(1-\alpha)} \int_0^t \frac{\dot{f}(\tau)}{(t-\tau)^\alpha} d\tau,\ 0 < \alpha < 1, \tag{10}$$

where $\dot{f}(\tau) = \frac{df(\tau)}{d\tau}$ and $\Gamma(z) = \int_0^\infty t^{x-1} e^{-t} dt$, $\mathrm{Re}(x) > 0$ is the Euler gamma function.

Consider the fractional continuous-time linear system

$$\frac{d^\alpha x(t)}{dt^\alpha} = Ax(t) + Bu(t), \tag{11a}$$

$$y(t) = Cx(t), \tag{11b}$$

where $x(t) \in \Re^n$, $u(t) \in \Re^m$, $y(t) \in \Re^p$ are the state, input and output vectors and $A \in \Re^{n\times n}$, $B \in \Re^{n\times m}$, $C \in \Re^{p\times n}$.

Definition 3 [13, 18] The fractional system (11a), (11b) is called (internally) positive if $x(t) \in \Re_+^n$ and $y(t) \in \Re_+^p$, $t \geq 0$ for any initial conditions $x(0) \in \Re_+^n$ and all inputs $u(t) \in \Re_+^m$, $t \geq 0$.

Theorem 5 [17, 22] The fractional system (11a), (11b) is positive if and only if

$$A \in M_n,\ B \in \Re_+^{n\times m},\ C \in \Re_+^{p\times n}. \tag{12}$$

The fractional positive linear system (11a), (11b) is called asymptotically stable (and the matrix A Hurwitz) if

$$\lim_{t\to\infty} x(t) = 0 \text{ for all } x(0) \in \Re_+^n. \tag{13}$$

The positive fractional system (11a), (11b) is asymptotically stable if and only if the real parts of all eigenvalues s_k of the matrix A are negative, i.e. Re $s_k < 0$ for $k = 1, \ldots, n$ [13, 22, 23].

Theorem 6 [17, 22] The positive fractional system (11a), (11b) is asymptotically stable if and only if one of the following equivalent conditions is satisfied:

(1) All coefficients of the characteristic polynomial

$$\det[I_n s - A] = s^n + a_{n-1}s^{n-1} + \cdots + a_1 s + a_0 \tag{14}$$

are positive, i.e. $a_i > 0$ for $i = 0, 1, \ldots, n-1$.

(2) There exists strictly positive vector $\lambda = [\lambda_1 \cdots \lambda_n]$, $\lambda_k > 0$, $k = 1, \ldots, n$ such that

$$A\lambda < 0 \text{ or } \lambda^T A < 0. \tag{15}$$

The transfer matrix of the system (11a), (11b) is given by

$$T(s^\alpha) = C[I_n s^\alpha - A]^{-1} B. \tag{16}$$

Theorem 7 If the matrix $A \in M_n$ is Hurwitz and $B \in \Re_+^{n\times m}$, $C \in \Re_+^{p\times n}$ of the linear positive system (11), then all coefficients of the transfer matrix (16) are positive.

Proof is similar to the proof given in for the standard positive linear systems.

4 Fractional Positive Discrete-Time Linear Systems

Consider the autonomous fractional discrete-time linear system

$$\Delta^{\alpha} x_{i+1} = Ax_i, 0 < \alpha < 1, i \in Z_+, \tag{17a}$$

where

$$\Delta^{\alpha} x_i = \sum_{j=1}^{i} c_j x_{i-j}, \tag{17b}$$

$$c_j = (-1)^j \binom{\alpha}{j}, \binom{\alpha}{j} = \begin{cases} 1 & \text{for } j = 0 \\ \frac{\alpha(\alpha-1)\dots(\alpha-j+1)}{j!} & \text{for } j = 1, 2, \dots \end{cases} \tag{17c}$$

is the fractional α-order difference of x_i and $x_i \in \Re^n$, $u_i \in \Re^m$ are the state and input vectors and $A \in \Re^{n \times n}$. Substitution of (17b) into (17a) yields

$$x_{i+1} = A_{\alpha} x_i - \sum_{j=2}^{i+1} c_j x_{i-j+1}, i \in Z_+, \tag{18a}$$

where

$$A_{\alpha} = A + I_n \alpha. \tag{18b}$$

Lemma 2 If $0 < \alpha < 1$ then

$$(1) \quad -c_j > 0 \text{ for } j = 1, 2, \dots \tag{19a}$$

$$(2) \quad \sum_{j=1}^{n} c_j = -1. \tag{19b}$$

Proof is given in [17].

Definition 4 [17, 22] The fractional system (17a) is called (internally) positive if $x_i \in \Re_+^n$, $i \in Z_+$ for any initial conditions $x_0 \in \Re_+^n$.

Theorem 8 [17, 22] The fractional system (17a) is positive if and only if

$$A_{\alpha} \in \Re_+^{n \times n}. \tag{20}$$

Proof is given in [17, 22].

Definition 5 The fractional positive system (17a) is called asymptotically stable if

$$\lim_{i\to\infty} x_i = 0 \text{ for all } x_0 \in \Re_+^n. \tag{21}$$

Theorem 9 [17, 22] The fractional positive system (17a) is asymptotically stable if and only if one of the equivalent conditions is satisfied:

(1) All coefficient of the characteristic polynomial

$$p_A(z) = \det[I_n(z+1) - A] = z^n + a_{n-1}z^{n-1} + \cdots + a_1 z + a_0 \tag{22}$$

are positive, i.e. $a_k > 0$ for $k = 0, 1, \ldots, n-1$.

(2) There exists strictly positive vector $\lambda^T = [\lambda_1 \cdots \lambda_n]^T$, $\lambda_k > 0$, $k = 1, \ldots, n$ such that

$$[A - I_n]\lambda < 0 \text{ or } \lambda^T[A - I_n] < 0. \tag{23}$$

Proof is given in [17, 22].

Theorem 10 The positive system (17a) is asymptotically stable if the sum of entries of each column (row) of the matrix A is less than one.

Proof The proof follows from condition (23) for $\lambda^T = [1 \cdots 1]$ since $[1 \cdots 1]^T A < [1 \cdots 1]$ if the sum of entries of each column of the matrix A is less than 1. Proof for rows is similar. □

5 Descriptor Positive Discrete-Time Linear Systems

Consider the descriptor discrete-time linear system

$$Ex_{i+1} = Ax_i + Bu_i, i = 0, 1, \ldots \tag{24a}$$

$$y_i = Cx_i, \tag{24b}$$

where $x_i \in \Re^n$, $u_i \in \Re^m$, $y_i \in \Re^p$ are the state, input and output vectors and $E, A \in \Re^{n\times n}$, $B \in \Re^{n\times m}$, $C \in \Re^{p\times n}$. It is assumed that the pencil (E, A) of (24a) is regular, i.e.

$$\det[Ez - A] \neq 0, z \in C \text{ (the field of complex numbers)}. \tag{25}$$

Definition 6 The descriptor system (24a), (24b) is called (internally) positive if $x_i \in \Re_+^n$, $y_i \in \Re_+^p$, $i = 0, 1, \dots$ for every consistent nonnegative initial conditions $x_0 \in \Re_+^n$ and all inputs $u_i \in \Re^m$.

The transfer matrix of the system (24) is given by

$$T(z) = C[Ez - A]^{-1}B \in \Re^{p\times m}(z), \tag{25}$$

where $\Re^{p\times m}(z)$ is the set of $p \times m$ rational matrices in z. The transfer matrix (25) can be always decomposed into the strictly proper transfer matrix

$$T_{sp}(z) = C_1[I_{n_1}z - A_1]^{-1}B_1 \tag{26a}$$

and the polynomial matrix

$$P(z) = D_0 + D_1z + \cdots + D_qz^q \in \Re^{p\times m}[z], \tag{26b}$$

where $\Re^{p\times m}[z]$ is the set of $p \times m$ polynomial matrices in z and q is the index of E.

Theorem 11 [12, 17] The descriptor system (24a), (24b) is positive if and only if

$$A_1 \in \Re_+^{n_1\times n_1}, \quad B_1 \in \Re_+^{n_1\times m}, \quad C_1 \in \Re_+^{p\times n_1} \tag{27}$$

and

$$D_k \in \Re_+^{p\times m} \quad \text{for} \quad k = 0, 1, \dots, q. \tag{28}$$

It is assumed that the singular matrix E has only $n_1 < n$ linearly independent columns and the pencil (E, A) is regular. In this case by Weierstrass-Kronecker theorem [18] there exist nonsingular matrices $P \in \Re^{n\times n}$ and $Q \in \Re^{n\times n}$ monomial (in each row and in each column only one entry is positive and the remaining entries are zero) such that

$$PEQ = \begin{bmatrix} I_{n_1} & 0 \\ 0 & N \end{bmatrix}, \quad PAQ = \begin{bmatrix} A_1 & 0 \\ 0 & I_{n_2} \end{bmatrix}, \quad n = n_1 + n_2, \tag{29}$$

where $N \in \Re^{n_2\times n_2}$ is the nilpotent matrix such that $N^\mu = 0$, $N^{\mu-1} \neq 0$, μ is the nilpotency index, $A_1 \in \Re^{n_1\times n_1}$ and $n_1 = \deg\det[Es - A]$.

Premultiplying the Eq. (24a) by the matrix $P \in \Re^{n\times n}$ and defining the new state vector

$$\begin{bmatrix} x_{1i} \\ x_{2i} \end{bmatrix} = Q^{-1}x_i, \quad x_{1i} \in \Re^{n_1}, \quad x_{2i} \in \Re^{n_2}, \quad i = 0, 1, \dots \tag{30}$$

we obtain

$$x_{1,i+1} = A_1 x_{1i} + B_1 u_i, \tag{31}$$

$$N x_{2,i+1} = x_{2i} + B_2 u_i, \tag{32}$$

where $A_1 \in \Re^{n_1 \times n_1}$, $B_1 \in \Re^{n_1 \times m}$, $B_2 \in \Re^{n_2 \times m}$ and $\begin{bmatrix} B_1 \\ B_2 \end{bmatrix} = PB$.

Note that if $Q \in \Re_+^{n \times n}$ is monomial then $Q^{-1} \in \Re_+^{n \times n}$ and $x_{1i} \in \Re_+^{n_1}$ and $x_{2i} \in \Re_+^{n_2}$ for $i = 0, 1, \ldots$ if $x_i \in \Re_+^n$, $i = 0, 1, \ldots$. Defining $CQ = [\,C_1\ C_2\,]$, $C_1 \in \Re_+^{p \times n_1}$, $C_2 \in \Re_+^{p \times n_2}$ for any $C \in \Re_+^{p \times n}$ from (24b) we have

$$y_i = C_1 x_{1i} + C_2 x_{2i}. \tag{33}$$

It is easy to verify that

$$\begin{aligned} T(z) &= C[Ez - A]^{-1}B = CQ[P(Ez - A)Q]^{-1}PB \\ &= \begin{bmatrix} C_1 & C_2 \end{bmatrix} \begin{bmatrix} I_{n_1}z - A_1 & 0 \\ 0 & Nz - I_{n_2} \end{bmatrix}^{-1} \begin{bmatrix} B_1 \\ B_2 \end{bmatrix} \\ &= C_1[I_{n_1}z - A_1]^{-1}B_1 \\ &\quad - C_2[I_{n_2} + Nz + \cdots + N^{\mu-1}z^{\mu-1}]B_2. \end{aligned} \tag{34}$$

From (31), (32) and (33) we have the following theorem.

Theorem 12 [12, 17] The descriptor discrete-time system (24a), (24b) is positive if and only if

$$\begin{gathered} A_1 \in \Re_+^{n_1 \times n_1}, \quad B_1 \in \Re_+^{n_1 \times m}, \quad -B_2 \in \Re_+^{n_2 \times m}, \\ C_1 \in \Re_+^{p \times n_1}, \quad C_2 \in \Re_+^{p \times n_2}. \end{gathered} \tag{35}$$

Theorem 13 [12, 17] The positive linear discrete-time system (24a), (24b) is asymptotically stable (the matrix A_1 is Schur) if and only if one of the following equivalent conditions is satisfied:

(1) All coefficient of the characteristic polynomial

$$p_{n_1}(z) = \det[I_{n_1}(z+1) - A_1] = z^{n_1} + a_{n-1}z^{n_1-1} + \cdots + a_1 z + a_0 \tag{36}$$

are positive, i.e. $a_i > 0$ for $i = 0, 1, \ldots, n-1$.

(2) There exists strictly positive vector $\lambda^T = [\,\lambda_1 \cdots \lambda_n]^T$, $\lambda_k > 0$, $k = 1, \ldots, n$ such that

$$(A_1 - I_{n_1})\lambda < 0 \text{ or } \lambda^T(A_1 - I_{n_1}) < 0. \tag{37}$$

6 Stability of Fractional Interval Positive Continuous-Time Linear Systems

Consider the fractional interval positive continuous-time linear system

$$\frac{d^\alpha x}{dt^\alpha} = Ax, 0 < \alpha < 1, \tag{38}$$

where $x = x(t) \in \Re^n$ is the state vector and the matrix $A \in M_n$ is defined by

$$A_1 \le A \le A_2 \text{ or equivalently } A \in [A_1, A_2]. \tag{39}$$

Definition 7 The fractional interval positive system (38) is called asymptotically stable if the system is asymptotically stable for all matrices $A \in M_n$ satisfying the condition (39).

By condition (2) of Theorem 6 the positive system (38) is asymptotically stable if there exists strictly positive vector $\lambda > 0$ such that the condition (15) is satisfied.

For two fractional positive linear systems

$$\frac{d^\alpha x_1}{dt^\alpha} = A_1 x_1, A_1 \in M_n \tag{40a}$$

and

$$\frac{d^\alpha x_2}{dt^\alpha} = A_2 x_2, A_2 \in M_n \tag{40b}$$

there exists a strictly positive vector $\lambda \in \Re^n_+$ such that

$$A_1\lambda < 0 \text{ and } A_2\lambda < 0 \tag{41}$$

if and only if the systems (40a), (40b) are asymptotically stable.

Theorem 14 If the matrices A_1 and A_2 of fractional positive systems (40a), (40b) are asymptotically stable then their convex linear combination

$$A = (1 - k)A_1 + kA_2 \text{ for } 0 \le k \le 1 \tag{42}$$

is also asymptotically stable.

Proof By condition (2) of Theorem 6 if the fractional positive linear systems (40a), (40b) are asymptotically stable then by the condition (15) there exists strictly positive vector $\lambda \in \Re_+^n$ such that

$$A_1\lambda < 0 \text{ and } A_2\lambda < 0. \tag{43}$$

Using (42) and (43) we obtain $A\lambda = [(1-k)A_1+kA_2]\lambda = (1-k)A_1\lambda+kA_2\lambda < 0$ for $0 \leq k \leq 1$. Therefore, if the positive linear systems (40a), (40b) are asymptotically stable, then their convex linear combination (42) is also asymptotically stable. □

Theorem 15 The interval positive system (38) is asymptotically stable if and only if the positive linear systems (40a), (40b) are asymptotically stable.

Proof If the matrices $A_1 \in M_n$, $A_2 \in M_n$ are asymptotically stable, then by condition (2) of Theorem 6 there exists a strictly positive vector $\lambda \in \Re_+^n$ such that (43) holds. The convex linear combination (42) satisfies the condition $A\lambda < 0$ if and only if (43) holds. Therefore, the interval system (39) is asymptotically stable if and only if the positive linear system is asymptotically stable. □

Example 1 Consider the fractional interval positive continuous-time linear system (38) with the matrices

$$A_1 = \begin{bmatrix} -2 & 1 \\ 1 & -3 \end{bmatrix}, \quad A_2 = \begin{bmatrix} -3 & 2 \\ 2 & -4 \end{bmatrix}. \tag{44}$$

Using the condition (2) of Theorem 6 we choose $\lambda = [1 \quad 1]^T$ and we obtain

$$A_1\lambda = \begin{bmatrix} -2 & 1 \\ 1 & -3 \end{bmatrix}\begin{bmatrix} 1 \\ 1 \end{bmatrix} = \begin{bmatrix} -1 \\ -2 \end{bmatrix} < 0, \tag{45a}$$

and

$$A_2\lambda = \begin{bmatrix} -3 & 2 \\ 2 & -4 \end{bmatrix}\begin{bmatrix} 1 \\ 1 \end{bmatrix} = \begin{bmatrix} -1 \\ -2 \end{bmatrix} < 0. \tag{45b}$$

Therefore the matrices (44) are Hurwitz.

Consider the interval positive linear discrete-time system

$$x_{i+1} = Ax_i \tag{46}$$

where $x_i \in \Re^n$ is the state vector and the matrix $A \in \Re_+^{n \times n}$ is defined by

$$A_1 \leq A \leq A_2 \text{ or equivalently } A \in [A_1, A_2] \tag{47}$$

Definition 8 The interval positive system (46) is called asymptotically stable if the system is asymptotically stable for all matrices $A \in \Re_+^{n\times n}$ satisfying the condition (47).

By condition (2) of Theorem 9 the positive system (46) is asymptotically stable if and only if there exists strictly positive vector $\lambda > 0$ such that (23) holds.

For two positive linear systems

$$x_{1,i+1} = A_1 x_{1,i},\ A_1 \in \Re_+^{n\times n} \tag{48a}$$

and

$$x_{2,i+1} = A_2 x_{2,i},\ A_2 \in \Re_+^{n\times n} \tag{48b}$$

there exists a strictly positive vector $\lambda \in \Re_+^n$ such that

$$A_1\lambda < \lambda \text{ and } A_2\lambda < \lambda \tag{49}$$

if and only if the systems (48a), (48b) are asymptotically stable.

Definition 9 The matrix

$$A = (1-k)A_1 + kA_2,\ 0 \le k \le 1,\ A_1 \in \Re^{n\times n},\ A_2 \in \Re^{n\times n} \tag{50}$$

is called the convex linear combination of the matrices A_1 and A_2.

Theorem 16 The convex linear combination (50) is asymptotically stable if and only if the matrices $A_1 \in \Re^{n\times n}$ and $A_2 \in \Re^{n\times n}$ are asymptotically stable.

Proof By condition (2) of Theorem 9 if the fractional positive linear systems (48a), (48b) are asymptotically stable then there exists strictly positive vector $\lambda \in \Re_+^n$ such that

$$A_1\lambda < \lambda \text{ and } A_2\lambda < \lambda. \tag{51}$$

Using (50) and (51) we obtain

$$A\lambda = [(1-k)A_1 + kA_2]\lambda = (1-k)A_1\lambda + kA_2\lambda < \lambda \tag{52}$$

for $0 \le k \le 1$. Therefore, if the positive linear systems (48a), (48b) are asymptotically stable then their convex linear combination (50) is also asymptotically stable. □

Theorem 17 The interval positive system (47) is asymptotically stable if and only if the positive linear systems (48a), (48b) are asymptotically stable.

Proof By condition (2) of Theorem 9, if the matrices $A_1 \in \Re_+^{n\times n}$, $A_2 \in \Re_+^{n\times n}$ are asymptotically stable then there exists a strictly positive vector $\lambda \in \Re_+^n$ such that (23) holds. The convex linear combination (50) satisfies the condition $A\lambda < \lambda$ if and only if (51) holds. Therefore, the interval system (46) is asymptotically stable if and only if the positive linear systems (48a), (48b) are asymptotically stable. □

7 Global Stability of Nonlinear Feedback Systems with Positive Linear Parts

Consider the nonlinear feedback system shown in Fig. 1 which consists of the positive linear part, the nonlinear element with characteristic $u = f(e)$ and positive feedback with $h > 0$. The linear part is described by the equations

$$\dot{x} = Ax + Bu,$$
$$y = Cx, \tag{53}$$

where $x = x(t) \in \Re_+^n$, $u = u(t) \in \Re_+$, $y = y(t) \in \Re_+$ are the state, input and output vectors and $A \in M_n$, $B \in \Re_+^{n\times 1}$, $C \in \Re_+^{1\times n}$.

The characteristic of the nonlinear element is shown in Fig. 2 and it satisfies the condition

$$0 \le \frac{f(e)}{e} \le k < \infty. \tag{54}$$

It is assumed that the positive linear part is asymptotically stable (the matrix $A \in M_n$ is Hurwitz).

Definition 10 The nonlinear positive system is called globally stable if it is asymptotically stable for all nonnegative initial conditions $x(0) \in \Re_+$.

The following theorem gives sufficient conditions for the global stability of the positive nonlinear system.

Theorem 17 The nonlinear positive feedback system consisting of the positive linear part, the nonlinear element satisfying the condition (54) and feedback *h*, is globally stable if

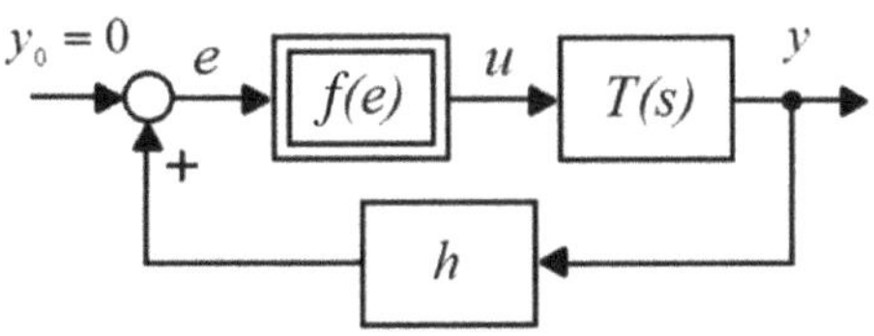

Fig. 1 The nonlinear feedback system

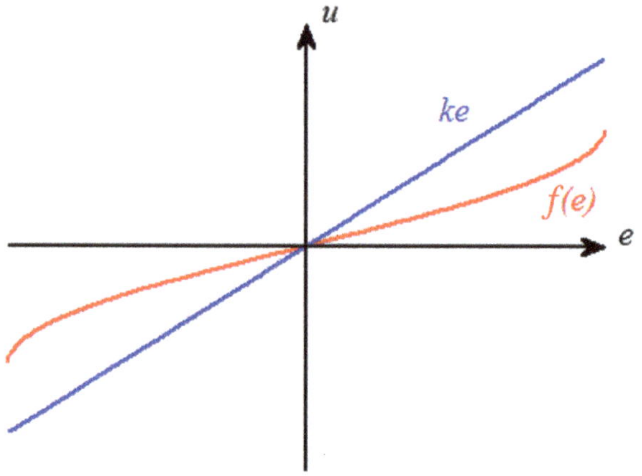

Fig. 2 Characteristic of the nonlinear element

$$A + khBC \in M_n. \tag{55}$$

Proof The proof will be accomplished by the use of the Lyapunov method [27, 28]. As the Lyapunov function $V(x)$ we choose

$$V(x) = \lambda^T x \geq 0 \text{ for } x \in \Re_+^n, \tag{56}$$

where λ is strictly positive vector, i.e. $\lambda_k > 0, k = 1, \ldots, n$. Using (56) and (53) we obtain

$$\dot{V}(x) = \lambda^T \dot{x} = \lambda^T (Ax + Bu) = \lambda^T (Ax + Bf(e)) \leq \lambda^T (A + khBC)x \tag{57}$$

since $u = f(e) \leq ke = khCx$. From (57) it follows that $\dot{V}(t) < 0$ if the condition (55) is satisfied and the nonlinear system is globally stable. □

Example 2 Consider the nonlinear system with the positive linear part with the matrices

$$A = \begin{bmatrix} -4 & 2 \\ 1 & -3 \end{bmatrix}, B = \begin{bmatrix} 1 \\ 1 \end{bmatrix}, C = [1\ 0] \tag{58}$$

and the nonlinear element satisfying the condition (54). The following two cases will be considered.

Case 1. $k = 1$.
Case 2. $k = 2$.

In Case 1 using (55) and (58) for $k = 1$ we obtain

$$A_1 = A + kBC = \begin{bmatrix} -4 & 2 \\ 1 & -3 \end{bmatrix} + \begin{bmatrix} 1 \\ 1 \end{bmatrix} [\, 1 \; 0 \,] = \begin{bmatrix} -3 & 2 \\ 2 & -3 \end{bmatrix} \in M_2. \tag{59}$$

The matrix (59) is Hurwitz since the characteristic polynomial

$$\det(I_2 s - A_1) = \begin{vmatrix} s+3 & -2 \\ -2 & s+3 \end{vmatrix} = s^2 + 6s + 5 \tag{60}$$

has the zeros $s_1 = -1$, $s_2 = -5$. The same result we obtain using Theorem 2 since for $\lambda^T = \begin{bmatrix} 1 & 1 \end{bmatrix}$ we have

$$A_1 \lambda = \begin{bmatrix} -3 & 2 \\ 2 & -3 \end{bmatrix} \begin{bmatrix} 1 \\ 1 \end{bmatrix} = -\begin{bmatrix} 1 \\ 1 \end{bmatrix} < \begin{bmatrix} 0 \\ 0 \end{bmatrix}. \tag{61}$$

In Case 2 for $k = 2$ we obtain

$$A_2 = A + kBC = \begin{bmatrix} -4 & 2 \\ 1 & -3 \end{bmatrix} + 2 \begin{bmatrix} 1 \\ 1 \end{bmatrix} [\, 1 \; 0 \,] = \begin{bmatrix} -2 & 2 \\ 3 & -3 \end{bmatrix}. \tag{62}$$

The matrix (62) is not Hurwitz since

$$\det(I_2 s - A_2) = \begin{vmatrix} s+2 & -2 \\ -3 & s+3 \end{vmatrix} = s(s+5) \tag{63}$$

and the nonlinear system for $k = 2$ does not satisfy the condition (55).

8 Global Stability of Fractional Nonlinear Feedback Continuous-Time Systems

Consider the fractional nonlinear feedback system shown in Fig. 3 which consists of the fractional positive linear part, the nonlinear element with the matrix characteristic $u = f(e)$ and the feedback with positive gain matrix H. The fractional linear part is described by the equations

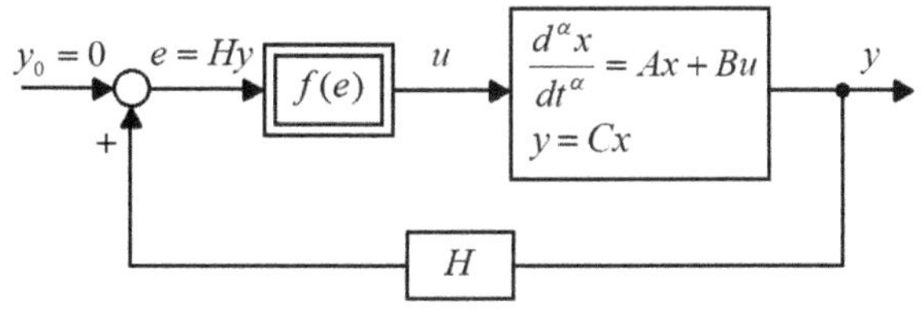

Fig. 3 The fractional nonlinear feedback system

$$\frac{d^\alpha x}{dt^\alpha} = Ax + Bu, \quad 0 < \alpha < 1,$$
$$y = Cx, \tag{64a}$$

where $x = x(t) \in \Re_+^n$, $u = u(t) \in \Re_+^m$, $y = y(t) \in \Re_+^p$ is the state, input and output vectors and

$$A \in [A_1, A_2] \in M_n, B \in [B_1, B_2] \in \Re_+^{n\times 2}, C \in [C_1, C_2] \in \Re_+^{2\times n}. \tag{64b}$$

The matrix characteristic of the nonlinear element satisfies the condition

$$u_i = f(e_i) \le k_{i1}e_1 + \cdots + k_{ip}e_p, i = 1, \ldots, m \text{ or } u \le Ke, \tag{65a}$$

where

$$u = \begin{bmatrix} u_1 \\ \vdots \\ u_m \end{bmatrix}, K = \begin{bmatrix} k_{11} & \cdots & k_{1p} \\ \vdots & \ddots & \vdots \\ k_{m1} & \cdots & k_{mp} \end{bmatrix}, e = \begin{bmatrix} e_1 \\ \vdots \\ e_p \end{bmatrix}. \tag{65b}$$

Definition 11 The fractional nonlinear positive system is called globally stable if it is asymptotically stable for all nonnegative initial conditions $x(0) \in \Re_+$.

The following theorem gives sufficient conditions for the global stability of the fractional positive nonlinear system.

Theorem 18 The fractional nonlinear system consisting of the positive asymptotically stable linear part described by (64a) with interval matrices (64b), the nonlinear element satisfying the condition (65a) and the feedback with positive gain matrix $H \in \Re_+^{m\times p}$ is globally stable if there exists a matrix K with positive entries such that the sum of entries of each column (row) of the matrix

$$(1-q)A_1 + qA_2 + BKHC = \begin{cases} A_1 + B_1K_1HC_1 \in M_n \text{ for } q = 0 \\ A_2 + B_2K_2HC_2 \in M_n \text{ for } q = 1 \end{cases} \tag{66}$$

is negative.

Proof The proof will be accomplished by use of the Lyapunov method [27, 28]. As the Lyapunov function $V(x)$ we choose

$$V(x) = \lambda^T x \ge 0 \text{ for } x \in \Re_+^n, \tag{67}$$

where λ is strictly positive vector, i.e. $\lambda_k > 0$, $k = 1, \ldots, n$. Using (67) and (64a) we obtain

$$\frac{d^{\alpha}V(x)}{dt^{\alpha}} = \lambda^T \frac{d^{\alpha}x}{dt^{\alpha}} = \lambda^T(Ax + Bu) = \lambda^T(Ax + Bf(e)) \le \lambda^T(A + BKHC)x \tag{68}$$

since $u = f(e) \le Ke = KHCx$. From (68) it follows that $\frac{d^{\alpha}V(x)}{dt^{\alpha}} < 0$ if the sum of entries of each column (row) of the matrix (66) is negative (Theorem 3) and the fractional nonlinear positive system is globally stable. □

To find the maximal K satisfying the condition (66) for the fractional nonlinear positive system the following procedure can be used.

Procedure 1.

Step 1. Using the matrices A_1, B_1, C_1 of the positive linear system and the matrix H compute the matrix K_1 such that the sum of all entries of each column (row) of the matrix

$$\hat{A} = A_1 + B_1K_1HC_1 \tag{69}$$

is negative.

If $m\ p > n$ then we choose $m\ p - n$ nonnegative entries of the matrix K_1 and its remaining entries (components of vector k) we compute as the solution of the linear matrix equation

$$Gk = h, \tag{70}$$

where the matrix G and the column vector h are defined by the sum of entries of each column (row) of the matrix (69).

Step 2. Using the matrices A_2, B_2, C_2 of the positive linear system and the matrix H compute the matrix K_2 such that the sum of all entries of each column (row) of the matrix

$$\overline{A} = A_2 + B_2K_2HC_2 \tag{71}$$

is negative.

Step 3. Find the desired K as such one that the matrices $\hat{A}$ and $\overline{A}$ are Hurwitz.

Remark 1 The conditions of Theorem 2 can be also used to compute the entries of the matrix K. Usually in this case the computations are more complicated.

Example 3 Consider the nonlinear feedback system shown in Fig. 3 with the interval matrices of the positive linear part

$$A_1 = \begin{bmatrix} -3 & 1 \\ 2 & -4 \end{bmatrix}, \quad A_2 = \begin{bmatrix} -4 & 2 \\ 3 & -5 \end{bmatrix}, \quad B_1 = \begin{bmatrix} 0.4 & 0.2 \\ 0.1 & 0.3 \end{bmatrix}, \quad B_2 = \begin{bmatrix} 0.5 & 0.3 \\ 0.2 & 0.4 \end{bmatrix},$$
$$C_1 = \begin{bmatrix} 0.8 & 0 \\ 0 & 0.8 \end{bmatrix}, \quad C_2 = \begin{bmatrix} 1 & 0 \\ 0 & 1 \end{bmatrix} \tag{72}$$

and the gain matrix

$$H = \begin{bmatrix} 0.4 & 0.6 \\ 0.2 & 0.8 \end{bmatrix}. \tag{73}$$

Using Procedure 1 we obtain:

Step 1. Using (69), (72) and (73) we obtain

$$\hat{A} = A_1 + B_1 K_1 H C_1$$
$$= \begin{bmatrix} 0.128k_{11} + 0.064k_{12} + 0.064k_{21} + 0.032k_{22} - 3 & 0.192k_{11} + 0.256k_{12} + 0.096k_{21} + 0.032k_{22} + 1 \\ 0.032k_{11} + 0.016k_{12} + 0.096k_{21} + 0.048k_{22} + 2 & 0.048k_{11} + 0.064k_{12} + 0.144k_{21} + 0.192k_{22} - 4 \end{bmatrix}. \tag{74}$$

The sum of entries of the first column of the matrix (74) is $0.16k_{11} + 0.08k_{12} + 0.16k_{21} + 0.08k_{11} - 1$ and the sum of entries of the second column of the matrix (74) is $0.24k_{11} + 0.32k_{12} + 0.24k_{21} + 0.32k_{11} - 3$. Assuming $k_{11} = k_{12} = 1$ and solving the system of linear inequalities

$$\begin{cases} 0.16k_{11} + 0.08k_{12} + 0.16k_{21} + 0.08k_{22} - 1 < 0 \\ 0.24k_{11} + 0.32k_{12} + 0.24k_{21} + 0.32k_{22} - 3 < 0 \end{cases}$$

we obtain $k_{21} < 1.5$ and $k_{22} < 6.5$. Therefore, the maximal K_1 for which the matrix (5.1) is Hurwitz is $K_1 = \begin{bmatrix} 1 & 1 \\ 1.5 & 6.5 \end{bmatrix}$.

Step 2. Using (72), (73) and (71) we obtain

$$\overline{A} = A_2 + B_2 K_2 H C_2$$
$$= \begin{bmatrix} 0.2k_{11} + 0.1k_{12} + 0.12k_{21} + 0.06k_{22} - 4 & 0.3k_{11} + 0.4k_{12} + 0.18k_{21} + 0.24k_{22} + 2 \\ 0.08k_{11} + 0.04k_{12} + 0.16k_{21} + 0.08k_{22} + 3 & 0.12k_{11} + 0.16k_{12} + 0.24k_{21} + 0.32k_{22} - 5 \end{bmatrix}. \tag{75}$$

The sum of entries of the first column of the matrix (75) is $0.28k_{11} + 0.14k_{12} + 0.28k_{21} + 0.14k_{22} - 1$ and the sum of entries of the second column of the matrix (75) is $0.42k_{11} + 0.56k_{12} + 0.42k_{21} + 0.56k_{22} - 3$. Assuming $k_{11} = k_{12} = 1$ and solving the system of linear inequalities

$$\begin{cases} 0.28k_{11} + 0.14k_{12} + 0.28k_{21} + 0.14k_{22} - 1 < 0 \\ 0.42k_{11} + 0.56k_{12} + 0.42k_{21} + 0.56k_{22} - 3 < 0 \end{cases}$$

we obtain $k_{21} < 0.43$ and $k_{22} < 3.29$. Therefore, the maximal K_2 for which the matrix (71) is Hurwitz is $K_2 = \begin{bmatrix} 1 & 1 \\ 0.428 & 3.286 \end{bmatrix}$.

Step 3. Therefore the maximal K for which the matrices (74) and (75) are Hurwitz is $K = \begin{bmatrix} 1 & 1 \\ 0.428 & 3.286 \end{bmatrix}$.

9 Global Stability of Nonlinear Feedback Discrete-Time Systems

Consider the nonlinear multi-input multi-output feedback system shown in Fig. 4. which consists of the nonlinear element with matrix characteristic $u = f(e)$, positive linear part with interval matrices and feedback with matrix gain H. The linear part is described by the equations

$$x_{i+1} = Ax_i + Bu_i, \quad i \in Z_+ = \{0, 1, \ldots\}, \tag{76a}$$

$$y_i = Cx_i, \tag{76b}$$

where $x_i \in \Re^n$, $u_i \in \Re^m$, $y_i \in \Re^p$ are the state, input and output vectors of the system $A \in \Re^{n\times n}$, $B \in \Re^{n\times m}$, $C \in \Re^{p\times n}$ are interval

$$A_1 \le A \le A_2, \quad B_1 \le B \le B_2, \quad C_1 \le C \le C_2. \tag{76c}$$

The matrix characteristic of the nonlinear element satisfies the condition

$$u \le Ke, u_i = f(e_i) \le k_{i1}e_1 + \cdots + k_{ip}e_p, i = 1, \ldots, m, \tag{77a}$$

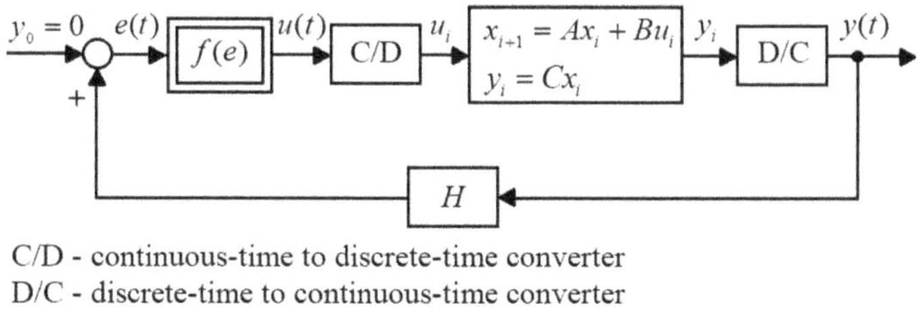

Fig. 4 The nonlinear feedback system

where

$$u = \begin{bmatrix} u_1 \\ \vdots \\ u_m \end{bmatrix}, K = \begin{bmatrix} k_{11} & \cdots & k_{1p} \\ \vdots & \ddots & \vdots \\ k_{m1} & \cdots & k_{mp} \end{bmatrix}, e = \begin{bmatrix} e_1 \\ \vdots \\ e_p \end{bmatrix}. \tag{77b}$$

In general case the feedback matrix H is not square.

Theorem 19 The multi-input multi-output nonlinear discrete-time system consisting of the positive linear part with interval matrices (76c), the matrix nonlinear element satisfying the condition (77a) and feedback with the matrix $H \in \Re_+^{m\times p}$ is globally stable if there exists a matrix K with positive entries such that the sum of entries of each column (row) of the matrix

$$(1-q)A_1 + qA_2 + khBC = \begin{cases} A_1 + B_1K_1HC_1 \in \Re_+^{n\times n} \text{ for } q = 0 \\ A_2 + B_2K_2HC_2 \in \Re_+^{n\times n} \text{ for } q = 1 \end{cases} \tag{78}$$

is less than one.

Proof As the Lyapunov [27, 28] function $V(x_i)$ we choose

$$V(x_i) = \lambda^T x_i \geq 0 \quad \text{for} \quad x_i \in \Re_+^n, \quad i \in Z_+, \tag{79}$$

where $\lambda \in \Re_+^n$ is a strictly positive vector. Using (79) and (76) we obtain

$$\begin{aligned} \Delta V(x_i) &= V(x_{i+1}) - V(x_i) = \lambda^T(Ax_i + Bu_i) - \lambda^T x_i \\ &= \lambda^T(Ax_i + Bf(e_i)) - \lambda^T x_i \leq \lambda^T[(A - I_n) + BKHC]x_i \end{aligned} \tag{80}$$

since $u \leq Ke = KHCx_i$. From (80) it follows that $\Delta V(x_i) < 0$ if the sum of entries of each column (row) of the matrix (78) is less than one. □

To find the maximal matrix K for which the nonlinear system shown in Fig. 4 is globally stable the following procedure can be used.

Procedure 2.

Step 1. Find the matrix K_1 such that the sum of entries of each column of the matrix

$$A_1 + B_1K_1HC_1 \in \Re_+^{n\times n} \tag{81}$$

is less than one. If $m\ p > n$ then we choose $m\ p - n$ nonnegative entries of the matrix K_1 and the its remaining entries (components of vector k) we compute as the solution of the linear matrix equation

$$Gk = h, \tag{82}$$

where the matrix G and the column vector h are defined by the sum of entries of each column (row) of the matrix (81).

Step 2. In a similar way as in Step 1 find the matrix K_2 such that the sum of entries of each column of the matrix

$$A_2 + B_2 K_2 H C_2 \in \Re_+^{n \times n} \tag{83}$$

is less than one.

Step 3. Knowing K_1 and K_2 find the desired matrix K which satisfies the conditions (81) and (83).

Example 4 Consider the nonlinear system with the positive linear part with the interval matrices

$$A_1 = \begin{bmatrix} 0.3 & 0.2 \\ 0.2 & 0.4 \end{bmatrix}, A_2 = \begin{bmatrix} 0.5 & 0.3 \\ 0.3 & 0.5 \end{bmatrix}, \quad B_1 = \begin{bmatrix} 0.4 \\ 0.3 \end{bmatrix}, B_2 = \begin{bmatrix} 0.6 \\ 0.5 \end{bmatrix}$$
$$C_1 = \begin{bmatrix} 0.5 & 0.2 \\ 0.1 & 0.3 \end{bmatrix}, \quad C_2 = \begin{bmatrix} 0.6 & 0.3 \\ 0.2 & 0.5 \end{bmatrix}. \tag{84}$$

The matrix K of the nonlinear element satisfies the condition (77) and the matrix

$$H = \begin{bmatrix} 0.5 & 0.4 \\ 0.3 & 0.5 \end{bmatrix}. \tag{85}$$

Find the maximal matrix K for which the nonlinear system is globally stable. Using Procedure 2 we obtain.

Step 1. Using (81) and (84) we obtain

$$\begin{aligned} A_1 + B_1 K_1 H C_1 &= \begin{bmatrix} 0.3 & 0.2 \\ 0.2 & 0.4 \end{bmatrix} + \begin{bmatrix} 0.4 \\ 0.3 \end{bmatrix} \begin{bmatrix} k_{11} & k_{12} \end{bmatrix} \begin{bmatrix} 0.5 & 0.4 \\ 0.3 & 0.5 \end{bmatrix} \begin{bmatrix} 0.5 & 0.2 \\ 0.1 & 0.3 \end{bmatrix} \\ &= \begin{bmatrix} 0.116k_{11} + 0.08k_{12} + 0.3 & 0.088k_{11} + 0.084k_{12} + 0.2 \\ 0.087k_{11} + 0.06k_{12} + 0.2 & 0.066k_{11} + 0.063k_{12} + 0.4 \end{bmatrix}. \end{aligned} \tag{86}$$

The sum of entries of the first column of the matrix (86) is $0.203k_{11}+0.14k_{12}+0.5$ and the sum of entries of the second column of the matrix (86) is $0.154k_{11}+0.147k_{12}+0.63$. Solving the system of linear inequalities

$$\begin{cases} 0.203k_{11} + 0.14k_{12} + 0.5 < 1 \\ 0.154k_{11} + 0.147k_{12} + 0.63 < 1 \end{cases} \tag{87}$$

we obtain $k_{11} < 2.113$ and $k_{12} < 0.507$. Therefore, the maximal K_1 for which the matrix (81) is Schur is $K_1 = \begin{bmatrix} 2.113 & 0.507 \end{bmatrix}$.

Step 2. Using (83) and (84) we obtain

$$\begin{aligned} A_2 + B_2K_2HC_2 &= \begin{bmatrix} 0.5 & 0.3 \\ 0.3 & 0.5 \end{bmatrix} + \begin{bmatrix} 0.6 \\ 0.5 \end{bmatrix} \begin{bmatrix} k_{12} & k_{21} \end{bmatrix} \begin{bmatrix} 0.5 & 0.4 \\ 0.3 & 0.5 \end{bmatrix} \begin{bmatrix} 0.6 & 0.3 \\ 0.2 & 0.5 \end{bmatrix} \\ &= \begin{bmatrix} 0.228k_{11} + 0.168k_{12} + 0.5 & 0.21k_{11} + 0.204k_{12} + 0.3 \\ 0.19k_{11} + 0.14k_{12} + 0.3 & 0.175k_{11} + 0.17k_{12} + 0.5 \end{bmatrix}. \end{aligned} \tag{88}$$

The sum of entries of the first column of the matrix (88) is $0.418k_{11}+0.308k_{12}+0.8$ and the sum of entries of the second column of the matrix (85) is $0.385k_{11}+0.374k_{12}+0.8$. Solving the system of linear inequalities

$$\begin{cases} 0.418k_{11} + 0.308k_{12} + 0.8 < 1 \\ 0.385k_{11} + 0.374k_{12} + 0.8 < 1 \end{cases} \tag{89}$$

we obtain $k_{11} < 0.349$ and $k_{12} < 0.174$. Therefore, the maximal K_2 for which the matrix (83) is Schur is $K_2 = \begin{bmatrix} 0.349 & 0.174 \end{bmatrix}$.

Step 3. Using the results obtained in Steps 1 and 2 we obtain that the maximal K for which the matrices (86) and (88) are Schur is $K = \begin{bmatrix} 0.349 & 0.174 \end{bmatrix}$. Therefore, the nonlinear system is globally stable for given maximal matrix $K = \begin{bmatrix} 0.349 & 0.174 \end{bmatrix}$.

These considerations can be easily extended to fractional feedback nonlinear discrete-time systems.

10 Global Stability of Descriptor Nonlinear Feedback Discrete-Time Systems

Consider the nonlinear feedback system shown in Fig. 5. which consists of the descriptor positive linear part, the nonlinear element with characteristic $u = f(e)$, positive scalar gain feedback h and interval state matrices. The descriptor linear part

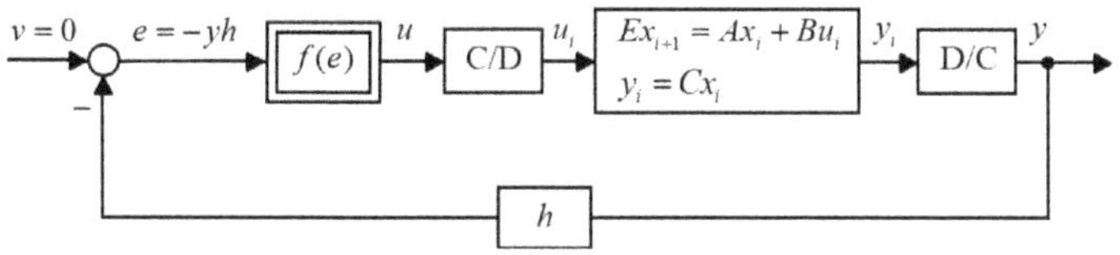

C/D - continuous-time to discrete-time converter
D/C - discrete-time to continuous-time converter

Fig. 5 Nonlinear system

is described by the equations

$$Ex_{i+1} = Ax_i + Bu_i, i = 0, 1, \ldots \tag{90a}$$

$$y_i = Cx_i, \tag{90b}$$

where $x_i \in \Re^n$, $u_i \in \Re$, $y_i \in \Re$ are the state vector, input and output of the system $E, A \in \Re^{n\times n}$, $B \in \Re^{n\times 1}$, $C \in \Re^{1\times n}$. The characteristic $f(e)$ of the nonlinear element (Fig. 6) satisfies the condition

$$0 < f(e) < ke, \; 0 < k < \infty. \tag{91}$$

It is assumed that:

(1) the pencil (E, A) is regular (the condition (25) is satisfied),

Fig. 6 Characteristic of nonlinear element

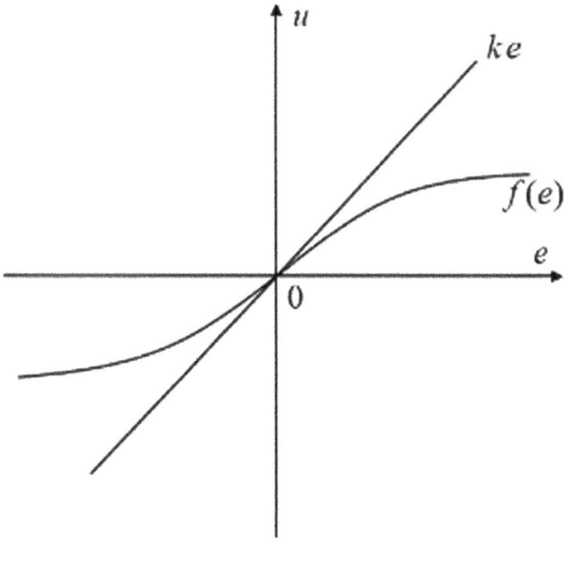

(2) the matrix E has n_1 linearly independent columns,
(3) rank $E = \deg\det[Ez - A] = n_1$.

If the assumptions are satisfied then by Weirstrass-Kronecker theorem, there exists nonsingular matrices $P \in \Re^{n\times n}$ and $Q \in \Re^{n\times n}$ monomial such that

$$PEQ = \begin{bmatrix} I_{n_1} & 0 \\ 0 & 0 \end{bmatrix}, \quad PAQ = \begin{bmatrix} \overline{A}_1 & 0 \\ 0 & I_{n_2} \end{bmatrix}, \tag{92}$$
$$n = n_1 + n_2,$$

where $\overline{A}_1 \in \Re^{n_1\times n_1}$, $A_1 \le \overline{A}_1 \le A_2$ and $n_1 = \deg\det[Es - A]$. Premultiplying the Eq. (90a) by the matrix $P \in \Re^{n\times n}$ and defining new state vector

$$\begin{bmatrix} x_{1i} \\ x_{2i} \end{bmatrix} = Q^{-1}x_i, \quad x_{1i} \in \Re^{n_1}, \quad x_{2i} \in \Re^{n_2} \tag{93}$$

we obtain

$$x_{1,i+1} = \overline{A}_1 x_{1i} + B_1 u_i, \tag{94a}$$

$$Nx_{2,i+1} = x_{2i} + B_2 u_i, \tag{94b}$$

where $\overline{A}_1 \in \Re^{n_1\times n_1}$, $B_1 \in \Re^{n_1}$, $B_2 \in \Re^{n_2}$ and

$$\begin{bmatrix} B_1 \\ B_2 \end{bmatrix} = PB. \tag{94c}$$

Note that if $Q \in \Re_+^{n\times n}$ is monomial then $Q^{-1} \in \Re_+^{n\times n}$ and $x_{1i} \in \Re_+^{n_1}$ and $x_{2i} = 0$ for $i = 0, 1, \ldots$, $B_2 = 0$ since $N = 0$ (assumptions (2) and (3)). In this case defining $CQ = [\, C_1 \; C_2 \,]$, $C_1 \in \Re_+^{1\times n_1}$, $C_2 \in \Re^{1\times n_2}$ for any $C \in \Re_+^{1\times n}$ we have

$$y(t) = C_1 x_1(t). \tag{95}$$

Definition 12 The nonlinear positive system is called globally stable if it is asymptotically stable for all nonnegative initial conditions $x_0 \in \Re_+^n$.

The following theorem gives sufficient conditions for the global stability of the descriptor positive nonlinear system.

Theorem 20 The nonlinear system consisting of the positive linear part with interval state matrix (92) satisfying the assumptions (1), (2), (3), the nonlinear element satisfying the condition (91) and the gain feedback h is globally stable if the matrix

$$A_i + k_i h B_1 C_1 \in \Re_+^{n_1\times n_1} \quad \text{for} \quad i = 1, 2 \tag{96}$$

is asymptotically stable.

Proof The proof will be accomplished by the use of the Lyapunov method [27, 28]. As the Lyapunov function $V(x_{1i})$ for each system we choose

$$V(x_{1i}) = \lambda^T x_{1i} \geq 0 \quad \text{for} \quad x_{i1} \in \Re_+^{n_1}, \tag{97}$$

where λ is strictly positive vector, i.e. $\lambda_k > 0$, $k = 1, \ldots, n_1$. Using (97) and (94a) we obtain

$$\begin{aligned} \Delta V(x_{1i}) &= V(x_{1,i+1}) - V(x_{1i}) = \lambda^T(x_{1,i+1} - x_{1i}) \\ &= \lambda^T(A_i - I_{n_1})x_{1i} + B_1 h f(e) = \lambda^T(A_i + k_i h B_1 C_1)x_{1i} \text{ for } i = 1, 2 \end{aligned} \tag{98}$$

since $\lambda^T(A_i - I_{n_1}) < 0$ and $(I_{n_1} - A_i) > B_1 h f(e)$ for $i = 1, 2$. From (98) it follows that $\Delta V(x_{1i}) < 0$ if the matrix (96) is asymptotically stable and the nonlinear system is globally stable. □

To find the maximal value of k for which the nonlinear systems is globally stable, the following procedure can be used.

Procedure 3.

Step 1. Find the value of k_1 for which the matrix

$$A_1 + k_1 h B C \in \Re_+^{n_1 \times n_1} \tag{99}$$

is asymptotically stable.

Step 2. Find the value of k_2 for which the matrix

$$A_2 + k_2 h B C \in \Re_+^{n_1 \times n_1} \tag{100}$$

is asymptotically stable.

Step 3. Find the desired value of k as

$$k = \min(k_1, k_2). \tag{101}$$

From Theorem 20 we have the following conclusions.

Conclusion 1. The nonlinear positive feedback system is asymptotically stable only if the sum of all entries in every rows (columns) of the matrix $A_1, A_2 \in \Re_+^{n_1 \times n_1}$ is less than 1.

Conclusion 2. To check the global stability of the nonlinear system it sufficient to check the condition (96) only for the matrix A_1 (A_2) with greater sum of all its entries.

Example 5 Consider the nonlinear feedback system with the descriptor linear part with the interval matrix

$$A_1 = \begin{bmatrix} 0.3 & 0.1 \\ 0.2 & 0.4 \end{bmatrix}, A_2 = \begin{bmatrix} 0.4 & 0.2 \\ 0.25 & 0.5 \end{bmatrix}, B_1 = \begin{bmatrix} 0.5 \\ 0.6 \end{bmatrix}, C_1 = \begin{bmatrix} 0.2 & 0.4 \end{bmatrix}, \tag{102}$$

the nonlinear element satisfies the condition (91) and $h = 0.5$. Find the maximal value of k for which the nonlinear system is globally stable. Using the Procedure 1 and (102) we obtain.

Step 1. Using (96) and (102) we obtain

$$\begin{aligned} A_1 + k_1 hBC &= \begin{bmatrix} 0.3 & 0.1 \\ 0.2 & 0.4 \end{bmatrix} + 0.5k_1 \begin{bmatrix} 0.5 \\ 0.6 \end{bmatrix} \begin{bmatrix} 0.2 & 0.4 \end{bmatrix} \\ &= \begin{bmatrix} 0.3 + 0.05k_1 & 0.1 + 0.1k_1 \\ 0.2 + 0.06k_1 & 0.4 + 0.12k_1 \end{bmatrix} \end{aligned} \tag{103}$$

and the maximal value of k_1 for which the matrix (103) is asymptotically stable is $k_1 < 2.857$, since for this value the coefficients of the polynomial

$$\det[I_2(z+1) - A_1 - k_1 h B_1 C_1] = z^2 + (1.3 - 0.17k_1)z + 0.4 - 0.14k_1 \tag{104}$$

are positive.

Step 2. Using (96) and (100) we obtain

$$\begin{aligned} A_2 + k_2 hBC &= \begin{bmatrix} 0.4 & 0.2 \\ 0.25 & 0.5 \end{bmatrix} + 0.5k_2 \begin{bmatrix} 0.5 \\ 0.6 \end{bmatrix} \begin{bmatrix} 0.2 & 0.4 \end{bmatrix} \\ &= \begin{bmatrix} 0.4 + 0.05k_2 & 0.2 + 0.1k_2 \\ 0.25 + 0.06k_2 & 0.5 + 0.12k_2 \end{bmatrix} \end{aligned} \tag{105}$$

and the maximal value of k_2 for which the matrix is asymptotically stable is $k_2 < 1.866$, since for this value the coefficients of the polynomial

$$\det[I_2(z+1) - A_2 - k_2hB_1C_1] = z^2 + (1.1 - 0.17k_2)z + 0.25 - 0.134k_2 \quad (106)$$

are positive.

Step 3. Using (101) and the results of Steps 1 and 2 we obtain

$$k = \min(k_1, k_2) = \min(2.857, 1.866) = 1.866. \quad (107)$$

Therefore the nonlinear system is globally stable for $k < 1.866$.

11 Global Stability of Positive Nonlinear Electrical Circuits

Consider the linear continuous-time electrical circuit described by the state equations

$$\dot{x}(t) = Ax(t) + Bu(t), \quad (108a)$$

$$y(t) = Cx(t) + Du(t), \quad (108b)$$

where $x(t) \in \Re^n$, $u(t) \in \Re^m$, $y(t) \in \Re^p$ are the state, input and output vectors and $A \in \Re^{n\times n}$, $B \in \Re^{n\times m}$, $C \in \Re^{p\times n}$, $D \in \Re^{p\times m}$. It is well-known [22] that any linear electrical circuit composed of resistors, coils, capacitors and voltage (current) sources can be described by the state Eqs. (108a), (108b). Usually as the state variables $x_1(t),\ldots, x_n(t)$ (the components of the state vector $x(t)$) the currents in the coils and voltages on the capacitors are chosen.

Definition 13 [22] The electrical circuit (108a), (108b) is called (internally) positive if $x(t) \in \Re^n_+$ and $y = y(t) \in \Re^p_+$, $t \in [0, +\infty]$ for any $x_0 = x(0) \in \Re^n_+$ and every $u(t) \in \Re^m_+$, $t \in [0, +\infty]$.

Theorem 21 [22] The electrical circuit (108a), (108b) is positive if and only if

$$A \in M_n,\ B \in \Re^{n\times m}_+,\ C \in \Re^{p\times n}_+,\ D \in \Re^{p\times m}_+. \quad (109)$$

Theorem 22 [22] The linear electrical circuit composed of resistors, coils and voltage sources is positive for any values of the resistances, inductances and source voltages if the number of coils is less than or equal to the number of its linearly independent meshes and the direction of the mesh currents are consistent with the directions of the mesh source voltages.

Theorem 23 [22] The linear electrical circuit composed of resistors, capacitors and voltage sources is not positive for all values of its resistances, capacitances and source voltages if each of its branches contain resistor, capacitor and voltage source.

Theorem 24 [22] The *R*, *L*, *C*, *e* electrical circuits are not positive for any values of its resistances, inductances, capacitances and source voltages if at least one of its branches contain coil and capacitor.

Definition 14 [22] The positive electrical circuit is called asymptotically stable if

$$\lim_{t\to\infty} x(t) = 0 \text{ for any } x_0 \in \Re_+^n. \tag{110}$$

Theorem 25 [22] The positive electrical circuit (108a), (108b) is asymptotically stable if and only if one of the equivalent conditions is satisfied:

(1) All coefficient of the characteristic polynomial

$$\det[I_n s - A] = s^n + a_{n-1}s^{n-1} + \cdots + a_1 s + a_0 \tag{111}$$

are positive, i.e. $a_k > 0$ for $k = 0, 1, \ldots, n-1$.

(2) All principal minors $\overline{M}_i$, $i = 1, \ldots, n$ of the matrix $-A$ are positive, i.e.

$$\overline{M}_1 = |-a_{11}| > 0, \overline{M}_2 = \begin{vmatrix} -a_{11} & -a_{12} \\ -a_{21} & -a_{22} \end{vmatrix} > 0, \ldots, \overline{M}_n = \det[-A] > 0. \tag{112}$$

(3) There exists strictly positive vector $\lambda^T = [\lambda_1 \cdots \lambda_n]^T$, $\lambda_k > 0$, $k = 1, \ldots, n$ such that

$$A\lambda < 0 \text{ or } A^T\lambda < 0. \tag{113}$$

Consider the nonlinear feedback electrical circuit shown in Fig. 7 which consists of the positive linear part, the nonlinear element with the characteristic $u = f(e)$ and the feedback with positive gain *h*. The linear part is described by the equations

$$\begin{aligned} \dot{x} &= Ax + Bu, \\ y &= Cx, \end{aligned} \tag{114}$$

where $x = x(t) \in \Re_+^n$, $u = u(t) \in \Re_+$, $y = y(t) \in \Re_+$ is the state vector, input and output and $A \in M_n$, $B \in \Re_+^{n\times 1}$, $C \in \Re_+^{1\times n}$.

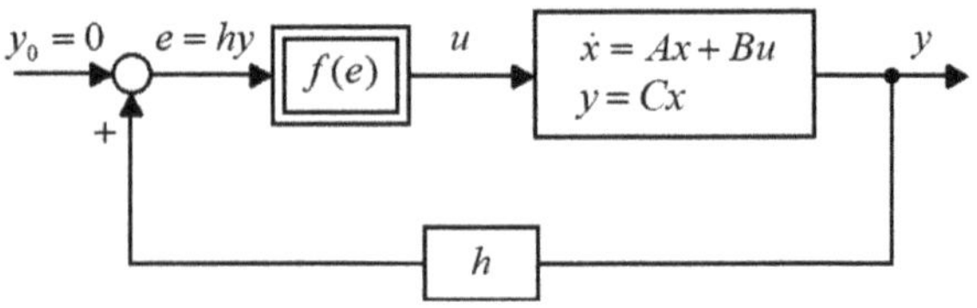

Fig. 7 The nonlinear feedback electrical circuit

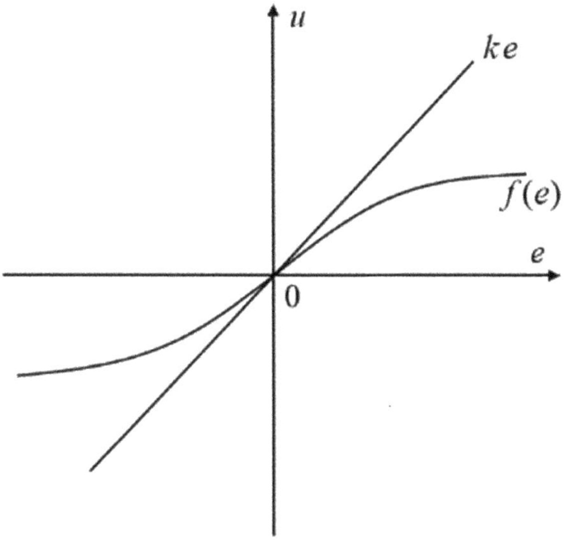

Fig. 8 Characteristic of the nonlinear element of the electrical circuit

The characteristic of the nonlinear element is shown in Fig. 8 and it satisfies the condition

$$0 \leq \frac{f(e)}{e} \leq k < \infty. \tag{115}$$

It is assumed that the positive linear part is asymptotically stable (the matrix $A \in M_n$ is Hurwitz).

Definition 15 The nonlinear positive electrical circuit is called globally stable if it is asymptotically stable for all nonnegative initial conditions $x(0) \in \Re_+^n$.

The following theorem gives sufficient conditions for the global stability of the positive nonlinear electrical circuit.

Theorem 26 The nonlinear electrical circuit consisting of the positive linear part, the nonlinear element satisfying the condition (115) and the feedback with positive gain h is globally stable if

$$A + khBC \in M_n \tag{116}$$

is asymptotically stable.

Proof The proof will be accomplished by use of the Lyapunov method [27, 28]. As the Lyapunov function $V(x)$ we choose

$$V(x) = \lambda^T x \geq 0 \text{ for } x \in \Re_+^n, \tag{117}$$

where λ is strictly positive vector, i.e. $\lambda_k > 0$, $k = 1, \ldots, n$. Using (117) and (114) we obtain

$$\dot{V}(x) = \lambda^T \dot{x} = \lambda^T (Ax + Bu) = \lambda^T (Ax + Bf(e)) \leq \lambda^T (A + khBC)x \quad (118)$$

since $u = f(e) \leq ke = khCx$. From (118) it follows that $\dot{V}(t) < 0$ if the condition (116) is satisfied and the nonlinear positive electrical circuit is globally stable. □

To find the maximal value of k satisfying the condition (115) for the nonlinear positive electrical circuit the following procedure can be used.

Procedure 4.

Step 1. Using the matrices A, B, C of the positive electrical circuit and h compute the matrix

$$A_1 = A + khBC. \quad (119)$$

Step 2. Compute the characteristic polynomial of the matrix (119) with coefficient depending on k

$$\det[I_n s - A_1]. \quad (120)$$

Step 3. Using the condition (1) of Theorem 25 find the maximal k for which all coefficients of (120) are positive.

Example 6 Consider the nonlinear positive electrical circuit shown in Fig. 9 with given resistances $R_1 = 3$, $R_2 = 2$, inductances $L_1 = L_2 = 1$, the characteristic $u = f(i)$ satisfying the condition

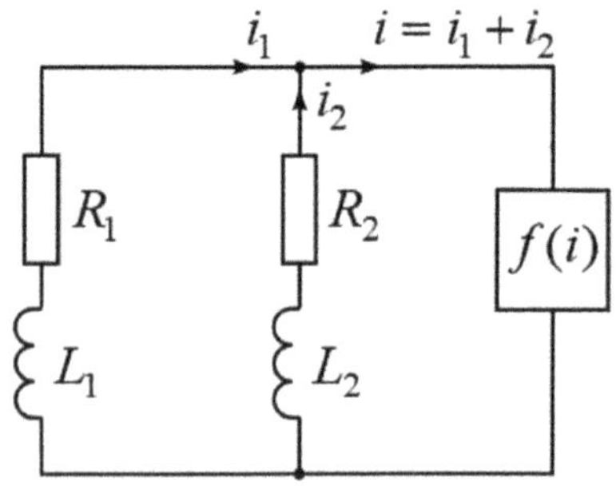

Fig. 9 Nonlinear electrical circuit of Example 6

$$0 \leq \frac{f(i)}{i} \leq k \tag{121}$$

and with the feedback gain $h = 0.5$.

Find the maximal value of k for which the nonlinear electrical circuit is globally stable. Using Procedure 4 we obtain the following.

Step 1. The positive linear part is described by the equations

$$\frac{d}{dt}\begin{bmatrix} i_1 \\ i_2 \end{bmatrix} = A\begin{bmatrix} i_1 \\ i_2 \end{bmatrix} + Bu, \quad y = C\begin{bmatrix} i_1 \\ i_2 \end{bmatrix}, \tag{122a}$$

where

$$A = \begin{bmatrix} -\frac{R_1}{L_1} & 0 \\ 0 & -\frac{R_2}{L_2} \end{bmatrix} = \begin{bmatrix} -3 & 0 \\ 0 & -2 \end{bmatrix}, \quad B = \begin{bmatrix} \frac{1}{L_1} \\ \frac{1}{L_2} \end{bmatrix} = \begin{bmatrix} 1 \\ 1 \end{bmatrix}, \quad C = \begin{bmatrix} 1 & 1 \end{bmatrix}. \tag{122b}$$

The matrix (119) has the form

$$A_1 = A + khBC = \begin{bmatrix} -3 & 0 \\ 0 & -2 \end{bmatrix} + k\begin{bmatrix} 0.5 & 0.5 \\ 0.5 & 0.5 \end{bmatrix}. \tag{123}$$

Step 2. The characteristic polynomial of the matrix (123) has the form

$$\det[I_2 s - A_1] = \begin{vmatrix} s + 3 - 0.5k & -0.5k \\ -0.5k & s + 2 - 0.5k \end{vmatrix} = s^2 + (5 - k)s + 6 - 2.5k. \tag{124}$$

Step 3. Therefore, the maximal value of k for which all the coefficient of (124) are positive is $k < 2.4$.

Example 7 Consider the nonlinear positive electrical circuit shown in Fig. 10 with given resistances $R_1 = R_2 = 2$, capacitances $C_1 = 0.2$, $C_2 = 0.1$, the characteristic $i = f(u)$, $u = u_1 + u_2$ satisfying the condition

$$0 \leq \frac{f(i)}{i} \leq k \tag{125}$$

and the feedback gain $h = 0.2$.

Check if the positive nonlinear electrical circuit is globally stable for $k = 0.5$. The positive linear part of the nonlinear circuit is described by the equations

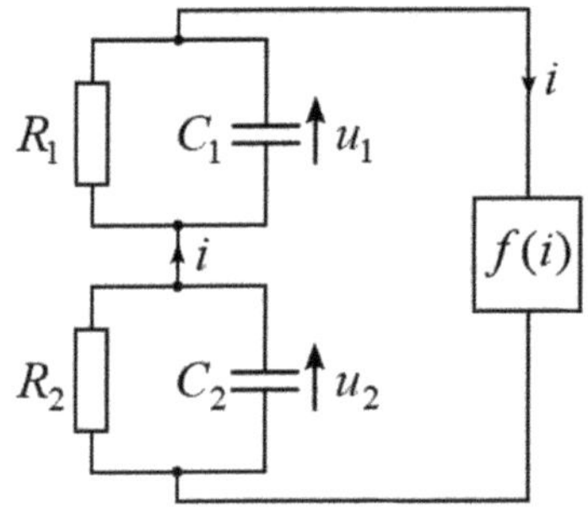

Fig. 10 Nonlinear electrical circuit of Example 7

$$\frac{d}{dt}\begin{bmatrix} u_1 \\ u_2 \end{bmatrix} = A\begin{bmatrix} u_1 \\ u_2 \end{bmatrix} + Bu, \quad y = C\begin{bmatrix} u_1 \\ u_2 \end{bmatrix}, \tag{126a}$$

where

$$A = \begin{bmatrix} -\frac{1}{R_1C_1} & 0 \\ 0 & -\frac{1}{R_2C_2} \end{bmatrix} = \begin{bmatrix} -2.5 & 0 \\ 0 & -5 \end{bmatrix}, \quad B = \begin{bmatrix} \frac{1}{C_1} \\ \frac{1}{C_2} \end{bmatrix} = \begin{bmatrix} 5 \\ 10 \end{bmatrix}, \quad C = [\,1\ 1\,]. \tag{126b}$$

In this case the matrix (119) has the form

$$A_1 = A + khBC = \begin{bmatrix} -2.5 & 0 \\ 0 & -5 \end{bmatrix} + 0.1\begin{bmatrix} 5 \\ 10 \end{bmatrix}[\,1\ 1\,] = \begin{bmatrix} -2 & 0.5 \\ 1 & -4 \end{bmatrix} \tag{127}$$

and its characteristic polynomial

$$\det[I_2 s - A_1] = \begin{vmatrix} s+2 & -0.5 \\ -1 & s+4 \end{vmatrix} = s^2 + 6s + 7.5. \tag{128}$$

Therefore, the positive nonlinear electrical circuit is asymptotically stable.

12 Summary

The global stability of multi-input multi-output feedback continuous-time and discrete-time nonlinear systems with interval state matrices of the positive linear parts has been analyzed. New sufficient conditions for the global stability of different classes of nonlinear systems with positive linear parts have been established. Procedures for computation of the coefficient matrices K satisfying the condition (77) have been proposed. The new stability conditions have been demonstrated on simple examples of standard and fractional nonlinear systems with interval state matrices of the positive linear parts. As an example of application of the new global stability conditions, the positive nonlinear electrical circuits have been given. The considerations can be extended to different orders fractional nonlinear systems with the positive linear parts.

Acknowledgements This work was supported by National Science Centre in Poland under work No. 2017/27/B/ST7/02443.

References

1. Berman, A., Plemmons, R.J.: Nonnegative Matrices in the Mathematical Sciences. SIAM (1994).
2. Borawski, K.: Modification of the stability and positivity of standard and descriptor linear electrical circuits by state feedbacks. Electr. Rev. **93**(11), 176–180 (2017)
3. Busłowicz, M., Kaczorek, T.: Simple conditions for practical stability of positive fractional discrete-time linear systems. Int. J. Appl. Math. Comput. Sci. **19**(2), 263–169 (2009)
4. Farina, L., Rinaldi, S.: Positive Linear Systems; Theory and Applications. Wiley, New York (2000)
5. Kaczorek, T.: Absolute stability of a class of fractional positive nonlinear systems. Int. J. Appl. Math. Comput. Sci. **29**(1), 93–98 (2019)
6. Kaczorek, T.: Analysis of positivity and stability of discrete-time and continuous-time nonlinear systems. Comput. Probl. Electr. Eng. **5**(1), 11–16 (2015)
7. Kaczorek, T.: Analysis of positivity and stability of fractional discrete-time nonlinear systems. Bull. Pol. Acad.: Tech. **64**(3), 491–494 (2016)
8. Kaczorek, T.: Global stability of nonlinear feedback systems with fractional positive linear parts. Int. J. Appl. Math. Comput. Sci. **30**(3), 493–500 (2020)
9. Kaczorek, T.: Global stability of positive standard and fractional nonlinear feedback systems. Bull. Pol. Acad.: Tech. **68**(2), 285–288 (2020)
10. Kaczorek, T.: Global stability of nonlinear feedback systems with positive linear parts, Int. J. Nonlinear Sci. Numer. Simul. **20**(5), 575–579 (2019)
11. Kaczorek, T.: Positivity and stability of descriptor continuous-time linear systems with interval state matrices. In: Nolle, L., Burger, A., Tholen, C., Werner, J., Wellhausen, J., (eds.) Proceedings 32nd European Conference on Modelling and Simulation ©ECMS. ISBN: 978–0–9932440–6–3/ ISBN: 978–0–9932440–7–0 (CD)
12. Kaczorek, T.: Positivity and stability of descriptor linear systems with interval state matrices. Comput. Probl. Electr. Eng. **8**(1), 7–17 (2018)
13. Kaczorek, T.: Positive 1D and 2D Systems. Springer, London (2002)
14. Kaczorek, T.: Positive linear systems with different fractional orders. Bull. Pol. Acad.: Tech.**58**(3), 453–458 (2010)
15. Kaczorek, T.: Positive linear systems consisting of n subsystems with different fractional orders. IEEE Trans. Circuits Syst. **58**(7), 1203–1210 (2011)
16. Kaczorek, T.: Positive fractional continuous-time linear systems with singular pencils. Bull. Pol. Acad.: Tech. **60**(1), 9–12 (2012)
17. Kaczorek, T.: Selected Problems of Fractional Systems Theory. Springer, Berlin (2011)
18. Kaczorek, T.: Stability of fractional positive nonlinear systems. Arch. Control Sci. **25**(4), 491–496 (2015)
19. Kaczorek, T., Sajewski, Ł.: The pointwise completeness and the pointwise degeneracy of linear discrete-time different fractional order systems. Bull. Pol. Acad.: Tech. **68**(6), 1513–1516 (2020)

20. Kaczorek, T.: Superstabilization of positive linear electrical circuit by state-feedbacks. Bull. Pol. Acad.: Tech. **65**(5), 703–708 (2017)
21. Kaczorek, T., Borawski, K.: Stability of Positive Nonlinear Systems. In: 22nd International Conference on Methods and Models in Automation and Robotics. Międzyzdroje, Poland (2017)
22. Kaczorek, T., Rogowski, K.: Fractional Linear Systems and Electrical Circuits. Springer, Cham (2015)
23. Kaczorek, T., Ruszewski, A.: Global stability of discrete-time nonlinear systems with descriptor standard and fractional positive linear parts and scalar feedbacks. Arch. Control Sci. **30**(4), 667–681 (2020)
24. Kaczorek, T., Ruszewski, A.: Global stability of positive discrete-time standard and fractional nonlinear systems with scalar feedbacks. In: Bartoszewicz, A., Kabziński, J., Kacprzyk, J. (eds.) Advanced, Contemporary Control. Advances in Intelligent Systems and Computing, vol. 1196, pp. 1167–1175. Springer, Cham (2020)
25. Kaczorek, T., Sajewski, Ł: Pointwise completeness and pointwise degeneracy of fractional standard and descriptor linear continuous-time systems with different fractional orders. Int. J. Appl. Math. Comput. Sci. **30**(4), 641–647 (2020)
26. Kudrewicz, J.: Ustoicivost nieliniejnych sistem z obratnoj swjazju. Avtomatika i Telemechanika **25**(8) (1964)
27. Lyapunov, A.M.: Obscaja zadaca ob ustoicivosti dvizenija. Gostechizdat, Moskwa (1963)
28. Leipholz, H.: Stability Theory. Academic Press, New York (1970)
29. Mitkowski, W.: Dynamical properties of Metzler systems. Bull. Pol. Acad.: Tech. **56**(4), 309–312 (2008)
30. Ostalczyk, P.: Discrete Fractional Calculus. World Scientific, River Edgle (2016)
31. Podlubny, I.: Fractional Differential Equations. Academic Press, San Diego (1999)
32. Ruszewski, A.: Practical and asymptotic stabilities for a class of delayed fractional discrete-time linear systems. Bull. Pol. Acad.: Tech. **67**(3), 509–515 (2019)
33. Ruszewski, A.: Stability of discrete-time fractional linear systems with delays. Arch. Control Sci. **29**(3), 549–567 (2019)
34. Sajewski, Ł.: Decentralized stabilization of descriptor fractional positive continuous-time linear systems with delays. In: 22nd International Conference on Methods and Models in Automation and Robotics, pp. 482–487. Międzyzdroje, Poland (2017)
35. Sajewski, Ł.: Stabilization of positive descriptor fractional discrete-time linear systems with two different fractional orders by decentralized controller. Bull. Pol. Acad.: Tech. **65**(5), 709–714 (2017)

Controllability of Fractional Linear Systems with Delays in Control

Jerzy Klamka

Abstract In the chapter linear, fractional, continuous time, finite-dimensional, dynamical control systems with multiple variable point delays and distributed delay in admissible control described by linear ordinary differential state equations are considered. Using notations, theorems and methods taken directly from functional analysis and linear controllability theory, necessary and sufficient conditions for global relative controllability in a given finite time interval are formulated and proved. The main result of the chapter is to show, that global relative controllability of fractional linear systems with different types of delays in admissible control is equivalent to non-singularity of a suitably defined relative controllability matrix. In the proofs of the main results, methods and concepts taken from the theory of linear bounded operators in Hilbert spaces are used. Applying a relative controllability matrix for relative controllable systems steering admissible control is proposed, which steers the fractional system from a given initial complete state to the desired final relative state. Some remarks and comments on the existing controllability results for linear fractional dynamical system with delays are also presented.

1 Introduction

Controllability is one of the fundamental concepts in mathematical control theory and plays an important role both in traditional and fractional control theory (see e.g. monographs [3, 14, 16, 18, 20]), and survey papers [21, 23, 25]. Controllability is a qualitative property of dynamical control systems and is of particular importance in different, mainly theoretical problems in control theory. Systematic study of controllability began at the beginning of the sixties, when the theory of controllability, based on the description in the form of state space for both time-invariant and time-varying linear control systems was presented in [20]. Roughly speaking,

---

J. Klamka (✉)
Institute of Theoretical and Applied Informatics, Polish Academy of Sciences, Bałtycka 5, 44-100 Gliwice, Poland
e-mail: jerzy.klamka@iitis.pl

P. Kulczycki et al. (eds.), *Fractional Dynamical Systems: Methods, Algorithms and Applications*, Studies in Systems, Decision and Control 402,
https://doi.org/10.1007/978-3-030-89972-1_11

controllability generally means that it is possible to steer a dynamical control system from an arbitrary initial state to an arbitrary final state using the set of admissible controls.

In the literature there are many different definitions of controllability, both for linear and nonlinear or semilinear dynamical systems [4, 5, 26, 30, 31, 36]. The concept of controllability strongly depends on the class of dynamical control systems and on the set of admissible controls, [11–13, 15, 33–35]. Therefore, for linear and nonlinear or semilinear dynamical systems and fractional systems there exists many different necessary and sufficient conditions for global and local controllability [8, 9, 19, 20]. These conditions are proved using different methods of linear algebra, functional analysis and theory of differential equations and difference equations. Using theory of difference equations and pure algebraic methods, controllability of different discrete time linear fractional control systems is discussed in [7, 13, 14].

The control processes frequently involve different types of delays in state variables or in admissible controls [20]. It should be pointed out, that delay is one of the general phenomenon in a real dynamical system which has a crucial effect on the system properties, for example on the controllability, observability and stability.

Delayed systems constitute a very important class of mathematical models of real phenomena. Delays are inherent in many physical and engineering systems. In particular, pure delays are often used to ideally represent the effects of transmission and transportations. Many applications of delayed systems in engineering, mechanics and economics are presented in the monograph [10]. For dynamical systems with delays in control and/or state variables, two fundamental concepts of states are considered, namely: finite-dimensional instantaneous or relative state and infinite-dimensional complete or functional state [21, 23, 25]. However it should be stressed, that relative state does not provide full information about the trajectory of a control system. Hence it is necessary to introduce at least two different concepts of controllability, namely: relative controllability connected with relative states and complete controllability connected with complete states. Moreover taking into account possible constraints posed on the state variables and on admissible controls [12], local controllability and global controllability are also discussed.

On the other hand, fractional order continuous and discrete mathematical models express the behavior of many real processes more precisely than integer order ones. The various types of fractional differential equations have many applications in different fields of technique for example signal processing, theory of visco-elastic materials [1, 37], supercapacitors [29] filter description and design, circuit theory, computer networks, and bioengineering [15–17]. Recently different controllability problems have been discussed both for linear and nonlinear fractional infinite dimensional control systems defined in Hilbert spaces. Stochastic boundary controllability of nonlinear fractional systems defined in infinite dimensional Hilbert space is considered in paper [27] using methods of stochastic differential equations. Approximation results for linear fractional diffusion wave equation are presented and discussed in paper [28]. Moreover, the existence and properties of solutions and the initial Cauchy problem for abstract linear differential fractional equations are formulated and discussed in paper [39].

In the present chapter we shall study global relative controllability in a given finite time interval for fractional, linear, continuous time dynamical systems with multiple time variable point delays and distributed delay in admissible control. There are natural generalizations of controllability concepts, which are rather well known in the theory of finite dimensional linear control systems [21, 23, 25] without delays in state variables or in admissible control. Using techniques and methods similar to those presented in monographs [20, 24] and in the series of papers [10, 14, 15] and [19] we shall formulate and prove necessary and sufficient conditions for global relative controllability of linear fractional systems in a prescribed time interval.

This chapter is organized as follows: Sect. 2 contains a mathematical model of a linear, stationary fractional dynamical system with multiple time variable point delays in admissible controls. Moreover, in this section, a basic solution, of a fractional linear finite dimensional differential equation is presented in compact integral form and its properties are also discussed. In Sect. 3 definition of global relative controllability in a given time interval is recalled and discussed. Next, using the results and methods taken directly from linear functional analysis, a global relative controllability problem is mathematically stated and considered. Moreover, using a suitably defined relative controllability matrix, the necessary and sufficient condition and rank condition for global relative controllability in a finite time interval is formulated and proved. The next Sect. 4, is devoted to a study of a popular special case, i.e., relative controllability of fractional systems with multiple constant point delays in admissible control. Necessary and sufficient condition for relative controllability of this system is formulated using results presented in Sect. 3. In Sect. 5, which may be treated as an illustrative example, a linear fractional system with one constant delay in admissible control is considered. In Sect. 6 controllability results for a linear fractional system with distributed delay in admissible control are given. Finally, Sect. 7 contains concluding remarks, and proposes some open controllability problems for more general fractional systems.

2 System Description

Let us consider linear, fractional, delay dynamical systems containing multiple lumped time varying delays in admissible controls, described by the following differential state equation [35, 38, 40]

$$D^{\alpha}x(t) = Ax(t) + \sum_{i=0}^{i=M} B_i(t)u(v_i(t)) \tag{1}$$

for $0 < \alpha \leq 1$, and $t \in [t_0 - h, t_1]$.

with initial complete state

$$x(t_0) = x_0 \in R^n, \quad u(t) = u_{t_0}(t), \quad t \in [t_0 - h, t_0] \tag{2}$$

where.

$D^{\alpha}(t)$ denotes a fractional Caputo derivative,

$x(t) \in R^n$, is the relative state,

A is $n \times n$ dimensional constant matrix with real coefficients,

B_i, for $i = 0,1,...,M$ are given nxp dimensional constant matrices with real coefficients.

admissible controls $u \in U_{ad} = L^2([t_0, t_1], R^p)$.

Initial data $\{x_0, u_{t_0}\}$ forms complete state of the fractional delayed system (1) at initial time t_0.

The strictly increasing and twice continuously differentiable functions $v_i(t):[t_0,t_1] \to R$, $i = 0,1,...,M$, represent deviating arguments in the admissible controls, i.e. $v_i(t) = t\text{-}h_i(t)$, where $h_i(t) \geq 0$ are lumped time varying delays for $i = 0,1,...M$.

Hence, $v_i(t) \leq t$ for $t \in [t_0,t_1]$, and $i = 0,1,...,M$, and we assume that $v_0(t) = t$ for $t \in [t_0,t_1]$, and $i = 0,1,...,M$.

Let us introduce the time-lead functions $r_i(t):[v_i(t_0),v_i(t_1)] \to [t_0,t_1]$, $i = 0,1,...,M$, such that $r_i(v_i(t)) = t$ for $t \in [t_0,t_1]$. Furthermore only for simplicity and compactness of notation, let us assume that $v_0(t) = t$ and for a given t_1 the functions $v_i(t)$ satisfy the following inequalities [20].

$$\begin{aligned} h = v_M(t_1) \leq v_{M-1}(t_1) \leq \ldots v_{m+1}(t_1) \leq t_0 = v_m(t_1) < v_{m-1}(t_1) \leq \ldots \\ \ldots \leq v_1(t_1) \leq v_0(t_1) = t_1 \end{aligned} \tag{3}$$

Let us observe, that without loss of generality it may be assumed that $t_0 = v_m(t_1)$.

It is well known (see e.g. [14, 15, 17, 32]), that for given initial conditions (2) and any admissible control $u \in U_{ad}$, there exists unique solution $x(t; x_0, u) \in L^2([t_0, t_1], R^n)$ of the linear fractional differential state Eq. (1), which can be represented in the integral form. In order to do that it is convenient to introduce many notations.

The strictly increasing and twice continuously differentiable functions

$$v_i(t) : [t_0, t_1] \to R, \quad i = 0, 1, \ldots, M$$

represent deviating arguments in the admissible control, i.e., $v_i(t) = t - h_i(t)$, where $h_i(t) > 0$ are point, lumped, time varying delays for $i = 0,1,...,M$ and

$$v_i(t) \leq t, t \in [t_0, t_1], i = 0, 1, \ldots, M$$

Let us introduce the time-lead functions

$$r_i(t) : [v_i(t_0), v_i(t_1)] \to [t_0, t_1] \text{ for } i = 0, 1, \ldots, M$$

such that

$$r_i(v_i(t)) = t, \quad t \in [t_0, t_1], \quad i = 0, 1, \ldots, M$$

Moreover for a given admissible control function

$$u : [v_M(t_0), t_1] \to R^p$$

where symbol u_t denotes the function defined by the equality

$$u_t(s) = u(t+s) \quad \text{for} \quad s \in [v_M(t), t)$$

For example u_{t_0} denotes the initial admissible control function defined on time interval $[v_M(t_0), t_0)$.

It is well known (see e.g. [14, 15, 19, 20]), that for given initial conditions (2) and any admissible control $u \in U_{ad}$, there exists unique solution $x(t; x_0, u) \in L^2([t_0, t_1], R^n)$ of the linear fractional differential state Eq. (1), which can be represented in the integral form.

Furthermore, taking into account linearity of the mathematical model (1) and using Laplace transform method [17], solution $x(t, x_0, u_{t_0}, u)$ of the linear fractional differential Eq. (1) with the given initial complete state $\{x_0, u_{t_0}\}$ (2) is represented by [14, 16]:

$$\begin{aligned} x(t, x_0, u_{t_0}, u) &= F_0(t - t_0)x_0 \\ &+ \int_{t_0}^{t} F(t-s) \sum_{i=0}^{i=M} B_i u(v_i(s))ds, \qquad t \in [t_0, t_1] \end{aligned} \tag{4}$$

where

$$F_0(t) = E_\alpha(At) = \sum_{k=0}^{\infty} \frac{A^k t^{k\alpha}}{\Gamma(k\alpha + 1)}$$

is the Mittag–Leffler $n \times n$ dimensional matrix function for At, [14, 16, 17], where A is $n \times n$ dimensional constant matrix and symbol Γ denotes the Euler gamma function. Similarly we have

$$F(t) = \sum_{k=0}^{\infty} \frac{A^k t^{(k+1)\alpha - 1}}{\Gamma((k+1)\alpha)}$$

Matrix functions $F_0(t)$ and $F(t)$ are used to find the compact integral form of the solution of Eq. (1) (see e.g. [17], for more details).

Since in this chapter global relative controllability will be considered, let us recall the definition of global relative controllability in a given finite time interval.

Definition 1 The system (1) is said to be globally relatively controllable over time interval $[t_0, t_1]$ if for each pair of vectors $x_0, x_1 \in R^n$ there exists an admissible control $u \in L^2([t_0, t_1], R^m)$ such that the solution of (1) with initial conditions (2) satisfies $x(t_1) = x_1$.

Now let us separate from the solution (4) all components which depend on the given initial complete state $\{x(t_0); u_{t_0}\}$. For given final time t_1, using set of inequalities (3) and properties of integrals, it is possible to transform equality (4) as follows:

$$\begin{aligned} x(t_0, t_1, x_0, u_{t_0}, u) &= F_0(t_1 - t_0)x_0 \\ &+ \sum_{i=0}^{i=m} \int_{v_i(t_0)}^{t_0} F(t_1 - r_i(s))B_i r_i'(s) u_{t_0}(s) ds \\ &+ \sum_{i=m+1}^{i=M} \int_{v_i(t_0)}^{v_i(t_1)} F(t_1 - r_i(s))B_i r_i'(s) u_{t_0}(s) ds \\ &+ \sum_{i=0}^{i=m} \int_{t_0}^{v_i(t_1)} \sum_{j=0}^{j=m-i} F(t_1 - r_i(s))B_j r_j'(s)) u(s) ds \end{aligned} \tag{5}$$

We can divide the right hand side of the formula (5) into two sets. Let us observe that the first three terms on the right-hand side of formula (5) depend only on the initial complete state $\{x_0, u_0\}$ and in fact, do not depend at all on the admissible control $u \in L^2([t_0, t_1], R^m)$. Therefore we can separate these terms and denote shortly as follows:

$$\begin{aligned} q(t_1, t_0, x_0, u_{t_0}) &= F_0(t_1 - t_0)x_0 \\ &+ \sum_{i=0}^{i=m} \int_{v_i(t_0)}^{t_0} F(t_1 - r_i(s))B_i r_i'(s) u_{t_0}(s) ds \\ &+ \sum_{i=m+1}^{i=M} \int_{v_i(t_0)}^{v_i(t_1)} F(t_1 - r_i(s))B_i r_i'(s) u_{t_0}(s) ds \end{aligned} \tag{6}$$

Thus for given initial data $q(t_1, t_0, x_0, u_{t_0}) \in R^n$ is a constant vector.

Furthermore substituting (6) into (5) we obtain

$$x(t_0, t_1, x_0, u_{t_0}, u) = q(t_1, t_0, x_0, u_{t_0})$$

$$+\sum_{i=0}^{i=m}\int_{t_0}^{v_i(t_1)}\sum_{j=0}^{j=m-i}F(t_1-r_j(s))B_jr_j'(s)u(s)ds$$

$$= q(t_1,t_0,x_0,u_{t_0})$$

$$+\sum_{i=0}^{i=m-1}\int_{v_{i+1}(t_1)}^{v_i(t_1)}\sum_{j=0}^{j=m-i-1}F(t_1-r_j(s))B_jr_j'(s)u(s)ds$$

In order to use results and methods taken directly from the theory of bounded linear operators in Hilbert spaces, let us define linear relative controllability operator [20] as follows:

$$C_\alpha : L^2([t_0,t_1],R^p)\to R^n \tag{7}$$

$$C_\alpha u = \sum_{i=0}^{i=m}\int_{t_0}^{v_i(t_1)}\sum_{j=0}^{j=m-i}F(t_1-r_j(s))B_jr_j'(s))u(s)ds$$

$$= \sum_{i=0}^{i=m-1}\int_{v_{i+1}(t_1)}^{v_i(t_1)}\sum_{j=0}^{j=m-i-1}F(t_1-r_j(s))B_jr_j'(s)u(s)ds \tag{8}$$

The range of the relative controllability operator is finite dimensional and since matrix $F(t_1-r_i(s))$ is bounded for every $t\in[t_0,t_1]$, $i=0,1,\ldots,M$, then C_α is a linear bounded operator.

From relative controllability definition follows, that admissible control $u(t)$ steers on the time interval $[t_0,t_1]$, fractional system (1) from the given initial state x_0 to the final state x_1. In fact the relative controllability of (1) is equivalent to finding admissible control $u(t)$, $t\in[t_0,t_1]$, such that for any x_0 and x_1 the following equality holds

$$x_1 - q(t_0,t_1x_0,u_{t_0})$$

$$= \sum_{i=0}^{i=m}\int_{t_0}^{v_i(t_1)}\sum_{j=0}^{j=m-i}F(t_1-r_j(s))B_ir_i'(s)u(s)ds$$

$$= \sum_{i=0}^{i=m-1}\int_{v_{i+1}(t_1)}^{v_i(t_1)}\sum_{j=0}^{j=m-i-1}F(t_1-r_j(s))B_jr_j'(s)u(s)ds = C_\alpha(u) \tag{9}$$

Taking into account the second sum of integrals we see that relative controllability operator C_α is a sum of integral linear bounded operators defined on disjoint sets.

Then the adjoint relative controllability operator C_α^* is also a sum of linear and bounded operators.

In order to find adjoint operator C_α^* let us consider set of integral operators C_α^i, $i = 0, 1, \ldots, m-1$ given in equality (8). Hence

$$C_\alpha^i u = \int_{v_{i+1}(t_1)}^{v_i(t_1)} \sum_{j=0}^{j=m-i-1} F(t_1 - r_j(s)) B_j r_j'(s) u(s) ds$$

for $i = 0, 1, \ldots, m-1$.

and the following relation for scalar products in Hilbert spaces:
R^n and $L^2([v_{i+1}(t_1), v_i(t_1)], R^p) = V$

$$\begin{aligned}
&\langle C_\alpha^i u, y\rangle_{R^p} \\
&= \left\langle \int_{v_{i+1}(t_1)}^{v_i(t_1)} \sum_{j=0}^{j=m-i-1} F(t_1 - r_j(s)) B_j r_j'(s) u(s) ds, y \right\rangle_{R^p} \\
&= \int_{v_{i+1}(t_1)}^{v_i(t_1)} \left\langle u(s), \left(\sum_{j=0}^{j=m-i-1} F(t_1 - r_j(s)) B_j r_j'(s) \right)^* y \right\rangle_V ds \\
&= \int_{v_{i+1}(t_1)}^{v_i(t_1)} \left\langle \sum_{j=0}^{j=m-i-1} B_j^* F^*(t_1 - r_j(s)) r_j'(s)) u(s), y \right\rangle_V ds \\
&= \langle u(s), C_\alpha^{i*} y\rangle_V
\end{aligned}$$

for $s \in (v_{i+1}(t_1), v_i(t_1)]$ and $i = 0, 1, \ldots, m-1$.

Hence every adjoint operator C_α^{i*} for $i = 0,1,\ldots,m\text{-}1$, which corresponds to C_α^i is defined in different disjoint time intervals as follows

$$\begin{aligned}
C_\alpha^{i*} y &= \sum_{j=0}^{j=m-i-1} (F(t_1 - r_j(s)) B_j r_j'(s))^* y \\
&= \sum_{j=0}^{j=m-i-1} B_j^* F^*(t_1 - r_j(s)) r_j'(s)) y
\end{aligned} \tag{10}$$

for $s \in (v_{i+1}(t_1), v_i(t_1)]$ and $i = 0, 1, \ldots, m-1$.

Therefore, adjoint operator C_α^* is in fact family of operators C_α^{i*} defined on disjoint time intervals $(v_{i+1}(t_1), v_i(t_1)]$ and $i = 0, 1, \ldots, m-1$, which covers whole time interval $[t_0, t_1]$.

As was mentioned in the introduction, in this chapter functional analysis methods are used, thus operators C_α and C_α^* defined above, play a crucial role in global relative controllability discussion presented in the next sections.

3 Controllability Conditions

Using the relative controllability operator C_α and its adjoint operator C_α^* let us define the $n \times n$ dimensional relative controllability matrix $W(t_0, t_1)$ for the linear fractional control system

$$W(t_0, t_1) = C_\alpha C_\alpha^* \tag{11}$$

Taking into account relations (8), (9), and (11), relative controllability matrix $W(t_0,t_1)$ is an $n \times n$ dimensional symmetric matrix generally with real coefficients and is defined by the equality:

$$\begin{aligned}
W(t_1, t_0) &= C_\alpha C_\alpha^* \\
&= \sum_{i=0}^{i=m-1} \int_{v_{i+1}(t_1)}^{v_i(t_1)} \left(\sum_{j=0}^{j=m-i-1} F(t_1 - r_j(s)) B_j r_j'(s) \right) \\
&\times \sum_{j=0}^{j=m-i-1} F(t_1 - r_j(s)) B_j r_j'(s))^* ds \\
&= \sum_{i=0}^{i=m-1} \int_{v_{i+1}(t_1)}^{v_i(t_1)} \left(\sum_{j=0}^{j=m-i-1} F(t_j - r_i(s)) B_j r_j'(s) \right) \\
&\times \left(\sum_{j=0}^{j=m-i-1} B_j^* F^*(t_1 - r_j(s)) r_j'(s) \right) ds
\end{aligned} \tag{12}$$

Using a relative controllability matrix, it is possible to formulate and prove the main result of the paper given the following theorem, which presents necessary and sufficient conditions for global relative controllability in a given time interval.

Theorem 1 The following statements are equivalent

(1) Fractional system (1) is globally relatively controllable over $t \in [t_0, t_1]$.
(2) Relative controllability operator $C_\alpha : L^2([t_0, t_1], R^p) \to R^n$ is onto.
(3) Adjoint relative controllability operator $C_\alpha^* : R^n \to L^2([t_0, t_1], R^m)$ is invertible i.e., it is one to one operator.
(4) The bounded linear operator $C_\alpha C_\alpha^* : R^n \to R^n$ is onto and may be realized by nxn nonsingular matrix.

Proof In the proof of Theorem 1, relative controllability linear bounded operator C_α and its adjoint operator C_α^* play an important role. Hence linear functional analysis theory may be applied to prove the theorem. More precisely, we shall use methods and results taken directly from theory of linear bounded operators in Hilbert spaces.

Firstly, let us use, the range of the relative controllability operator C_α that is finite dimensional, and since matrix function *F(t)* is bounded for every $t \in [t_0, t_1]$, then operator C_α is a bounded linear operator. Moreover, as was mentioned before, from the definition 1 and integral formula (8) it immediately follows that the global relative controllability property is equivalent and that relative controllability operator C_α for relatively controllable fractional system (1) is a surjective operator. Hence equivalence (1) and (2) follows.

From the theory of linear operators it follows that surjectivity of the operator C_α. implies (see e.g. [9, 11]) that its adjoint operator

$$C_\alpha^* : R^n \to L^2([t_0, t_1], R^m)$$

is also a linear and bounded operator and moreover it is invertible, i.e., "one to one" operator. Hence equivalence (2) and (3) follows

Similarly, from theory of linear bounded operators it follows that invertibility of the selfadjoint operator $C_\alpha C_\alpha^*$ means that there exists inverse operator $(C_\alpha C_\alpha^*)^{-1}$ and is equivalent to surjectivity of the operator C_α. Therefore, for relatively controllable fractional system (1), relative controllability matrix

$$W(t_0, t_1) = C_\alpha C_\alpha^* : R^n \to R^n$$

is invertible, i.e., it is a full rank matrix. Hence equivalence (4) and (1) follows. This statement completes proof of Theorem 1.

From Theorem 1 it follows, that relative controllability matrix $W(t_0, t_1)$ plays a crucial role in relative controllability investigations and moreover, it is also used in admissible control, which transfers initial complete state $x(t_0)$ to the final desired relative state x^1 at time t_1.

Let us define admissible control

$$\begin{aligned} u^0(t) &= C_\alpha^*(C_\alpha C_\alpha^*)^{-1}(x_1 - q(t_0, t_1, x_0, u_{t_0}) \\ &= C_\alpha^* W(t_0, t_1)^{-1}(x_1 - q(t_0, t_1, x_0, u_{t_0}) \\ &= \sum_{j=0}^{j=m-i-1} B_j^* F^*(t_1 - r_j(s)) W(t_0, t_1)^{-1}(x_1 - q(t_0, t_1, x_0, u_{t_0}) \end{aligned} \tag{13}$$

for $s \in (v_{i+1}(t_1), v_i(t_1)]$ and $i = 0, 1, \ldots, m-1$.

Corollary 1 Admissible control $u^0(t)$ given by formula (13) steers globally relatively controllable fractional system (1) from the given initial complete state $\{x(t_0);u_{t_0}\}$ to the desired final relative state x_1 at time t_1.

Proof Substituting equality (13) into solution formula (5) we obtain.

$$x(t_0, t_1, x_0, u) = q(t_1, t_0, x_0, u_{t_0}) + \sum_{i=0}^{i=m} \int_{t_0}^{v_i(t_1)} \sum_{j=0}^{j=m-i} F(t_1 - r_j(s))B_j r_j'(s)u^0(s)ds = C_\alpha C_\alpha^*(C_\alpha C_\alpha^*)^{-1}(x_1 - q(t_0, t_1, x_0, u_{t_0}) = x_1 \tag{14}$$

Therefore, Corollary 1 is proved.

Remark 1 With the wide class of dynamical systems and especially for dynamical systems with different types of delays, the length of time interval $[t_0, t_1]$ is essential in controllability discussion. It should be pointed out that if the fractional delayed system (1) is globally relatively controllable on the time interval $[t_0, t_1]$, it is also globally relatively controllable on every longer time interval $[t_0, t_2]$, where $t_1 < t_2$ However even for dynamical systems without delays in admissible controls, the opposite statement is not always true, i.e. there are dynamical systems, which are globally controllable on a longer time interval, but are not controllable on a shorter time interval.

Remark 2 Let us observe, that from definition 1 it directly follows that the trajectory of the dynamical system between vectors x_1, and x_2 generally is not prescribed. Therefore, for globally relatively controllable systems generally, there are infinitely many different admissible controls defined on time interval $[t_0, t_2]$, which steer the dynamical system from initial vector x_1, to final vector x_2. However, admissible control u^0 (t) defined by formula (13) is optimal in the sense that it has minimum value of energy (see e.g. monographs [2, 24]), so it is called minimum energy control.

4 Fractional Systems with Multiple Constant Delays in Control

General results presented in the previous sections may be applied to formulate and to prove necessary and sufficient global relative controllability conditions for systems with multiple constant delays in admissible control. Therefore, now let us consider the special case of a fractional control system (1), i.e. a fractional system with constant multiple delays in admissible control, described by the following equation

$$D^\alpha x(t) = Ax(t) + \sum_{i=0}^{i=M} B_i u(t - h_i). \tag{15}$$

$$0 < \alpha \leq 1, \quad t \in [t_0, t_1]$$

$$0 = h_0 < h_1 < h_2 < \cdots < h_i < h_{i+1} < \cdots < h_{M-1} < h_M$$

In this case

$$v_i(t) = t - h_i \quad and \quad r_i(t) = t + h_i, \quad r_j'(t) = 1 \quad for \quad i = 0, 1, \ldots, M$$

hence,

$$t_1 - h_M < t_1 - h_{M-1} < \ldots < t_1 - h_m = t_0 < \cdots < t_1 - h_1 < t_1 \tag{16}$$

In this case the solution of the linear fractional Eq. (15) in integral form is given by the following equality

$$\begin{aligned} x(t_0, t_1, x_0, u_{t_0}, u) &= F_0(t_1 - t_0)x_0 \\ &+ \sum_{i=0}^{i=m} \int_{v_i(t_0)}^{t_0} F(t_1 - s - h_i) B_i u_{t_0}(s) ds \\ &+ \sum_{i=m+1}^{i=M} \int_{v_i(t_0)}^{v_i(t_1)} F(t_1 - s - h_i) B_i u_{t_0}(s) ds \\ &+ \sum_{i=0}^{i=m} \int_{t_0}^{v_i(t_1)} \sum_{j=0}^{j=m-i} F(t_1 - s - h_j) B_j u(s) ds \end{aligned} \tag{17}$$

Moreover operator

$$C_\alpha(u) = \sum_{i=0}^{i=m-1} \int_{v_{i+1}(t_1)}^{v_i(t_1)} \sum_{j=0}^{j=m-i-1} F(t_1 - s - h_j)) B_j u(s) ds$$

Similarly, as in the previous section, the linear bounded adjoint operator C_α^* is defined as family adjoint operators C_α^{i*} for $i = 0,1,...,m\text{-}1$, which are defined on a different time interval as follows

$$C_\alpha^{i*} y = \sum_{j=0}^{j=m-i-1} (F(t_1 - s - h_j) B_j)^* y$$

$$= \sum_{j=0}^{j=m-i-1} B_j^* F^*(t_1 - s - h_j)y \tag{18}$$

for $s \in (t_1 - h_{i+1}, t_1 - h_i)]$ and $i = 0, 1, \ldots, m-1$.

Therefore relative controllability matrix $W(t_1, t_0)$ has the form

$$\begin{aligned} W(t_1, t_0) &= C_\alpha C_\alpha^* \\ &= \sum_{i=0}^{i=m-1} \int_{t_1-h_{i+1}}^{t_1-h_i} \left(\sum_{j=0}^{j=m-i-1} F(t_1 - s - h_j) B_j \right) \\ &\times \sum_{j=0}^{j=m-i-1} (F(t_1 - s - h_j) B_j)^* ds \\ &= \sum_{i=0}^{i=m-1} \int_{t_1-h_{i+1}}^{t_1-h_i} \left(\sum_{j=0}^{j=m-i-1} (F(t_1 - s - h_j) B_j \right) \\ &\times \sum_{j=0}^{j=m-i-1} B_j^* F^*(t_1 - s - h_j) ds \end{aligned} \tag{19}$$

Let us define admissible control

$$\begin{aligned} u^0(t) &= C_\alpha^* (C_\alpha C_\alpha^*)^{-1} (x_1 - q(t_0, t_1, x_0, u_{t_0}) \\ &= C_\alpha^* W(t_0, t_1)^{-1} (x_1 - q(t_0, t_1, x_0, u_{t_0})) \\ &= \sum_{j=0}^{j=i} B_j^* F^*(t_1 - t - h_j) \\ &\times W(t_0, t_1)^{-1} (x_1 - q(t_0, t_1, x_0, u_{t_0})) \end{aligned} \tag{20}$$

for $s \in (t_1 - h_{i+1}, t_1 - h_i)]$ and $i = 0, 1, \ldots, m$.

Substituting admissible control $u^0(t)$ into solution (17) we obtain

$$\begin{aligned} x(t_0, t_1, x_0, u) &= q(t_1, t_0, x_0, u_{t_0}) \\ &+ \sum_{i=0}^{i=m} \int_{t_0}^{v_i(t_1)} \sum_{j=0}^{j=m-i} F(t_1 - s - h_j)) B_j u^0(s) ds \\ &= C_\alpha C_\alpha^* (C_\alpha C_\alpha^*)^{-1} (x_1 - q(t_0, t_1, x_0, u_{t_0}) = x_1 \end{aligned}$$

In the next section, as an illustrative example, relative controllability of a fractional linear control system with only one constant point delay in admissible control will be considered.

5 Fractional System with a Single Constant Point Delay in Control

In this section let us consider the special case of a fractional control system which is linear. More precisely, we shall discuss the global relative controllability problem for the fractional systems containing only single point constant delay in admissible controls, described by the following fractional differential state equation:

$$D^{\alpha}(t) = Ax(t) + B_0(t)u(t) + B_1(t)u(t-h) \tag{21}$$

$$\text{for } 0 < \alpha \le 1, \quad t \in [t_0, t_1].$$

with initial complete state

$$x(t_0) = x_0 \in R^n, \quad u(t) = u_{t_0}(t), \quad t \in [t_0 - h, t_0] \tag{22}$$

In this case we have

$$v_0(t) = t, \quad r_0(t) = t, \quad r_1'(t) = 1, \quad r_1(t) = t + h, \quad r_1'(t) = 1$$

Since for $t \in [t_0, t_1]$ and $t_1 \le t_0 + h$ fractional system (21) in fact works as a system without delays in control, let us assume that the final time t_1 satisfies the following inequality, $t_0 + h < t_1$. Similarly, as in the previous sections, in this special case it is also more convincing to present integral relative controllability operator $C_{\alpha}(u)$ in two equivalent simple integral forms:

$$\begin{aligned} C_{\alpha}(u) &= \int_{t_0}^{t_1} F(t_1 - s)B_0 u(s)ds + \int_{t_0+h}^{t_1} F(t_1 - s - h)B_1 u(s)ds \\ &= \int_{t_0}^{t_0+h} F(t_1 - s)B_0 u(s)ds + \int_{t_0+h}^{t_1} (F(t_1 - s)B_0 + F(t_1 - s - h)B_1)u(s)ds \end{aligned} \tag{23}$$

However from a computations point of view, it is better to take into consideration the second part of the integral formula (23). Thus we obtain adjoint relative controllability operator $C_{\alpha}^*(u)$ defined by two equalities in two different separated time intervals.

$$
\begin{aligned}
C_\alpha^* y &= ((F(t_1 - t)B_0)^* + (F(t_1 - t - h)B_1)^*)y \\
&= (B_0^* F^*(t_1 - t) + B_1^* F^*(t_1 - t - h))y
\end{aligned}
$$

for $t \in (t_0 + h, t_1]$.

and

$$
C_\alpha^* y = (F(t_1 - t)B_0)^* y = B_0^* F^*(t_1 - t)y
$$

for $t \in (t_0, t_0 + h]$

Thus the relative controllability matrix is given by the following equality

$$
\begin{aligned}
W(t_1, t_0) &= C_\alpha C_\alpha^* \\
&= \int_{t_0}^{t_0+h} F(t_1 - s)B_0 B_0^* F^*(t_1 - s)ds \\
&\quad + \int_{t_0+h}^{t_1} (F(t_1 - s)B_0 + F(t_1 - s - h)B_1) \\
&\quad \times (F(t_1 - s)B_0 + F(t_1 - s - h)B_1)^* ds \\
&= \int_{t_0}^{t_0+h} F(t_1 - s)B_0 B_0^* F^*(t_1 - s)ds \\
&\quad + \int_{t_0+h}^{t_1} (F(t_1 - s)B_0 + F(t_1 - s - h)B_1) \\
&\quad \times (B_0^* F^*(t_1 - s) + B_1^* F^*(t_1 - s - h))ds
\end{aligned} \tag{24}
$$

Using formulas defining operators C_α, C_α^* and matrix $W(t_0, t_1)$ it is possible to find admissible control $u^0(t)$, which steers the initial complete $\{x_0, u_{t_0}\}$ state to the desired final relative state $x_1 \in R^n$ at time t_1.

$$
\begin{aligned}
u^0(t) &= C_\alpha^*(C_\alpha C_\alpha^*)^{-1}(x_1 - q(t_0, t_1, x_0, u_{t_0})) \\
&= C_\alpha^* W(t_0, t_1)^{-1}(x_1 - q(t_0, t_1, x_0, u_{t_0})) \\
&= B_0^* F^*(t_1 - t)W(t_0, t_1)^{-1}(x_1 - q(t_0, t_1, x_0, u_{t_0}))
\end{aligned} \tag{25}
$$

for $t \in (t_0, t_0 + h]$.

and

$$\begin{aligned} u^0(t) &= C_\alpha^*(C_\alpha C_\alpha^*)^{-1}(x_1 - q(t_0, t_1, x_0, u_{t_0}) \\ &= C_\alpha^* W(t_0, t_1)^{-1}(x_1 - q(t_0, t_1, x_0, u_{t_0}) \\ &= (B_0^* F^*(t_1 - t) + B_1^* F^*(t_1 - t - h)) \\ &\quad \times W(t_0, t_1)^{-1}(x_1 - q(t_0, t_1, x_0, u_{t_0}) \end{aligned} \tag{26}$$

for $t \in (t_0 + h, t_1]$.

In order to prove the above statement, it is enough to substitute admissible control $u^0(t)$ given by (26) into integral solution (17) of the fractional control system (1). Thus we verify that

$$\begin{aligned} x(t_0, t_1, x_0, u) &= q(t_1, t_0, x_0, u_{t_0}) \\ &+ \int_{t_0}^{t_0+h} F(t_1 - s)B_0 B_0^* F^*(t_1 - s) \\ &\times W(t_0, t_1)^{-1}(x_1 - q(t_0, t_1, x_0, u_{t_0})ds \\ &+ \int_{t_0+h}^{t_1} (F(t_1 - s)B_0 + F(t_1 - s - h)B_1) \\ &\times (B_0^* F^*(t_1 - s) + B_1^* F^*(t_1 - s - h)) \\ &= \times W(t_0, t_1)^{-1}(x_1 - q(t_0, t_1, x_0, u_{t_0})ds \\ &\times C_\alpha C_\alpha^*(C_\alpha C_\alpha^*)^{-1}(x_1 - q(t_0, t_1, x_0, u_{t_0}) = x_1 \end{aligned} \tag{27}$$

Hence the desired final relative state is reached.

6 Example

Let us consider the fractional control system (21) with the following data:

$$x(t) = \begin{bmatrix} x_1(t) \\ x_2(t) \end{bmatrix} \in R^2, \quad A = \begin{bmatrix} 1 & 0 \\ 0 & 1 \end{bmatrix}, \quad t \in [t_0, t_1] = [0, 2] \tag{28}$$

$$u(t) \in R, \quad u_0(t) = 0, \quad t \in [t_0 - h, t_0] = [-1, 0]$$

$$M = 1, \quad B_0 = \begin{bmatrix} 1 \\ 0 \end{bmatrix}, \quad B_1 = \begin{bmatrix} 0 \\ 1 \end{bmatrix}$$

For the diagonal matrix A given above, matrix

$$F(t) = \sum_{k=0}^{\infty} \frac{A^k t^{(k+1)\alpha-1}}{\Gamma((k+1)\alpha)}$$

is also a diagonal matrix

$$F(t_1 - s) = \begin{bmatrix} f_{11}(t_1 - s) & 0 \\ 0 & f_{22}(t_1 - s) \end{bmatrix},$$

with positive elements $f_{11}(t_1 - s) > 0$ and $f_{22}(t_1 - s) > 0$ on the main diagonal for $s \in [t_0 - h, t_1] = [-1, 2]$.

Let us consider two cases. The first case is for given final time $t_1 \in [t_0, t_0 + h] = [0, 1]$.

Since both matrices $F(t_1 - s)$ and $F^*(t_1 - s)$ are diagonal matrices, then in this case relative controllability matrix $W(t_1, t_0)$ given by (24) has the following form

$$W(t_1, t_0) = C_\alpha C_\alpha^* = \int_{t_0}^{t_0+h} F(t_1 - s) B_0 B_0^* F^*(t_1 - s) ds$$

$$= \int_0^1 F(t_1 - s) \begin{bmatrix} 1 & 0 \\ 0 & 0 \end{bmatrix} F^*(t_1 - s) ds = \int_0^1 \begin{bmatrix} f_{11}^2(t_1 - s) & 0 \\ 0 & 0 \end{bmatrix} ds \tag{29}$$

and of course is singular, so taking into account Theorem 1, fractional system (25) is not relatively controllable on the time interval [0, 1].

Now let us assume that final time $t_1 > t_0 + h = 1$. In this case relative controllability matrix $W(t_1, t_0)$ is as follows

$$\begin{aligned} W(t_1, t_0) &= C_\alpha C_\alpha^* \\ &= \int_0^1 F(t_1 - s) B_0 B_0^* F^*(t_1 - s) ds \\ &+ \int_1^{t_1} (F(t_1 - s) B_0 + F(t_1 - s - h) B_1) \\ &\times (B_0^* F^*(t_1 - s) + B_1^* F^*(t_1 - s - h)) ds \end{aligned} \tag{30}$$

Substituting parameters of the fractional system (28) we obtain

$$W(t_0,t_1)=\int_0^1\begin{bmatrix} f_{11}^2(t_1-s) & 0\\ 0 & 0\end{bmatrix}ds$$

$$+\int_1^{t_1}\begin{bmatrix} f_{11}(t_1-s)\\ f_{22}(t_1-s)\end{bmatrix}\begin{bmatrix} f_{11}(t_1-s) & f_{22}(t_1-s)\end{bmatrix}ds$$

$$=\int_0^1\begin{bmatrix} f_{11}^2(t_1-s) & 0\\ 0 & 0\end{bmatrix}ds$$

$$+\int_1^{t_1}\begin{bmatrix} f_{11}(t_1-s)\\ f_{22}(t_1-s)\end{bmatrix}\begin{bmatrix} f_{11}(t_1-s) & f_{22}(t_1-s)\end{bmatrix}ds$$

Therefore, the relative controllability matrix is given by the following formula

$$W(t_0,t_1)=\int_0^1\begin{bmatrix} f_{11}^2(t_1-s) & 0\\ 0 & 0\end{bmatrix}ds$$

$$+\int_1^{t_1}\begin{bmatrix} f_{11}^2(t_1-s) & f_{11}(t_1-s)f_{22}(t_1-s)\\ f_{11}(t_1-s)f_{22}(t_1-s) & f_{22}^2(t_1-s)\end{bmatrix}ds$$

Hence in this case the relative controllability matrix $W(t_1, t_0)$ is nonsingular and system (28) is globally relatively controllable.

From this example it directly follows that global relative controllability of fractional control systems with delays in admissible controls, strongly depends on the length of the time interval $[t_0, t_1]$.

7 Fractional Systems with Distributed Delays in Control

In this section linear, fractional control systems with distributed delays in admissible control are considered. These control systems are extensions of systems with lumped point delays in control and are represented by the following fractional differential state equation [2, 19, 26]

$$D^{\alpha}x(t)=Ax(t)+\int_{-h}^{0}d_{\tau}B(t,\tau)u(t+\tau)\qquad t\in[t_0,t_1] \tag{31}$$

where

$x(t_0) = x_0 \in R^n$ is the given initial condition.

$su_{t_0}(t), sst \in [t_0 - h, t_0]$ is the given initial admissible control.

integral term is in the Lebesque-Stieltjes sense [6, 8, 9] with respect to τ,

$B(t, \tau)$ is $n \times p$ dimensional matrix continuous in t for fixed τ and of bounded variation in τ on *[-h,0]* for each $t \in [t_0, t_1]$ and continuous from left in τ on the interval *(-h,0)*.

Using matrices $F_0(t)$ and $F(t)$, which are dependent on α, the solution of differential Eq. (1) can be expressed in integral form as follows

$$x(t_0, t_1, x_0, u_{t_0}, u) = F_0(t_1)x_0 + \int_{t_0}^{t_1} F(t_1 - s)\left[\int_{-h}^{0} d_\tau B(s, \tau)u(s + \tau)\right]ds \tag{32}$$

Now, using unsymmetric Fubini theorem (see e.g. [6, 8] for more details) and changing the order of integration in the last term we have [2, 19, 26]

$$\begin{aligned} x(t_0, t_1, x_0, u_{t_0}, u) &= F_0(t_1)x_0 \\ &+ \int_{-h}^{0} dB_\tau\left[\int_{t_0}^{t_1} F(t_1 - s))B(s, \tau)u(s)ds\right] \\ &= F_0(t_1)x_0 + \int_{-h}^{0} dB_\tau\left[\int_{\tau}^{t_0} F(t_1 - (s - \tau))B(s - \tau, \tau)u_{t_0}(s)ds\right] \\ &+ \int_{-h}^{0} dB_\tau\left[\int_{t_0}^{t_1+\tau} F(t_1 - (s - \tau))B(s - \tau, \tau)u(s)ds\right] \\ &= F_0(t_1)x_0 + \int_{-h}^{0} dB_\tau\left[\int_{\tau}^{t_0} F(t_1 - (s - \tau))B(s - \tau, \tau)u_{t_0}(s)ds\right] \\ &+ \int_{t_0}^{t_1}\left[\int_{-h}^{0} F(t_1 - (s - \tau))d_\tau B_{t_1}(s - \tau, \tau)\right]u(s)ds \end{aligned} \tag{33}$$

where

$$B_{t_1}(s, \tau) = \begin{cases} B(s, \tau), & s \le t_1, \\ 0, & s > t_1 \end{cases}$$

The first two terms in formula (33) are dependent on the given initial relative state $\{x_0, u_{t_0}\}$ and in fact do not depend on admissible control $u(t), \quad t \ge t_0$. Therefore let us introduce the following notation

$$q(t_1, t_0, x_0, u_{t_0}) = F_0(t_1)x_0 + \int_{-h}^{0} dB_\tau \left[\int_{t_0+\tau}^{t_0} F(t_1 - (s - \tau))B(s - \tau, \tau)u_{t_0}(s)ds \right] \tag{34}$$

where dB_τ denotes the Lebesque-Stieltjes integration [6, 8, 9] with respect to the variable τ in the matrix function $B(t, \tau)$.

Changing variables in the integral term

$$\left[\int_{-h}^{0} d_\tau B(s, \tau)u(s + \tau) \right]$$

and taking into account the form of solution (34) we obtain

$$x(t_0, t_1, x_0, u_{t_0}, u) = q(t_1, t_0, x_0, u_{t_0}) + \int_{t_0}^{t_1} \left[\int_{-h}^{0} F(t_1 - (s - \tau))d_\tau B_{t_1}(s - \tau, \tau) \right] u(s)ds \tag{35}$$

Similarly, as in the previous sections, let us introduce relative controllability operator $C_\alpha(t_1)$ and its adjoint operator $C_\alpha^*(t_1)$

$$C_\alpha(t_1)u = \int_{t_0}^{t_1} (\int_{-h}^{0} F(t_1 - (s - \tau))d_\tau B_{t_1}(s - \tau, \tau))u(s)ds \tag{36}$$

$$C_\alpha^*(t_1)y = (\int_{-h}^{0} F(t_1 - (s - \tau))d_\tau B_{t_1}(s - \tau, \tau))^* y \tag{37}$$

Finally let us define $n \times n$ dimensional relative controllability matrix

$$W(t_0, t_1) = C_\alpha(t_1)C_\alpha^*(t_1) = \int_{t_0}^{t_1} (\int_{-h}^{0} F(t_1 - (s - \tau))d_\tau B_{t_1}(s - \tau, \tau))(\int_{-h}^{0} F(t_1 - (s - \tau))d_\tau B_{t_1}(s - \tau, \tau))^* ds \tag{38}$$

Corollary 2 Fractional system (31) with distributed delay in admissible control is globally relatively controllable on time interval $[t_0, t_1]$ if and only if the relative controllability matrix (38) is nonsingular.

Proof From the global relative controllability definition it directly follows, that for relatively controllable fractional system (31) the operator relative controllability operator $C_\alpha(t_1)$ is onto. On the other hand by Theorem 1 this is equivalent, that relative controllability matrix $W(t_0, t_1)$ is nonsingular. Therefore Corollary 2 follows.

For a globally relative controllability fractional system with distributed delay (31) it is possible to find an admissible control, which transforms the given initial complete state to any final relative state at time t_1. Since relative controllability matrix $W(t_0, t_1)$ is a nonsingular matrix then its inverse is well defined. Therefore let us define admissible control as follows

$$\begin{aligned} u^0(t) &= C^*_\alpha(t_1)W^{-1}(t_0, t_1)(x_1 - F_0(t_1)x_0 \\ &\quad - \int_{-h}^{0} dB_\tau \left[\int_{t_0+\tau}^{t_0} F(t_1 - (s - \tau))B(s - \tau, \tau)u_{t_0}(s)ds \right]) \\ &= C^*_\alpha(t_1)W^{-1}(t_0, t_1)(x_1 - q(t_0, t_1, x_0, u_{t_0})) \end{aligned} \tag{39}$$

where complete initial state and the final relative state vector are chosen arbitrarily.

Inserting $u^0(t)$ given by (39) into solution formula (33) and taking into account equalities (36), (37) and (38) we have

$$\begin{aligned} x(t_0, t_1, x_0, u_{t_0}, u) &= q(t_1, t_0, x_0, u_{t_0}) \\ &\quad + \int_{t_0}^{t_1} \left[\int_{-h}^{0} F(t_1 - (s - \tau))d_\tau B_{t_1}(s - \tau, \tau) \right] u^0(s)ds \\ &= q(t_1, t_0, x_0, u_{t_0}) \\ &\quad + C_\alpha(t_1)C^*_\alpha(t_1)W^{-1}(t_0, t_1)(x_1 - q(t_1, t_0, x_0, u_{t_0})) \\ &= q(t_1, t_0, x_0, u_{t_0}) \\ &\quad + W(t_0, t_1)W^{-1}(t_0, t_1)(x_1 - q(t_1, t_0, x_0, u_{t_0})) = x_1 \end{aligned} \tag{40}$$

Thus the admissible control $u^0(t)$ transfers the initial complete state to the desired final vector at time t_1.

8 Conclusions

The main result of this chapter is to show and thus prove that global relative controllability of fractional control systems with delays in admissible control is equivalent to non-singularity of a suitably defined square relative controllability matrix.

Using a suitably defined relative controllability matrix for global relatively controllable systems steering admissible control is proposed, which steers the system from the given initial complete state to the desired final relative state. Moreover, at the beginning of the chapter some remarks and comments on the existing literature on controllability results for different types of linear continuous-time and discrete-time fractional dynamical system are also presented.

Using a functional analysis approach, the controllability results presented in this chapter may be extended in many different ways. First of all, using a relative controllability matrix, relative controllability problems for semilinear fractional control systems with different types of delays not only in admissible controls but also in the state variables recently considered in papers [26, 30, 31, 36].

The second possibility is to formulate and prove the necessary and sufficient conditions for relative controllability of fractional control systems with different orders of derivatives, applying methods and concepts proposed in paper [15].

The third direction is to consider infinite dimensional control systems by applying functional analysis methods and concepts (see monographs [20, 24]). Since in this case, relative state space is infinite dimensional space, then several additional concepts of controllability should be introduced, namely: approximate absolute controllability and exact absolute controllability, approximate relative controllability and exact relative controllability.

In last few years nonlinear or semilinear fractional control systems have been discussed in the literature, e,g. in papers [26, 27, 36]. However so far, only little known reports on global or local relative controllability have been published. It follows from the fact that for nonlinear or semilinear fractional systems, we do not know the exact form of the solution for the nonlinear state equation. Relative controllability conditions for semilinear fractional systems with dominated linear parts are discussed in the paper [36] under the assumption that the linear part is relatively controllable and the nonlinear part satisfies certain inequality.

Generally in the case of semilinear or nonlinear fractional control systems, different techniques are used. The most popular is the fixed-point technique. For example, it is possible to use Banach fixed point theorem, Schauder fixed point theorem, Schaefer fixed point theorem or Darbou fixed point theorem based on measures of noncompactness in Banach, spaces, [12, 14]. It strongly depends on the form of the nonlinear part of the fractional state equation.

Minimum energy control problems similarly as for standard linear systems, are strongly connected with the controllability concept, (see e.g., [16, 20, 24] for more details). First of all, let us observe that for a relatively controllable linear control system there exists generally, many different admissible controls transferring the given initial state complete state to the desired final relative state. Therefore, we may ask which of these possible admissible controls are the optimal one according to given a priori criterion.

For quadratic criterion and relatively controllable linear fractional systems (1), (21), [22] or (31) the solution to this problem can be found using a relative controllability matrix. Moreover, the minimum energy value may be computed in rather simple form. However it should be mentioned, that this method requires many additional restrictive assumptions [20] for example, that state variables and admissible controls are unbounded in the whole time interval.

Acknowledgements The research was founded by Polish National Research Centre under grant "The use of fractional order controllers in congestion control mechanism of Internet", grant number UMO-2017/27/B/ST6/00145.

References

1. Bagley, R.L., Calico, R.A.: Fractional order state equations for the control of viscoelastically damped structure. J. Guid. Control Dyn. **14**(2), 304–311 (1991)
2. Balachandran, K., Zhou, Y., Kokila, J.: Relative controllability of fractional dynamical systems with distributed delays in control. Comput. Math. Appl. **64**(10), 3201–3209 (2012)
3. Balachandran, K., Kokila, J.: On the controllability of fractional dynamical systems, International J. Appl. Math. Comput. Sci. (AMCS) **22**(3), 523–531 (2012)
4. Balachandran, K., Govindaraj, V., Rdriguez-Germa, L., Trujillo, J.J.: Controllability results for nonlinear fractional order dynamical systems. J. Optim. Theory Appl. (JOTA) **156**(1), 33–44 (2013)
5. Balachandran, K., Govindaraj, V., Rdriguez-Germa, L., Trujillo, J.J.: Controllability of nonlinear higher order fractional dynamical systems. Nonlinear Dyn. **71**(4), 605–612 (2013)
6. Billingsley, P.: Probability and Measure, Chapter 3 Integration, Section 18 Product Measure and Fubini's Theorem, pp. 231–240. Wiley, New York (1995)
7. Busłowicz, M.: Controllability, reachability minimum energy control of fractional discrete-time linear systems with multiple delays in state. Bull. Pol. Acad. Sci., Tech. Sci. **62**(2), 233–239 (2014)
8. Cameron, R.H., Martin, W.T.: An unsymmetric Fubini theorem. Bull. Am. Math. Soc. **47**(2), 121–125 (1941)
9. Dunford, N., Schwartz T..J.: Linear Operators. Part 1. Chapter 8, Operators and Their Applications. Wiley, New York (1988)
10. Górecki, H.: Analysis and Synthesis of Time Delay Systems. Wiley, New York (1989)
11. Kaczorek, T.: Rechability and controllability to zero of cone fractional linear systems. Arch. Control Sci. **17**(3), 357–367 (2007)
12. Kaczorek, T.: Fractional positive continuous-time linear systems and their reachability. Int. J. Appl. Math. Comput. Sci. **18**(2), 223–228 (2008)
13. Kaczorek, T.: Positive fractional 2D continuous-discrete time linear systems. Bull. Pol. Acad. Sci., Tech. Sci. **59**(4), 575–579 (2011)
14. Kaczorek, T.: Selected Problems of Fractional Systems Theory. Springer, Berlin (2012)
15. Kaczorek, T.: Positive linear systems with different fractional orders. Bull. Pol. Acad. Sci., Tech. Sci. **58**(3), 453–458 (2010)
16. Kaczorek, T., Rogowski, K.: Fractional Linear Systems and Electrical Circuits, Studies in Systems, Decision and Control, vol. 13. Springer (2015)
17. Kexue, L., Jigen, P.: Laplace transform and fractional differential equations. Appl. Math. Lett. **24**(12), 2019–2023 (2011)

18. Kilbas, A.A., Srivastava, H.M., Trujillo, J.J.: Theory and Applications of Fractional Differential Equations. Elsevier, Amsterdam (2006)
19. Klamka, J.: Relative controllability and minimum energy control of linear systems with distributed delays in control. IEEE Trans. Autom. Control **21**(4), 594–595 (1976)
20. Klamka, J.: Controllability of Dynamical Systems. Kluwer Academic, Dordrecht (1991)
21. Klamka, J.: Controllability of dynamical systems - a survey. Arch. Control Sci. **2**(3–4), 281–307 (1993)
22. Klamka, J.: Controllability and minimum energy control problem of fractional discrete-time systems, chapter in monograph. In: Baleanu, D., Guvenc, Z.B., Tenreiro Machado, J.A. (eds.) New Trends in Nanotechnology and Fractional Calculus, pp. 503–509. Springer, New York (2010)
23. Klamka, J.: Controllability of dynamical systems. A survey. Bull. Polish. Acad. Sci., Tech. Sci. **61**(2), 221–229 (2013)
24. Klamka, J.: Controllability and Minimum Energy Control, Studies in Systems, Decision and Control, vol. 162, pp. 1–175. Springer (2018)
25. Klamka, J.: Controllability of dynamical systems - a survey. Arch. Control Sci. **2**(3/4), 281–307 (1993)
26. Kumar, P., S., Balachandran, K., Annapoorani, N.: Relative controllability of nonlinear fractional Longevin systems with delays in control. Vietnam J. Math. https://doi.org/10.1007/s10013-019-00356-4 (2019)
27. Lizzy, R.M., Balachandran, K.: Boundary controllability of nonlinear stochastic fractional systems in Hilbert spaces. Int. J. Appl. Math. Comput. Sci. (AMCS) **28**(1), 123–133 (2018)
28. Mitkowski, W.: Approximation of fractional diffusion wave equation. Acta Mechanica et Automatica **5**(2), 65–68 (2011)
29. Mitkowski, W., Skruch, P.: Fractional-order models of the supercapacitors in the form of RC ladder networks. Bull. Pol. Acad. Sci., Tech. Sci. **61**(3), 581–587 (2013)
30. Nirmala, R.J., Balachandran, K., Rdriguez-Germa, L., Trujillo, J.J.: Controllability of nonlinear fractional delay dynamical systems. Rep. Math. Phys. **77**(1), 87–104 (2016)
31. Nirmala, R.J., Balachandran, K.: The control of nonlinear implicit fractional delay dynamical systems. Int. J. Appl. Math. Comput. Sci. (AMCS) **27**(3), 501–513 (2017)
32. Ostalczyk, P.: The non-integer difference of the discrete-time function and its application to the control system synthesis. Int. J. Syst. Sci. **31**(12), 1551–1561 (2000)
33. Sikora, B., Klamka, J.: Constrained controllability of fractional linear systems with delays in control. Syst. Control Lett. **106**(1), 9–15 (2017)
34. Sikora, B., Klamka, J.: Cone-type constrained relative controllability of semilinear fractional systems with delays. Kybernetika **53**(2), 370–381 (2017)
35. Sikora, B., Klamka, J.: New controllability criteria for fractional systems with varying delays. In: Artur, B., Adam, C., Jerzy, K., Michał, N. (eds.) Theory and Applications of Non-integer Order Systems. Lecture Notes in Electrical Engineering, vol. 407. Springer (2017)
36. Sivabalan, M., Sathiyanathan, K.: Relative controllability results for nonlinear higher order fractional delay integrodifferential systems with time varying delay in control, Communications Faculty of Sciences, University of Ankara, series A1. Math. Stat. **68**(1), 889–906 (2019)
37. Torvik, P.J., Bagley, R.L.: On the appearance of the fractional derivative in the behavior of real materials. J. Appl. Mech. **51**(2), 294–298 (1984)
38. Venkatesan, G., George, R.: Controllability of fractional dynamical systems. A functional analytic approach. Math. Control Relat. Fields **7**(4), 537–562 (2017)
39. Wang, J.R., Zhou, Y., Feckan, M.: Abstract Cauchy problem for fractional differential equations. Nonlinear Dyn. **71**(4), 685–700 (2013)
40. Wei, J.: The controllability of fractional control systems with control delay. Comput. Math. Appl. **64**(10), 3153–3159 (2012)

Applications

Selected Engineering Applications of Fractional-Order Calculus

Wojciech Mitkowski, Marek Długosz, and Paweł Skruch

Abstract In this chapter several examples of using fractional-order calculus in selected engineering applications are presented. It is shown that some real systems can be better mathematically described with fractional-order differential equations. The chapter focuses on ladder network structures with fractional-order elements to model both electrical and nonelectrical systems with distributed parameters. Examples of modeling of supercapacitors, batteries, a chain of vehicles functioning in adaptive cruise control mode, and thermal processes inside buildings are provided. The effectiveness of the proposed modeling approach is verified by both simulation and experimental results.

1 Introduction

In engineering applications the construction of appropriate mathematical models plays an important role. Fractional class models better describe the behavior of real physical processes, as evidenced by numerous examples. It should be also noted that typically, a mathematical model refers only to some specific aspects of the phenomenon and two models of the same phenomenon may be essentially different. This statement is valid for the structure of the model as well as its formal description. Such flexibility in mathematical modeling is visible in ladder network structures that have been intensively investigated in recent decades. Ladder networks are typically described as networks formed by numerous repetitions of an elementary cell. In the case of an electric ladder network, the elementary cell may consist of resistors, inductance coils and capacitors connected in series or in parallel. Mechanical ladder

---

W. Mitkowski · M. Długosz · P. Skruch (✉)
AGH University of Science and Technology, al. A. Mickiewicza 30, 30-059 Cracow, Poland
e-mail: pawel.skruch@agh.edu.pl

W. Mitkowski
e-mail: wojciech.mitkowski@agh.edu.pl

M. Długosz
e-mail: mdlugosz@agh.edu.pl

P. Kulczycki et al. (eds.), *Fractional Dynamical Systems: Methods, Algorithms and Applications*, Studies in Systems, Decision and Control 402,
https://doi.org/10.1007/978-3-030-89972-1_12

networks consist of masses, springs and dampers. Ladder networks may be applied to model high-order dynamics, systems with distributed parameters, integrated interconnection problems, complex nonlinear phenomena, coupled mechanical systems, analog neural nets and so on. An additional degree of freedom to model a specific system behavior can be achieved by introduction to the network, fractional-order elements.

This chapter is organized as follows: the first section discusses RC ladder systems. In the second section, a supercapacitor model is built. The third section deals with the analysis of the behavior of a series of vehicles. The next section discusses the problems of modeling the temperature distribution inside residential buildings. Examples of other applications and concluding remarks are given at the end of this chapter.

2 RC-Ladder Networks with Supercapacitors

There is extensive literature on different types of ladder systems, see, e.g., the literature list in Mitkowski [22]. The authors of the works often emphasize the close relationship between ladder systems and distributed systems described by means of partial differential equations. Models with non-integral-order derivatives are introduced to describe real objects more accurately. RC ladder systems are used for the analysis and modeling of microelectronic circuits, supercapacitors, biological systems, temperature distribution in the modeling of spatial structures and in problems of electricity transmission. The dynamic properties are characterized by the eigenvalues of the respective ladder systems. Integrated RC, RCR, RRC and RC ladders of flat and spatial RRCRR networks are considered. Such structures have recently been used, for example, for modeling supercapacitors. The study also considered RCR ring systems and RC exponentially convergent. Dynamic properties of the networks are characterised by eigenvalues of Jacobi cyclic state matrix. See for example Mitkowski [22] and Bauer et al. [4].

2.1 *Basic RC-Ladder Network*

Let us now consider the basic ladder RC system depicted in Fig. 1. For simplicity we choose $n = 3$. The capacitance of supercapacitors is denoted with C_k, respectively. The current $i(t)$ through capacitor C_k is equal to $C_k \frac{\mathrm{d}^\alpha x_k}{\mathrm{d}t^\alpha}$, where $x_k(t)$ denotes the voltage for $C_k, \alpha \in (0, 2]$ is the non-integer order derivative [4] and $u(t)$ is the voltage source. The basic RC-ladder system shown in Fig. 1 is described (for any n) by the following equations

$$\frac{\mathrm{d}^\alpha x_i(t)}{\mathrm{d}t^\alpha} = a_i x_{i-1}(t) + b_i x_i(t) + c_i x_{i+1}(t)\,, x_0(t) = u(t)\,, x_{n+1}(t) = 0\,, \quad (1)$$

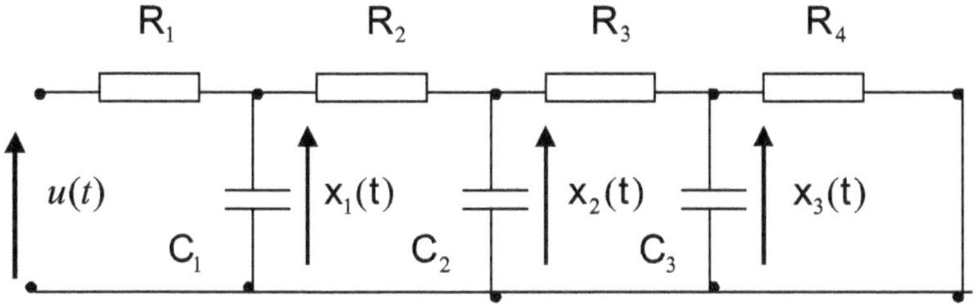

Fig. 1 Basic RC-ladder network for $n = 3$

$$a_i = 1/(R_i C_i)\,, \quad c_i = 1/(R_{i+1} C_i)\,, \quad b_i = -(a_i + c_i)\,. \tag{2}$$

In this section, firsty, the following Caputo definition of the fractional derivative will be used [16], ([17], p. 92)

$$\frac{\mathrm{d}^\alpha f(t)}{\mathrm{d}t^\alpha} = \frac{1}{\Gamma(n-\alpha)} \int_0^t f^{(n)}(\tau)/(t-\tau)^{\alpha+1-n} \mathrm{d}\tau\,, \tag{3}$$

where $n - 1 < \alpha < n$, $n = 1, 2, \ldots,$

$$\Gamma(x) = \int_0^\infty t^{x-1} \mathrm{e}^{-t} \mathrm{d}t \tag{4}$$

is the gamma Euler function and

$$f^{(n)}(\tau) = \frac{\mathrm{d}^n f(\tau)}{\mathrm{d}\tau^n}\,. \tag{5}$$

The system of equations (1)–(2) can be written in the vector-matrix form with the following form:

$$\frac{\mathrm{d}^\alpha x(t)}{\mathrm{d}t^\alpha} = Ax(t) + Bu(t)\,, \tag{6}$$

where $x(t) = [x_1(t)\, x_2(t)\, \ldots\, x_n(t)]^{\mathrm{T}}$ and

$$B = \left[a_1\ 0\ 0\ 0\ 0\right]^{\mathrm{T}}\,, \tag{7}$$

$$A = \begin{bmatrix} b_1 & c_1 & 0 & 0 & a_1 \\ a_2 & b_2 & c_2 & 0 & 0 \\ 0 & a_3 & b_3 & c_3 & 0 \\ 0 & 0 & a_4 & b_4 & c_4 \\ c_5 & 0 & 0 & a_5 & b_5 \end{bmatrix}\,. \tag{8}$$

In (7) the symbol T indicates transposition. The vector (7) and matrix (8) for simplification of notation has been shown for $n = 5$. The basic RC ladder is described by the equation (6) with the matrix (8), where $a_1 = c_5 = 0$ (generally $a_1 = c_n = 0$). If $a_1 = c_n = 0$ and $a_{i+1}c_i > 0$ for $i = 1, 2, \ldots, n-1$, then A is called **tridiagonal Jacobi matrix** dimension $n \times n$. Tridiagonal real Jacobi matrix has only single (distinct) real eigenvalues $\lambda_1, \lambda_2, \ldots, \lambda_n$ (see for example [22]).

The tridiagonal Jacobi matrix A is similar to the diagonal canonical Jordan form $J = \text{diag}(\lambda_1, \lambda_2, \ldots, \lambda_n)$. That is to say, there exists P, $\det P \neq 0$ such that $P^{-1}AP = J$. In other words, the matrix A is diagonalizable. Additionally, if $a_i > 0$, $c_i > 0$ and $b_i = -(a_i + c_i)$, then $\lambda_k \in [-m\ 0)$, $k = 1, 2, \ldots, n$, where $m = 2 \cdot \max_k(a_k + c_k)$. Thus, the matrix A is asymptotically stable.

The Jacobi matrix (8) is diagonalizable. Therefore, the system (6) is diagonalizable and can be decomposed to the n independent differential one-dimensional equation:

$$\frac{\mathrm{d}^\alpha z_k(t)}{\mathrm{d}t^\alpha} = \lambda_k z_k(t) + w_k u(t)\,, \tag{9}$$

where λ_k is the eigenvalue of matrix A and $P^{-1}AP = J$, P is an invertible real matrix $n \times n$. Jordan canonical form is the real matrix $J = \text{diag}(\lambda_1, \lambda_2, \ldots, \lambda_n)$. The vector $[z_1(t) \ \ldots \ z_n(t)]^{\mathrm{T}} = z(t) = P^{-1}x(t)$. The w_k factor in equation (9) is the k-th coordinate of the vector $w = P^{-1}B$, vector B is given in (7).

The general solution of the scalar equation (9) is generally known and is given by the following formula ([16], p. 92):

$$z_k(t) = E_\alpha(\lambda_k t^\alpha) b z_k(0) + \int_0^t \Phi(t-\tau) w_k u(\tau)\mathrm{d}\tau\,, \tag{10}$$

where

$$E_\alpha(\lambda_k t^\alpha) = \sum_{i=0}^{\infty} (\lambda_k t^\alpha)^i / \Gamma(i\alpha + 1)\,, \tag{11}$$

$$\Phi(t) = \sum_{i=0}^{\infty} \lambda_k^i t^{(i+1)\alpha - 1} / \Gamma((i+1)\alpha)\,. \tag{12}$$

In (11) the symbol E_α means the Mittag-Leffler function. See for example, Huseynov et al. [14], Kaczorek [17], Pillai [32]. The gamma function Γ is given in (4).

Consider $n \times n$ real **cyclic Jacobi matrix** A given by (8) with $a_i = a$, $c_i = c$, $b_i = b$. In this particular case, the **eigenvalues of the cyclic Jacobi matrix** A have the following form (see for example ([15], p. 159), [22]):

$$\lambda_k = b + 2\sqrt{ac}\cos\phi_k \quad k = 1, 2, \ldots, n\,, \tag{13}$$

$$\phi_k = k2\pi/n. \tag{14}$$

A special case of a $n \times n$ cyclic Jacobi matrix is a $n \times n$ **tridiagonal Jacobi matrix**. Matrix A is a tridiagonal Jacobi matrix, for example, if in A $a_i = c_i = 1$, $b_i = b$ and $a_1 = c_n = 0$. In this particular case, the **eigenvalues of the tridiagonal matrix** A have the following form given in (13), but now

$$\phi_k = k\pi/(n+1)\,. \tag{15}$$

Example

Consider a homogeneous RC-type ladder system described by equality (1) and shown in Fig. 1 for $n = 3$. The electric ladder system (1) is homogeneous (uniform), which means that $R_i = R$, $C_i = C$. In this case matrix (8) is a tridiagonal Jacobi matrix and from (6) we have

$$RC\frac{\mathrm{d}^\alpha x(t)}{\mathrm{d}t^\alpha} = Ax(t) + Bu(t)\,, \tag{16}$$

where $x(t) = [x_1(t)\,x_2(t)\,\ldots\,x_n(t)]^{\mathrm{T}}$ and

$$B = \begin{bmatrix}1\;0\;0\cdots 0\end{bmatrix}^{\mathrm{T}}\,, \tag{17}$$

$$A = \begin{bmatrix} -2 & 1 & 0 & \cdots & 0 \\ 1 & -2 & 1 & \cdots & 0 \\ 0 & 1 & -2 & \cdots & 0 \\ \vdots & \vdots & \vdots & \ddots & \vdots \\ 0 & 0 & 0 & \cdots & -2 \end{bmatrix}. \tag{18}$$

Consider the transformation matrix P by the similarity of the following form ([15], p. 159):

$$P = \sqrt{\frac{2}{n+1}}\begin{bmatrix} \sin\phi_1 & \sin\phi_2 & \sin\phi_3 & \cdots & \sin\phi_n \\ 2\sin\phi_1 & 2\sin\phi_2 & 2\sin\phi_3 & \cdots & 2\sin\phi_n \\ 3\sin\phi_1 & 3\sin\phi_2 & 3\sin\phi_3 & \cdots & 3\sin\phi_n \\ \vdots & \vdots & \vdots & \ddots & \vdots \\ n\sin\phi_1 & n\sin\phi_1 & n\sin\phi_1 & \cdots & n\sin\phi_1 \end{bmatrix}, \tag{19}$$

where ϕ_k, $k = 1, 2, \ldots, n$, is given in (15).

The matrix (19) has an interesting property: $P^2 = I$, $P^{-1} = P$. The matrix (18) is diagonalizable, that is,

$$P^{-1}AP = PAP = \mathrm{diag}(\lambda_1,\,\lambda_2,\,\ldots\,\lambda_n)\;\lambda_k = -\frac{2}{RC}(1-\cos\phi_k)\,, \tag{20}$$

$$\lambda_k = -\frac{2}{RC}(1 - \cos\phi_k) = -\frac{4}{RC}\sin^2\frac{\phi_k}{2}\,, \quad k = 1, 2, ..., n\,. \tag{21}$$

The Jacobi matrix (18) is also diagonalizable. Therefore, the system (16) is diagonalizable too and can be decomposed to the n independent differential one-dimensional equation (9), where

$$w_k = \frac{1}{RC}\sqrt{\frac{2}{n+1}}\,k\sin\phi_1\,, k = 1, 2, ..., n\,. \tag{22}$$

The solution of Eq. (9) with the parameters (21), (22) has the form (10).

Below we consider two cases for $\alpha = 1$ and $\alpha = 2$.

For $\alpha = 1$ we have $E_\alpha(\lambda_k t^\alpha) = \mathrm{e}^{\lambda_k t}$ and from (10)

$$z_k(t) = \mathrm{e}^{\lambda_k t} z_k(0) + \int_0^t \mathrm{e}^{\lambda_k(t-\tau)} w_k u(\tau)\,\mathrm{d}\tau. \tag{23}$$

Let $Z_k(s)$ be the Laplace transform of the function $z_k(t)$, that is $L(z_k(t)) = Z_k(s)$. The transmittance of (9) for $\alpha = 1$ is

$$G(s) = \frac{w_k}{s - \lambda_k} = \frac{Z_k(s)}{U(s)}\,. \tag{24}$$

The impulse characteristic of the system $G(s)$ has the form

$$y_k(t) = L^{-1}[G(s)] = w_k \mathrm{e}^{\lambda_k t}\,. \tag{25}$$

For $\alpha = 2$ we have $E_{\alpha=2}\lambda_k t^{\alpha=2} = \cos(\sqrt{\lambda_k} t)$. The solution of (9) for $\alpha = 2$ is of the form:

$$z_k(t) = \cos(\omega_k t) z_k(0) + (\omega_k)^{-1}\sin(\omega_k t) z_k^{(1)}(0) + f(t), \tag{26}$$

$$f(t) = (\omega_k)^{-1}\int_0^t \sin(\omega_k(t-\tau)) w_k u(\tau)\mathrm{d}\tau\,, \quad \omega_k = \sqrt{\lambda_k}\,. \tag{27}$$

In this example an interesting phenomenon can be observed. Notice that the RC system at $\alpha = 2$ behaves like the LC system. The LC circuit is obtained from the circuit shown in Fig. 1 by converting the resistance R to the inductance L.

The dynamic properties of the system (9) are well characterized by the impulse characteristics of the tested system. Figures 2 and 3 show the $z_k(t)$ time series of the system (9) with $n = 3$ and for $\alpha = 0.7$ and $\alpha = 1.5$, respectively (see [4]).

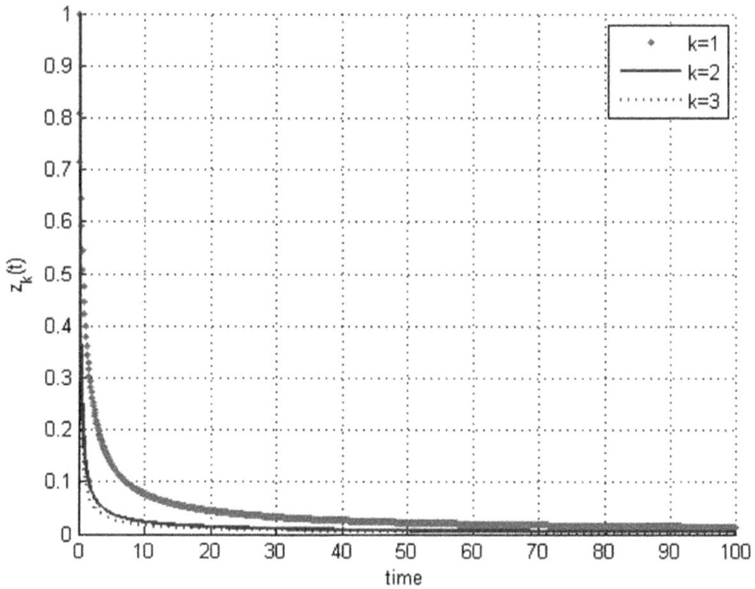

Fig. 2 Impulse time characteristics of the tested RC system for alpha = 0.7

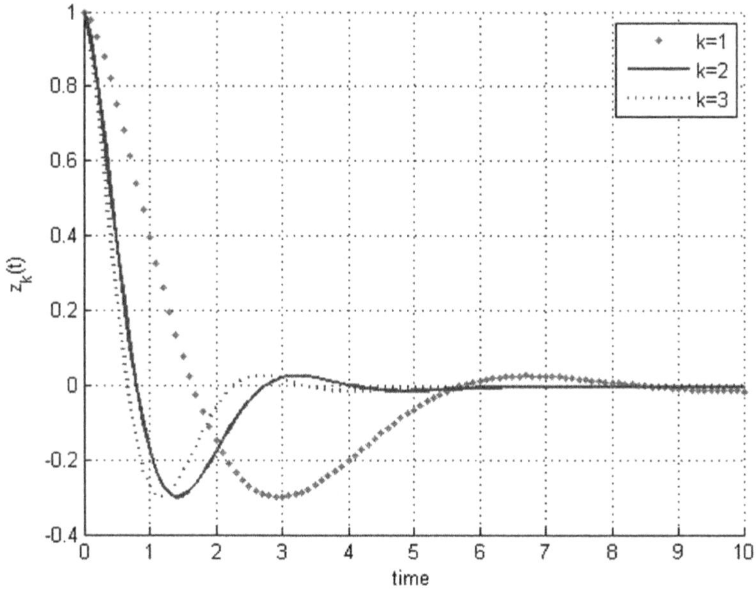

Fig. 3 Impulse time characteristics of the tested RC system for $\alpha = 1.5$

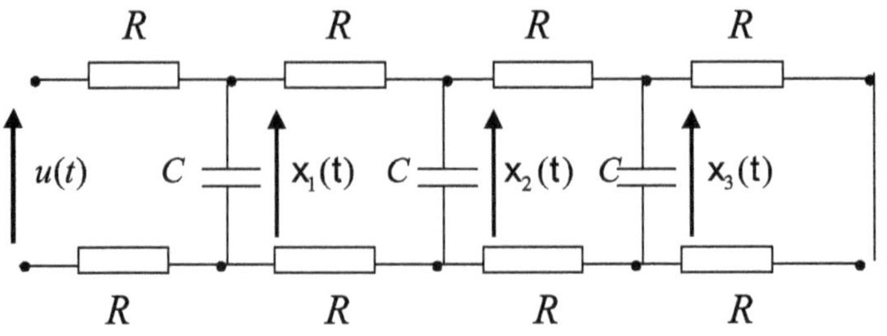

Fig. 4 RCR-uniform ladder network for n = 3

2.2 RCR-Uniform Ladder Network

Now we consider a RCR-uniform ladder network. For $n = 3$ the RCR–ladder network is shown in Fig. 4. The RCR–uniform ladder network can be described by the following equation (compare with (16) and Fig. 1 with Fig. 4):

$$2RC\,\frac{d^{\alpha}x(t)}{dt^{\alpha}} = Ax(t) + Bu(t)\ ,\ \ B = [1\,0\,0...0]^T\ , \tag{28}$$

where $x(t) = [x_1(t)\,x_2(t)\,\ldots\,x_n(t)]^{\mathrm{T}}$ and tridiagonal Jacobi matrix A is given in the pattern (18).

The system (28) after diagonalization takes the form (9) in which (compare with (21) and (22)):

$$\lambda_k = -\frac{1}{RC}(1-\cos\phi_k) = -\frac{2}{RC}\sin^2\frac{\phi_k}{2}\,,\quad k = 1, 2, \ldots, n\,, \tag{29}$$

$$w_k = \frac{1}{2RC}\sqrt{\frac{2}{n+1}}k\sin\phi_1\,,\quad \phi_k = \frac{k\pi}{n+1}\,,\quad k = 1, 2, \ldots, n\,. \tag{30}$$

2.3 RCR-Ring Uniform Ladder Network

Consider the (fundamental) basic RC–ladder system (1) and let additionally

$$x_n(t) = x_0(t)\,,\quad x_{n+1}(t) = x_1(t)\quad \text{and}\quad R_{n+1} = R_1\,. \tag{31}$$

In this case, the system (1), (31) is called an electric RC–ring network.

Let $R_i = R$ and $C_i = C$. Then the RCR–ring system (see Fig. 5 for $n = 6$) can be described by the equation

$$2RC\frac{\mathrm{d}^{\alpha}x(t)}{\mathrm{d}t^{\alpha}} = A_c x(t)\,, \tag{32}$$

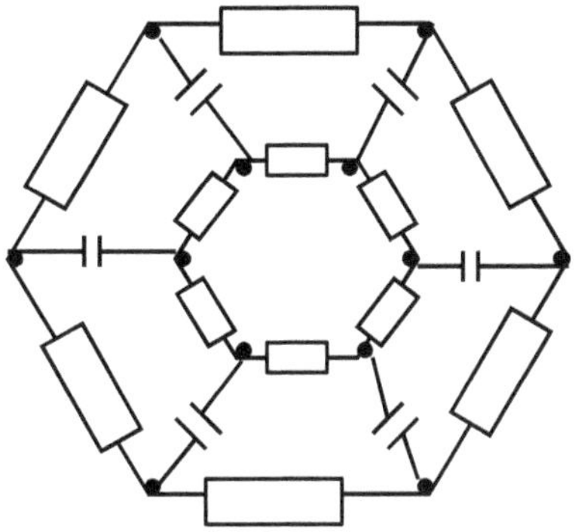

Fig. 5 RCR-ring uniform ladder network for n = 6

where $x(t) = [x_1(t)\, x_2(t)\, \ldots\, x_n(t)]^{\mathrm{T}}$ and cyclic Jacobi matrix A_c is given (to simplify the notation for $n = 5$) in the pattern (8) with $a_i = 1$, $c_i = 1$, $b_i = -2$.

The system (32) after diagonalization takes the following form:

$$2RC \frac{\mathrm{d}^{\alpha} z_k(t)}{\mathrm{d}t^{\alpha}} = \lambda_k\, z_k(t)\,, \tag{33}$$

where (see and compare with (13) and (14))

$$\lambda_k = -\frac{1}{RC}(1 - \cos\phi_k) = -\frac{2}{RC}\sin^2\frac{\phi_k}{2}\,, \quad \phi_k = \frac{k2\pi}{n}\,, \quad k = 1, 2, \ldots, n\,. \tag{34}$$

2.4 RRCr-Uniform Ladder Network

Let us consider an RRCr–uniform ladder network (see Fig. 6 for $n = 3$). If $r = 0$, then the ladder network is called an electric RRC–uniform ladder network.

Let

$$J(n; b) = \begin{bmatrix} b & 1 & 0 & \cdots & 0 \\ 1 & b & 1 & \cdots & 0 \\ 0 & 1 & b & \cdots & 0 \\ \vdots & \vdots & \vdots & \ddots & \vdots \\ 0 & 0 & 0 & \cdots & b \end{bmatrix}. \tag{35}$$

For example, $J(n; -2) = A$, where A is given in (18). Let us denote by the symbol $\lambda_k(A)$ the k-th eigenvalue of the matrix A. The following equality is true ([15], p. 159):

$$\lambda_k = b + 2\cos\phi_k\,, \quad \phi_k = k\pi/(n+1)\,, \quad k = 1, 2, \ldots, n\,. \tag{36}$$

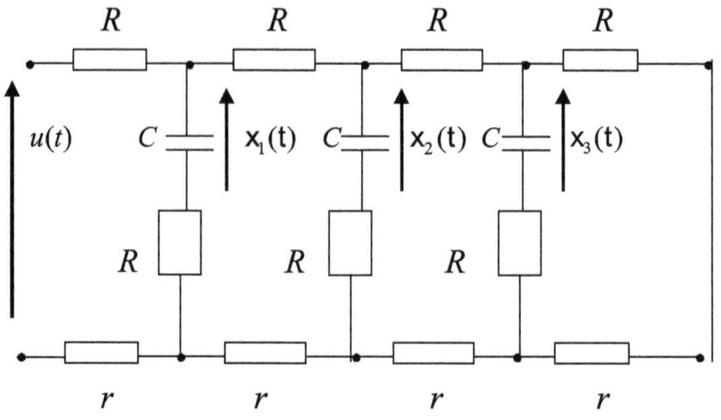

Fig. 6 RRCr-uniform ladder network for n = 3

The following equality is also true (for example [4, 22]):

$$P^{-1}J(n;b)P = PJ(n;b)P = \mathrm{diag}\,(\lambda_1, \lambda_2, \ldots, \lambda_k)\,, \tag{37}$$

where the matrix P is given by equation (19).

The RRC-uniform ladder network (see Fig. 6 with $n = 3$) can be described in general by equation:

$$-\,RCJ(n;-3)\frac{\mathrm{d}^\alpha x(t)}{\mathrm{d}t^\alpha} = J(n;-2)\,x(t) + Bu(t)\,, \quad B = [1\,0\,0\,\ldots 0]^{\mathrm{T}}\,, \tag{38}$$

where $x(t) = [x_1(t)\,x_2(t)\,\ldots\,x_n(t)]^{\mathrm{T}}$ and tridiagonal Jacobi matrix $J(n;b)$ is given in the pattern (35).

The system (38) is diagonalizable. After diagonalization, the system (38) takes the form

$$\frac{\mathrm{d}^\alpha z_k(t)}{\mathrm{d}t^\alpha} = \lambda_k z_k(t) + w_k u(t)\,, \quad k = 1, 2, \ldots, n\,, \tag{39}$$

where

$$\lambda_k = -\frac{4\sin^2(\phi_k/2)}{RC(1+4\sin^2(\phi_k/2))}\,, \quad w_k = \frac{\sqrt{2/(n+1)}\,\sin\phi_k}{RC(1+4\sin^2(\phi_k/2))}\,. \tag{40}$$

If $r = R$ (see Fig. 6), then the ladder network is called a RRCR–uniform ladder network. The RRCR-ladder network can be described by the equation

$$-\,RCJ(n;-4)\frac{\mathrm{d}^\alpha x(t)}{\mathrm{d}t^\alpha} = J(n;-2)x(t) + Bu(t)\,, \quad B = [1\,0\,0\ldots 0]^{\mathrm{T}}\,, \tag{41}$$

where $x(t) = [x_1(t)\,x_2(t)\,\ldots\,x_n(t)]^{\mathrm{T}}$ and tridiagonal Jacobi matrix $J(n;b)$ is given in the pattern (35).

The system (41) is diagonalizable. After diagonalization, the system (41) takes the form (39) with

$$\lambda_k = -\frac{2\sin^2(\phi_k/2)}{RC(1+2\sin^2(\phi_k/2))}, \quad w_k = \frac{\sqrt{2/(n+1)}\sin\phi_k}{2RC(1+2\sin^2(\phi_k/2))}. \tag{42}$$

2.5 Exponential RC-Ladder Network

Consider the long line of heterogeneous parameters R and C. Let the length of the line be equal to 1. Let $h = 1/(n+1)$ be a step discretization spatial variable $\xi \in [0, 1]$. The heterogeneous, exponentially convergent transmission line has the following parameters given by the formulas: $R(\xi) = R\exp(\alpha\xi)$ and $C(\xi) = C\exp(-\alpha\xi)$, where α is the convergence parameter. In this case, a suitable RC-ladder system similar to that is shown in Fig. 1 (see also (2)) with the parameters:

$$R_i = k^i R \quad C_i = k^{-1}C \quad k > 0. \tag{43}$$

The system (6) with $a_1 = c_n = 0$ and (43) is called an exponential RC–ladder network. The state $n \times n$ matrix A_e of our exponential system has the form [22]

$$A_e = \frac{1}{RC}\begin{bmatrix} -(1+1/k) & 1/k & 0 & \cdots & 0 \\ 1 & -(1+1/k) & 1/k & \cdots & 0 \\ \vdots & \ddots & \ddots & \ddots & \vdots \\ 0 & \cdots & 1 & -(1+1/k) & 1/k \\ 0 & \cdots & 0 & 1 & -(1+1/k) \end{bmatrix}. \tag{44}$$

The exponential RC–ladder network is diagonalizable. Eigenvalues of matrix (44) are given by

$$\lambda_i = -\frac{1}{RC}(1 + \frac{1}{k} - 2\sqrt{\frac{1}{k}}\cos\phi_i), \quad \phi_i = \frac{i\pi}{n+1}, \tag{45}$$

where $i = 1, 2, 3, ..., n$. It is evident that the exponential RC–ladder network may be represented in the form (39).

2.6 RC-Plane Network

Details of the RC–uniform plane network in space $\mathbb{R}^2$ is shown in Fig. 7. The scheme of a flat RC–network is shown in Fig. 8 [4, 22].

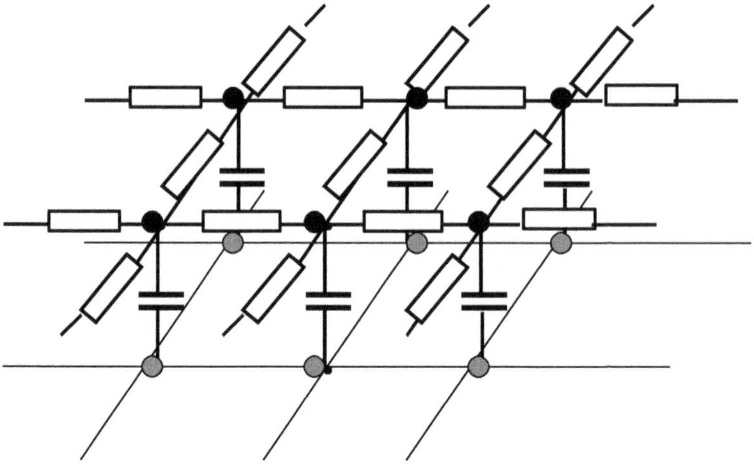

Fig. 7 RC–plane network

Fig. 8 The scheme of a flat RC–network

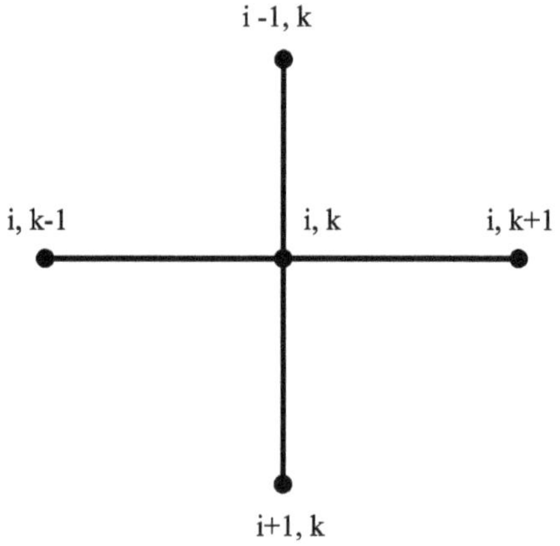

The uniform RC–plane network with supercapacitors C can be described by equations

$$RC\frac{\mathrm{d}^{\alpha}}{\mathrm{d}t^{\alpha}}x_{i,k} = x_{i,k-1} - 2x_{i,k} + x_{i,k+1} + x_{i-1,k} - 2x_{i,k} + x_{i+1,k}\,, \tag{46}$$

where $x_{i,k}$ is the voltage on the capacitance C associated with the node (i, k), $i = 1, 2, \ldots, m$ and $k = 1, 2, \ldots, n$.

The shape of the flat network and the network boundary conditions are important. We consider a rectangular network with Dirichlet boundary conditions. In this case, the equality (46) can be written in vector-matrix form (details see [21])

$$RC\frac{\mathrm{d}^{\alpha}}{\mathrm{d}t^{\alpha}}x(t) = A_{pl}x(t) + Bu(t)\,. \tag{47}$$

Method power supply determines the matrix B. The matrix B is dependent on the boundary control. Now we consider a rectangular network with Dirichlet boundary conditions and $u(t) = 0$. The dimension of the vector $x(t)$ is $m \times n$. For example, if $m = 2$, $n = 3$, then real block-matrix A_{pl} is given by the equation

$$A_{pl} = \begin{bmatrix} -4 & 1 & 0 & 1 & 0 & 0 \\ 1 & -4 & 1 & 0 & 1 & 0 \\ 0 & 1 & -4 & 0 & 0 & 1 \\ 1 & 0 & 0 & -4 & 1 & 0 \\ 0 & 1 & 0 & 1 & -4 & 1 \\ 0 & 0 & 1 & 0 & 1 & -4 \end{bmatrix}, \tag{48}$$

or

$$A_{pl} = J_{bloc}[2;\, J(3;-4)] = \begin{bmatrix} J(3;-4) & I_{3x3} \\ I_{3x3} & J(3;-4) \end{bmatrix}, \tag{49}$$

where $J(n; b)$ is given in (35). In the general case we have $A_{pl}{=}J_{block}[m;\, J(n;b)]$.

The eigenvalues $\lambda_{i,k}(A_{pl})$ of the block matrix A_{pl} are given by following formulas [21]:

$$\lambda_{i,k}(A_{pl}) = b + 2(\cos\psi_i + \cos\phi_k), \tag{50}$$

where $i = 1, 2, \ldots, m$, $k = 1, 2, \ldots, n$ and

$$\psi_i = i\pi/(m+1)\,, \quad \phi_k = k\pi/(n+1). \tag{51}$$

Formulas for the eigenvectors of the block-matrix A_{pl} are given in Mitkowski [21]. The system (47) is diagonalizable (see eigenvalues of the block-matrix A_{pl} given in Eq. (50)). The eigenvalues of A_{pl}, are not necessarily single. This is important for the study of controllability of the system (47).

2.7 Conclusion and Remarks

In this section, the dynamic properties characterized by the eigenvalues of the following structures are considered: ladder systems with supercapacitors of RC, RCR, RRC, RRCRR types and a flat RC network. The study also consideres exponentially convergent ladder networks and RCR ring systems.

The ladder and ring networks can be applied in the approximation of processes such as: hot mills in metallurgy, long transmission lines, multicomponent rectification in a distillation column, or when we consider an approximation of some distributed parameter systems.

Similarly to integer order systems, it may be decomposed into n scalar subsystems (39) which simplifies the analysis.

Depending on the order of the differential equation, the parameter α changes the dynamic properties of the time signals. See Figs. 2 and 3.

Depending on the order α of the differential equation, it becomes either an RC or LC ladder system. See (23) and (26).

We proved that the analytical approach to complex RC-ladder systems with supercapacitors is possible. The RC structure can be transformed to (39).

In this section, flat RC networks with supercapacitors are considered. Considerations can be generalized to the RC spatial networks discussed in Mitkowski [21].

3 Modeling of Supercapacitors

Supercapacitors (also called ultracapacitors or electrochemical double-layer capacitors) can be considered as good representatives of physical systems, which mathematical models can effectively and efficiently describe using fractional-order calculus. A transmission line model in the form of the RC ladder network presented in Fig. 9 is one of the possible modeling approaches that aims to capture specific supercapacitor physical phenomena such as the distributed double-layer capacitance and the distributed electrolyte resistance that extends the depth of the pore. The dynamic behavior of the RC ladder network illustrated in Fig. 9 can be described by the following fractional-order differential equations (see [26])

$$C_s R_s D \frac{d^\alpha x(t)}{dt^\alpha} = Ax(t) + Bu(t), \quad x(0) = x_0 \,, \tag{52}$$

where $x(t) = [x_1(t)\ x_2(t)\ \ldots\ x_n(t)]^T \in \mathbb{R}^n$ is an n-dimensional vector which elements $x_i(t)$, $i = 1, 2, \ldots, n$ denote the voltages across the plates of the capacitors C_s, $x_0 \in \mathbb{R}^n$ is the given initial condition, $u(t) \in \mathbb{R}$ is the control voltage, $\alpha \in (0, 1]$ denotes the order of the fractional derivative according to the Caputo definition [33], $t > 0$,

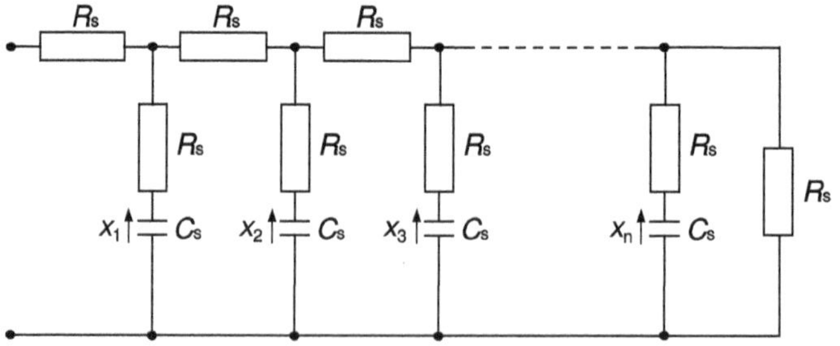

Fig. 9 Electrical circuit model of a supercapacitor in the form of a RC ladder network

$$D = \begin{bmatrix} 3 & -1 & 0 & \dots & 0 & 0 & 0 \\ -1 & 3 & -1 & \dots & 0 & 0 & 0 \\ 0 & -1 & 3 & \dots & 0 & 0 & 0 \\ \vdots & \vdots & \vdots & \ddots & \vdots & \vdots & \vdots \\ 0 & 0 & 0 & \dots & -1 & 3 & -1 \\ 0 & 0 & 0 & \dots & 0 & -1 & 3 \end{bmatrix}_{n\times n}, \tag{53}$$

$$A = \begin{bmatrix} -2 & 1 & 0 & \dots & 0 & 0 & 0 \\ 1 & -2 & 1 & \dots & 0 & 0 & 0 \\ 0 & 1 & -2 & \dots & 0 & 0 & 0 \\ \vdots & \vdots & \vdots & \ddots & \vdots & \vdots & \vdots \\ 0 & 0 & 0 & \dots & 1 & -2 & 1 \\ 0 & 0 & 0 & \dots & 0 & 1 & -2 \end{bmatrix}_{n\times n}, \quad B = \begin{bmatrix} 1 \\ 0 \\ \vdots \\ 0 \end{bmatrix}_{n\times 1}. \tag{54}$$

As shown by Mitkowski and Skruch [26] the system (52) is diagonalizable, that is, it can be broken down into n scalar systems. Let $x(t) = \widetilde{P}z(t)$, where $\widetilde{P}$ is given by

$$\widetilde{P} = \sqrt{\frac{2}{n+1}} \begin{bmatrix} \sin\varphi_1 & \sin 2\varphi_1 & \dots & \sin n\varphi_1 \\ \sin\varphi_2 & \sin 2\varphi_2 & \dots & \sin n\varphi_2 \\ \vdots & \vdots & \ddots & \vdots \\ \sin\varphi_n & \sin 2\varphi_n & \dots & \sin n\varphi_n \end{bmatrix}, \tag{55}$$

$\varphi_k = k\pi/(n+1)$, $k = 1, 2, \dots, n$. It can be checked that $\det \widetilde{P} \neq 0$ and $\widetilde{P}^{-1} = \widetilde{P}$. Thus $z(t) = \widetilde{P}x(t)$ and from (52) we have

$$C_s R_s \widetilde{P} D \widetilde{P} \frac{d^\alpha z(t)}{dt^\alpha} = \widetilde{P} A \widetilde{P} z(t) + \widetilde{P} B u(t)\,. \tag{56}$$

Finally,

$$C_s R_s s_k(D) \frac{d^\alpha z_k(t)}{dt^\alpha} = s_k(A) z_k(t) + \sqrt{\frac{2}{n+1}} \sin\varphi_k u(t)\,, \tag{57}$$

where $k = 1, 2, \dots, n$ and

$$s_k(D) = 1 + 4\sin^2\frac{\varphi_k}{2} > 0, \quad s_k(A) = -4\sin^2\frac{\varphi_k}{2} < 0\,. \tag{58}$$

Figures 10 and 11 show experimental data obtained during the discharging process of two supercapacitors: 10F/2.5V and 2F/2.5V. The discharging process has been mathematically described by an RC ladder network with $n = 10$ elementary cells containing fractional-order elements. Simulated data for different fractional-order values have been plotted on the same figures for comparison purposes.

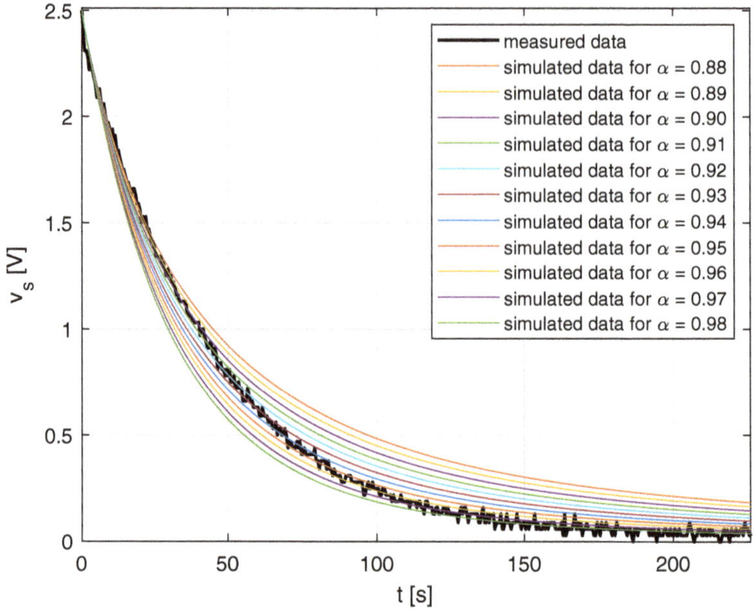

Fig. 10 Comparison of the simulated and real data obtained during discharging process of the 10F/2.5V supercapacitor. The simulation data is based on the model (52) for $n = 10$

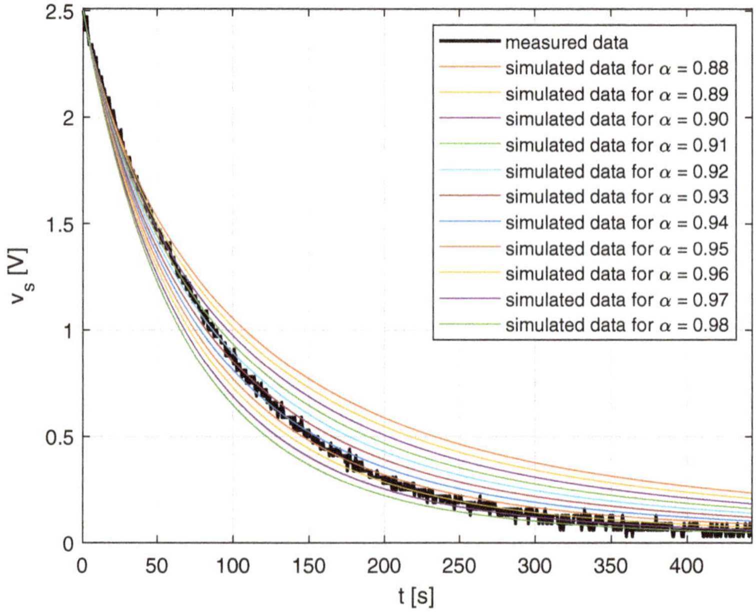

Fig. 11 Comparison of the simulated and real data obtained during discharging process of the 22F/2.5V supercapacitor. The simulation data is based on the model (52) for $n = 10$

4 A Chain of Vehicles Driving in Adaptive Cruise Control Mode

Modern vehicles are often equipped with the adaptive cruise control (ACC) system. Using this system, the driver can set a desired speed and time interval to the vehicle ahead. When the system detects a slower vehicle, the speed is automatically adjusted, so the vehicle ahead remains being followed at the setup distance. Once the road is clear again, the vehicle returns to the selected speed. Consider then a chain of vehicles following each other and functioning in the ACC mode (see Fig. 12). The system can be represented by an equivalent structure containing masses, springs and dampers as depicted in Fig. 13 (see [36]). The masses $m_i, i = 1, 2, \ldots, n$ correspond to the masses of the vehicles with m_0 the mass of the leading car. The coefficients k_i, $i = 1, 2, \ldots, n$ are related to gains of the controllers responsible to keep a set safe distance to the vehicle ahead. The parameters c_i, $i = 1, 2, \ldots, n$ are related to aerodynamic or other types of friction forces that in general are modeled by nonlinear functions. Here we propose to use fractional-order calculus to model this physical phenomenon. The dynamic behaviour of n-cars driving in adaptive cruise control mode can be represented by a matrix-vector linear differential equation of fractional-order of the following form

$$E\frac{\mathrm{d}^2 x(t)}{\mathrm{d}t^2} + F\frac{\mathrm{d}^\alpha x(t)}{\mathrm{d}t^\alpha} + Ax(t) = 0\,, \quad x(0) = x_0\,, \quad \dot{x}(0) = x_{\mathrm{d}0}, \tag{59}$$

where A, E and F are matrices of the form:

$$A = \begin{bmatrix} k_1+k_2 & -k_2 & 0 & \ldots & 0 & 0 & 0 \\ -k_2 & k_2+k_3 & -k_3 & \ldots & 0 & 0 & 0 \\ 0 & -k_3 & k_3+k_4 & \ldots & 0 & 0 & 0 \\ \vdots & \vdots & \vdots & \ddots & \vdots & \vdots & \vdots \\ 0 & 0 & 0 & \ldots & k_{n-2}+k_{n-1} & -k_{n-1} & 0 \\ 0 & 0 & 0 & \ldots & -k_{n-1} & k_{n-1}+k_n & -k_n \\ 0 & 0 & 0 & \ldots & 0 & -k_n & k_n \end{bmatrix}_{n\times n}, \tag{60}$$

$$E = \mathrm{diag}\,(m_1, m_2, m_3, \ldots, m_{n-2}, m_{n-1}, m_n)\,, \tag{61}$$

$$F = \begin{bmatrix} c_1+c_2 & -c_2 & 0 & \ldots & 0 & 0 & 0 \\ -c_2 & c_2+c_3 & -c_3 & \ldots & 0 & 0 & 0 \\ 0 & -c_3 & c_3+c_4 & \ldots & 0 & 0 & 0 \\ \vdots & \vdots & \vdots & \ddots & \vdots & \vdots & \vdots \\ 0 & 0 & 0 & \ldots & c_{n-2}+c_{n-1} & -c_{n-1} & 0 \\ 0 & 0 & 0 & \ldots & -c_{n-1} & c_{n-1}+c_n & -c_n \\ 0 & 0 & 0 & \ldots & 0 & -c_n & c_n \end{bmatrix}_{n\times n}, \tag{62}$$

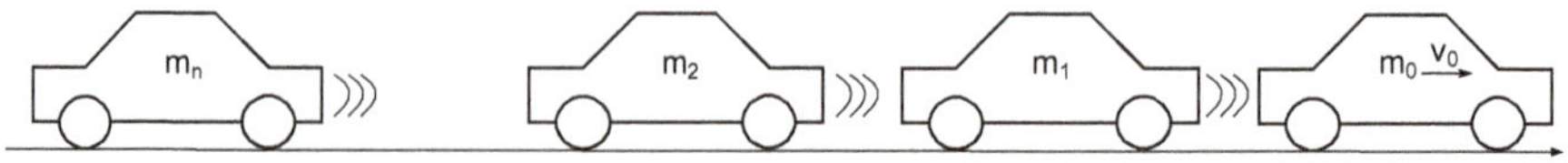

Fig. 12 Series of n-cars following each other in adaptive cruise control mode

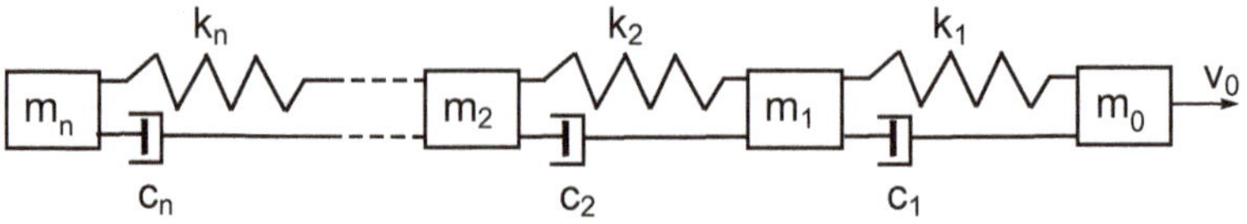

Fig. 13 Mechanical equivalent model of a series of n-cars driving in adaptive cruise control mode

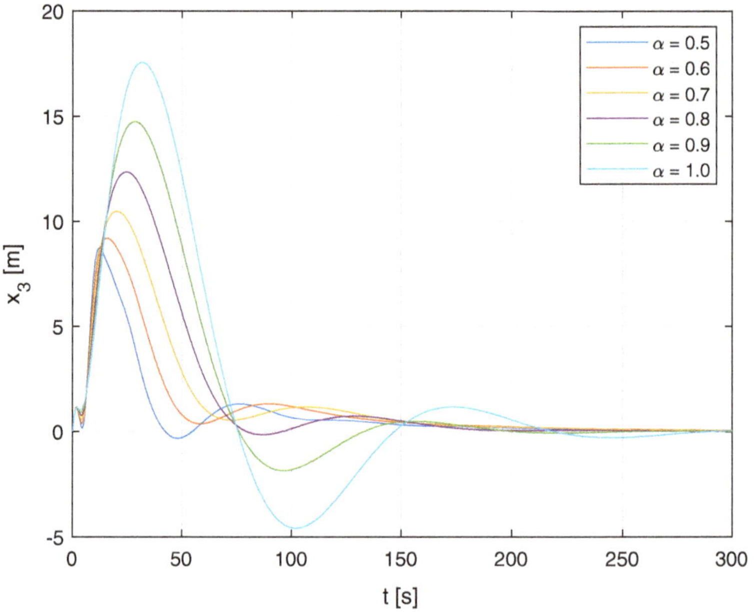

Fig. 14 Displacement of the 1st vehicle from the equilibrium point for different fractional-order damping elements

with the parameters $c_i > 0$, $k_i > 0$, $m_i > 0$ for $i = 1, 2, \ldots, n$, $\alpha \in (0, 1]$ denotes the order of the fractional derivative according to the Caputo definition [33], $t > 0$,

Figures 14, 15 and 16 present simulation results for a chain of three vehicles with the ACC mode enabled and following a lead vehicle. On the figures are plotted displacements for the vehicles from the equilibrium points obtained for different fractional-orders of damping elements. It is easy to observe that the fractional-order derivatives significantly influence the vehicles' dynamics.

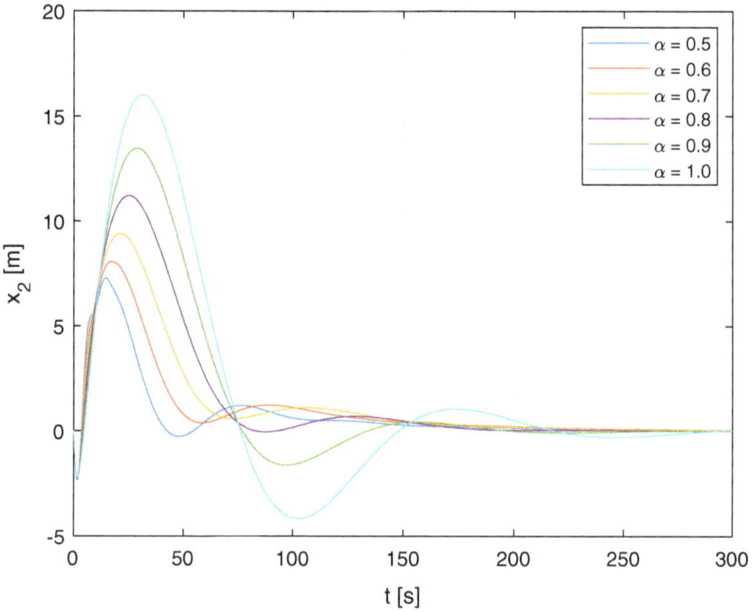

Fig. 15 Displacement of the 2nd vehicle from the equilibrium point for different fractional-order damping elements

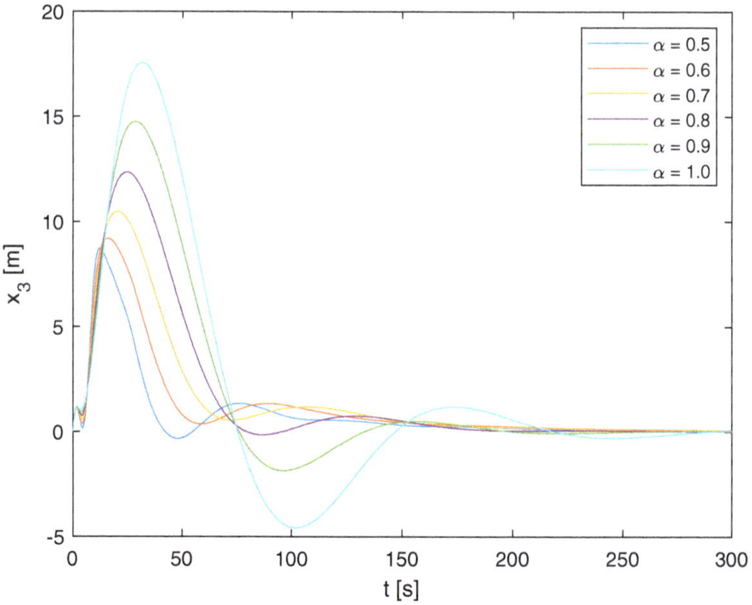

Fig. 16 Displacement of the 3rd vehicle from the equilibrium point for different fractional-order damping elements

5 Modeling of Thermal Processes Inside Buildings

The dynamics of thermal processes and in particular flow of the heat flux, is a very complex physics problem and can be modeled by using partial differential equations. Heat flow phenomenon is modeled very accurately by partial differential equations but in real applications, these type of equations are very hard to use. In practice applications, simpler mathematical models are used which are less or more accurate, e.g., the model of lumped heat capacity [7, 8, 13, 39]. The idea of extending a classical mathematical model with integer derivatives to a model with fractional derivatives is not new. One of the practical applications of fractional derivative calculus is its use in modeling heat flow processes [28–30]. It turns out that due to their specific properties, models with fractional derivatives of the heat flux transfer or diffusion process are very often more accurate than the approximate models [2, 12, 34, 35].

Very interesting results of practical use of fractional derivative models can be found in the article [11]. In this paper two types of fractional order models are considered with the following transmittances which are given by the equations:

$$G(s) = \frac{b_0}{a_1 s^{\alpha} + a_0}, \quad \alpha \in [0.1;\ 2] \tag{63}$$

$$G(s) = \frac{b_0}{a_2 s^{\alpha_2} + a_1 s^{\alpha_1}}, \quad \alpha_1, \alpha_2 \in [0.1;\ 2] \tag{64}$$

where $a_0, a_1, a_2 \in \mathbb{R}$ and $b_0 \in \mathbb{R}$ are transmittance parameters, $\alpha, \alpha_1, \alpha_2 \in \mathbb{R}$ mean order values for fractional derivatives (based on the Caputo definition). The input signal for both transmittances is the radiator temperature and the output signal is the air temperature at a given point in the room. Equations (63) and (64) were used to model temperature change in a standard room. The measurement data required for identification was recorded during a practical experiment. A thermal imaging camera was used to measure the temperature of the heater and the air in the room. The camera took a picture every 10 s. Figure 17 shows the location of the heater air temperature measurements at three different points (T_1, T_2, T_3) and the thermal imaging camera itself.

The experiment scenario was as follows: after stabilizing the air temperature in the room, the valve of the radiator was opened to the maximum value. The thermal imaging camera recorded the air temperature in the room at three points and the temperature of the radiator surface. Figure 18 presents an example of a picture made by a thermal imaging camera.

The input signal for the models (63) and (64) was the radiator temperature $u(t)$ and the output signal was air temperature inside room T_o. As an optimisation method, the non-linear least-squares algorithm was used to find the coefficient of equations (63) and (64). Additionally, it was assumed that the air in the tested room could change temperature only as a result of the adiabatic process (high thermal resistance of external walls). The experiment time was relatively short, so the outside air temperature did not have any effect on the air temperature in the tested room. Figure 19 presents comparison of the output signal of the two types of fractional models (63) and (64) for which parameters were obtained during numerical optimisation.

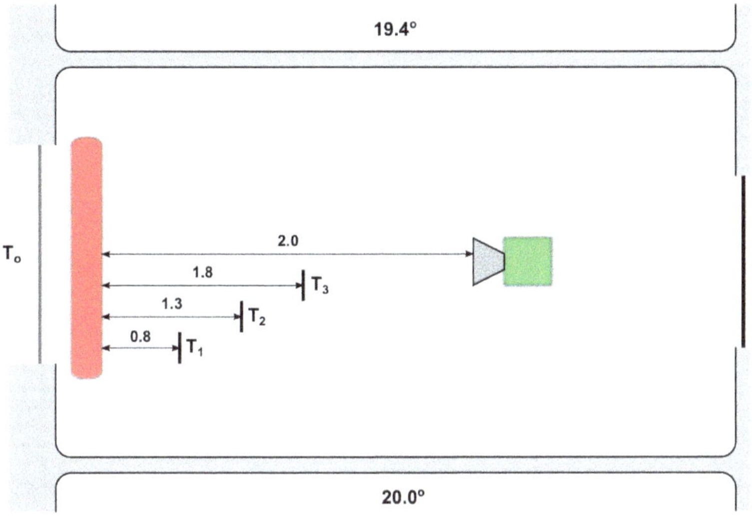

Fig. 17 Measurement system diagram

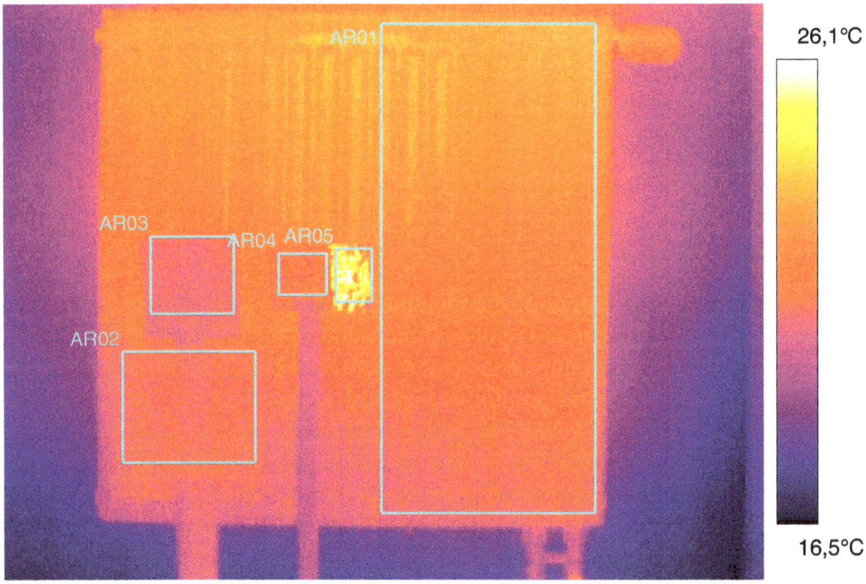

Fig. 18 Example image from a thermal imaging camera with marked measurement areas

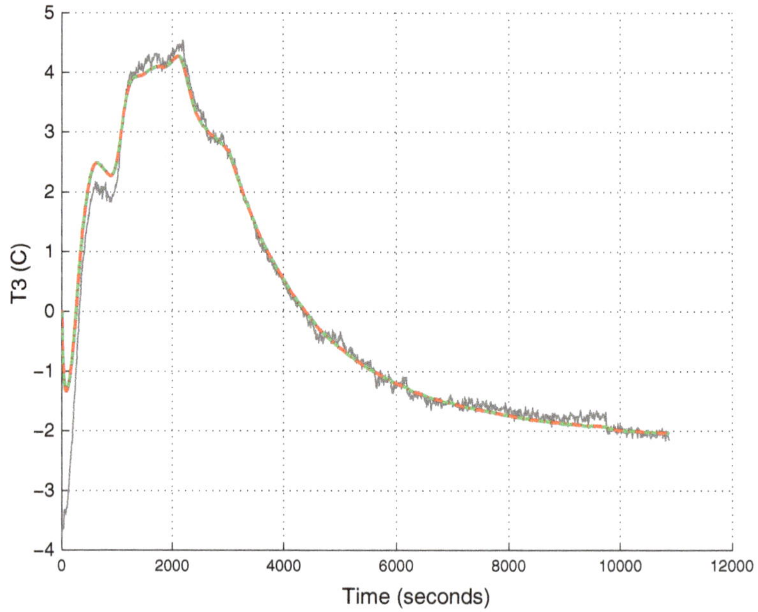

Fig. 19 Responses of the identified models (63) (green line) and (64) (red line) for best parameter fit

Data recorded during the practical experiment described above were adopted as the input $u(t)$ and output T_o signals. As can be seen, the response of both models (63) and (64) is almost identical. A more accurate comparison of the given models can be done by using an algebraic quality indicator. One of those quality indicators can be defined as [19]:

$$\text{FIT} = \left(1 - \frac{|y - \hat{y}|}{|y - \bar{y}|}\right) \times 100\% \tag{65}$$

where y—the measured output of the identified system, $\hat{y}$—model output, $\bar{y}$—the average value for the measured output y. If the value of (65) is closer to 100% then the identified model is the same as the real system. For both identified models, the matching value ratio (65) has an almost identical value of 83.8%. In effect, both model (64) and the simpler model (63) yield comparable results.

MATLAB software with the FOMCON (Fractional-order Modeling and Control, http://fomcon.net) toolbox was used to make all calculations related to fractional-order dynamical systems. The FOMCON toolbox allows us to make such computations as modeling, simulation and identification for fractional-order systems and is very useful. It was created by Aleksei Tepljakov and the software itself was developed as the outcome of an interdisciplinary scientific project at the University of Talin [37].

6 Other Models

6.1 *Battery Modeling*

Supercapacitors are primarily used for energy storage purposes in a variety of commercial applications. They can act as short-term backup supplies to retain data in digital components with memory in case of a short interruption of the power supply. Supercapacitors offer a long performance lifetime and therefore do not need to be replaced regularly as is the case with batteries. Using a supercapacitor in combination with a battery can relieve the battery of the most severe load demands by meeting the peak power requirements and allowing the battery to supply the average load. The reduction in pulsed current drawn from the battery can result in an extended battery lifetime in portable electronic devices such as laptops and mobile phones.

The ability of supercapacitors to deliver high electrical performance can resolve the limitation of lead-acid and lithium-ion batteries in the automotive industry [6]. This issue is especially challenging in hybrid and electric vehicles which are becoming increasingly popular, as well as in vehicles equipped with automated Start & Stop systems. The use of supercapacitors can support the cold cranking condition and therefore extend the battery life. The support of the warm cranking condition in Start & Stop systems can improve fuel efficiency. Supercapacitors also have applications in KERS (Kinetic Energy Recovery System) systems for regenerative energy capture during braking and coasting. Promising areas of application are distributed power systems where supercapacitors can play an important role in reducing wiring size, weight and consequently cost.

Consider a battery model that is based on a simple electrical circuit shown in Fig. 20. The circuit consists of a bulk capacitance C_{cb}, a surface capacitance C_{cs}, an internal resistance R_{i} and a polarization resistance R_{t}. The bulk capacitor characterizes the ability of the battery to store charge and the surface capacitor represents battery diffusion effects. The voltages across the bulk capacitor and the surface capacitor are denoted by V_{cb} and V_{cs}, respectively. The current and voltage observed at the terminals of the battery are represented in the circuit by I and U, respectively. The current I is taken as positive in case of charging and negative otherwise.

The dynamic behavior of the model in Fig. 20 can be governed by the following equations:

$$\frac{\mathrm{d}^{\alpha} V_{\mathrm{cb}}(t)}{\mathrm{d}t^{\alpha}} = \frac{1}{C_{\mathrm{cb}}} I(t) , \tag{66}$$

$$\frac{\mathrm{d}^{\alpha} V_{\mathrm{cs}}(t)}{\mathrm{d}t^{\alpha}} = -\frac{1}{R_{\mathrm{t}} C_{\mathrm{cs}}} V_{\mathrm{cs}}(t) + \frac{1}{C_{\mathrm{cs}}} I(t) , \tag{67}$$

$$U(t) = V_{\mathrm{cs}}(t) + V_{\mathrm{cb}}(t) + R_{\mathrm{i}} I(t) . \tag{68}$$

By defining the following state variables

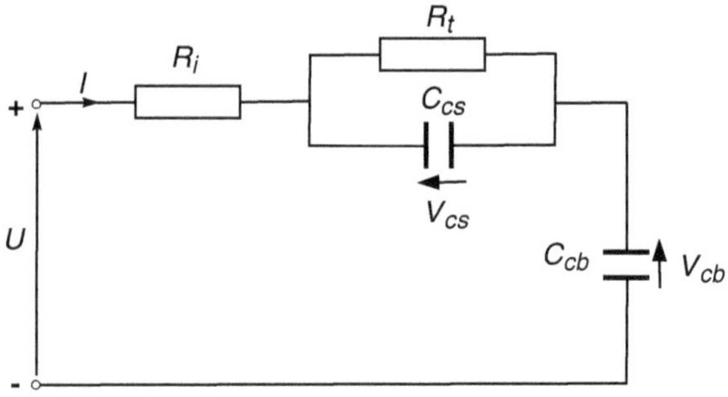

Fig. 20 Electrical circuit model of a battery

$$x_1(t) = V_{\text{cb}}(t)\,, \quad x_2(t) = V_{\text{cs}}(t)\,, \tag{69}$$

and denoting the input and output as

$$u(t) = I(t)\,, \quad y(t) = U(t)\,, \tag{70}$$

we can formulate a fractional-order state space model of the battery

$$\frac{\mathrm{d}^\alpha x(t)}{\mathrm{d}t^\alpha} = Ax(t) + Bu(t)\,, \tag{71}$$

$$y(t) = Cx(t) + Du(t)\,, \tag{72}$$

where $x(t) = [x_1(t)\ x_2(t)]^{\mathrm{T}}$, $\alpha \in (0, 1]$, $t > 0$ and

$$A = \begin{bmatrix} 0 & 0 \\ 0 & -\frac{1}{R_{\text{t}} C_{\text{cs}}} \end{bmatrix}, \quad B = \begin{bmatrix} C_{\text{cb}}^{-1} \\ C_{\text{cs}}^{-1} \end{bmatrix}, \quad C = \begin{bmatrix} 1 & 1 \end{bmatrix}, \quad D = \begin{bmatrix} R_{\text{i}} \end{bmatrix}. \tag{73}$$

Figure 21 presents the results of the simulation experiment where the fractional-order electrical circuit model from Fig. 20 was used to describe the charging process of the battery.

6.2 *Other Possible Applications*

Fractional PID controllers are increasingly used in automatic control systems, for example [3, 23, 25, 38].

In engineering applications, we encounter fractional models with distributed parameters [1, 5, 25, 27].

The problems of identifying the parameters of the appropriate mathematical models are also important [11, 24].

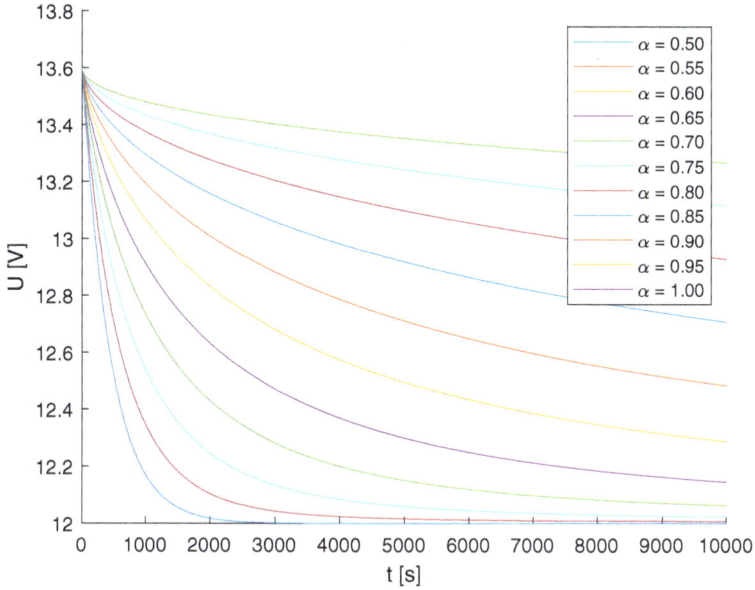

Fig. 21 Discharging of the battery

Appropriate numerical methods must be used to determine the solutions of fractional models [9, 10, 18, 20, 31, 40].

7 Final Remarks

In this chapter, selected applications of fractional-order calculus in engineering are presented. It was shown that ladder network structures can be successfully used to model high-order dynamics and complex nonlinear phenomena with few coefficients, that can be achieved by numerous repetitions of an elementary cell in the network. Moreover, by introducing in the elementary cell an element which, of behavior is described by fractional-order differential equations, an additional degree of freedom can be obtained. This additional degree of freedom very often allows us to capture a specific system behavior which is usually neglected by integer-order calculus. The following final conclusions can be drawn based on the results of this research effort:

1. The inherent complexity and lack of efficient implementation methods, in particular on embedded platforms with resource constraints, somewhat limit broader applications of fractional-order calculus in engineering;
2. Experimental and simulation results obtained for supercapacitors suggest using variable fractional-order derivatives when modeling charging and discharging processes to obtain better accuracy;

3. The example of a series of vehicles functioning in adaptive cruise control mode is a good evidence that fractional-order calculus can be successfully applied in the design process of robust fractional-order controllers which is nowadays a key requirement expected from advanced control systems;
4. One of the main advantages of fractional-order calculus is visible when modeling distributed parameter systems (such as thermal processes in buildings) as the complexity of the model in terms of the number of parameters, implementation and computation effort is disproportionately small compared to partial differential equations;
5. It has to be highlighted that the fractional system is not a dynamical system according to the formal definition. Thus the analysis of such systems can not be carried out using properties of dynamical systems;
6. Fractional-order systems are infinite-dimensional so their implementation on digital computers is actually an integer-higher-order (and finite dimensional) approximation.

References

1. Agrawal, O.: Solution for a fractional diffusion-wave equation defined in a bounded domain. Nonlinear Dyn. **29**(1–4), 145–155 (2002)
2. Aoki, Y., Sen, M., Paolucci, S.: Approximation of transient temperatures in complex geometries using fractional derivatives. Heat Mass Transf. **44**(7), 771–777 (2008). ISSN 0947-7411. https://doi.org/10.1007/s00231-007-0305-0
3. Bauer, W., Baranowski, J.: Fractional $PI^{\lambda}D$ controller design for a magnetic levitation system. Electronics **9**(12), 2135 (2020)
4. Bauer, W., Mitkowski, W., Zagorowska, M.: RC-ladder networks with supercapacitors. Arch. Electr. Eng. (2018)
5. Bohaienko, V., Bulavatsky, V.: Fractional-fractal modeling of filtration-consolidation processes in saline saturated soils. Fractal Fract. **4**(4), 59 (2020)
6. Burke, A.: Ultracapacitor technologies and application in hybrid and electric vehicles. Int. J. Energy Res. **34**(2), 133–151 (2010)
7. Clarke, J.: Energy Simulation in Building Design, 2 edn. Butterworth-Heinemann (2001). ISBN 9780750650823. http://books.google.com/books?id=ksNIQ4kx6UIC
8. Crabb, J.A., Murdoch, N., Penman, J.M.: A simplified thermal response model. Build. Serv. Eng. Res. Technol. **8**(1), 13–19 (1987)
9. Diethelm, K., Ford, N., Freed, A.: A predictor-corrector approach for the numerical solution of fractional differential equations. Nonlinear Dyn. **29**(1–4), 3–22 (2002)
10. Diethelm, K., Walz, G.: Numerical solution of fractional order differential equations by extrapolation. Numer. Algorithms **16**(3–4), 231–253 (1997)
11. Dlugosz, M., Skruch, P.: The application of fractional-order models for thermal process modelling inside buildings. J. Build. Phys. **39**(5), 440–451 (2016)
12. Gabano, J.D., Poinot, T.: Fractional modelling applied to heat conductivity and diffusivity estimation. Physica Scripta **2009**(T136), 014015 (2009). http://stacks.iop.org/1402-4896/2009/i=T136/a=014015
13. Gouda, M., Danaher, S., Underwood, C.: Low-order model for the simulation of a building and its heating system. Build. Serv. Eng. Res. Technol. **21**(3), 199–208 (2000)
14. Huseynov, I., Ahmadovay, A., Ojo, G., Mahmudov, N.: A natural extension of Mittag-Leffler function associated with a triple infinite series. arXiv preprint arXiv:2011.03999 (2020)

15. Ilin, W., Kuznyetsow, Y.: Tridiagonal matrices and their applications. Science, Moscow (1985)
16. Kaczorek, T.: Positive fractional linear systems. Pomiary Automatyka Robotyka **2**, 91–112 (2011)
17. Kaczorek, T.: Selected Problems of Fractional Systems Theory, vol. 411. Springer (2011)
18. Lakeb, A., Kaisserli, Z., Bouagada, D.: Computing the h_2-norm of a fractional-order system using the state-space linear model. Kragujevac J. Math. **47**(4), 531–538 (2023)
19. Ljung, L.: System Identification Toolbox. The MathWorks, Inc., 3 Apple Hill Drive, Natick, MA 01760-2098 (2013)
20. Lopes, A., Machado, J.: A review of fractional order entropies. Entropy **22**(12), 1374 (2020)
21. Mitkowski, W.: Uniform spatial networks RC. Arch. Electrotech. **22**(2), 398–405 (1973)
22. Mitkowski, W.: Finite-dimensional approximations of distributed RC networks. Bull. Pol. Acad. Sci. Tech. Sci. 263–269 (2014)
23. Mitkowski, W., Bauer, W.: Comparison of non-integer PID, PD and PI controllers for DC motor. In: Advanced, Contemporary Control, pp. 904–913. Springer (2020)
24. Mitkowski, W., Obraczka, A.: Simple identification of fractional differential equation. In: Solid State Phenomena, vol. 180, pp. 331–338. Trans Tech Publications (2012)
25. Mitkowski, W., Oprzedkiewicz, K.: Modelling and control of heat conduction processes. In: Automatic Control, Robotics, and Information Processing, pp. 767–789. Springer (2021)
26. Mitkowski, W., Skruch, P.: Fractional-order models of the supercapacitors in the form of RC ladder networks. Bull. Pol. Acad. Sci. Tech. Sci. **61**(3) (2013)
27. Obraczka, A., Mitkowski, W.: The comparison of parameter identification methods for fractional, partial differential equation. In: Solid State Phenomena, vol. 210, pp. 265–270. Trans Tech Publications (2014)
28. Oprzedkiewicz, K.: Non integer order, state space model of heat transfer process using Atangana-Baleanu operator. Bull. Pol. Acad. Sci. Tech. Sci. 43–50 (2020)
29. Oprzedkiewicz, K., Dziedzic, K., Wieckowski, L.: Non integer order, discrete, state space model of heat transfer process using Grünwald-Letnikov operator. Bull. Pol. Acad. Sci. Tech. Sci. 905–914 (2019)
30. Oprzedkiewicz, K., Podsiadlo, M., Dziedzic, K.: Integer order vs fractional order temperature models in the forced air heating system
31. Pellegrino, E., Pezza, L., Pitolli, F.: A collocation method based on discrete spline quasi-interpolatory operators for the solution of time fractional differential equations. Fractal Fract. **5**(1), 5 (2021)
32. Pillai, R.: On Mittag-Leffler functions and related distributions. Ann. Inst. Stat. Math. **42**(1), 157–161 (1990)
33. Podlubny, I.: Fractional Differential Equation. Rinehart, and Winston, San Diego, USA, SHolt (1999)
34. Sierociuk, D., Dzielinski, A., Sarwas, G., Petras, I., Podlubny, I., Skovranek, T.: Modelling heat transfer in heterogeneous media using fractional calculus. Philos. Trans. R. Soc. A: Math. Phys. Eng. Sci. **371**, 2013 (1990)
35. Sierociuk, D., Petráš, I.: Modeling of heat transfer process by using discrete fractional-order neural networks. In: 2011 16th International Conference on Methods and Models in Automation and Robotics, MMAR 2011, pp. 146–150 (2011)
36. Skruch, P., Dlugosz, M., Mitkowski, W.: Stability analysis of a series of cars driving in adaptive cruise control mode. In: Trends in Advanced Intelligent Control, Optimization and Automation, vol. 577, pp. 168–177. Springer (2017)
37. Tepljakov, A., Petlenkov, E., Belikov, J.: Fomcom: a matlab toolbox for fractional-order system identification and control. Int. J. Microelectron. Comput. Sci. **2**(2), 51–62 (2011)
38. Tufenkci, S., Senol, B., Matusu, R., Alagoz, B.: Optimal V-plane robust stabilization method for interval uncertain fractional order PID control systems. Fractal Fract. **5**(1), 3 (2021)
39. Underwood, C., Yik, F.: Modelling Methods for Energy in Buildings. Blackwell Science (2004). ISBN 0-632-05936-2. http://books.google.com/books?id=bzdV3iSRhsUC
40. Weilbeer, M.: Efficient Numerical Methods for Fractional Differential Equations and Their Analytical Background. Papierflieger (2005)

Fractional Order State Space Models of the One-Dimensional Heat Transfer Process

Krzysztof Oprzędkiewicz and Wojciech Mitkowski

Abstract The chapter presents the fractional order, state space models of the one dimensional heat transfer process. The proposed models are based on a known semi-group state space model of a parabolic system with distributed control and observation. The first model presented is the time continuous model using the Caputo definition of the fractional derivative over time and the Riesz definition to express the spatial fractional derivative. The second proposed model is the discrete time model, employing the discrete Grünwald-Letnikov operator. The last discrete time fractional order model uses a new, memory-effective method, proposed by the authors as a method of solution for the discrete fractional state equation. The proposed method uses the CFE approximant to express the fractional derivative. Elementary properties (stability, convergence and accuracy) for each proposed model are analysed and results are verified using real experimental data.

1 Introduction

Mathematical models of distributed parameter systems based on partial differential equations can be described in an infinite-dimensional state space, usually in a Hilbert space, but a Sobolev space can also be applied. This problem has been analyzed by many authors. Fundamentals have been drawn in [6] and [38], they are also recalled in [16] and [17, 20]. Distributed RC networks are discussed in [19]. Analysis of a hyperbolic system in a Hilbert space is presented in [2]. This paper also gives a broad overview of literature.

The modeling of processes and phenomena that are hard to analyse with the use of other tools is one of main areas where non integer order calculus is applied. Fundamentals of fractional calculus with respect to modeling and control are given,

K. Oprzędkiewicz (✉) · W. Mitkowski
AGH University, al. A Mickiewicza 30, 30-059 Kraków, Poland
e-mail: kop@agh.edu.pl

W. Mitkowski
e-mail: wojciech.mitkowski@agh.edu.pl

P. Kulczycki et al. (eds.), *Fractional Dynamical Systems: Methods, Algorithms and Applications*, Studies in Systems, Decision and Control 402,
https://doi.org/10.1007/978-3-030-89972-1_13

for example in books [4] and [41]. Fractional Order (FO) models of different physical phenomena are presented by many authors. The amount of FO models of various processes is collected in the book [7]. The book [4] presents fractional order models of chaotic systems and Ionic Polymer Metal Composites (IPMC). FO models of ultracapacitors are given for example in [10]. The use of fractional calculus to model diffusion processes is considered in [11, 18, 42, 44]. A collection of recent results employing the new Atangana-Baleanu operator can be found in [12]. In this book for example the FO blood alcohol model, the Christov diffusion equation and fractional advection-dispersion equation for a groundwater transport process are presented.

As it is mentioned in [16], the heat transfer processes can also be described using non integer order approach. For example the temperature–heat flux relationship for heat flow in a semi-infinite conductor is presented in [7], the beam heating problem is given in [10], the FO transfer function temperature models in a room are presented in [8], temperature models in a three dimensional solid body are given in [21, 23, 24]. The use of a fractional order approach to the modeling and control of heat systems is also presented in [43]. The paper [28] proposes the use of the Caputo-Fabrizio operator to state space modeling for the heat transfer process.

2 Preliminaries

A presentation of elementary ideas is started with a definition of a non integer-order, integro-differential operator. It is given for example in [7, 14, 15, 41]:

Definition 1 (*The elementary non integer order operator*) The non integer-order integro-differential operator is defined as follows:

$$
{}_aD_t^{\alpha} f(t) = \begin{cases} \frac{d^{\alpha} f(t)}{dt^{\alpha}} & \alpha > 0 \\ f(t) & \alpha = 0 \\ \int\limits_a^t f(\tau)(d\tau)^{\alpha} & \alpha < 0 \end{cases} \tag{1}
$$

where a and t denote time limits for operator calculation, $\alpha \in \mathbb{R}$ denotes the non integer order of the operation.

The fractional-order, integro-differential operator can be described by different definitions, given by Grünwald and Letnikov, Riemann and Liouville (RL) and Caputo (C). In this chapter the C nd GL definitions are used (see e.g. [7, 14, 15, 41]):

Definition 2 (*The Caputo definition of the FO operator*)

$$
{}_0^C D_t^{\alpha} f(t) = \frac{1}{\Gamma(M-\alpha)} \int\limits_0^{\infty} \frac{f^{(M)}(\tau)}{(t-\tau)^{\alpha+1-M}} d\tau \tag{2}
$$

where $M-1<\alpha<M$ denotes the non integer order of operation and $\Gamma(..)$ is the Gamma function.

The GL definition follows [4, 36]:

Definition 3 (*The Grünwald-Letnikov definition of the FO operator*)

$$ {}_0^{GL}D_t^{\alpha} f(t) = \lim_{h\to 0} h^{-\alpha} \sum_{l=0}^{[\frac{t}{h}]} (-1)^l \binom{\alpha}{j} f(t-lh) \tag{3} $$

In (3) $\binom{\alpha}{l}$ is the binomial coefficient:

$$ \binom{\alpha}{l} = \left\{ \begin{matrix} 1, & l=0 \\ \frac{\alpha(\alpha-1)\ldots(\alpha-l+1)}{l!}, & l>0 \end{matrix} \right\} \tag{4} $$

The GL definition is limit case for $h\to 0$ of the Fractional Order Backward Difference (FOBD), commonly employed in discrete FO calculations (see for example [37], p. 68):

Definition 4 (*The Fractional Order Backward Difference-FOBD*)

$$ (\Delta^{\alpha}x)(t) = \frac{1}{h^{\alpha}} \sum_{l=0}^{L} (-1)^l \binom{\alpha}{l} x(t-lh) \tag{5} $$

Denote coefficients $(-1)^l\binom{\alpha}{l}$ by d_l:

$$ d_l = (-1)^l \binom{\alpha}{l} \tag{6} $$

The coefficients (6) can be also calculated with the use of the following, equivalent recursive formula (see for example [4], p. 12), useful in numerical calculations:

$$ \begin{aligned} d_0 &= 1 \\ d_l &= \left(1-\frac{1+\alpha}{l}\right) d_{l-1}, \quad l=1,\ldots,L \end{aligned} \tag{7} $$

It is proven in [3] that:

$$ \sum_{l=1}^{\infty} d_l = 1-\alpha \tag{8} $$

From (7) and (8) we obtain at once that:

$$ \sum_{l=2}^{\infty} d_l = 1 \tag{9} $$

In (5) L denotes the memory length necessary for correct approximation of a non integer order operator. Unfortunately high accuracy of approximation requires the use of long memory L which can cause difficulties during implementation.

For the Caputo operator the Laplace transform can be defined (see for example [13]):

Definition 5 (*The Laplace transform for the Caputo operator*)

$$\mathcal{L}({}_0^C D_t^\alpha f(t)) = s^\alpha F(s), \quad \alpha < 0$$
$$\mathcal{L}({}_0^C D_t^\alpha f(t)) = s^\alpha F(s) - \sum_{k=0}^{M-1} s^{\alpha-k-1}{}_0 D_t^k f(0), \ \alpha > 0, \quad M-1 < \alpha \leq M \in \mathbb{Z} \tag{10}$$

Consequently, the inverse Laplace transform for non integer order function is expressed as follows [15]:

$$\mathcal{L}^{-1}[s^\alpha F(s)] ={}_0 D_t^\alpha f(t) + \sum_{k=0}^{M-1} \frac{t^{k-1}}{\Gamma(k-\alpha+1)} f^{(k)}(0^+) \ M-1<\alpha < M, \quad M \in \mathbb{Z} \tag{11}$$

The non integer order spatial derivative is given by Riesz and it has the following form (see for example [48]):

Definition 6 (*The Riesz definition of FO spatial derivative*)

$$\frac{\partial^\beta \Theta(x,t)}{\partial x^\beta} = -r_\beta \left({}_0 D_x^\beta \Theta(x,t) +_x D_1^\beta \Theta(x,t)\right) \tag{12}$$

where:

$$r_\beta = \frac{1}{2\cos(\frac{\pi\beta}{2})} \tag{13}$$

In (12) ${}_0 D_x^\beta$ and ${}_x D_1^\beta$ denote left- and right-side Riemann-Liouville derivatives, defined as below:

$${}_0 D_x^\beta \Theta(x,t) = \frac{1}{\Gamma(2-\beta)} \frac{\partial}{\partial x} \int_0^x \frac{\Theta(\xi,t) d\xi}{(x-\xi)^{\beta-1}} \tag{14}$$

$${}_x D_1^\beta \Theta(x,t) = \frac{1}{\Gamma(2-\beta)} \frac{\partial}{\partial x} \int_x^1 \frac{\Theta(\xi,t) d\xi}{(\xi-x)^{\beta-1}} \tag{15}$$

In (14) and (15) $\Gamma(..)$ denotes the Gamma function, $\beta > 1$ is the non integer derivative order with respect to length.

A fractional-order linear state space system is described as:

$$\begin{cases} {}_0D_t^{\alpha}x(t) = Ax(t) + Bu(t) \\ y(t) = Cx(t) \end{cases} \tag{16}$$

where $\alpha \in (0, 1)$ denotes the fractional order of the state equation, $x(t) \in \mathbb{R}^N$, $u(t) \in \mathbb{R}^L$, $y(t) \in \mathbb{R}^P$ are the state, control and output vectors respectively, A, B, C are the state, control and output matrices respectively.

Analogically the discrete, fractional order state equation using definition (5) is written as follows (see for example [9, 22]):

$$\begin{cases} (\Delta_L^{\alpha}x)(t+h) = A^+x(t) + B^+u(t) \\ y(t) = C^+x(t) \end{cases} \tag{17}$$

where $x(t) \in \mathbb{R}^N$ is the state vector, $u(t) \in \mathbb{R}^P$ is the control, $y(t) \in \mathbb{R}^M$ is the output. A^+, B^+ and C^+ are state, control and output matrices respectively. If we shortly denote k-th time instant: hk by k, then Eq. (17) turns to:

$$\begin{cases} (\Delta_L^{\alpha}x)(k+1) = A^+x(k) + B^+u(k) \\ y(k) = C^+x(k) \end{cases} \tag{18}$$

where:

$$A^+ = h^{\alpha}A \tag{19}$$

$$B^+ = h^{\alpha}B \tag{20}$$

$$C^+ = C. \tag{21}$$

The solution of the discrete state equation (18) takes the form:

$$x(k+1) = P^+x(k) - \sum_{l=2}^{L} A_l^+x(k-l) + h^{\alpha}B^+u(k) \tag{22}$$

where:

$$P^+ = A^+ + \alpha I \tag{23}$$

$$A_l^+ = d_l I_{N\times N} \tag{24}$$

2.1 Final Value Theorem

Next the Final Value Theorem (FVT) needs to be recalled. It allows us to calculate the steady-state value of a time function described by the Laplace transform or the "z" transform. It is formulated as follows:

Theorem 1 (Final Value Theorem for continuous time) *Let $f(t)$ is a function of time t and $F(s)$ is its Laplace transform. Assume that $F(s)$:*

1. *has no poles in the right part of the complex plane,*
2. *has maximally one pole on the imaginary axis: $s = 0$.*

then:

$$\lim_{t\to\infty} f(t) = \lim_{s\to 0} sF(s) \tag{25}$$

Theorem 2 (Final Value Theorem for discrete time) *Let $f^+(k)$ is a discrete function of time, defined in k time moments and $F^+(z)$ is its z-transform. Assume that $F^+(z)$:*

1. *has no poles outside the unit circle,*
2. *has maximally one pole on the unit circle: $z = 1$.*

then:

$$\lim_{k\to\infty} f^+(k) = \lim_{z\to 1}(z-1)F^+(z) \tag{26}$$

2.2 CFE Approximation

The CFE method allows us to approximate the basic s^α element using a discrete IIR filter containing both poles and zeros. It is more rapidly covergent and easier to implement because its order is relatively low, typically not higher than 5. It is obtained via discretization of the elementary fractional order element s^α. This can be done using the so called generating function $s \approx \omega(z^{-1})$. The new operator raised to the power α has the following form (see [5, 39]):

$$\begin{array}{l} \left(\omega(z^{-1})\right)^\alpha = g_h CFE\{\left(\frac{1-z^{-1}}{1+az^{-1}}\right)^\alpha\}_{M,M} = \\ = \frac{P_{\alpha M}(z^{-1})}{Q_{\alpha M}(z^{-1})} = g_h \frac{CFE_N(z^{-1},\alpha)}{CFE_D(z^{-1},\alpha)} = \\ = g_h \frac{\sum\limits_{m=0}^{M} w_m z^{-m}}{\sum\limits_{m=0}^{M} v_m z^{-m}} \end{array} \tag{27}$$

where M is the order of approximation, g_h is the coefficient depending on sample time and type of approximation:

$$g_h = \left(\frac{1+a}{h}\right)^\alpha \tag{28}$$

In (28) h is the sample time and a is the coefficient depending on approximation type. For $a = 0$ and $a = 1$ we obtain the Euler and Tustin approximations respectively. For $a \in (0, 1)$ we arrive at the Al-Alaoui-based approximation, which is a linear combination of the Euler and Tustin approaches. Note that in this case the parameter a in Eq. (27) is equal to $a = \frac{1-\gamma}{1+\gamma}$, with γ being the Al-Alaoui weighting coefficient [1, 47]. Numerical values of coefficients w_m and v_m and various values of the parameter a can be calculated with the use of the MATLAB function written by I. Petras and available in [40]. If the Tustin approximation is considered ($a = 1$) then $CFE_D(z^{-1}, \alpha) = CFE_N(z^{-1}, -\alpha)$ and the polynomial $CFE_D(z^{-1}, \alpha)$ can be given in the direct form [5] see. Examples of the polynomial $CFE_D(z^{-1}, \alpha)$ for $M = 1, 3, 5$ are given in Table 1. The detailed analysis of various forms of CFE approximators has been presented by [47].

3 The CFE Based Method of Solution for a FO State Equation

The novel method of a memory-effective solution for a FO state equation is proposed and analysed with details in paper [35]. Its idea consists of replacing the continuous operator s^α in the Laplace transform of the FO state equation (16) by its discrete CFE approximant expressed by (27), with coefficients given in the Table 1. The CFE approximant is a function of discrete complex variable z^{-1}. This allows us to directly pass to the discrete time domain. Then the solution of the state equation takes the following form:

$$\sum_{m=0}^{M} E_m x^+(k-m) = \sum_{m=0}^{M} F_m u^+(k-m) + \sum_{m=-M}^{0} x_0(m) \tag{29}$$

Table 1 Coefficients of CFE polynomials $CFE_{N,D}(z^{-1}, \alpha)$ for Tustin approximation

Order M	w_m	v_m
$M = 1$	$w_1 = -\alpha$	$v_1 = \alpha$
	$w_0 = 1$	$v_0 = 1$
$M = 3$	$w_3 = -\frac{\alpha}{3}$	$v_3 = \frac{\alpha}{3}$
	$w_2 = \frac{\alpha^2}{3}$	$v_2 = \frac{\alpha^2}{3}$
	$w_1 = -\alpha$	$v_1 = \alpha$
	$w_0 = 1$	$v_0 = 1$
$M = 5$	$w_5 = -\frac{\alpha}{5}$	$v_5 = \frac{\alpha}{5}$
	$w_4 = \frac{\alpha^2}{5}$	$v_4 = \frac{\alpha^2}{5}$
	$w_3 = -\left(\frac{\alpha}{5} + \frac{2\alpha^3}{35}\right)$	$v_3 = -\left(\frac{-\alpha}{5} + \frac{-2\alpha^3}{35}\right)$
	$w_2 = \frac{2\alpha^2}{5}$	$v_2 = \frac{2\alpha^2}{5}$
	$w_1 = -\alpha$	$v_1 = \alpha$
	$w_0 = 1$	$v_0 = 1$

where matrices E_m and F_m are defined as follows:

$$\begin{cases} E_m = g_h w_m I_{N \times N} - v_m A \\ F_m = v_m B \\ m = 0, 1, ..., M \end{cases} \tag{30}$$

In (30) g_h is described by (28), w_m and v_m denote coefficients of CFE approximant given in the Table 1. From (29) the state vector x^+ can be directly calculated as follows:

$$\begin{array}{c} x^+(k) = -E_0^{-1} \sum_{m=1}^{M} E_m x^+(k-m) + \\ +E_0^{-1} \sum_{m=0}^{M} F_m u^+(k-m) + E_0^{-1} \sum_{m=-M}^{0} x_0(m) \end{array} \tag{31}$$

The Eq. (31) allows us to solve the discrete-time FO state equation using the CFE approximant. It has the form of the M-th order difference equation. Its solution requires M previous steps of state and control signals to be known.

The state equation (29) can also be written in the extended form:

$$\begin{cases} x_q^+(k+1) = A_q^+ x_q^+(k) + B_q^+ u_q^+(k) \\ y_q^+(k) = C_q^+ x_q^+(k) \end{cases} \tag{32}$$

where:

$$x_q^+(k) = \begin{bmatrix} x_1^+(k) \\ x_2^+(k) \\ \dots \\ x_M^+(k) \end{bmatrix}_{MN \times 1} \tag{33}$$

$$\begin{cases} x_1^+(k) = x(k) \\ x_2^+(k) = x(k-1) \\ \dots \\ x_M^+(k) = x(k+1-M) \end{cases} \tag{34}$$

$$u_q^+(k) = \begin{bmatrix} u_1^+(k) \\ u_2^+(k) \\ \dots \\ u_{M+1}^+(k) \end{bmatrix}_{M+1 \times 1} \tag{35}$$

$$\begin{cases} u_1^+(k) = u(k) \\ u_2^+(k) = u(k-1) \\ \dots \\ u_{M+1}^+(k) = u(k-M) \end{cases} \tag{36}$$

$$A_q^+ = \begin{bmatrix} -E_0^{-1}E_1, ..., -E_0^{-1}E_M \\ I_{N\times N}, 0, 0, ..., 0 \\ 0, I_{N\times N}, 0, ..., 0 \\ ..., ..., ..., ..., .. \\ 0, ..., I_{N\times N}, 0 \end{bmatrix}_{MN\times MN} \tag{37}$$

$$B_q^+ = \begin{bmatrix} E_0^{-1}F_0, ..., E_0^{-1}F_M \\ 0, 0, 0, ..., 0 \\ ..., ..., ..., ..., .. \\ 0, 0, 0, ..., 0 \end{bmatrix}_{MN\times M+1} \tag{38}$$

The output matrix C is as follows:

$$C_q^+ = \left[C^+, 0, ..., 0 \right]_{N\times MN} \tag{39}$$

The initial condition for state equation (32) also turns to the extended form:

$$x_{q0}^+ = \begin{bmatrix} x^+(M-1) \\ x^+(M-2) \\ \dots \\ x^+(0) \end{bmatrix}_{MN\times 1} \tag{40}$$

Notice that the summarized size of the proposed discrete, FO model is equal to NM. This size is significantly smaller than the size of the model using PSE approximation, analysed in [29, 34].

3.1 Stability of the CFE Based Discrete Model

Stability of the CFE based, discrete FO system is analysed by using the approach presented in [35], [37], pp. 202–223, [45] and [46]. Its idea consists of testing the location of the spectrum of the continuous system (before its discretization) in the complex plane, with respect to restricted area, limited by the form of the CFE approximant (27). Let us assume that the approximation $\omega(z^{-1})$ of (27) is stable and the term $\omega(e^{-j\varphi})$, $\varphi \in [-\pi, \pi]$, draws a simple closed curve in the complex plane. Then the stability/instability areas with respect to the spectrum of the continuous system are separated from each other by the contour defined as follows (see [45], [37], Theorem 7.4, page 205):

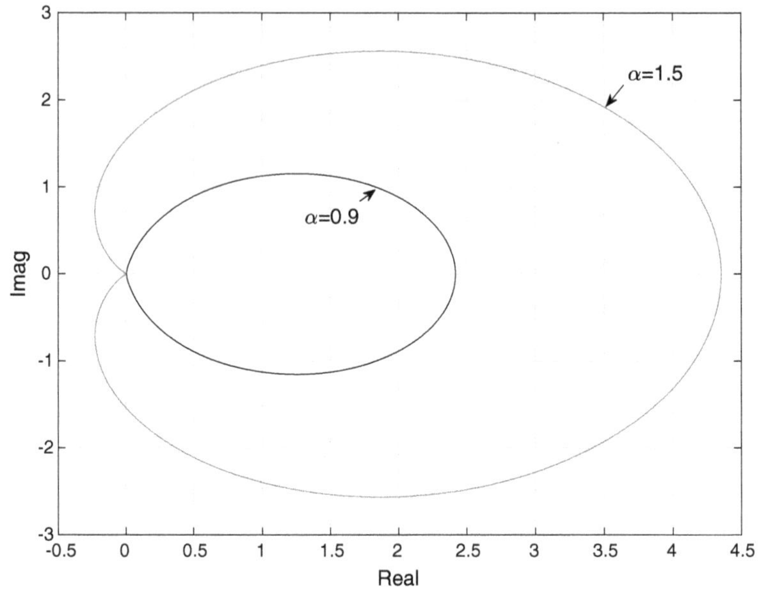

Fig. 1 Stability/instability areas for: $\alpha = 0.9$, $\alpha = 1.5$, $a = \frac{1}{7}$, $h = 1\,[\mathrm{s}]$, $M = 5$

$$S = \left\{\omega(e^{-j\varphi}), \varphi \in [-\pi, \pi]\right\} \tag{41}$$

where:

$$\omega(e^{-j\varphi}) = g_h \frac{\sum_{m=0}^{M} w_m (e^{-j\varphi})^m}{\sum_{m=0}^{M} v_m (e^{-j\varphi})^m} \tag{42}$$

An example of the stability/instability areas for the discrete-time approximator (27) and different ranges of order α is given in Fig. 1. In each case the "restricted" area is located inside the contour S.

It is important to notice that:

- For $0.0 < \alpha < 1.0$ the restricted area is located in the right semi plane only,
- for $1.0 < \alpha < 2.0$ the restricted area is located in the both semi planes, but it does not cover the negative part of real axis.

The above notes will be fundamental during stability analysis for the considered discrete, FO, CFE based model of the heat plant. This is due to the fact that the spectrum of the considered heat system has an unique location in the complex plane.

3.2 The Cost Function

Finally the cost function employed for accuracy estimation of all proposed models should be given. This is the known Mean Square Error (MSE) cost function:

$$MSE = \frac{1}{3K_s}\sum_{j=1}^{3}\sum_{k=1}^{K_s}\left(y^+_{p_j}(k) - y^+_j(k)\right)^2 \tag{43}$$

In (43) K_s denotes the number of collected samples for one sensor, $y^+ p_j(k)$ and $y^+_j(k)$ are responses of the plant and model in k-th time moment and at j-th output respectively. The parameter identification for all presented models is run via minimization of the cost function (43) using Matlab function *fminsearch*.

4 Experimental Heat Transfer System

Experiments were carried out using the experimental system shown with details in Fig. 2. The length of the rod is equal to 260 [mm]. The control signal in the system is the standard current 0–20 [mA] given from analog output of the PLC. This signal is amplified to the range 0–1.5 [A] and it is the input signal for the heater. The temperature distribution along the rod is measured by the standard RTD sensors Pt-100. Signals from the sensors are directly read by analog inputs of the PLC in Celsius degrees. Data from the PLC is collected by SCADA. The whole system is connected via the PROFINET industrial network. The temperature distribution with respect to time and length is shown in the Fig. 3.

The fundamental model describing the heat conduction in the rod is the partial differential equation of the parabolic type with the homogeneous Neumann boundary conditions at the ends, the homogeneous initial condition, the heat exchange along the length of rod and distributed control and observation. This equation with integer orders of both differentiations has been considered in many papers, for example in [25–27]. Its fractional order models using the Caputo operator, discrete GL operator and CFE approximation are presented in the next subsections.

5 The State Space Model Using the Caputo Operator

The state space model using the Caputo operator (C model) is proposed and analyzed with details in the papers: [31, 32]. The main results are recalled here. The simplified scheme of the considered heat plant is recalled in Fig. 4. It has a form of a thin copper rod heated with an electric heater of the length Δx_u located at one end of rod. Output

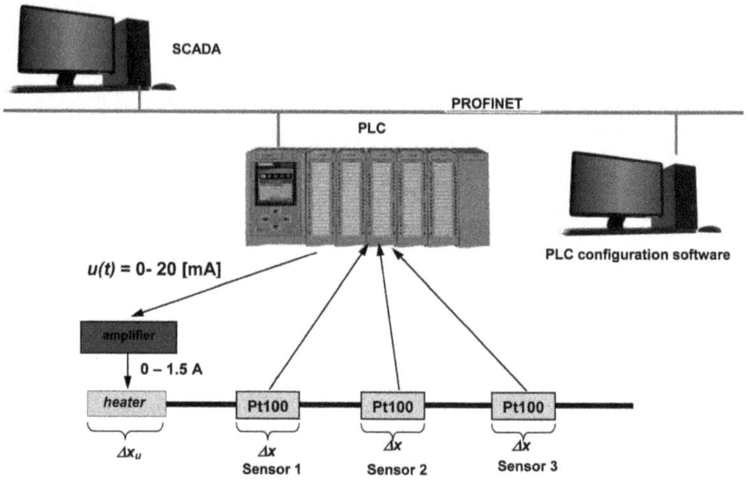

Fig. 2 The construction of the experimental system

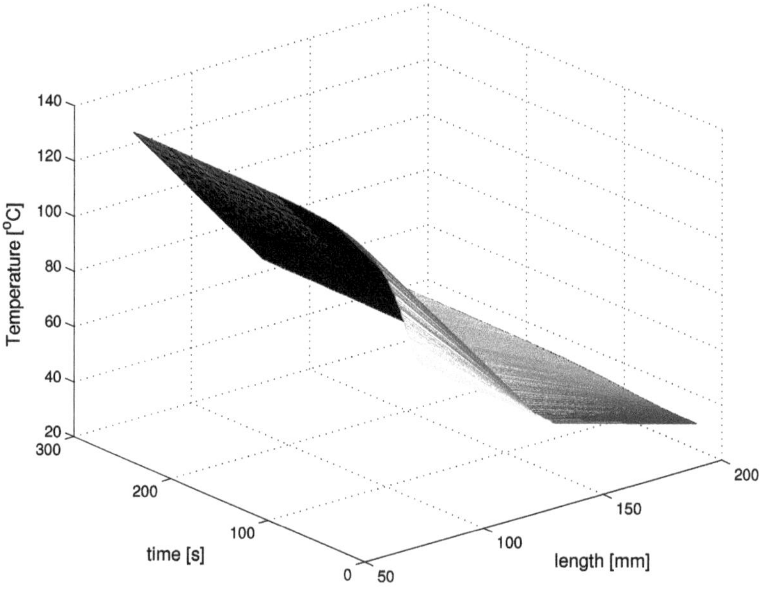

Fig. 3 The spatial-time temperature distribution in the plant

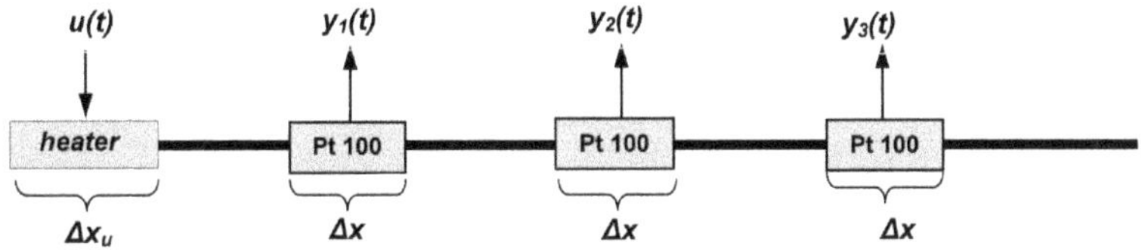

Fig. 4 The simplified scheme of the experimental system

temperature is measured using Pt-100 RTD sensors Δx long attached at points: 0.29, 0.50 and 0.73 of rod length. More details of the construction are given in the Sect. 4.

The non integer order model with respect to both time and space coordinates, employing the Caputo and Riesz operators takes the following form (see [31]):

$$\begin{cases} {}^C D_t^{\alpha} Q(x,t) = a_w \dfrac{\partial^{\beta} Q(x,t)}{\partial x^{\beta}} - R_a Q(x,t) + b(x)u(t) \\ \dfrac{\partial Q(0,t)}{dx} = 0, t \geq 0 \\ \dfrac{\partial Q(1,t)}{dx} = 0, t \geq 0 \\ Q(x,0) = Q_0, 0 \leq x \leq 1 \\ y(t) = k_0 \int_0^1 Q(x,t)c(x)dx \end{cases} \tag{44}$$

where $\alpha, \beta > 0$ denote non integer orders of the system, a_w, R_a denote coefficients of heat conduction and heat exchange, k_0 is a steady-state gain of the model. Now we can express (44) as the following, inifinite dimensional state equation:

$$\begin{cases} {}^C D_t^{\alpha} Q(t) = AQ(t) + Bu(t) \\ Q(0) = Q_0 \\ y(t) = k_0 C Q(t) \end{cases} \tag{45}$$

where:

$$\begin{cases} AQ = a_w \dfrac{\partial^{\beta} Q(x)}{\partial x^{\beta}} - R_a Q \\ \mathcal{D}(A) = \left\{ Q \in H^2(0,1) : Q'(0) = 0, Q'(1) = 0 \right\} \\ a_w, R_a > 0 \\ H^2(0,1) = \left\{ u \in L^2(0,1) : u', u'' \in L^2(0,1) \right\} \\ CQ(t) = \langle c, Q(t) \rangle, Bu(t) = \langle bu(t) \rangle \\ Q(t) = [q_1(t), q_2(t)..]^T \end{cases} \tag{46}$$

In (46) $\mathcal{D}(A)$ denotes the field of the state operator A, $'$, $''$ denotes the first and second derivative with respect to length, $< .. >$ is the standard scalar product.

The following set of the eigenvectors for the state operator A creates the orthonormal basis of the state space:

$$h_n = \begin{cases} 1, n = 0 \\ \sqrt{2}cos(n\pi x), n = 1, 2, ... \end{cases} \tag{47}$$

Eigenvalues of the state operator are expressed as underneath:

$$\lambda_n = -a_w \pi^\beta n^\beta - R_a, n = 0, 1, 2, ... \tag{48}$$

and consequently the state operator takes the form:

$$A = diag\{\lambda_0, \lambda_1, \lambda_2, ...\} \tag{49}$$

Next, the spectrum σ of the state operator A is expressed as underneath:

$$\sigma(A) = \{\lambda_0, \lambda_1, \lambda_2, ...\} \tag{50}$$

From (48) it follows at once that $\lambda_0 > \lambda_1 > \lambda_2 ...$
The input operator B has the following form:

$$B = [b_0, b_1, b_2, ...]^T \tag{51}$$

where $b_n = \langle b, h_n \rangle$, $b(x)$ denotes the heater function:

$$b(x) = \begin{cases} 1, x \in [0, x_0] \\ 0, x \notin [0, x_0] \end{cases} \tag{52}$$

With respect to (47) and (52) each element b_n takes the following form:

$$b_n = \begin{cases} x_u, \ n = 0, \\ \frac{\sqrt{2}\sin(n\pi x_u)}{n\pi}, \ n = 1, 2, ... \end{cases} \tag{53}$$

The output operator C is expressed as follows:

$$C = \begin{bmatrix} C_1 \\ C_2 \\ C_3 \end{bmatrix} \tag{54}$$

Rows of output operator C are as underneath:

$$C_j = \left[c_{j0}, c_{j1}, c_{j2}, ...\right] \ j = 1, 2, 3 \tag{55}$$

where $c_{jn} = \langle c, h_n \rangle$, $c(x)$ denotes the sensor function:

$$c(x) = \begin{cases} 1, x \in [x_1, x_2] \\ 0, x \notin [x_1, x_2] \end{cases} \quad (56)$$

With respect to (47) and (56) each element c_{jn} takes the form:

$$c_{jn} = \begin{cases} x_{j2} - x_{j1}, \ j = 1, 2, 3, \ n = 0 \\ \frac{\sqrt{2}(\sin(n\pi x_{j2}) - \sin(n\pi x_{j1}))}{n\pi}, \ j = 1, 2, 3, \ n = 1, 2, \ldots \end{cases} \quad (57)$$

5.1 The Spectrum Decomposition of the C Model

The state operator A has a discrete spectrum consisting of single eigenvalues λ_n associated to orthonormal eigenvectors (47) forming the base $L^2(0, 1)$. The spectrum decomposition of the system (46) is presented underneath.

$$^C D_t^\alpha Q(t) = AQ(t) + Bu(t)$$

$$Ah_n = \lambda_n h_n$$

$$< h_n, h_m >= \begin{cases} 1, n = m \\ 0, n \neq m \end{cases}$$

$$^C D_t^\alpha Q = a_w \frac{\partial^\beta Q}{\partial x^\beta} - R_a Q + Bu$$

$$^C D_t^\alpha \sum_{n=0}^{\infty} c_n h_n = a_w \frac{\partial^\beta \sum_{n=0}^{\infty} c_n h_n}{\partial x^\beta} - R_a \sum_{n=0}^{\infty} c_n h_n + \frac{Bu}{h_n}$$

$$^C D_t^\alpha c_n = a \frac{\partial^\beta c_n}{\partial x^\beta} - R_a c_n + b_n u \quad (58)$$

The form of Eq. (58) implies the decomposition of system (46) into subsystems related to the different eigenvalues λ_n, $n = 0, 1, 2, 3, \ldots$. This property is also mapped to discrete models presented in the next subsections.

5.2 The Step Responses of the C Model

Assume that the control is the Heaviside function: $u(t) = 1(t)$. Consequently the step response formula is as follows:

$$y_j(t) = k_{0_j} \sum_{n=0}^{\infty} \frac{\left(E_\alpha(\lambda_{\beta_n} t^\alpha) - 1(t)\right)}{\lambda_{\beta_n}} b_n c_{jn}, \quad j = 1, 2, 3 \tag{59}$$

The model (44)–(59) is infinite dimensional. Its use to modeling of the considered plant requires us to apply its finite dimensional approximation. This can be done by truncating further modes of solution (59). Consequently the operators A, B and C can be interpreted as matrices and the solution takes the form of the following finite sum:

$$y_j(t) = k_{0_j} \sum_{n=0}^{N} \frac{\left(E_\alpha(\lambda_{\beta_n} t^\alpha) - 1(t)\right)}{\lambda_{\beta_n}} b_n c_{jn}, \quad j = 1, 2, 3 \tag{60}$$

In (60) k_{0_j} is the coefficient necessary to fit the step response of the model to the experimental step response in j-th output, N denotes the order of finite approximation. Its estimation assuring stability and predefined convergence of the model can be done numerically or analytically. The numerical estimation is presented with details in [32]). The step responses of the above model are illustrated by Fig. 5.

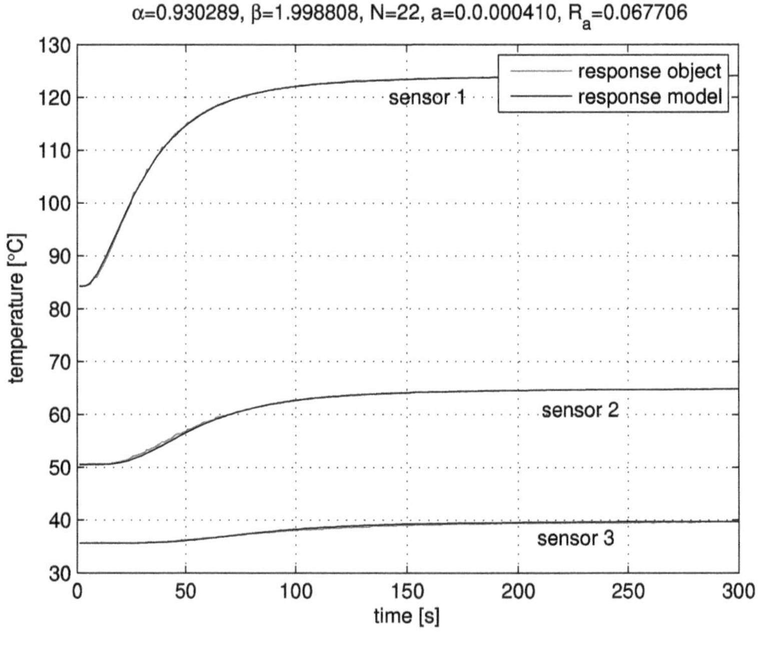

Fig. 5 Step response object and model 4 for 1–3 outputs with non-integer order ($\alpha = 0.930289$, $\beta = 1.998808$) and optimal order $N = 22$, $a = 0.000410$, $R_a = 0.0677066$. The cost function $MSE = 0.007108$ (see [32])

6 The Discrete State Space Model Using the Discrete GL Operator

The discrete time model using the discrete GL definition (the discrete GL model) follows directly from the continuous model (45) after use of FOBD (5). It is presented with details in [29]. The solution of the discrete state equation (22) takes the following form:

$$\begin{cases} Q^+(k+1) = P^+Q^+(k) - \sum_{l=1}^{L} A_l^+ Q^+(k-l) + B^+u(k) \\ y^+(k) = C^+Q^+(k) \end{cases} \tag{61}$$

In (61) A_l^+ is expressed by (24), P^+, B^+ and C^+ take the following form:

$$\begin{cases} P^+ = diag\{\lambda_{\beta_0}^+, \lambda_{\beta_1}^+ ... \lambda_{\beta_N}^+\} \\ B^+ = h^\alpha B \\ C^+ = C \end{cases} \tag{62}$$

where:

$$\lambda_{\beta_n}^+ = \alpha + h^\alpha \lambda_{\beta_n} = \alpha - h^\alpha \left(a_w \pi^\beta n^\beta + R_a\right) \tag{63}$$

The spectrum of the time-continuous system can be decomposed into single, separated eigenvalues (analogically as in the integer order case). This property is mapped to the discrete time system. Particularly, the solution (61) can be decomposed to separated "subsolutions" associated to the single eigenvalues (63).

6.1 Decomposition of the Discrete GL Model

The state vector $Q^+(k)$ of the discrete model (61) can be expressed as:

$$Q^+(k) = \begin{bmatrix} q_1^+(k) \\ ... \\ q_N^+(k) \end{bmatrix} \tag{64}$$

The matrices P^+ and A_l^+ describing the solution of the discrete system (61) are diagonal matrices. Consequently the solution (61) can be decomposed into N independent modes, associated with n-th state variable $Q_n^+(k)$ and described by n-th eigenvalue. The n-th mode of solution for fixed memory length L takes the form as follows:

$$q_n^{+L}(k+1) = \lambda_{\beta_n}^+ q_n^+(k) - \sum_{l=2}^{L} d_l q_n^+(k-l) + b_n^+ u^+(k), \quad n = 0, .., N \tag{65}$$

For each memory length the solution takes the following form:

$$q_n^{+\infty}(k+1) = \lambda_{\beta_n}^+ q_n^+(k) - \sum_{l=2}^{\infty} d_l q_n^+(k-l) + b_n^+ u^+(k), \quad n = 0, .., N \tag{66}$$

Between input of the system and the j-th output the discrete transfer function $G_j^+(z^{-1})$ can be defined:

$$G_j^{+L\infty}(z^{-1}) = \sum_{n=0}^{N} G_{nj}^{+L,\infty}(z^{-1}), \ j = 1, 2, 3 \tag{67}$$

The upper index "L" denotes the fixed memory length, index "∞" denotes each memory length. The transfer function $G_{nj}^{+L}(z^{-1})$ associated to n-th mode of solution for fixed memory length L is as follows:

$$G_{nj}^{+L}(z^{-1}) = \frac{c_{jn}^+ b_n^+ z^{-1}}{1 - z^{-1}\lambda_{\beta n}^+ + \sum_{l=2}^{L} d_l z^{-l-1}}, \ j = 1, 2, 3 \tag{68}$$

and analogically $G_{nj}^{+\infty}(z^{-1})$ can be defined:

$$G_{nj}^{+\infty}(z^{-1}) = \frac{c_{jn}^+ b_n^+ z^{-1}}{1 - z^{-1}\lambda_{\beta n}^+ + \sum_{l=2}^{\infty} d_l z^{-l-1}}, \ j = 1, 2, 3 \tag{69}$$

6.2 *Stability of the Discrete GL Model*

The stability conditions for the model (61)–(63) are proven in the paper [30]. The fundamental result is that too high an order N in the considered discrete model can cause its instability. Propositions describing the maximum permissible value of N assuring the preservation of the stability are recalled here.

Proposition (Maximum size of model N_{sL} assuring the stability of the discrete model for fixed memory length L) *Let us consider the discrete model of the heat transfer process described by* (61). *The size* N_{sL} *of finite-dimensional approximation assuring the stability of the discrete model* (61) *meets the following inequality:*

$$N_{sL} \leq Int\left(\left(\frac{1 + \alpha - h^{\alpha} R_a + \sum_{l=2}^{L} d_l}{h^{\alpha} a_w \pi^{\beta}}\right)^{\frac{1}{\beta}}\right) \tag{70}$$

Proposition (Maximum size of model $N_{s\infty}$ assuring the stability of the discrete model for each memory length) *Let us consider the discrete model of the heat transfer process described by* (61). *The size N_{sL} of finite-dimensional approximation assuring the stability of the discrete model* (61) *meets the following inequality:*

$$N_{s\infty} \leq Int\left(\left(\frac{2 + \alpha - h^{\alpha} R_a}{h^{\alpha} a_w \pi^{\beta}}\right)^{\frac{1}{\beta}}\right) \tag{71}$$

In (70) and (71) $Int(x)$ denotes an integer number nearest to x.

From (8) it turns out that condition (71) is the limit case of (70) for $L \to \infty$. Results of numerical calculations show that the both propositions give practically the same result [30].

6.3 Accuracy of the Discrete GL Model

The accuracy of the considered model can be described using the approach given in papers [33] and [35]. The steady-state error of the model we deal with is defined as follows:

$$\epsilon = y^{ss} - y^{+ss} \tag{72}$$

where y^{ss} and y^{+ss} are steady-state responses of continuous and discrete models respectively. They are equal:

$$y^{ss} = \lim_{t \to \infty} y(t) \tag{73}$$

$$y^{ss+} = \lim_{k \to \infty} y^{+}(k) \tag{74}$$

Both of the above responses can be calculated using the Final Value Theorem (FVT) (25) and (26) for continuous and discrete systems respectively.

For a time continuous system and the control in the form of the Heaviside function: $u(k) = 1(k)$ the steady state response is equal:

$$y^{ss} = -CA^{-1}B \tag{75}$$

With respect to (49)–(54) it is as follows:

$$y^{ss} = \left[y_1^{ss}, y_2^{ss}, y_3^{ss}\right]^T \tag{76}$$

where:

$$y_j^{ss} = \sum_{n=0}^{N} y_{nj}^{ss}, \ j = 1, 2, 3 \tag{77}$$

$$y_{nj}^{ss} = \frac{c_{jn} b_n}{\lambda_{\beta_n}}, \ j = 1, 2, 3 \tag{78}$$

Next the steady state response of the discrete system needs to be given. Fixed memory length L and each memory length need to be analysed separately. Using the FVT Theorem (26) and discrete transfer functions (68), (69) and with respect to (6), (7), (8) and (9) we obtain the steady-state response of the system $y^{+ss} = [y_1^{+ssL,\infty}, y_2^{+ssL,\infty}, y_3^{+ssL,\infty}]^T$ Upper index "L" denotes the fixed memory length, index "∞" denotes each memory length. For fixed memory length the steady state response is equal:

$$y_j^{+ssL} = \sum_{n=0}^{N} y_{nj}^{ssL}, \ j = 1, 2, 3 \tag{79}$$

where:

$$y_{nj}^{+ssL} = \frac{c_{jn}^{+} b_n^{+}}{1 - \lambda_{\beta n}^{+} + \sum_{l=2}^{L} d_l} \ j = 1, 2, 3 \tag{80}$$

Next the steady state response for each memory length equals to:

$$y_j^{+ss\infty} = \sum_{n=0}^{N} y_{nj}^{ss\infty} \ j = 1, 2, 3 \tag{81}$$

where:

$$y_{nj}^{+ss\infty} = \frac{c_{jn}^{+} b_n^{+}}{2 - \lambda_{\beta n}^{+}} \ j = 1, 2, 3 \tag{82}$$

With respect to (62) and (63) y_{nj}^{ssL} and $y_{nj}^{ss\infty}$ take the form:

$$y_{nj}^{+ssL} = \frac{h^{\alpha} c_{jn} b_n}{1 - \alpha + \sum_{l=2}^{L} d_l - h^{\alpha} \lambda_{\beta n}} \ j = 1, 2, 3 \tag{83}$$

$$y_{nj}^{+ss\infty} = \frac{h^{\alpha} c_{jn} b_n}{2 - \alpha - h^{\alpha} \lambda_{\beta n}} \ j = 1, 2, 3 \tag{84}$$

Finally the steady-state error with respect to (72) takes the following form:

$$\epsilon^{ssL,\infty} = \left[\epsilon_1^{ssL,\infty}, \epsilon_2^{ssL,\infty}, \epsilon_3^{ssL,\infty}\right]^T \tag{85}$$

where:

$$\epsilon_j^{ssL,\infty} = \left|y_j^{ss} - y_j^{+ssL,\infty}\right|, \ j = 1, 2, 3 \tag{86}$$

With respect to (83) and (84):

$$\epsilon_j^{ssL,\infty} = \left|\sum_{n=0}^{N} \epsilon_{jn}^{ssL,\infty}\right| \tag{87}$$

Each component of (87) is as follows:

$$\epsilon_{nj}^{ssL} = c_{jn} b_n \left(\frac{1 + \sum_{l=2}^{L} d_l - \alpha - 2h^{\alpha}\lambda_{\beta n}}{\lambda_{\beta n}\left(1 + \sum_{l=2}^{L} d_l - \alpha - h^{\alpha}\lambda_{\beta n}\right)} \right), \ j = 1, 2, 3 \tag{88}$$

$$\epsilon_{nj}^{ss\infty} = c_{jn} b_n \left(\frac{2 - \alpha - 2h^{\alpha}\lambda_{\beta n}}{\lambda_{\beta n}\left(2 - \alpha - h^{\alpha}\lambda_{\beta n}\right)} \right), \ j = 1, 2, 3 \tag{89}$$

6.4 *Convergence of the Discrete GL Model*

The convergence can be analyzed with respect to order N or with respect to memory length L. Unfortunately the analysis with respect to L can only be done numerically. However the analysis with respect to N is presented in paper [29]. It is recalled below.

The rate of convergence ROC_N for the model of the size N is defined as the absolute value of the steady state value of the N-th mode of the solution for the j-th output:

$$ROC_{Nj}^{L,\infty} = \left|y_{Nj}^{+ssL,\infty}\right| \tag{90}$$

where upper indices L and ∞ denote fixed memory length and each memory length respectively, $y_{Nj}^{+ss\infty}$ is calculated using (83) and (84) with $n = N$. The size of the model $N_{\Delta L}$ assuring the predefined value Δ_L of ROC for fixed memory length L can be estimated. It is described by the following proposition:

Proposition (Minimum size of model $N_{\Delta L}$ assuring the predefined Rate of Convergence Δ_L of the discrete model for fixed memory length L) *Let us consider the discrete model of the heat transfer process described by* (61). *The size $N_{\Delta L}$ of the model assuring the predefined value Δ_L of ROC meets the following inequality:*

$$N_{\Delta L} \geq Int\left(\frac{\sqrt{\sqrt{S_L^2 + \frac{8h^{2\alpha}a_w}{\Delta_L}} - S_L}}{\pi\sqrt{h^\alpha a_w}}\right) \tag{91}$$

where:

$$S_L = 1 - \alpha + \sum_{l=2}^{L} d_l + h^\alpha R_a \tag{92}$$

In (91) $Int(..)$ denotes the nearest integer value.

Proof The condition $ROC_{Nj}^L \leq \Delta_L$ with respect to (53) and (57) is equivalent to:

$$\Delta_L \leq \left|\frac{h^\alpha}{S_L + h^\alpha a_w \pi^\beta N_{\Delta L}^\beta}\right| \cdot P \tag{93}$$

where:

$$P = \frac{2}{N_{\Delta L}^2 \pi^2}\left|\sin\left(\frac{N\pi(x_{j2} - x_{j1})}{2}\right)\cos\left(\frac{N\pi(x_{j2} + x_{j1})}{2}\right)\cdot \sin\left(\frac{N\pi x_u}{2}\right)\right| \tag{94}$$

The factor P expressed by (94) is not greater than $\frac{2}{N_{\Delta L}^2 \pi^2}$, because the expression inside absolute value $|..|$ does not exceed one. It allows to assume that:

$$P \leq \frac{2}{N_{\Delta L}^2 \pi^2} \tag{95}$$

This gives the upper estimation of $N_{\Delta L}$, but (93) takes to simpler form:

$$\Delta_L \leq \left|\frac{2h^\alpha}{N_{\Delta L}^2 \pi^2 \left(S_L + h^\alpha a_w \pi^\beta N_{\Delta L}^\beta\right)}\right| \tag{96}$$

The expression inside $|...|$ is always positive for $S_L > 0$. This allows to ignore the absolute value. Next to simplify further calculations assume that $\beta = 2$. Then the (96) takes the form:

$$\Delta_L \leq \frac{2h^\alpha}{N_{\Delta L}^2 \pi^2 \left(S_L + h^\alpha a_w \pi^\beta N_{\Delta L}^\beta\right)} \Longleftrightarrow \\ \Longleftrightarrow \Delta_L \pi^4 h^\alpha a_w N_{\Delta L}^4 + \Delta_L \pi^2 S_L N_{\Delta L}^2 - 2h^\alpha \geq 0 \tag{97}$$

Solving the double quadratic inequality (97) we obtain directly the condition (91) and the proof is completed. □

The case of each memory length is the limit case for $L \to \infty$ and it is expressed as follows:

Proposition (Minimum size of model $N_{\Delta\infty}$ assuring the predefined Rate of Convergence Δ_∞ of the discrete model for each memory length) *Let us consider the discrete model of the heat transfer process described by* (61). *The size* $N_{\Delta\infty}$ *of the model assuring the predefined value* Δ_∞ *of ROC meets the following inequality:*

$$N_{\Delta\infty} \geq Int\left(\frac{\sqrt{\sqrt{S_\infty^2 + \frac{8h^{2\alpha}a_w}{\Delta_\infty}} - S_\infty}}{\pi\sqrt{h^\alpha a_w}}\right) \tag{98}$$

where:

$$S_\infty = 2 - \alpha + h^\alpha R_a \tag{99}$$

In (98) $Int(..)$ denotes the nearest integer value.

Proof Each memory length is a limit case: $L \to \infty$ of condition (91). This gives:

$$S_\infty = \lim_{L\to\infty} S_L \tag{100}$$

where S_L is expressed by (92). The rest of the proof is identical as (93)–(97). □

The use of both conditions (91) and (98) gives practically the same result. This can be explained by the fact that the factor S_∞ in (98) is the limit case of factor S_L in (91) and the sum (9) quickly goes to one. The simplification (95) gives the upper, "cautious" estimation of $N_{\Delta L,\infty}$. However it allows us to analyze the convergence independently of size and location of the heater and sensors.

Finally the recommended size N of the discrete model presented in this subsection is proposed as follows:

$$N_\Delta < N < N_s \tag{101}$$

6.5 Experimental Verification of the Discrete GL Model

Experiments were executed using the experimental system given in Sect. 4. The step response of the model was tested in time range from 0 to $T_f = 300\,[\mathrm{s}]$ with sample time $h = 1\,[\mathrm{s}]$, parameters of the model (61)–(63) were calculated via the minimization of the MSE (Mean Square Error) cost function (43) using the MATLAB *fminsearch* function. The parameters are given in the Table 2. The use of stability conditions (70) and (71) gives $N_{sL} \leq 20$, $N_{s\infty} \leq 20$. Next the steady state error is analysed. The error $\epsilon_{N2}^{ss\infty}$ as a function of size N, calculated with respect to (86)

Table 2 Parameters of the heat plant, $N = 13$, $L = 150$

Parameter	α	β	a	R_a
Value	0.9448	2.0336	0.0006	0.0531

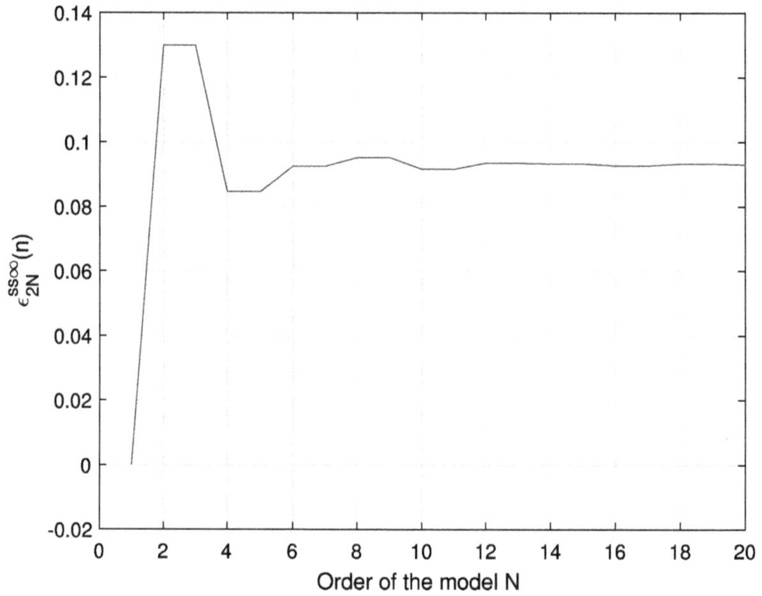

Fig. 6 The steady state error of the discrete GL model $\epsilon_{N2}^{ss\infty}$ as a function of N for each memory length

for each memory length is shown in Fig. 6. The comparison steady state errors for each memory length vs fixed memory length is shown in the Fig. 7. Furthermore the convergence of the presented model needs to be considered. It should be assumed that we need to find the size $N_{\Delta L}$ and $N_{\Delta\infty}$ of the model assuring the value $\Delta_N = 0.0005$. The use of conditions (91) and (98) gives the values: $N_L = N_\infty \geq 13$. This result is verified by diagrams shown in Figs. 8 and 9. Finally, the order N of the model assuring maintained stability and predefined rate of convergence is as follows:

$$13 \leq N \leq 20$$

To verify the above results, in Figs. 10 and 11 are shown exemplary spectra of standard systems, calculated as sets of eigenvalues (63) and, to verify results spectra of the whole discrete system, calculated as poles of transfer function (67). The comparison of the experimental step response to step response of the model with parameters given in the Table 2 and the order of the model $N = 13$ and the order of the discrete GL operator equal $L = 150$ is given in Fig. 12. For diagrams in this figure the cost function (43) equals to: $MSE = 0.0719$.

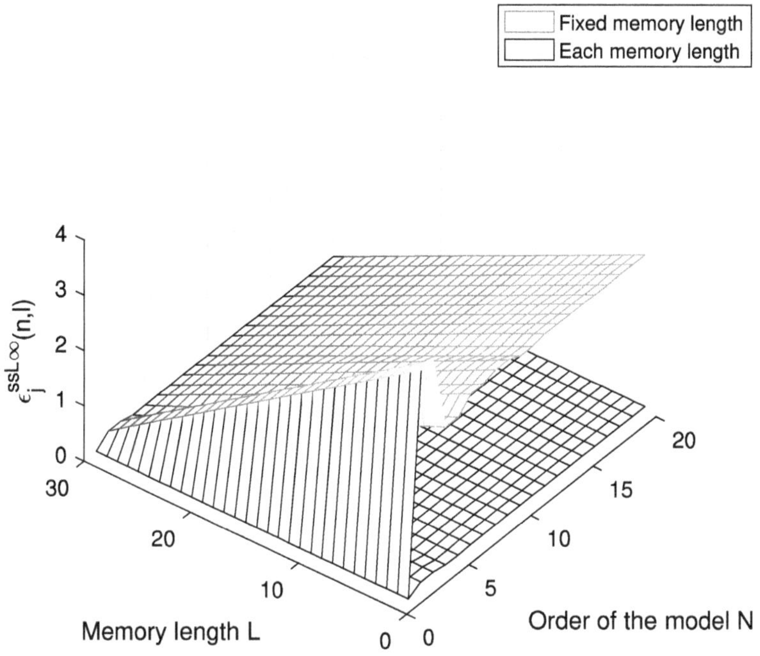

Fig. 7 The comparison of steady state error in the discrete GL model $\epsilon_{N2}^{ss\infty}$ for each memory length to fixed memory length

7 The Discrete State Space Model Using CFE Approximation

The discrete state equation using CFE approximation (31) for the model (44) will be called the discrete CFE model. It takes the following form:

$$\begin{gathered} Q^+(k) = -E_0^{-1}\sum_{m=1}^{M} E_m Q^+(k-m) + \\ +E_0^{-1}\sum_{m=0}^{M} F_m u^+(k-m) + E_0^{-1}\sum_{m=-M}^{0} q_0(m) \end{gathered} \tag{102}$$

The state vector has the form: $Q^+(k) = \left[q_1^+(k), q_2^+(k), ...\right]^T$, matrices $-E_0^{-1}E_m$ in (102) with respect to (49) and (48) take the form:

$$E_m^+ = E_0^{-1}E_m = diag\{e_{1m}, .., e_{Nm}\},\ m = 1, .., M \tag{103}$$

where:

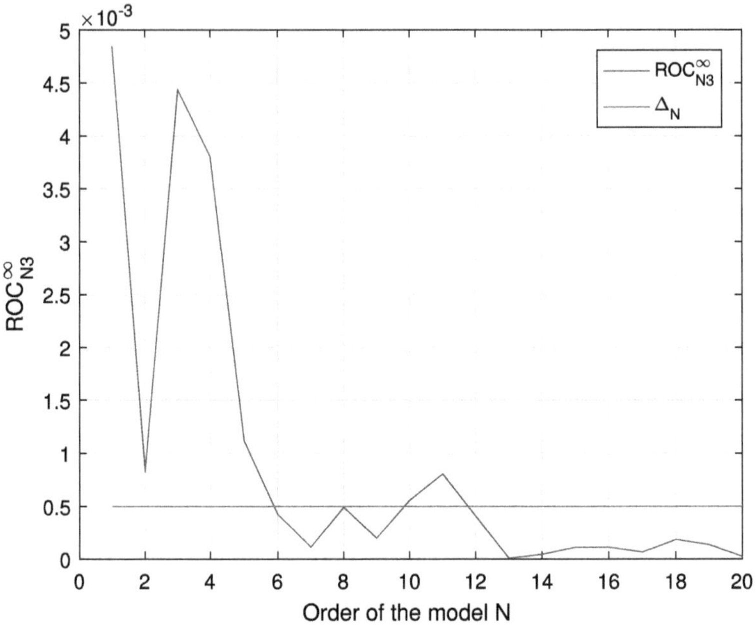

Fig. 8 The Rate of Convergence of the discrete GL model as a function of N for each memory length

$$e_{nm} = \frac{g_h w_m + v_m(a_w n^\beta \pi^\beta + R_a)}{v_0(a_w n^\beta \pi^\beta + R_a)} \tag{104}$$

and analogically:

$$F_m^+ = E_0^{-1} F_m = [f_{1m}, .., f_{Nm}]^T\ ,\ m = 1, .., M \tag{105}$$

where:

$$f_{nm} = \frac{v_m b_n}{v_0(a_w n^\beta \pi^\beta + R_a)} \tag{106}$$

The discrete system (102)–(106) can be expressed also in the 1'st order extended form (32)–(40). The matrices A_q^+, B_q^+ and C_q^+ for the extended system (32) are obtained using (104) and (106). The extended system is easy to use during simulations because it can be solved with the use of standard tools available on the MATLAB platform.

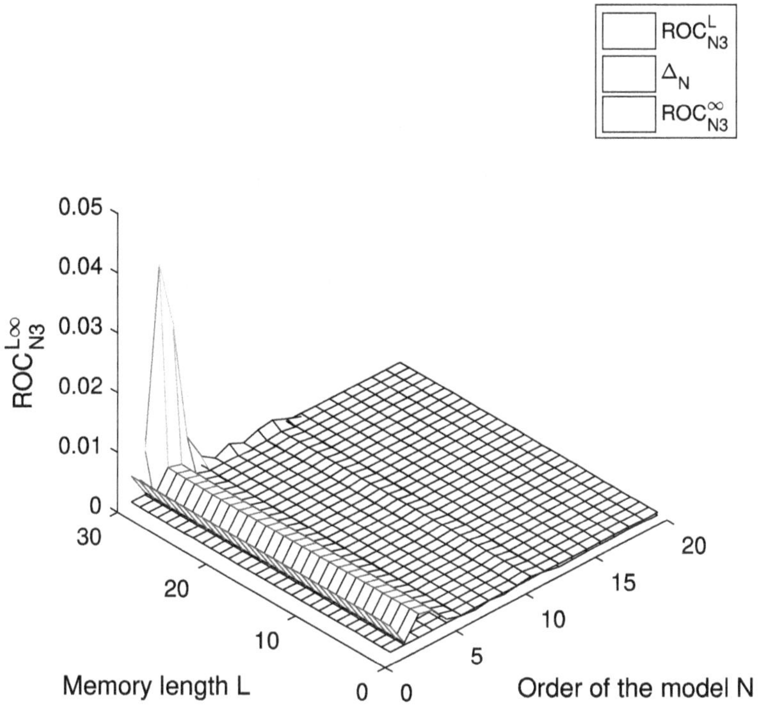

Fig. 9 The comparison of Rate of Convergence of the discrete GL model for each memory length to fixed memory length

7.1 Decomposition of the Discrete CFE Model

The form of discrete equation (102) with matrices (103) and (105) implies the possibility of its decomposition, analogically as it was done for time continuous case. The n-th mode of solution for the decomposed system is expressed as follows:

$$q_n^+(k) = \sum_{m=1}^{M} e_{nm} q_n^+(k-m) + \sum_{m=0}^{M} f_{nm} u^+(k-m) \tag{107}$$

The discrete transfer function $G_n^+(z^{-1})$ of the n-th mode is as underneath:

$$G_n^+(z^{-1}) = \frac{c_{jn} \sum_{m=1}^{M} f_{nm} z^{-m}}{1 - \sum_{m=1}^{M} e_{nm} z^{-m}} \tag{108}$$

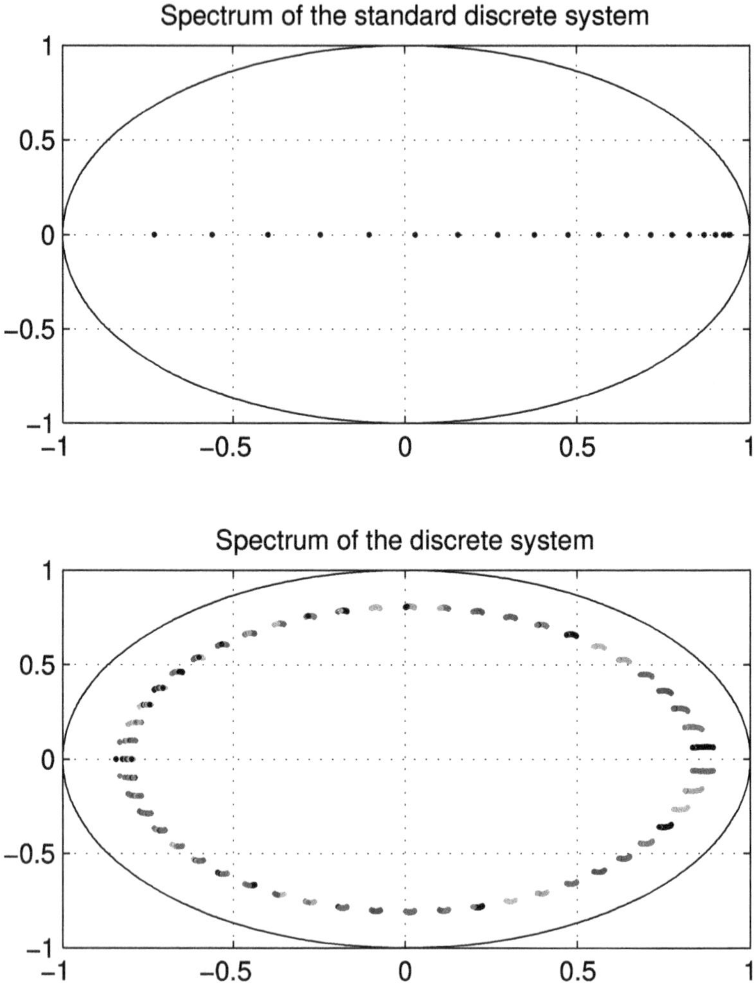

Fig. 10 The spectra of stable system: $h = 1[\mathrm{s}]$, $N = 20$

where e_{nm} and f_{nm} are expressed by (104) and (106) respectively. The characteristic polynomial associated with the n-th mode of solution (107) has the following form:

$$w_n^+(z^{-1}) = 1 - \sum_{m=1}^{M} e_{nm} z^{-m} \tag{109}$$

The steady-state response of the n-th mode y_{jn}^{ss} can be obtained with the use of the Final Value Theorem (FVT) for discrete systems. If the control signal is the Heaviside function: $u(k) = 1(k)$ then y_{jn}^{ss} is as follows:

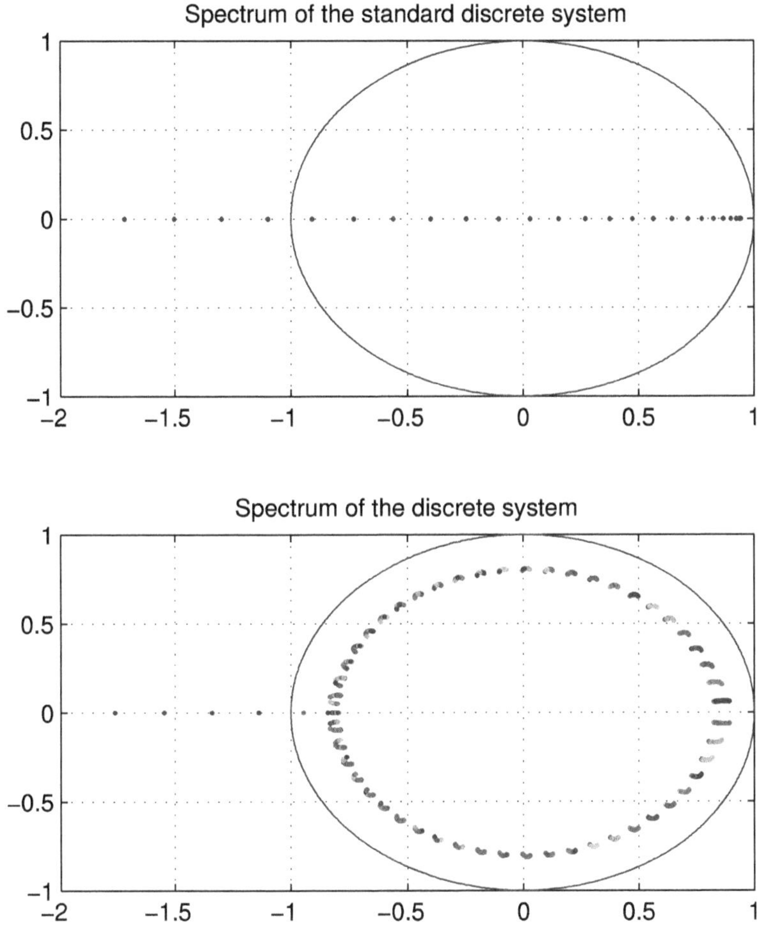

Fig. 11 The spectra of instable system: $h = 1[\text{s}]$, $N = 25$

$$y_{jn}^{ss} = \frac{c_{jn} \sum\limits_{m=1}^{M} f_{nm}}{1 - \sum\limits_{m=1}^{M} e_{nm}} = \frac{F c_{jn} b_n}{\lambda_{\beta_n} (F - \lambda_{\beta_n})} \tag{110}$$

where F is expressed as follows:

$$F = \frac{\sum_{m=0}^{M} w_m}{\sum_{m=0}^{M} v_m} \tag{111}$$

In (111) w_m and v_m denote coefficients of CFE approximation, given in Table 1.

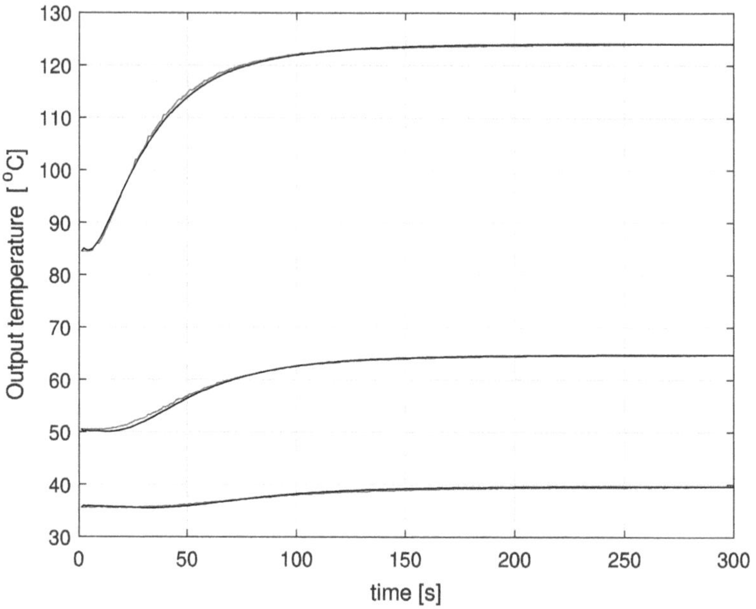

Fig. 12 Comparison of experimental step response to step response of the discrete GL model. Gray line is the experimental step response, black line denotes the step response of the model

With respect to (53), (57) and some elementary transformations we obtain the direct dependency between the steady-state response of the n-th mode and parameters of the plant:

$$y_{jn}^{ss} = \frac{4}{(F - \lambda_{\beta_n})\pi^2 n^2} \sin\left(\frac{n\pi(x_{j2} - x_{j1})}{2}\right) \cdot \cdot \cos\left(\frac{n\pi(x_{j2} + x_{j1})}{2}\right) \sin\left(\frac{n\pi x_u}{2}\right) \tag{112}$$

7.2 Stability of the Discrete CFE Model

Stability of this model was analysed using the frequency approach presented in Sect. 2. It is described by the following proposition:

Proposition 1 *Asymptotic stability of the discrete CFE model Consider the discrete model of heat transfer process described by* (44)–(55) *with non integer order* $0.0 < \alpha < 2.0$ *and its discrete, CFE based approximation* (102).

The discrete approximation (102) *is asymptotically stable for fractional order* $0.0 < \alpha < 2.0$, *and each approximation order* M, *sample time* h *and weight parameter* a.

Proof Recall that the spectrum $\sigma(A)$ of the considered heat plant is given by (50). It contains negative, single, separated, pure real eigenvalues. All these eigenvalues are located in the left semi plane, at the real axis.

Stability areas for the discrete, fractional order system described by CFE approximation and the different ranges of order α are given in Fig. 1. From this figure it can be noted immediately that:

- The instability area expressed by (41) for $0.0 < \alpha < 1.0$ is located in the right semi plane and for each M, h and a it does not exceed the imaginary axis.
- The instability area expressed by (41) for $1.0 < \alpha < 2.0$ is located in both semi planes, but for each M, h and a it does not cover the negative part of the real axis.

The above notes allow us to conclude that for $0.0 < \alpha < 2.0$ and each set of the other approximation parameters: M, h and s the spectrum $\sigma(A)$ is located outside the instability area. This finishes the proof. □

7.3 Accuracy of the Discrete CFE Model

Accuracy of the model we deal with can be estimated using the approach presented by [35] with the use of steady-state error of the considered model. This error can be estimated with respect to Proposition 1 in [35]:

$$\epsilon_{ss} = C\left(F - A\right)^{-1} FA^{-1}Bu_{ss} = [\epsilon_{ss1}, \epsilon_{ss2}, \epsilon_{ss3}]^T \tag{113}$$

where u_{ss} is the steady-state value of the control signal, F is defined by (111). The steady-state error given by (113) has the form of a vector. Each component of this vector describes error of suitable output. If the control is the Heaviside function: $u(t) = 1(t)$ then steady-state error at j-th output ($j = 1, 2, 3$) takes the following form:

$$\epsilon_{ssj} = F\sum_{n=0}^{N} \frac{c_{jn}b_n}{\lambda_{\beta_n}\left(F - \lambda_{\beta_n}\right)} \tag{114}$$

where λ_{β_n}, b_n and c_{jn} are described by (48), (53) and (57) respectively.

7.4 Convergence of the Discrete CFE Model

Convergence analysis is done by estimating the order N assuring a predefined value of Rate Of Convergence (ROC). In the considered case the ROC can be defined as the increment of the steady-state response y_{jn}^{ss} as a function of order N. This increment can be defined as the absolute value of N-th mode of steady-state response:

$$ROC_N = |y_{jN}^{ss}| \tag{115}$$

where y_{jn}^{ss} is expressed by (110) and (112). The order N assuring the keeping predefined value Δ_N of ROC_N is described by the following proposition:

Proposition 2 *The order of model N assuring the keeping of predefined value Δ_N by ROC of the model*
Consider the discrete model of the heat transfer process described by (44)–(55) *with non integer order* $0.0 < \alpha < 2.0$ *and its discrete, CFE based approximation* (102), *let the ROC of the discrete approximated model be defined by* (115).
The order N of the model assuring the predefined value Δ_N of ROC_N meets the following inequality:

$$N \geq \sqrt{\frac{-(F+R_a)+\sqrt{(F+R_a)^2+\frac{16a_w}{\Delta_N}}}{\pi^2 a_w}} \tag{116}$$

Proof The condition $ROC_N \leq \Delta_N$ is equivalent to:

$$\Delta_N \geq \left|\frac{4}{(F-\lambda_{\beta_N})\pi^2 N^2}\right| \cdot P \tag{117}$$

where:

$$P = \left|\sin\left(\frac{N\pi(x_{j2}-x_{j1})}{2}\right)\cos\left(\frac{N\pi(x_{j2}+x_{j1})}{2}\right)\cdot \right. \\ \left. \cdot \sin\left(\frac{N\pi x_u}{2}\right)\right| \tag{118}$$

Notice that P expressed by (118) is not greater than one. It allows us to assume that P is equal to one. It will give us the upper estimation of N, but (117) takes much a more simplier form:

$$\left|\frac{4}{(F-\lambda_{\beta_N})\pi^2 N^2}\right| \leq \Delta_N \tag{119}$$

With respect to (48) the expression (119) takes the form:

$$\left|\frac{4}{(F+a_w\pi^\beta N^\beta+R_a)\pi^2 N^2}\right| \leq \Delta_N \tag{120}$$

The expression inside the absolute value is always positive, consequently the absolute value can be ignored:

$$\frac{4}{(F+a_w\pi^\beta N^\beta+R_a)\pi^2 N^2} \leq \Delta_N \tag{121}$$

The left side of (121) will be called the non integer order limiter $L_{nio}(N)$:

$$L_{nio}(N) = \frac{4}{(F + a_w \pi^\beta N^\beta + R_a)\pi^2 N^2} \tag{122}$$

Next assume that $\beta = 2$ (we consider the integer order model with respect to length). Then the non integer order limiter (122) takes its integer order form $L_{io}(N)$:

$$L_{io}(N) = \frac{4}{(F + a_w \pi^2 N^2 + R_a)\pi^2 N^2} \tag{123}$$

Consequently the inequality (121) turns to:

$$\Delta_N \pi^4 a_w N^4 + \Delta_N \pi^2 (F + R_a) N^2 - 4 \geq 0 \tag{124}$$

The solution of double quadratic inequality (124) gives directly the condition (116). This finishes the proof. □

The condition (116) gives only the upper estimation of N, but for decreasing values of Δ_N the accuracy of the proposed estimation increases. This will be shown in the next subparagraph.

7.5 *Experimental Verification of the Discrete CFE Model*

Experiments have been executed using the experimental system presented in the Sect. 4. The parameters of the model are given in the Table 3. They were achieved using experimental results and via minimization the cost function (43).

The step response of the model compared to the step response of the plant is given in the Fig. 13. Next the accuracy of the proposed model was estimated using steady state error (113) and (114). Results are given in the Table 4. From Table 4 it can be noted that the steady state accuracy of the proposed model does not practically depend on order N.

Finally, the convergence has been tested using the Proposition 2. The predefined value of ROC was equal: $\Delta_N = 0.001$. Using condition (115) we obtain $N = 15$. The comparison limiters (122) and (123) to steady-state values of modes (112) is shown in Fig. 14. In this figure it can be noted that use of the non integer order limiter gives the exact estimation $N = 13$, which is slightly better than the integer order estimation proposed by Proposition 2.

Table 3 Parameters of the discrete CFE model

α	β	a_w	R_a	a	M	N	MSE (43)
0.9402	2.2054	0.0007	0.0336	0.7215	5	8	0.1366

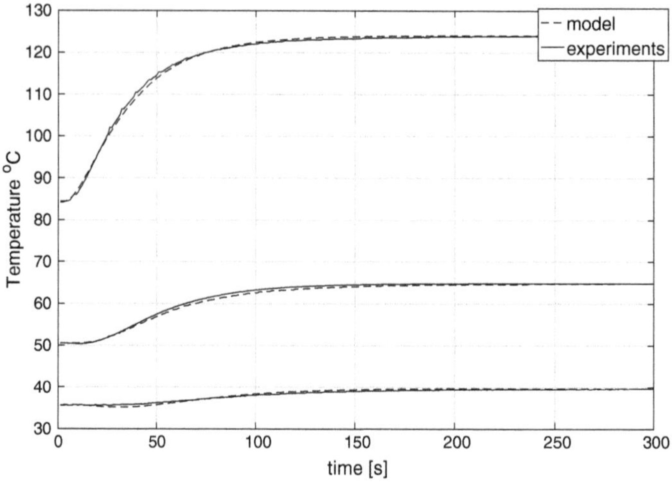

Fig. 13 The comparison experiment to discrete CFE model with parameters given in the Table 3

Table 4 Steady state error ϵ_{ss} for different N and all outputs

N	8	15	25
ϵ_{ss1}	−0.0404	−0.0405	−0.0406
ϵ_{ss2}	−0.0173	−0.0172	−0.0172
ϵ_{ss3}	−0.0045	−0.0043	−0.0043

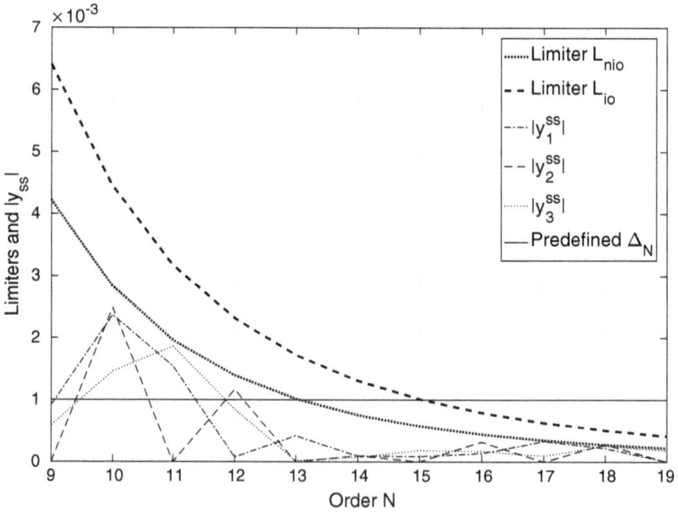

Fig. 14 Convergence estimation for $\Delta_N = 0.001$, and parameters of the model given in Table 3

8 Conclusions

The main, final conclusion from this chapter is that the use of the fractional order approach allows us to obtain models that are more accurate in the sense of MSE cost function than analogical integer order models. The proposed non integer order models are consistent with the integer order model due to fact that setting non integer orders: $\alpha = 1.0$ and $\beta = 2.0$ allows us to obtain the integer order model. Next, the spectrum decomposition property is fully mapped to each proposed non integer order model. The accuracy and convergence of all proposed models can be analyzed numerically or analytically. The most problematic is the stability of the discrete model using the GL operator, but the analytical stability condition is given and proved.

Acknowledgements This chapter is sponsored by AGH project no 16.16.120.773.

References

1. Al-Alaoui, M.A.: Novel digital integrator and differentiator. Electron. Lett. **29**(4), 376–378 (1993)
2. Bartecki, K.: A general transfer function representation for a class of hyperbolic distributed parameter systems. Int. J. Appl. Math. Comput. Sci. **23**(2), 291–307 (2013)
3. Buslowicz, M., Kaczorek, T.: Simple conditions for practical stability of positive fractional discrete-time linear systems. Int. J. Appl. Math. Comput. Sci. **19**(2), 263–269 (2009)
4. Caponetto, R., Dongola, G., Fortuna, L., Petras, I.: Fractional order systems: modeling and control applications. In: Chua, L.O. (ed.) World Scientific Series on Nonlinear Science, pp. 1–178. University of California, Berkeley (2010)
5. Chen, Y.Q., Moore, K.L.: Discretization schemes for fractional-order differentiators and integrators. IEEE Trans. Circuits Syst. I Fundam. Theory Appl. **49**(3), 263–269 (2002)
6. Curtain, R.F., Zwart, H.: An Introduction to Infinite-Dimensional Linear Systems Theory. Springer, New York (1995)
7. Das, S.: Functional Fractional Calculus for System Identification and Controls. Springer, Berlin (2010)
8. Dlugosz, M., Skruch, P.: The application of fractional-order models for thermal process modelling inside buildings. J. Build. Phys. **1**(1), 1–13 (2015)
9. Wyrwas, M., Mozyrska, D., Girejko, E.: Comparison of h-difference fractional operators. In: Mitkowski, W. et al. (ed.) Advances in the Theory and Applications of Non-integer Order Systems, pp. 1–178. Springer, Switzerland (2013)
10. Dzieliński, A., Sierociuk, D., Sarwas, G.: Some applications of fractional order calculus. Bull. Pol. Acad. Sci. Tech. Sci. **58**(4), 583–592 (2010)
11. Gal, C.G., Warma, M.: Elliptic and parabolic equations with fractional diffusion and dynamic boundary conditions. Evol. Eqn. Control Theory **5**(1), 61–103 (2016)
12. Gómez, J.F., Torres, L., Escobar, R.F. (eds.): Fractional derivatives with Mittag-Leffler kernel. Trends and applications in science and engineering. In: Kacprzyk, J. (ed.) Studies in Systems, Decision and Control, vol. 194, pp. 1–339. Springer, Switzerland (2019)
13. Kaczorek, T.: Selected Problems of Fractional Systems Theory. Springer, Berlin (2011)
14. Kaczorek, T.: Singular fractional linear systems and electrical circuits. Int. J. Appl. Math. Comput. Sci. **21**(2), 379–384 (2011)
15. Kaczorek, T., Rogowski, K.: Fractional Linear Systems and Electrical Circuits. Bialystok University of Technology, Bialystok (2014)

16. Kulczycki, P., Korbicz, J., Kacprzyk, J. (eds.): Control, Robotics and Information Processing (in Polish). PWN, Warsaw (2020)
17. Mitkowski, W.: Stabilization of Dynamic Systems (in Polish). WNT, Warszawa (1991)
18. Mitkowski, W.: Approximation of fractional diffusion-wave equation. Acta Mechanica et Automatica **5**(2), 65–68 (2011)
19. Mitkowski, W.: Finite-dimensional approximations of distributed RC networks. Bull. Pol. Acad. Sci. Tech. Sci. **62**(2), 263–269 (2014)
20. Mitkowski, W.: Outline of Control Theory (in Polish). Wydawnictwa AGH, Kraków (2019)
21. Mitkowski, W., Obraczka, A.: Simple identification of fractional differential equation. Solid State Phenom. **1**(180), 331–338 (2012)
22. Mozyrska, D., Pawluszewicz, E.: Fractional discrete-time linear control systems with initialisation. Int. J. Control **1**(1), 1–7 (2011)
23. Obraczka, A., Mitkowski, W.: The comparison of parameter identification methods for fractional, partial differential equation. Diffusion and defect data – solid state data. Part B. Solid State Phenom. **210**(2014), 265–270 (2014)
24. Obrączka, A.: Control of heat processes with the use of non-integer models. Ph.D. thesis, AGH University, Krakow, Poland (2014)
25. Oprzędkiewicz, K.: The interval parabolic system. Arch. Control Sci. **13**(4), 415–430 (2003)
26. Oprzędkiewicz, K.: A controllability problem for a class of uncertain parameters linear dynamic systems. Arch. Control Sci. **14**(1), 85–100 (2004)
27. Oprzędkiewicz, K.: An observability problem for a class of uncertain-parameter linear dynamic systems. Int. J. Appl. Math. Comput. Sci. **15**(3), 331–338 (2005)
28. Oprzędkiewicz, K.: Non integer order, discrete, state space model of heat transfer process using grünwald-letnikov operator. Bull. Pol. Acad. Sci. Tech. Sci. **66**(3), 249–255 (2018)
29. Oprzędkiewicz, K., Dziedzic, K., Więckowski, L.: Non integer order, discrete, state space model of heat transfer process using Grünwald-Letnikov operator. Bull. Pol. Acad. Sci. Tech. Sci. **67**(5), 905–914 (2019)
30. Oprzędkiewicz, K., Gawin, E.: The practical stability of the discrete, fractional order, state space model of the heat transfer process. Arch. Control Sci. **28**(3), 463–482 (2018)
31. Oprzędkiewicz, K., Gawin, E., Mitkowski, W.: Modeling heat distribution with the use of a non-integer order, state space model. Int. J. Appl. Math. Comput. Sci. **26**(4), 749–756 (2016)
32. Oprzędkiewicz, K., Gawin, E., Mitkowski, W.: Parameter identification for non integer order, state space models of heat plant. In: MMAR 2016 : 21th International Conference on Methods and Models in Automation and Robotics : 29 August–01 September 2016, Międzyzdroje, Poland, pp. 184–188 (2016)
33. Oprzędkiewicz, K., Mitkowski, W.: A memory efficient non integer order discrete time state space model of a heat transfer process. Int. J. Appl. Math. Comput. Sci. **28**(4), 649–659 (2018)
34. Oprzędkiewicz, K., Mitkowski, W., Gawin, E.: An accuracy estimation for a non integer order, discrete, state space model of heat transfer process. In: Automation 2017 : Innovations in Automation, Robotics and Measurement Techniques : 15–17 March, Warsaw, Poland, pp. 86–98 (2017)
35. Oprzędkiewicz, K., Stanisławski, R., Gawin, E., Mitkowski, W.: A new algorithm for a CFE approximated solution of a discrete-time non integer-order state equation. Bull. Pol. Acad. Sci. Tech. Sci. **65**(4), 429–437 (2017)
36. Ostalczyk, P.: Equivalent descriptions of a discrete-time fractional-order linear system and its stability domains. Int. J. Appl. Math. Comput. Sci. **22**(3), 533–538 (2012)
37. Ostalczyk, P.: Discrete Fractional Calculus. Applications in Control and Image Processing. World Scientific, New Jersey, London, Singapore (2016)
38. Pazy, A.: Semigroups of Linear Operators and Applications to Partial Differential Equations. Springer, New York (1983)
39. Petras, I.: Fractional order feedback control of a DC motor. J. Electr. Eng. **60**(3), 117–128 (2009)
40. Petras, I.: http://people.tuke.sk/igor.podlubny/usu/matlab/petras/dfod2.m (2009)
41. Podlubny, I.: Fractional Differential Equations. Academic Press, San Diego (1999)

42. Popescu, E.: On the fractional Cauchy problem associated with a feller semigroup. Math. Rep. **12**(2), 181–188 (2010)
43. Rauh, A., Senkel, L., Aschemann, H., Saurin, V.V., Kostin, G.V.: An integrodifferential approach to modeling, control, state estimation and optimization for heat transfer systems. Int. J. Appl. Math. Comput. Sci. **26**(1), 15–30 (2016)
44. Sierociuk, D., Skovranek, T., Macias, M., Podlubny, I., Petras, I., Dzielinski, A., Ziubinski, P.: Diffusion process modeling by using fractional-order models. Appl. Math. Comput. **257**(1), 2–11 (2015)
45. Stanisławski, R., Latawiec, K.: Stability analysis for discrete-time fractional-order LTI state-space systems. Part i: new necessary and sufficient conditions for asymptotic stability. Bull. Pol. Acad. Sci. Tech. Sci. **61**(2), 353–361 (2013)
46. Stanisławski, R., Latawiec, K.: Stability analysis for discrete-time fractional-order LTI state-space systems. Part ii: stability criterion for FD-based systems. Bull. Pol. Acad. Sci. Tech. Sci. **61**(2), 362–370 (2013)
47. Stanisławski, R., Latawiec, K., Łukaniszyn, M.: A comparative analysis of Laguerre-based approximators to the Grünwald-Letnikov fractional-order difference. Hindawi Publ. Corpor. Math. Prob. Eng. **2015**(1), 1–10 (2015)
48. Yang, Q., Liu, F., Turner, I.: Numerical methods for fractional partial differential equations with Riesz space fractional derivatives. Appl. Math. Model. **34**(1), 200–218 (2010)

Lightning Source UK Ltd.
Milton Keynes UK
UKHW020625240222
399172UK00001B/2